AF543498

FISH AND FISHERY PRODUCTS

WORLD APPARENT CONSUMPTION STATISTICS BASED ON FOOD BALANCE SHEETS (1961-1995)

Compiled by
Edmonodo Laureti
Fishery Information, Data and Statistics Unit
FAO, Fishery Department

Reprinted by arrangement with the
Food and Agriculture Organization of the United Nations

by

Daya Publishing House®
A Division of
Astral International Pvt. Ltd.
New Delhi - 110 002

"This book was originally published by the Food and Agriculture Organization of the United Nations (FAO) as Pond construction for freshwater fish culture. Pond-farm structures and layouts, FAO Training Series, 20/2".

"The designations employed and the presentation of material in this publication do not imply the expression of any opinion, whatsoever on the part of the Food and Agriculture Organization of the United Nations concerning the legal status of any country, territory, city or area or of its authorities, or concerning the delimitations of its frontiers or boundaries."

'The designations 'developed' and 'developing' economies are intended for statistical convenience and do not necessarily express a judgment about the stage reached by a particular country, territory or area in the development process.'

First Indian Edition, 1996
Reprinted, 2016
ISBN: 978-81-7035-157-3

Published by : **Daya Publishing House®**
A Division of
Astral International Pvt. Ltd.
– ISO 9001:2008 Certified Company –
4760-61/23, Ansari Road, Darya Ganj
New Delhi-110 002
Ph. 011-43549197, 23278134
E-mail: info@astralint.com
Website: www.astralint.com

Printed at : **Kirpal Print-N-Pack, Delhi**

Laureti, E. (comp.)
Fish and fishery products: world apparent consumption statistics based on food balance sheets (1961-1995).
***FAO Fisheries Circular.* No. 821, Rev. 4. Rome, FAO. 1998. 253p.**

ABSTRACT

The document summarizes over three decades of statistics of apparent consumption of fish and fishery products for 222 countries, 7 continental aggregates, the group of the low-income food-deficit countries and the world total. Data are given for total and *per caput* supply in live weight on a yearly basis. Indicative nutritional values in terms of animal and total proteins are also provided. A section comprising descriptive and analytical graphs supplements the data.

FOREWORD

The central role of apparent consumption statistics for fish and fishery products in consideration of food security and nutrition issues is illustrated by the fact that previous issues of this publication were prepared for the International Conference on Nutrition (Rome, 5-11 December 1992) and the World Food Summit (Rome, 13-17 November 1996).

From its foundation, FAO has recognised the importance of fish as a food source, particularly for its role in protein supply. In its first report on Fisheries[1] published in 1945, FAO stated that "... fish should be regarded as one of the more important sources of food in any program for raising the nutritional levels of peoples throughout the world". The same report contained the first estimate of average world fish consumption per caput published by FAO which was 5 pounds of edible fish meat per year (or about 5.4 kg live weight per year) and this probably referred to years before the Second World War. However, no consumption information for individual countries was provided and the report commented that national consumption figures were available are often not comparable due to the use of different methodologies. Further, it states that "Accurate comparison must await the preparation of more complete and standardised international fishery statistics".

International fishery statistics are still far from perfect but they have improved during the last four decades and the promotion of standard definitions, classifications and methodologies by FAO and other organisations has made them more comparable. Food balance sheets have been constructed from statistics on fish production, disposition of landings, commodity production and trade on the basis of eight major groups of species and nine commodity groups. The present publication presents summary food balance sheets by country and for some aggregations for years between 1961 and 1995 which allow trends in elements such as per caput fish supply and its contribution to protein supply to be examined.

The role of fish in world protein supply is becoming increasingly important. Average per caput supply of fish grew from 8.9 kg in 1961 to 15.3 kg in 1995. Compared to an average annual compound growth rate in animal protein supply per caput of 0.9% during the same period, fish protein supply per caput showed a rate of 1.4%. Some developing countries are particularly dependent on fish for food and all but four of the top 30 fish consuming countries (in terms of fish as a proportion of all protein) are from the developing world. With aquaculture the fastest growing food producing sector, the contribution of fish to world nutrition looks sets to continue increasing in importance.

---

[1] FAO 1945. Five Technical Reports on Food and Agriculture: Fisheries. Report of the Technical Committee on Fisheries, submitted to the United Nations Interim Commission on Food and Agriculture. Washington. Pp 175-216

The contribution of national fishery statistical offices to the construction of these food balance sheets through the provision of basic statistics is gratefully acknowledged. Where such basic statistics are lacking or considered unreliable, data from other sources have been utilised, when available, and FAO estimated used when there are no other usable data.

FAO welcomes comments on these statistics and suggestions as to how they can be improved. Comments may be sent to the address below, or by e-mail to: FIDI-Inquiries@fao.org.

R. J. Grainger
Chief
Fishery Information, Data and Statistics Unit
Fisheries Department
FAO
Viale delle Terme di Caracalla
00100 Rome, Italy
Tel: (06) 57054828
Fax: (06) 57052476
E-mail: Richard.Grainger@fao.org
http://www.fao.org/fi

CONTENTS

INTRODUCTION

This document presents a compendium of more than three decades of statistics of apparent consumption for fish and fishery products for 222 countries, 7 continental aggregates, the group of the low-income food-deficit countries and the world total based on the Supply/Utilization Accounts (SUAs) maintained in FAOSTAT, the statistical component on the FAO World Agricultural Information Centre (WAICENT).

The system currently includes SUAs for eight groups of primary fishery commodities and nine groups of processed products derived from them for almost all the countries and territories in the world from 1961 to 1995. SUAs contain the estimates of supplies from different sources matched against estimates of different forms of utilization of each product. These series are regularly updated and, in addition, revised in the light of any new information. Accordingly, the food balance sheets derived from the SUAs of food products are consistent internally. In FAO's work these data are required to meet the requests of its statutory bodies to keep the world's food and nutritional situation under constant review, to update FAO's analytical work in the field of food and population and to provide the statistical base for the projections of demand and other assessment studies.

The food balance sheets included in this document have been extracted from individual SUA's series for each primary and processed commodity, prepared on a calendar-year basis. In constructing the SUA's and the food balance sheets derived from them, both official and unofficial data available in the Fishery Information, Data and Statistics Unit, mainly through FISHDAB, the Fishery Statistical Database of the Fisheries Department, have been used.

In order to facilitate descriptions and analyses of data, graphs have been provided. Section 1 presents long-term series of detailed food balance sheets on a calendar-year basis starting from 1961 and ending with 1995, in live weight, for continental aggregates, the group of the least economically developed and food-deficit countries and for the world total. On each food balance sheet the fish contribution to animal and total protein supply is indicated as a percentage. Section 2 gives, in live weight, the same data as described in Section 1 for 222 countries, which represent more than 90% of all countries in the world.

The derived consumption data are only as reliable as the basic trade and production data on which they are based; therefore, trends in food availability may reflect improved primary data rather than real changes in food intake. For a full description of concepts used and the major shortcomings of food balance sheet data, the reader may refer to publications indicated as reference. The nutritional values have been calculated utilizing appropriate standard food composition factors.

For detailed information on other food commodities and the role of fish in national diets, see the domain Food Balance Sheets Report at:

http://www.fao.org

Comments on this document and suggestions for its improvement are invited.

GENERAL NOTES

Data -	The data shown in this document were calculated on the basis on information available to the Fishery Information, Data and Statistics Unit in April 1998. Therefore, in certain instances, discrepancies may occur with fishery data held in the Fishery Statistical Database (FISHDAB) and published in the latest Volumes of the FAO Yearbooks of Fishery Statistics.
	Where necessary, the data published in preceding Revisions have been revised.
Country Names -	Country names are listed in alphabetical order and preceded by the appropriate FAOSTAT Code.
Population -	Refers to the present-in-area (*de facto*) population, i.e. includes all persons physically present within the geographical boundaries of countries, at the mid point of the reference period.
	The source of the data is the 1996 Revision of World Population Prospects, published by the Population Division, UN.
Live Weight -	The weight of Finfish and Shellfish at the time of their capture or harvest. Calculated on the basis of conversion factors from the landed weight to nominal weight and on rates prevailing among national industries for each type of processing.
Production -	Refers to catch and culture of all fish, crustaceans, molluscs and aquatic organisms, excluding mammals and aquatic plants.
Non-Food Uses -	Includes reduction to meal and oil; utilization of aquatic products for feed and bait; ornamental purposes; withdrawals from markets and any other non-food use of fish production.
Imports/Exports -	In accordance with the internationally-recommended practice, fishery import statistics include fish caught by foreign fishing vessels and landed in domestic ports, and export statistics include fish caught by domestic fishing vessels and landed directly in foreign ports.
Stocks -	Information on changes in stocks occurring between the production and the retail levels, or in levels of inventories, is incomplete. In most instances data indicated are the minimum required to avoid a negative balance.
Total Food Supply -	The total fish available for apparent human consumption is derived by using the following equation:

total food supply equals production less reduction to meal and other non-food uses, plus imports, less exports, plus or less variation in stocks.

All the calculations have been made in terms of live weight equivalent.

Per Caput Food Supply - The estimate of the total supply available for human consumption divided by the unrounded population total.

Percentages - Percentages have been calculated on unrounded figures, though such figures are rounded in the tabulations due to publication space requirements.

Standard Symbols -

...	=	data not available/unavailable
0/0.0	=	magnitude know to be nil or more than zero, but less than half the unit used.

Groups of Species - The system provides statistics of apparent consumption for Finfish and Shellfish divided into 8 broad groups of species, such as:

Freshwater:	including Carps, Barbels, Tilapias, etc.
Diadromous:	including Sturgeons, Eels, Salmons, Trouts, Shad, etc.
Demersal:	including Flounders, Cods, Redfishes, Sharks, etc.
Pelagic:	including Jacks, Herrings, Tunas, Mackerels, etc.
Marine unspecified:	including unidentified marine fish
Crustaceans:	including Crabs, Lobsters, Shrimps, Krill, etc.
Molluscs:	including Abalones, Oysters, Mussels, Scallops, Clams, etc.
Cephalopods:	including Squids, Cuttlefishes, Octopuses, etc.
Aquatic Animals:	including Frogs, Turtles, Sea-urchins, etc.

For more detailed information on the catch and culture, trade composition and other variables, please refer to the FAO Yearbooks of Fishery Statistics. For information on food consumption in general please refer to relevant publications included in the reference list (page 253).

NOTES ON GRAPHS

This section includes statistical graphs based on (a) 1995 data and (b) data from the whole time-series contained in this report and analysed for the world and/or geo-economic aggregates.

NOTES ON SECTION 1

This section provides yearly food balance sheets for fish and fishery products in live weight from 1961 to 1995, for the world as a whole, 7 tables showing totals for continents and 1 table showing the total for the least economically developed and food-deficit countries (for the composition of this group see index on page 18). In addition, yearly calculation of percentages of fish contribution to animal and total proteins are provided in each of the above tables.

The quality of data in the time-series varies.

NOTES ON SECTION 2

This section provides yearly food balance sheets for fish and fishery products in live weight for 222 countries from 1961 to 1995. As for Section 1, yearly calculation of percentages of fish contribution to animal and total proteins are provided.

The quality of data in the time-series varies.

FISH AND FISHERY PRODUCT

STATISTICAL GRAPHS ON APPARENT CONSUMPTION BY WORLD AND GEOGRAPHICAL AREAS

1995 AND 1961-1995

GRAPHS	- INDEX OF GRAPHS -	FAO-FIDI

GRAPHS	PAGE

Note: for the list of low-income food-deficit countries see page 18.

FIGURE 1

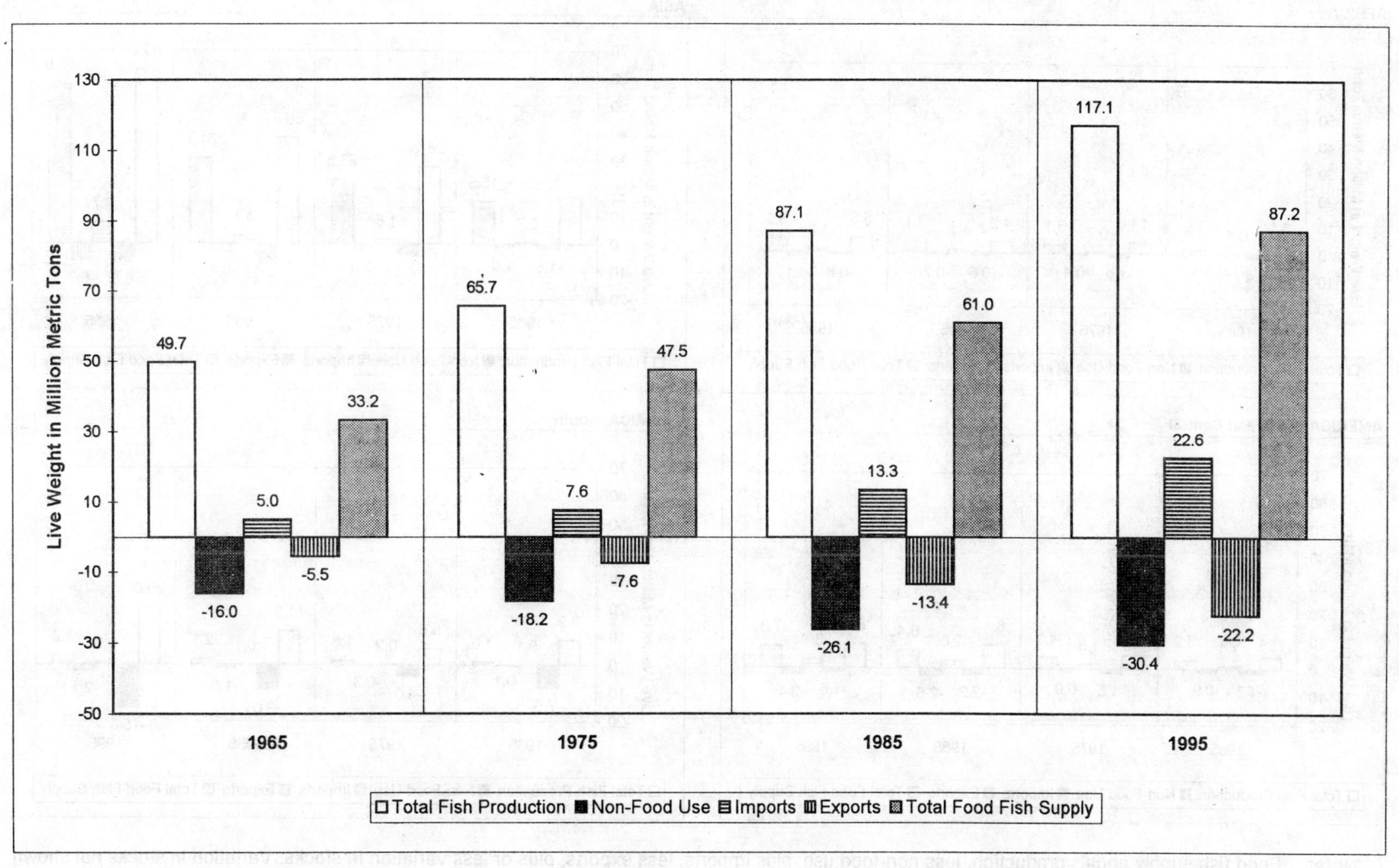

Note: Food fish supply equals production, less non-food use, plus imports, less exports, plus or less variation in stocks. Variation in stocks not shown.

FIGURE 2

Food Fish Balance in Live Weight

by Geographical Areas (1965, 1975, 1985, 1995)

AFRICA

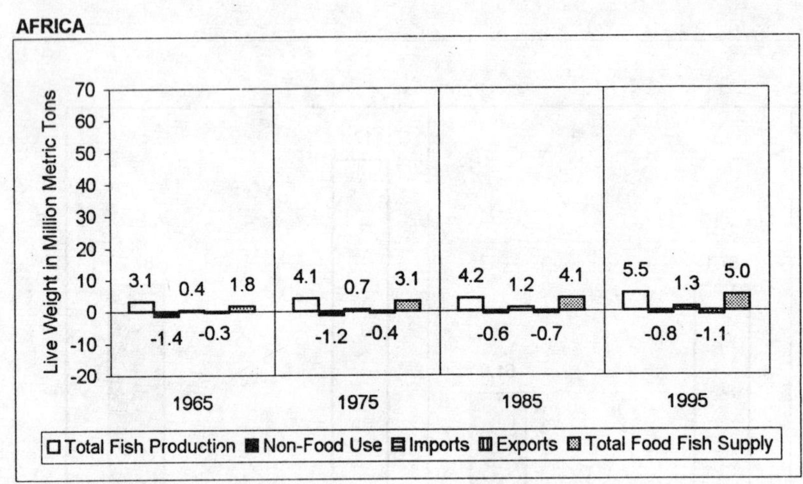

ASIA

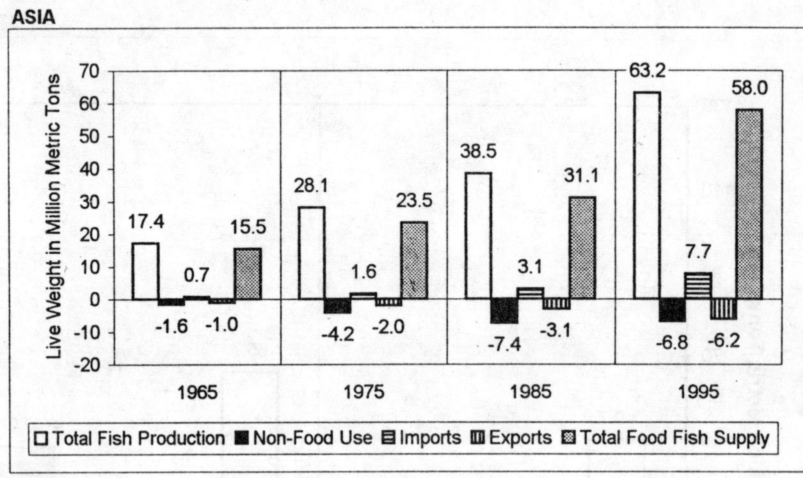

AMERICA, North and Central

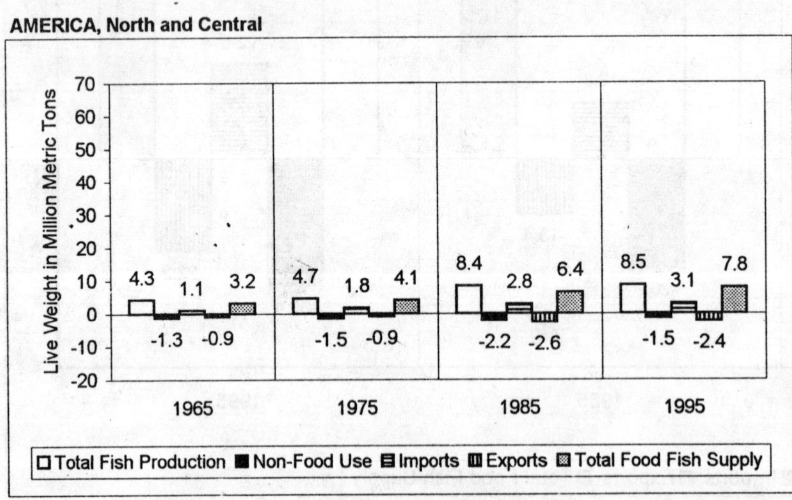

AMERICA, South

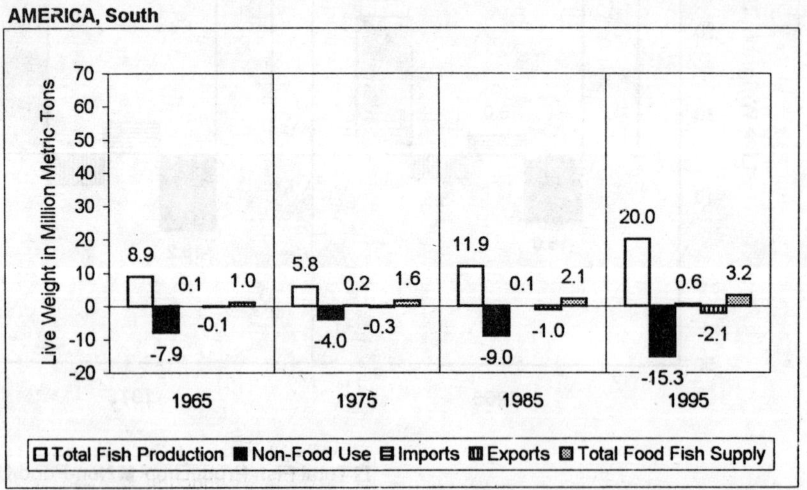

Note: Food fish supply equals production, less non-food use, plus imports, less exports, plus or less variation in stocks. Variation in stocks not shown.

Food Fish Balance in Live Weight

by Geographical Areas (1965, 1975, 1985, 1995)

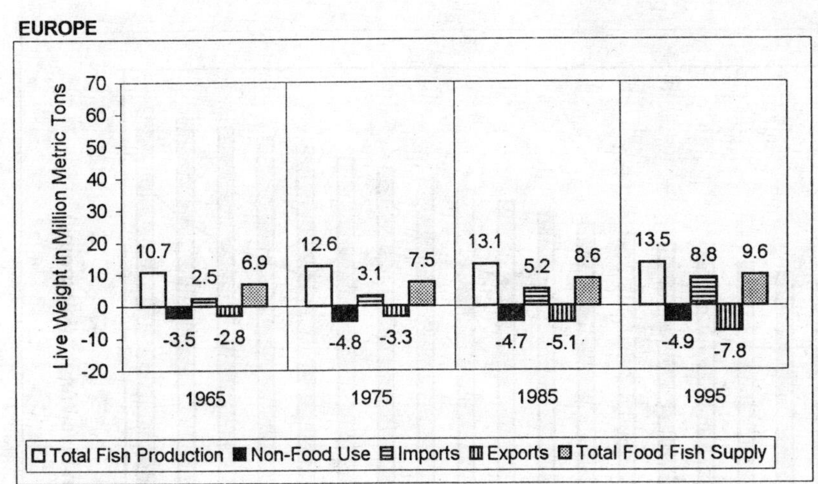

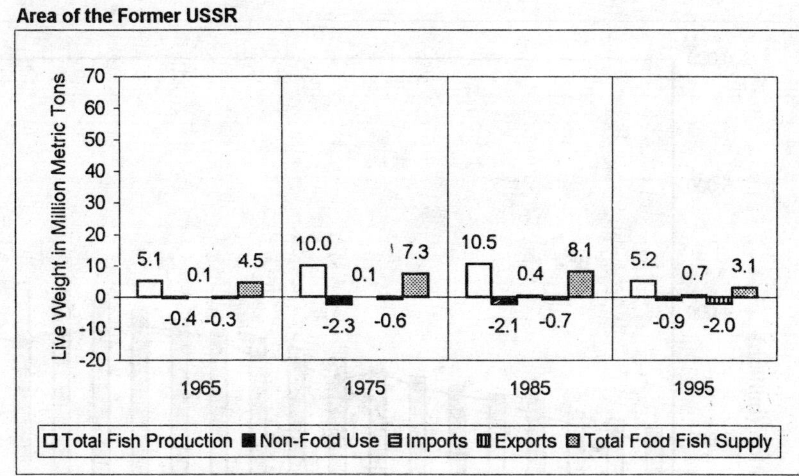

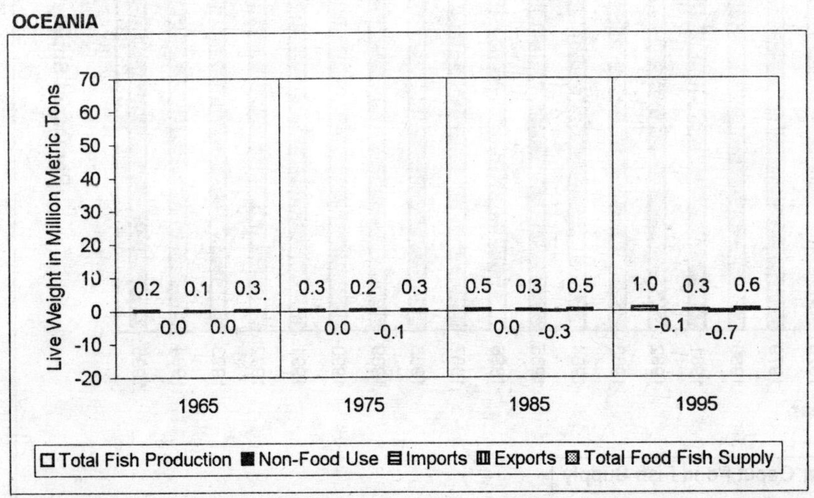

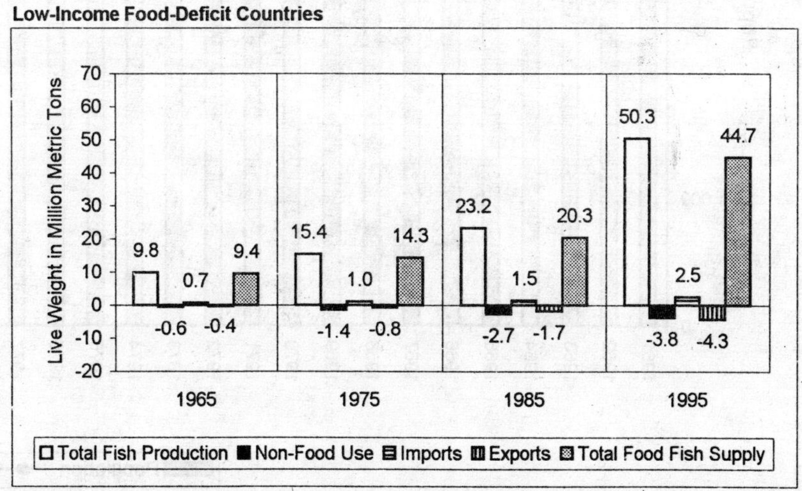

Note: Food fish supply equals production, less non-food use, plus imports, less exports, plus or less variation in stocks. Variation in stocks not shown.

FIGURE 4

Population and Per Caput Food Fish Supply in Live Weight

World (1961-1995)

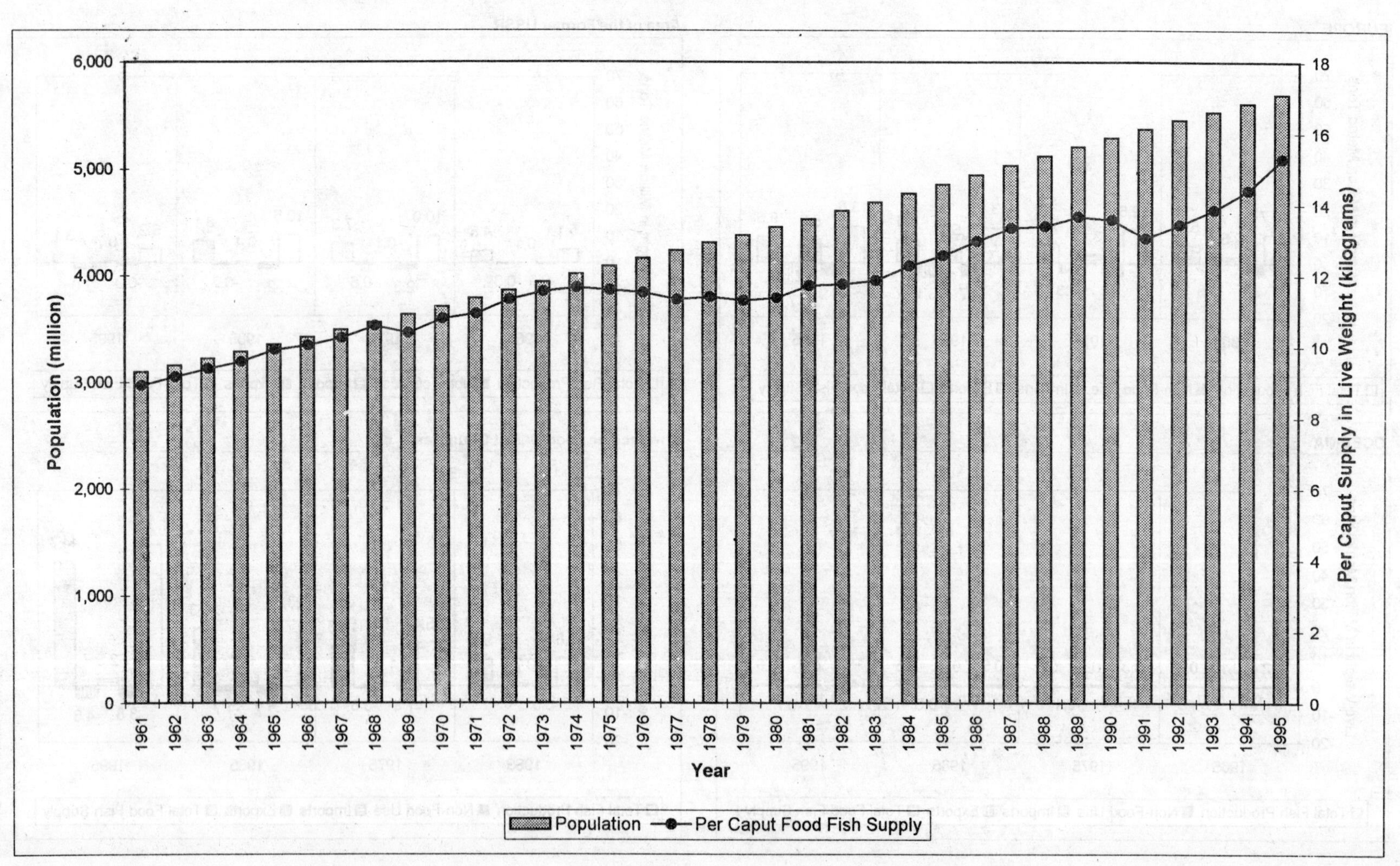

FIGURE 5

Role of Fish in Animal and Total Protein Supply

World (1961-1995)

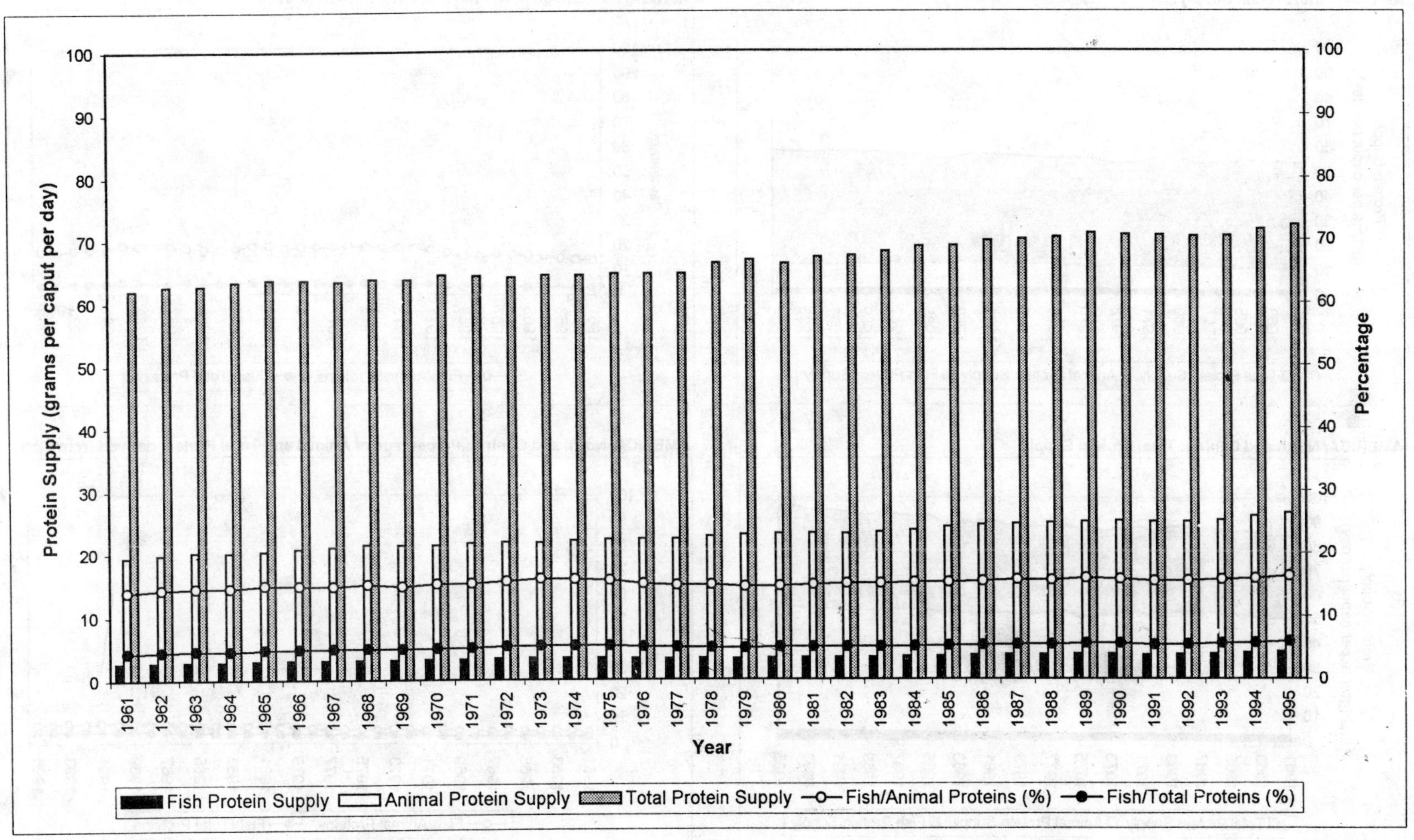

FIGURE 6

Role of Fish in Animal and Total Protein Supply

by Geographical Areas (1961-1995)

AFRICA: Total Protein Supply

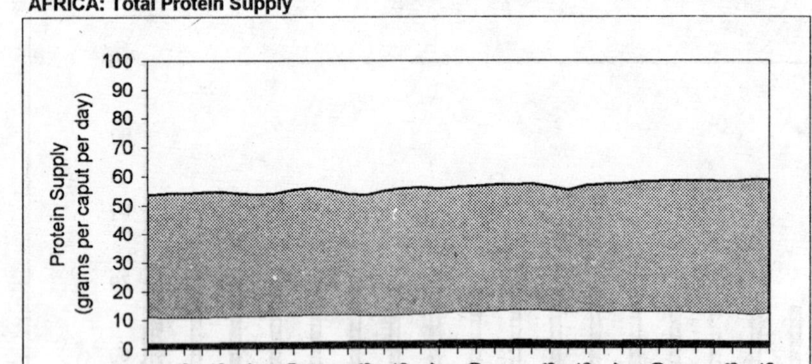

AFRICA: Percentage of Animal and Total Protein derived from Fish

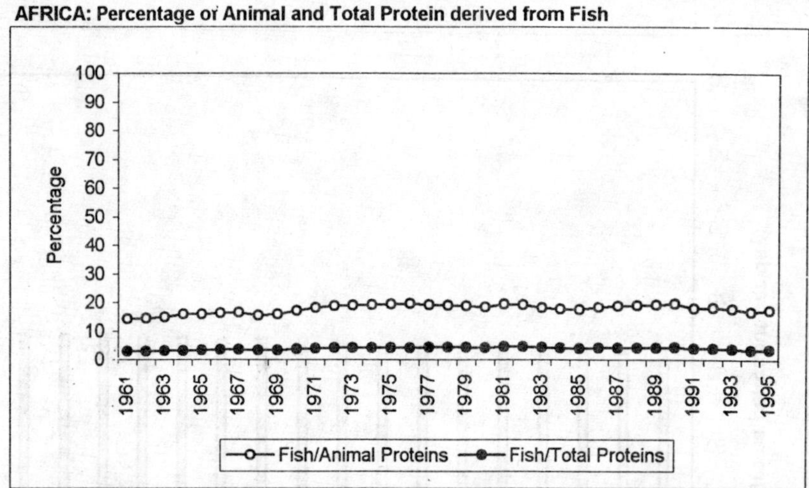

AMERICA, North and Central: Total Protein Supply

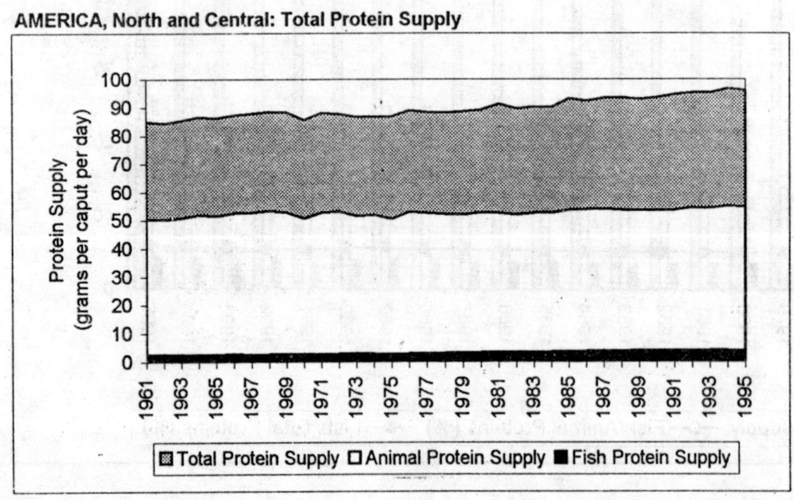

AMERICA, North and Central: Percentage of Animal and Total Protein derived from Fish

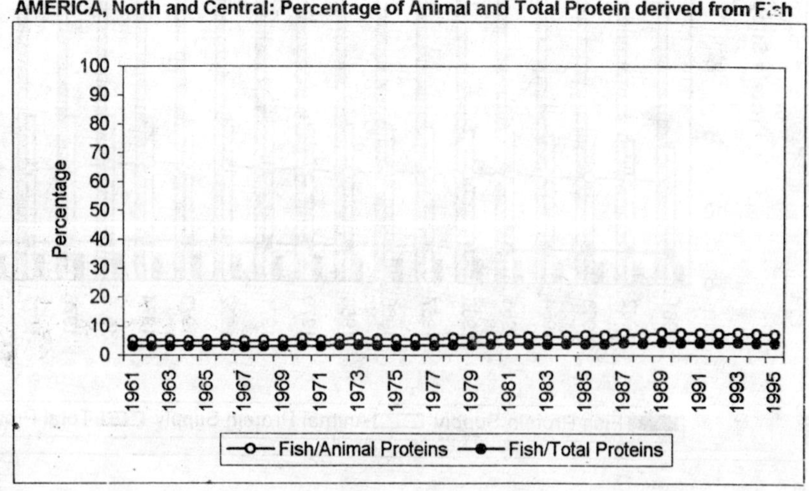

FIGURE 7

Role of Fish in Animal and Total Protein Supply

by Geographical Areas (1961-1995)

AMERICA, South: Total Protein Supply

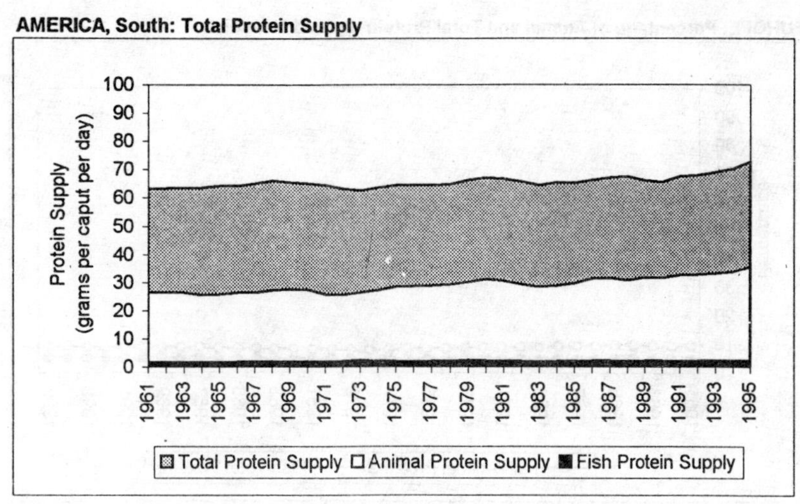

AMERICA, South: Percentage of Animal and Total Protein derived from Fish

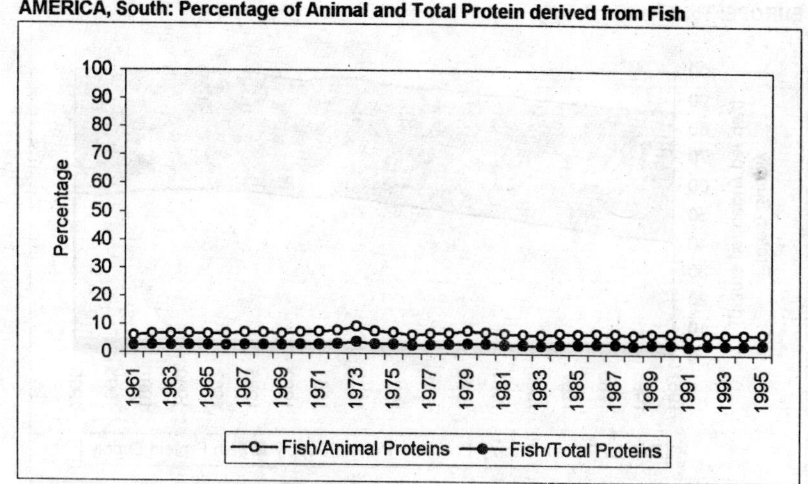

ASIA: Total Protein Supply

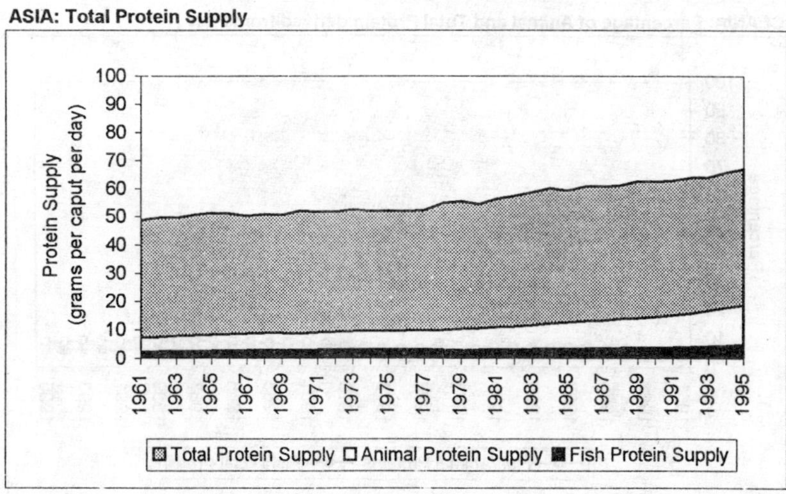

ASIA: Percentage of Animal and Total Protein derived from Fish

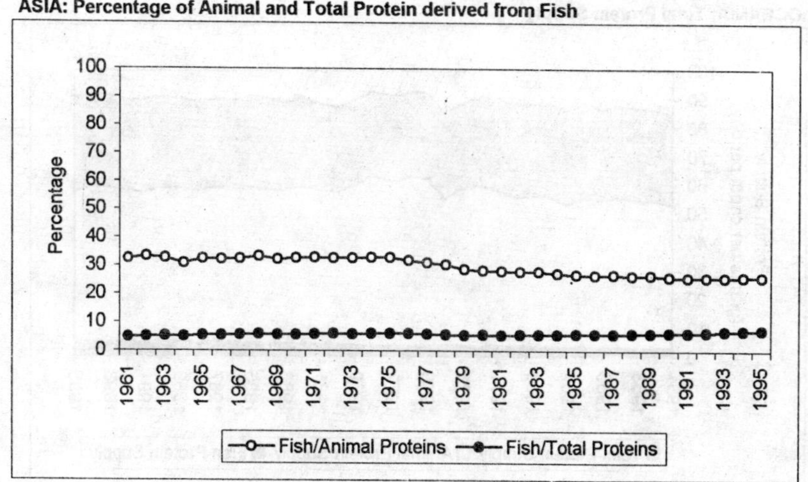

FIGURE 8

Role of Fish in Animal and Total Protein Supply

by Geographical Areas (1961-1995)

EUROPE: Total Protein Supply

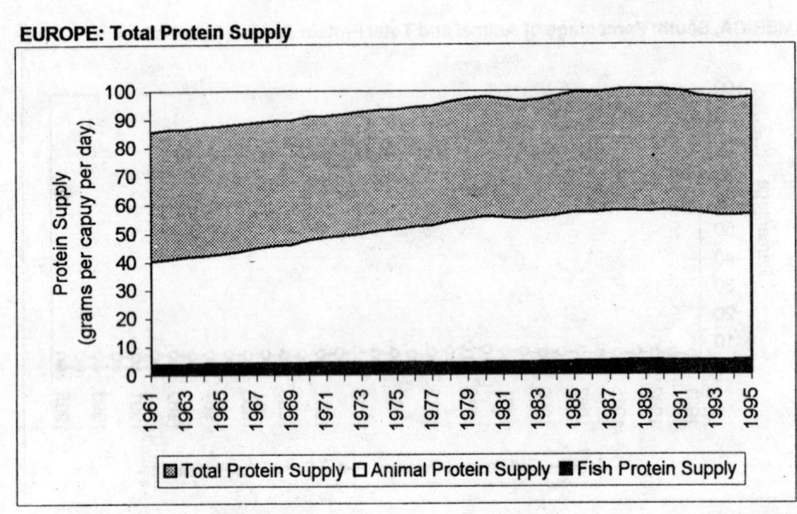

EUROPE: Percentage of Animal and Total Protein derived from Fish

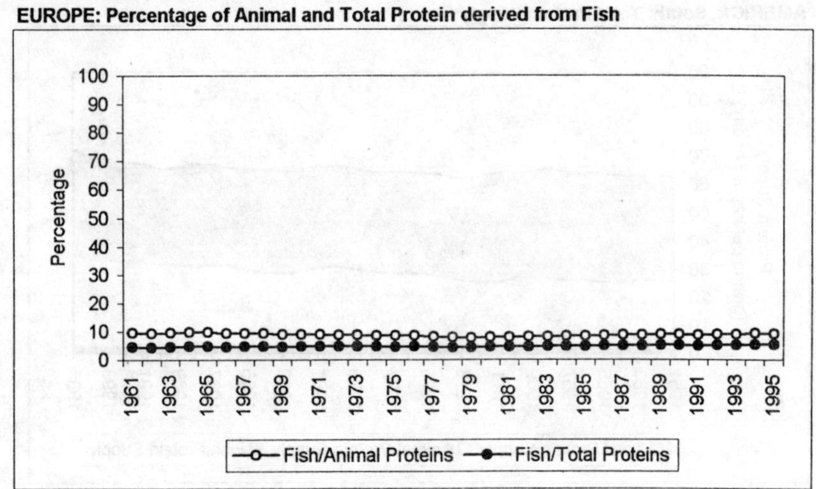

OCEANIA: Total Protein Supply

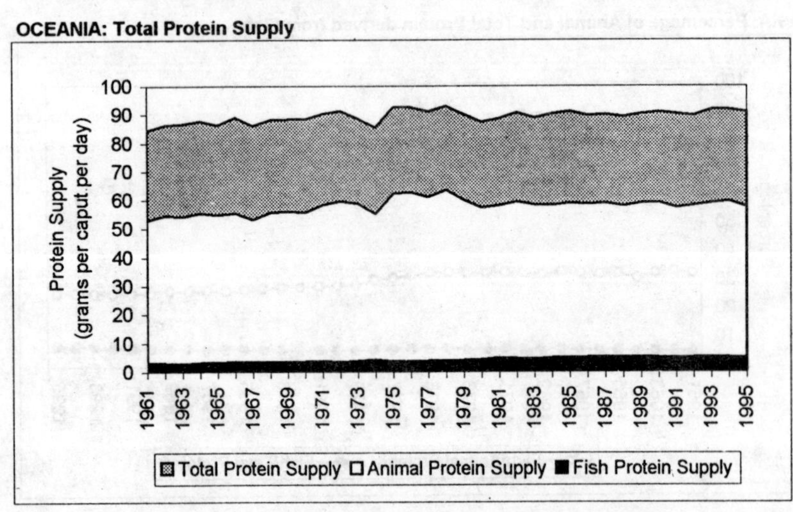

OCEANIA: Percentage of Animal and Total Protein derived from Fish

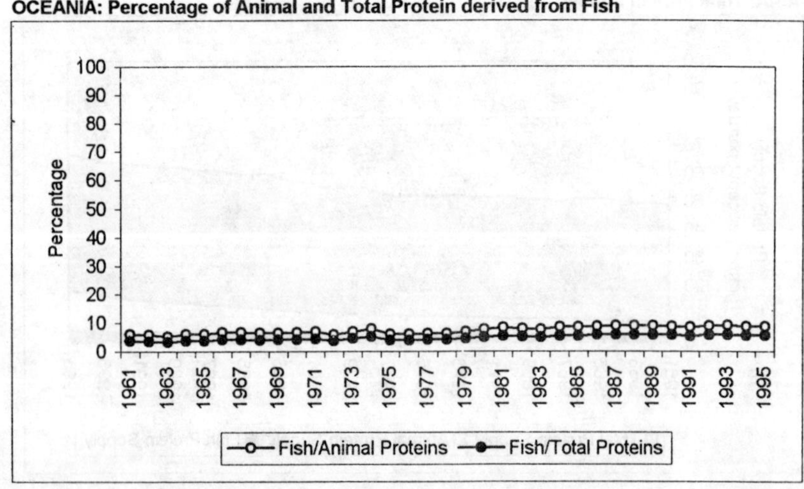

FIGURE 9

Role of Fish in Animal and Total Protein Supply

by Geographical Areas (1961-1995)

Area of the former USSR: Total Protein Supply

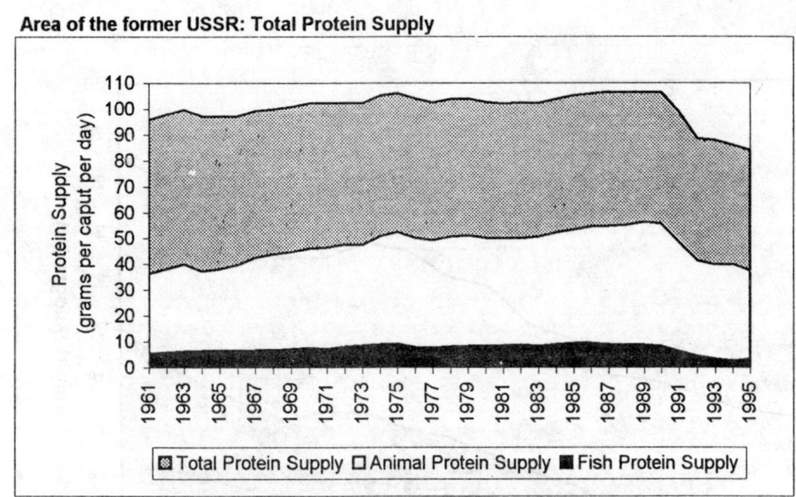

Area of the former USSR: Percentage of Animal and Total Protein derived from Fish

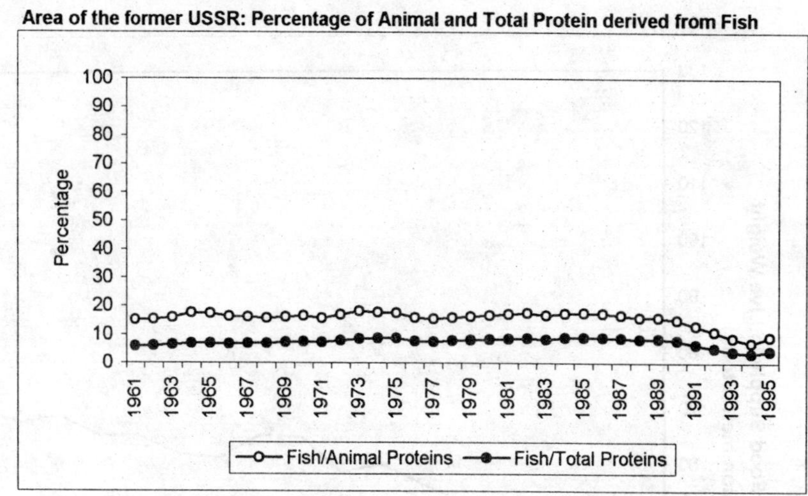

Low-Income Food-Deficiy Countries: Total Protein Supply

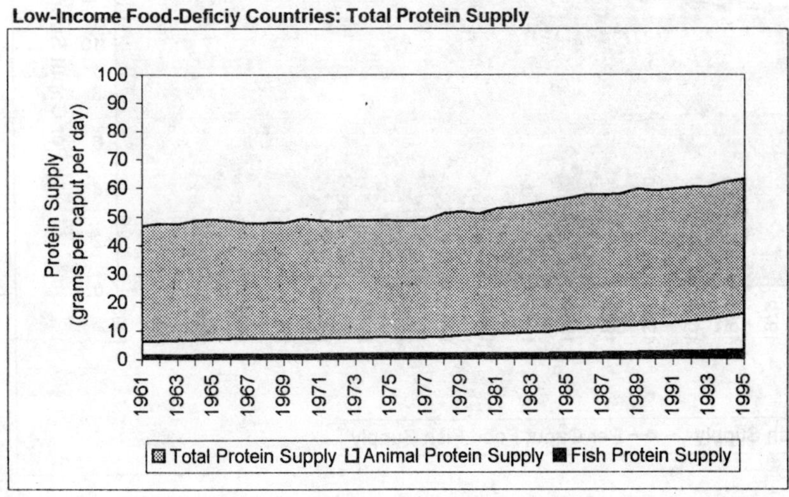

Low-Income Food-Deficit Countries: Percentage of Animal and Total Protein derived from Fish

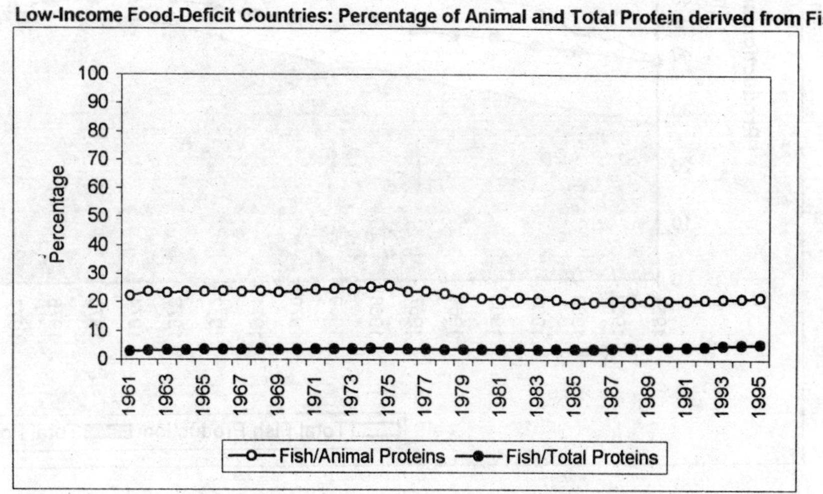

FIGURE 10

Total Fish Production, Food Fish Supply and Per Caput Food Fish Supply in Live Weight

World (1961-1995)

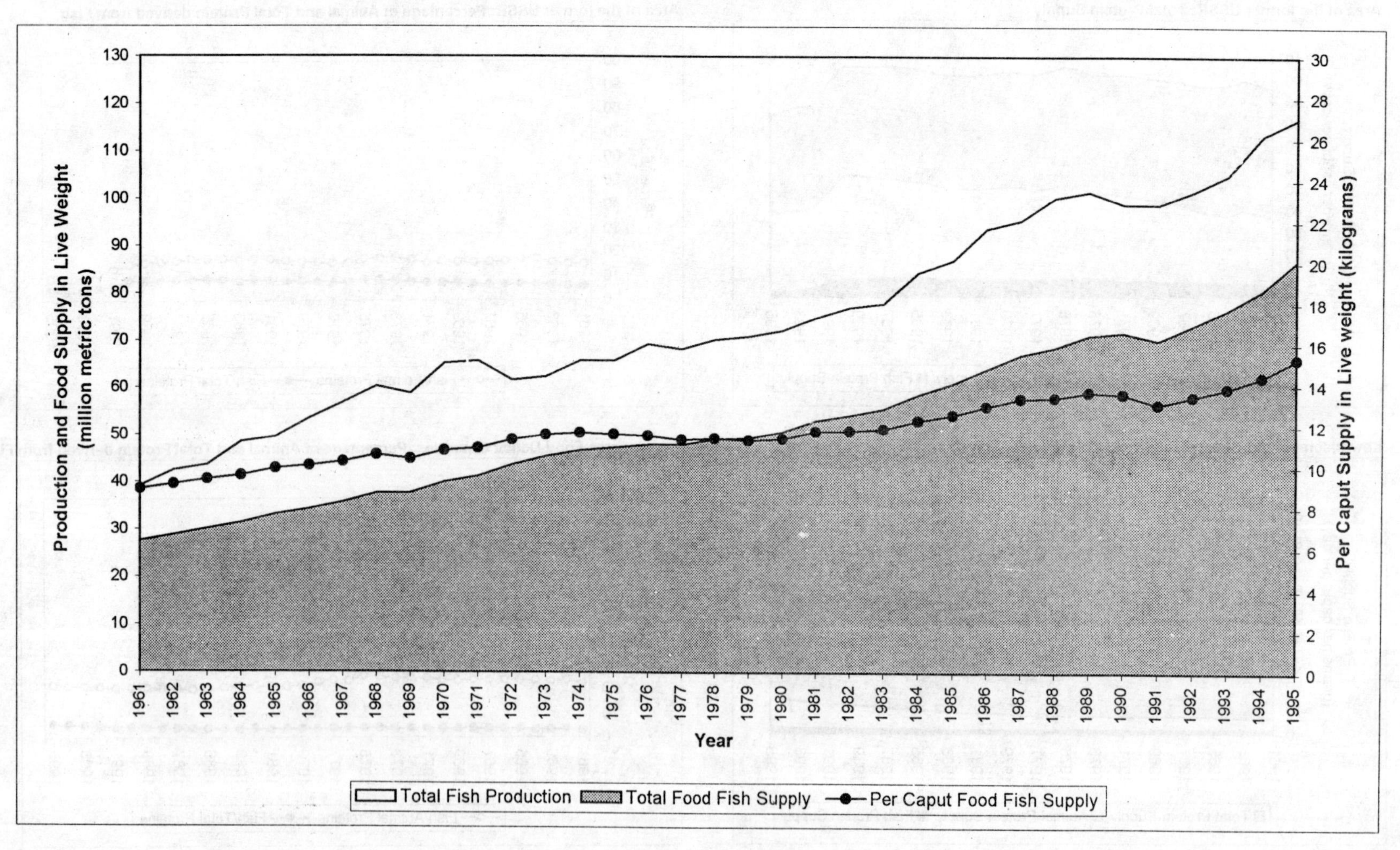

FIGURE 11

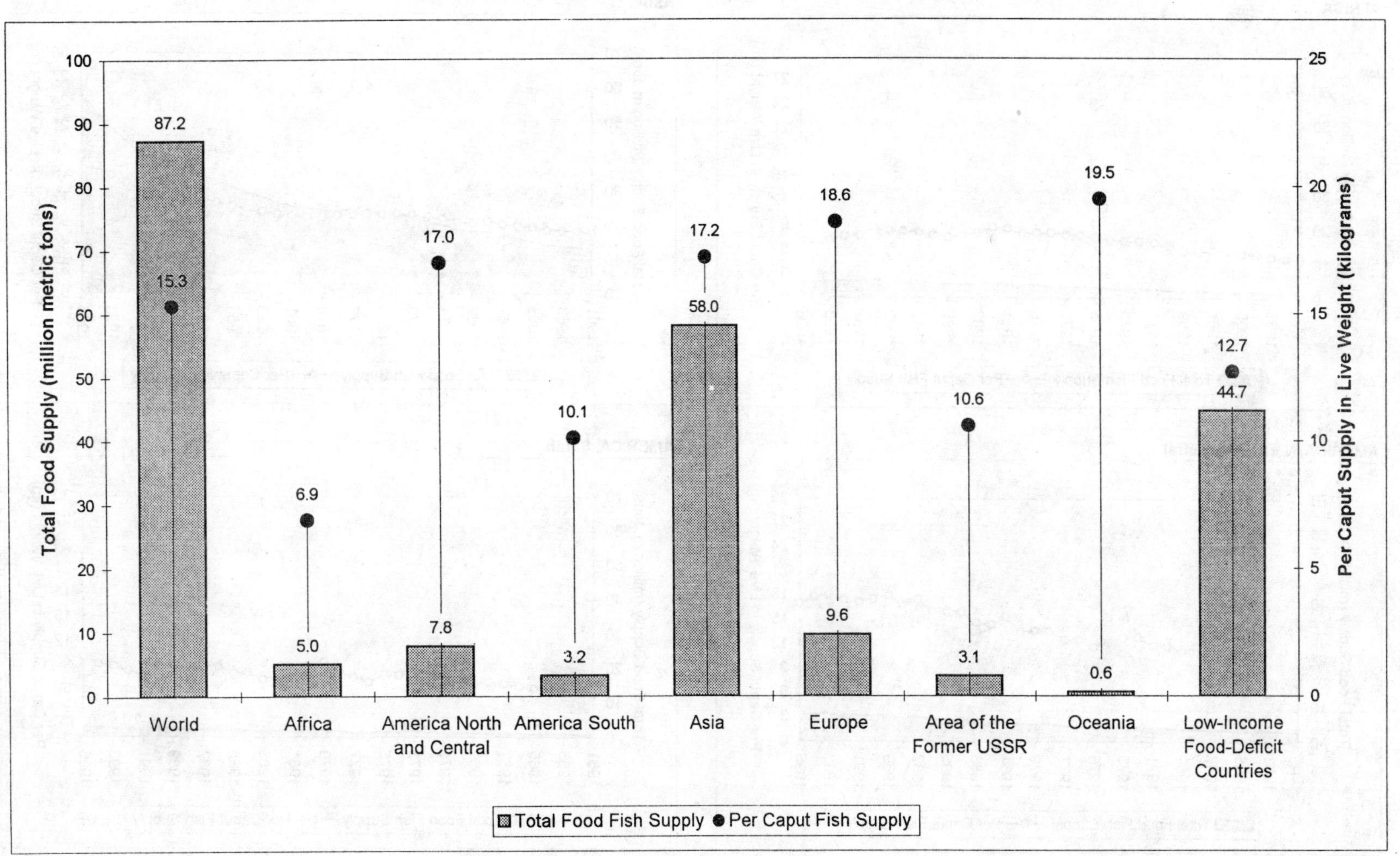
Food Fish Supply and Per Caput Food Fish Supply in Live Weight
by World and Geographical Areas in 1995
Total Food Supply (million metric tons)
Per Caput Supply in Live Weight (kilograms)
100
90
80
70
60
50
40
30
20
10
0
25
20
15
10
5
0
87.2
15.3
6.9
5.0
17.0
7.8
10.1
3.2
17.2
58.0
18.6
9.6
10.6
3.1
19.5
0.6
12.7
44.7
World
Africa
America North and Central
America South
Asia
Europe
Area of the Former USSR
Oceania
Low-Income Food-Deficit Countries
Total Food Fish Supply
Per Caput Fish Supply

FIGURE 12

Food Fish Supply and Per Caput Food Fish Supply in Live Weight

by Geographical Areas (1961-1995)

AFRICA

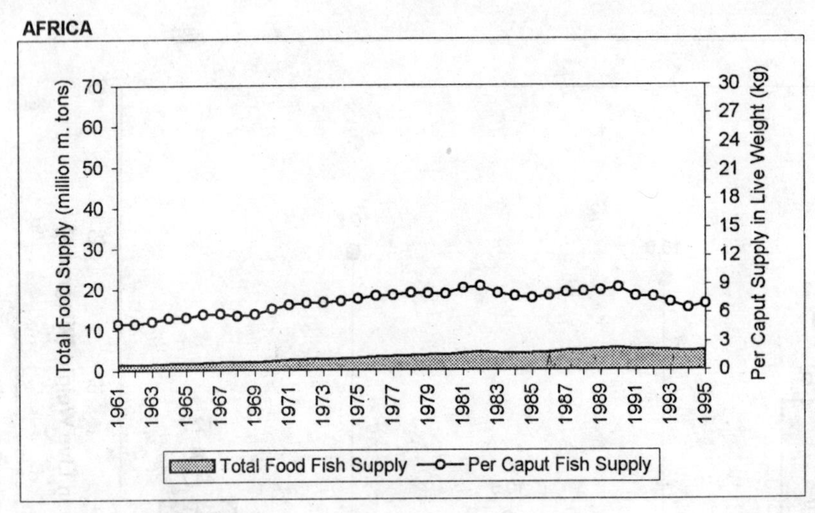

ASIA

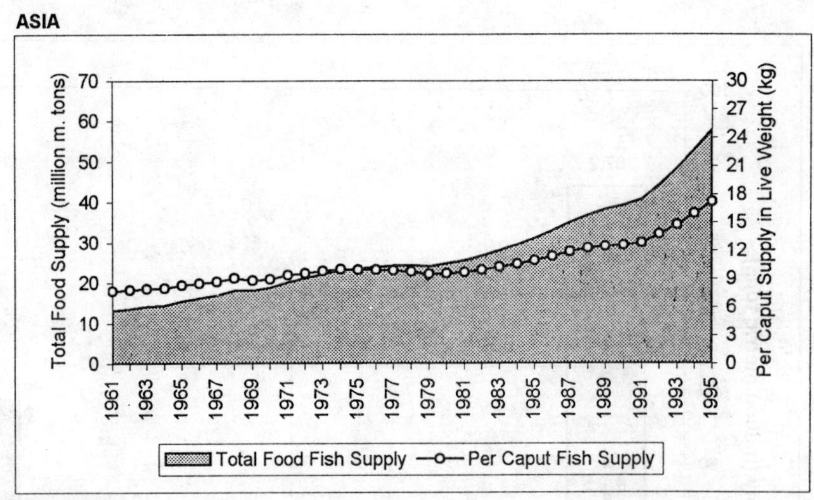

AMERICA, North and Central

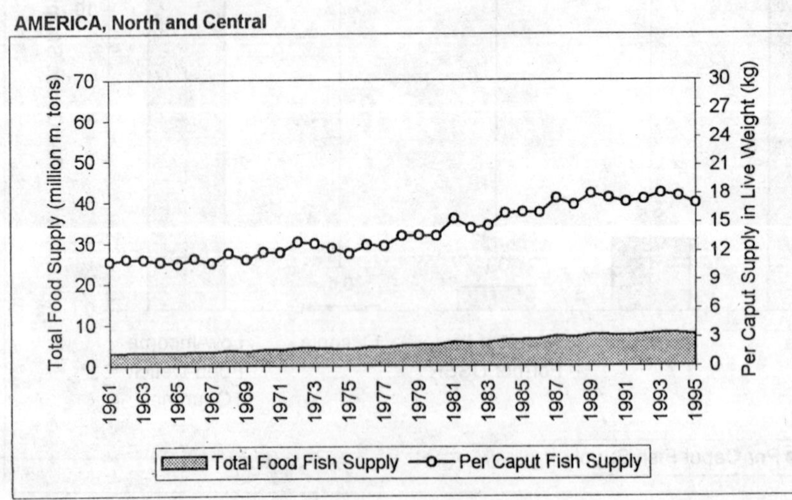

AMERICA, South

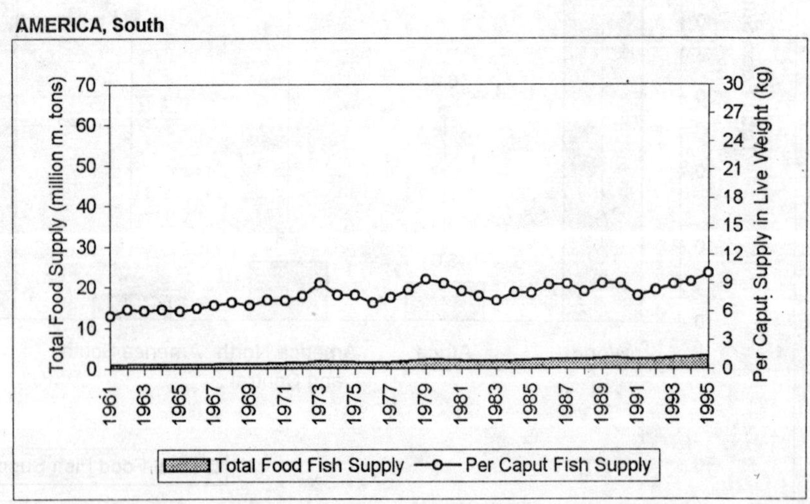

Food Fish Supply and Per Caput Food Fish Supply in Live Weight

by Geographical Areas (1961-1995)

EUROPE

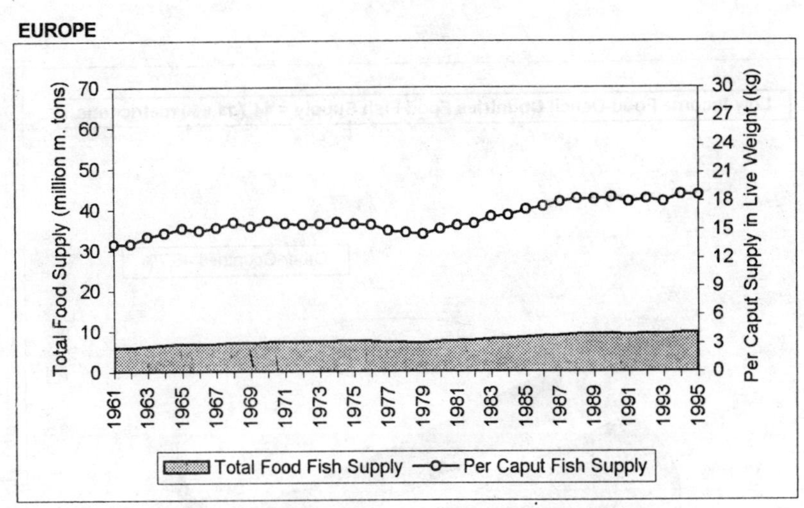

Area of the Former USSR

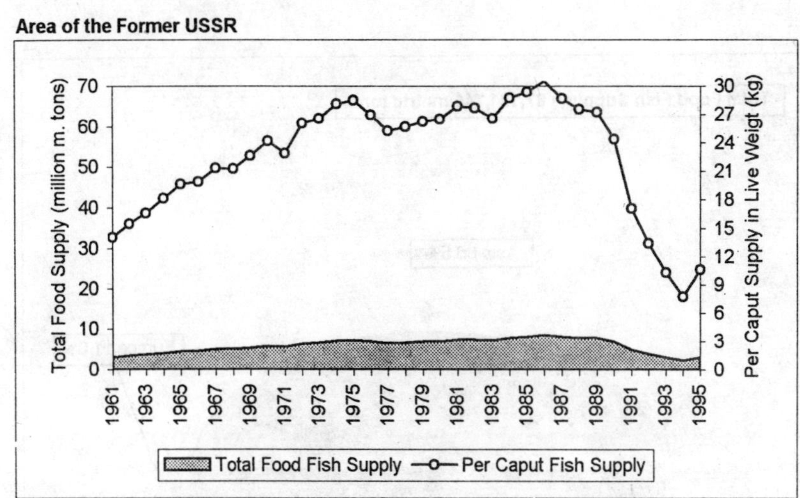

OCEANIA

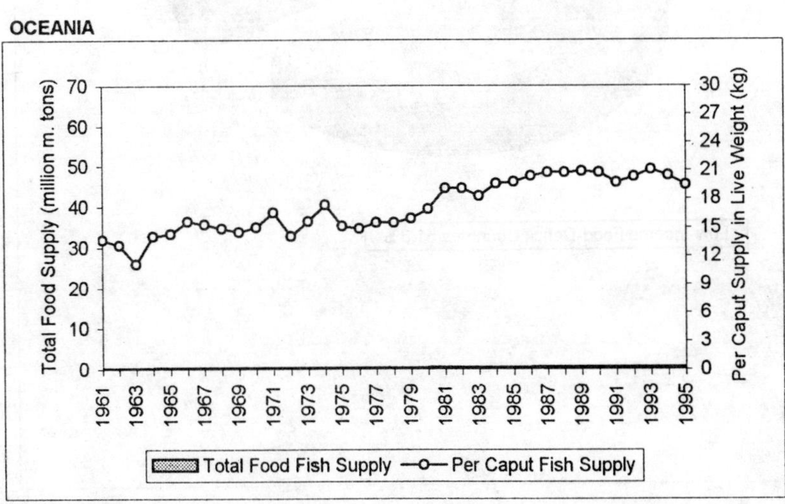

Low-Income Food-Deficit Countries

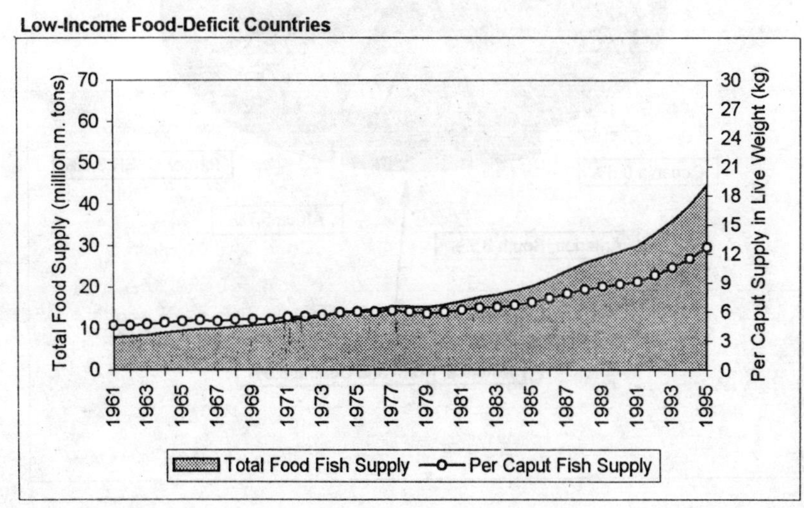

Food Fish Supply in Live Weight as a Percentage of World Total

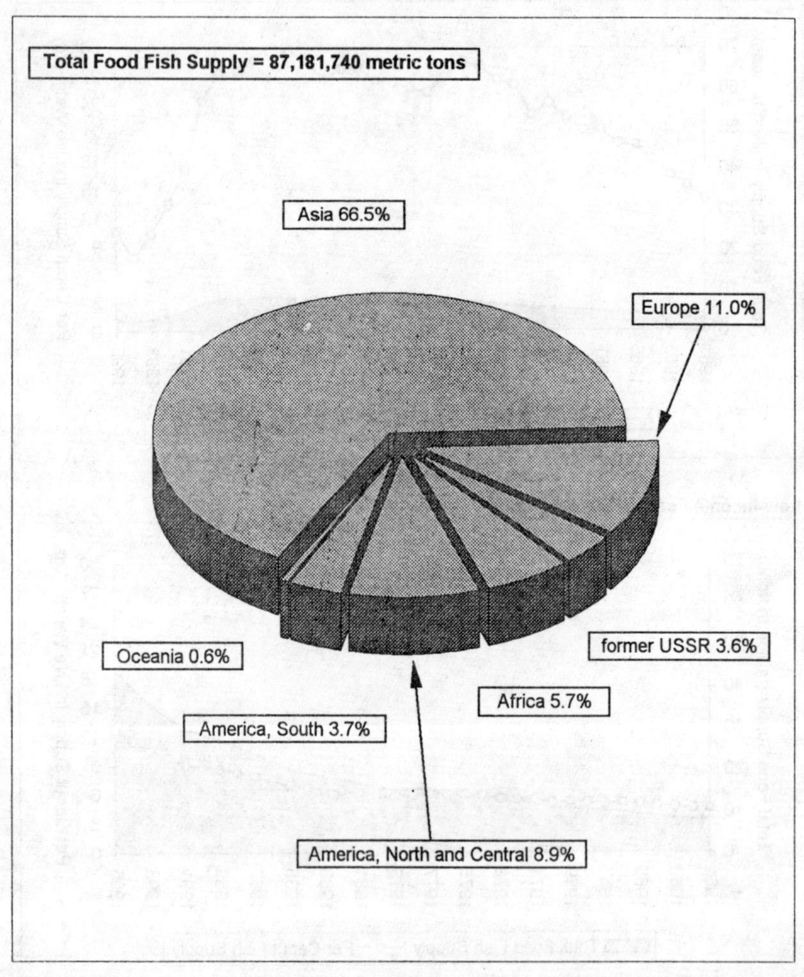

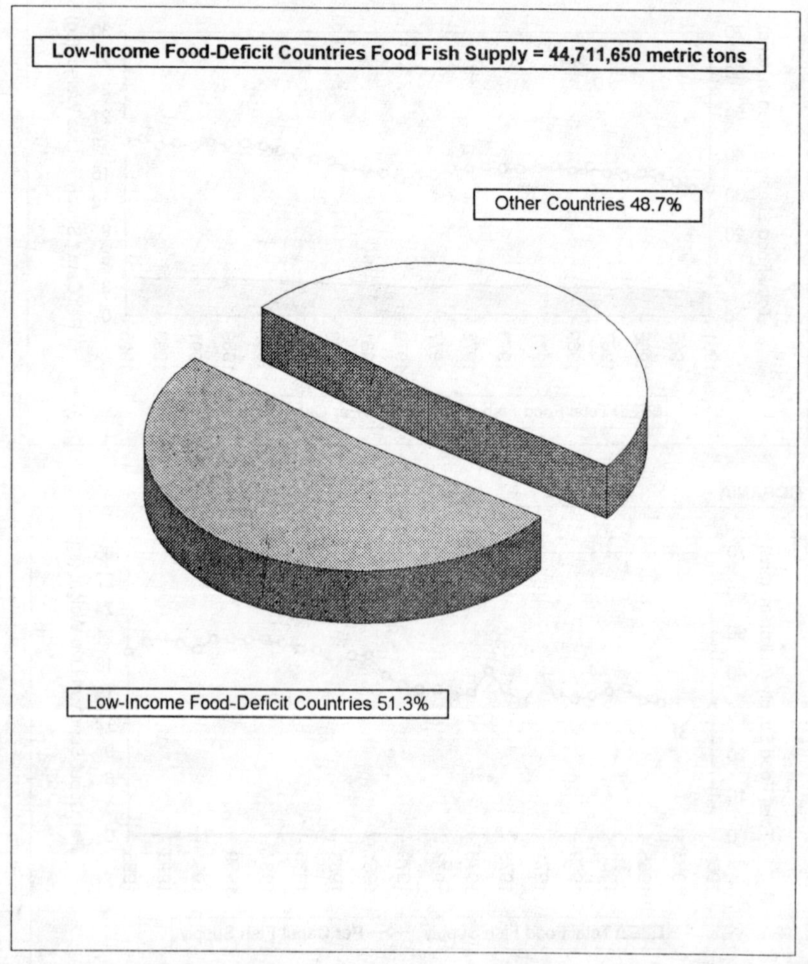

SECTION 1

FISH AND FISHERY PRODUCTS

PROVISIONAL FOOD BALANCE SHEETS AND FISH CONTRIBUTION TO PROTEIN SUPPLY BY WORLD AND GEOGRAPHICAL AREAS

FROM 1961 TO 1995

(Live Weight)

SECTION 1	- INDEX FOR FOOD BALANCE SHEETS BY GEOGRAPHICAL AREA AND COUNTRY GROUPS COMPOSITION -	FAO-FIDI

INDEX BY AREA	PAGE

GROUP COMPOSITION: LOW-INCOME FOOD-DEFICIT COUNTRIES AS OF MARCH 1998

AFRICA (42)

Angola
Benin
Burkina Faso
Burundi
Cameroon
Cape Verde
Central African Republic
Chad
Comoros
Congo, Democratic Rep.
Congo, Republic
Côte d'Ivoire
Djibouti
Egypt
Equatorial Guinea
Eritrea
Ethiopia
Gambia
Ghana
Guinea
Guinea-Bissau
Kenya
Lesotho
Liberia
Madagascar
Malawi
Mali
Mauritania
Morocco
Mozambique
Niger
Nigeria
Rwanda
Sao Tome and Principe
Senegal

Africa (continued)

Sierra Leone
Somalia
Sudan
Swaziland
Tanzania
Togo
Zambia

ASIA (24)

Afghanistan
Armenia
Azerbaijan
Bangladesh
Bhutan
Cambodia
China
Georgia
India
Indonesia
Korea, Dem. Peop. Rep.
Kyrgyzstan
Laos
Maldives
Mongolia
Nepal
Pakistan
Philippines
Sri Lanka
Syria
Tajikistan
Turkmenistan
Uzbekistan
Yemen

AMERICA (7)

Bolivia
Cuba
Ecuador
Guatemala
Haiti
Honduras
Nicaragua

EUROPE (3)

Albania
Bosnia and Herzegovina
Macedonia

OCEANIA (7)

Kiribati
Papua New Guinea
Samoa
Solomon Islands
Tokelau
Tuvalu
Vanuatu

TOTAL LIFDC (83)

341 WORLD	- FOOD BALANCE SHEET OF FISH AND FISHERY PRODUCTS IN LIVE WEIGHT AND FISH CONTRIBUTION TO PROTEIN SUPPLY -	FAO-FIDI

YEAR	PRODUCTION	NON-FOOD USES	IMPORTS	EXPORTS	STOCKS VARIATIONS	TOTAL FOOD SUPPLY	POPULATION	PER CAPUT SUPPLY	FISH PROTEINS	ANIMAL PROTEINS	TOTAL PROTEINS	FISH/ANIMAL PROTEINS	FISH/TOTAL PROTEINS
	(..		metric tons in live weight			....)	(thousands)	(kilograms)		(grams per caput per day)		(%)	(%)
1961	39,167,110	11,124,061	4,272,468	4,465,716	-275,909	27,574,900	3,095,380	8.9	2.7	19.4	62.0	13.9	4.4
1962	42,561,060	13,298,493	4,391,545	4,830,873	56,355	28,880,300	3,156,704	9.2	2.8	19.8	62.7	14.3	4.5
1963	43,716,600	13,185,511	4,633,660	4,936,386	-37,423	30,191,150	3,220,044	9.4	2.9	20.2	62.8	14.5	4.7
1964	48,562,280	16,832,907	4,878,572	5,257,407	119,069	31,469,610	3,285,344	9.6	2.9	20.1	63.4	14.5	4.6
1965	49,661,850	15,952,827	4,994,630	5,487,666	13,987	33,229,980	3,352,514	9.9	3.1	20.4	63.7	15.0	4.8
1966	53,549,800	18,758,534	5,440,280	5,760,484	-86,386	34,384,660	3,421,479	10.1	3.1	20.8	63.6	14.9	4.9
1967	56,917,460	20,733,134	5,384,133	5,600,060	-154,007	35,814,400	3,492,126	10.3	3.1	21.1	63.4	14.9	4.9
1968	60,319,970	22,590,365	5,805,184	5,820,473	13,470	37,727,790	3,564,256	10.6	3.2	21.4	63.8	15.1	5.1
1969	58,946,700	20,896,775	5,816,796	6,116,578	79,703	37,829,860	3,637,634	10.4	3.2	21.4	63.7	14.8	5.0
1970	65,270,380	24,819,800	6,284,198	6,730,137	92,452	40,097,120	3,712,019	10.8	3.3	21.4	64.5	15.3	5.1
1971	65,780,980	24,009,030	6,517,366	6,830,711	-64,871	41,393,740	3,787,438	10.9	3.3	21.7	64.3	15.3	5.2
1972	61,763,670	17,969,083	7,322,160	7,352,703	66,295	43,830,340	3,863,739	11.3	3.4	21.9	64.0	15.6	5.3
1973	62,419,420	16,750,837	7,649,700	7,840,742	57,906	45,535,470	3,940,294	11.6	3.5	21.8	64.3	16.1	5.4
1974	65,777,050	19,114,223	7,674,768	7,400,298	-82,891	46,854,420	4,016,311	11.7	3.5	22.1	64.4	15.9	5.5
1975	65,742,780	18,237,707	7,641,620	7,595,994	-74,871	47,475,860	4,091,282	11.6	3.5	22.2	64.6	15.8	5.5
1976	69,276,400	21,354,417	8,300,318	8,105,674	-106,784	48,009,850	4,164,920	11.5	3.4	22.4	64.6	15.2	5.3
1977	68,260,560	19,971,057	8,441,638	8,681,948	22,506	48,071,700	4,237,502	11.3	3.4	22.4	64.6	15.0	5.2
1978	70,231,900	20,900,292	9,344,169	9,514,274	983	49,162,460	4,309,805	11.4	3.4	22.7	66.3	15.0	5.2
1979	71,001,020	21,459,802	10,311,070	10,249,820	-49,289	49,553,180	4,382,940	11.3	3.4	23.0	66.7	14.8	5.1
1980	72,253,100	20,859,418	10,174,430	10,764,970	-62,594	50,740,530	4,457,742	11.4	3.4	23.2	66.2	14.8	5.2
1981	74,909,780	21,193,830	10,839,310	11,341,560	-31,580	53,182,110	4,534,299	11.7	3.5	23.2	67.2	15.0	5.2
1982	77,180,620	22,777,606	11,234,420	11,233,900	-99,520	54,304,020	4,612,424	11.8	3.5	23.2	67.5	15.2	5.2
1983	77,982,270	22,141,027	11,492,180	11,748,050	58,761	55,644,180	4,692,297	11.9	3.6	23.4	68.1	15.2	5.2
1984	84,511,490	25,779,146	12,065,230	12,185,050	28,466	58,640,990	4,774,058	12.3	3.7	23.7	68.9	15.4	5.3
1985	87,126,700	26,094,385	13,308,400	13,371,900	73,626	61,042,440	4,857,718	12.6	3.7	24.2	69.0	15.4	5.4
1986	93,775,790	29,190,187	14,330,810	14,876,530	97,666	64,137,600	4,943,488	13.0	3.8	24.5	69.8	15.6	5.5
1987	95,592,340	28,577,936	16,151,700	15,790,120	-207,337	67,168,660	5,031,106	13.4	3.9	24.7	70.0	15.8	5.6
1988	100,357,900	31,967,985	17,279,440	16,854,910	-141,862	68,672,240	5,119,456	13.4	3.9	24.9	70.3	15.7	5.6
1989	101,649,300	31,826,556	18,255,250	16,900,750	28,582	71,205,470	5,207,050	13.7	4.0	25.0	71.0	16.1	5.7
1990	99 003,000	28,384,010	19,156,250	17,808,790	-88,759	71,877,160	5,292,826	13.6	4.0	25.2	70.8	15.9	5.7
1991	98,901,490	30,258,740	19,638,540	18,207,020	117,996	70,191,510	5,376,136	13.1	3.9	25.0	70.7	15.5	5.5
1992	101,738,300	29,977,146	19,975,670	18,577,980	247,070	73,404,640	5,457,622	13.5	3.9	25.0	70.6	15.6	5.5
1993	105,234,800	29,262,518	20,508,760	19,883,740	179,361	76,776,120	5,537,747	13.9	4.0	25.3	70.6	15.8	5.7
1994	113,503,600	33,245,402	22,192,250	21,669,360	158,494	80,938,290	5,617,508	14.4	4.1	25.9	71.6	16.0	5.8
1995	117,076,100	30,415,827	22,623,900	22,230,070	129,459	87,181,740	5,697,656	15.3	4.3	26.4	72.4	16.4	6.0

338 AFRICA	- FOOD BALANCE SHEET OF FISH AND FISHERY PRODUCTS IN LIVE WEIGHT AND FISH CONTRIBUTION TO PROTEIN SUPPLY -	FAO-FIDI

YEAR	PRODUCTION	NON-FOOD USES	IMPORTS	EXPORTS	STOCKS VARIATIONS	TOTAL FOOD SUPPLY	POPULATION	PER CAPUT SUPPLY	FISH PROTEINS	ANIMAL PROTEINS	TOTAL PROTEINS	FISH/ANIMAL PROTEINS	FISH/TOTAL PROTEINS
	(........ metric tons in live weight ....)						(thousands)	(kilograms)	(grams per caput per day)			(%)	(%)
1961	2,430,000	1,023,000	431,285	415,474	0	1,422,812	289,284	4.9	1.6	11.1	53.6	14.3	2.9
1962	2,598,400	1,091,000	380,415	409,359	-17,147	1,461,309	296,706	4.9	1.6	10.9	54.1	14.3	2.9
1963	2,738,714	1,216,403	413,009	379,952	7,146	1,562,515	304,384	5.1	1.6	10.9	53.9	14.9	3.0
1964	3,061,306	1,352,955	435,890	424,344	10,000	1,729,898	312,286	5.5	1.7	10.9	54.4	15.8	3.2
1965	3,128,599	1,385,109	416,295	344,943	-16,789	1,798,059	320,391	5.6	1.8	11.1	54.6	16.0	3.2
1966	3,379,419	1,534,373	481,205	388,205	-6,409	1,931,636	328,707	5.9	1.8	11.3	53.9	16.3	3.4
1967	3,701,709	1,758,364	428,597	354,299	-13,663	2,003,981	337,247	5.9	1.9	11.4	53.5	16.4	3.5
1968	4,177,323	2,200,000	369,288	380,158	5,318	1,971,771	345,997	5.7	1.8	11.5	54.0	15.4	3.3
1969	4,160,797	2,129,264	393,633	348,897	7,632	2,083,901	354,943	5.9	1.8	11.6	55.2	15.9	3.3
1970	3,684,741	1,491,882	472,898	297,134	-14,294	2,354,330	364,083	6.5	2.0	11.8	55.8	17.2	3.6
1971	3,755,355	1,359,045	490,341	310,662	-5,827	2,570,162	373,406	6.9	2.1	11.8	55.0	18.1	3.9
1972	4,111,544	1,568,336	512,979	361,387	13,406	2,708,206	382,944	7.1	2.2	11.7	53.9	18.7	4.1
1973	4,403,554	1,724,918	571,633	460,453	17,489	2,807,305	392,792	7.2	2.2	11.6	53.5	19.0	4.1
1974	4,332,693	1,570,197	647,934	507,067	16,548	2,919,911	403,075	7.2	2.2	11.6	54.9	19.2	4.0
1975	4,065,143	1,236,642	727,825	403,689	-39,966	3,112,670	413,885	7.5	2.3	11.8	55.7	19.5	4.1
1976	3,833,181	1,066,965	937,133	369,184	-16,970	3,317,196	425,259	7.8	2.4	12.0	56.2	19.6	4.2
1977	3,808,027	902,643	875,684	379,646	47,160	3,448,582	437,180	7.9	2.4	12.4	55.4	19.1	4.3
1978	3,834,504	996,312	1,153,013	333,642	6,056	3,663,618	449,605	8.2	2.4	12.8	56.1	18.9	4.3
1979	3,770,148	896,806	1,254,343	371,841	-10,980	3,744,864	462,467	8.1	2.4	12.8	56.4	18.8	4.3
1980	3,687,924	822,758	1,320,002	385,094	7,370	3,807,444	475,713	8.0	2.4	13.0	56.8	18.4	4.2
1981	3,838,867	876,761	1,753,251	476,485	-964	4,237,909	489,334	8.7	2.6	13.3	56.8	19.5	4.6
1982	3,828,528	715,155	1,807,046	470,726	-7,864	4,441,829	503,344	8.8	2.6	13.4	57.2	19.3	4.5
1983	4,151,014	906,686	1,481,433	559,562	10,793	4,176,991	517,745	8.1	2.4	13.2	56.2	18.1	4.3
1984	4,074,288	677,780	1,317,448	592,393	8,586	4,130,148	532,542	7.8	2.3	12.8	55.0	17.8	4.1
1985	4,199,333	621,793	1,233,513	663,118	351	4,148,286	547,739	7.6	2.3	12.9	56.6	17.5	4.0
1986	4,555,499	760,112	1,265,080	666,656	-361	4,393,450	563,338	7.8	2.3	12.8	57.0	18.3	4.1
1987	5,290,548	1,283,167	1,430,013	674,565	-7,145	4,755,685	579,327	8.2	2.4	12.9	57.3	18.8	4.2
1988	5,362,111	1,203,943	1,417,955	653,024	-15,979	4,907,122	595,684	8.2	2.4	12.8	58.0	19.0	4.2
1989	4,985,978	740,163	1,604,666	681,551	-22,295	5,146,636	612,378	8.4	2.5	12.8	58.2	19.1	4.2
1990	5,137,410	535,065	1,713,005	785,607	-60,858	5,468,886	629,388	8.7	2.5	12.8	58.2	19.6	4.3
1991	4,850,944	489,975	1,484,416	877,160	54,900	5,023,126	646,705	7.8	2.2	12.4	58.2	18.0	3.8
1992	5,293,089	743,816	1,427,908	885,370	11,078	5,102,890	664,339	7.7	2.3	12.4	58.1	18.2	3.9
1993	5,197,057	735,833	1,315,719	886,907	762	4,890,800	682,323	7.2	2.2	12.1	58.0	17.7	3.7
1994	5,166,583	766,116	1,174,360	1,015,730	30,374	4,589,470	700,696	6.6	2.0	11.8	58.5	16.8	3.4
1995	5,539,600	768,064	1,268,503	1,062,489	14,918	4,992,468	719,495	6.9	2.1	12.0	58.5	17.4	3.6

336 AMERICA, NORTH and CENTRAL	- FOOD BALANCE SHEET OF FISH AND FISHERY PRODUCTS IN LIVE WEIGHT AND FISH CONTRIBUTION TO PROTEIN SUPPLY -	FAO-FIDI

YEAR	PRODUCTION	NON-FOOD USES	IMPORTS	EXPORTS	STOCKS VARIATIONS	TOTAL FOOD SUPPLY	POPULATION	PER CAPUT SUPPLY	FISH PROTEINS	ANIMAL PROTEINS	TOTAL PROTEINS	FISH/ANIMAL PROTEINS	FISH/TOTAL PROTEINS
	(.. metric tons in live weight					....)	(thousands)	(kilograms)	(grams per caput per day)			(%)	(%)
1961	4,187,885	1,445,687	996,367	700,085	-1,008	3,038,473	279,216	10.9	2.6	50.5	85.0	5.1	3.0
1962	4,324,523	1,485,716	1,103,780	770,625	-263	3,172,399	284,486	11.2	2.6	50.3	84.4	5.3	3.1
1963	4,214,875	1,330,356	1,118,298	780,915	473	3,222,575	289,703	11.1	2.6	51.0	85.4	5.1	3.0
1964	4,105,932	1,193,096	1,172,032	874,761	-667	3,209,440	294,833	10.9	2.6	51.9	86.6	5.0	3.0
1965	4,276,826	1,279,502	1,137,188	933,215	1,013	3,202,310	299,858	10.7	2.6	51.5	86.2	5.1	3.0
1966	4,241,651	1,173,566	1,337,359	965,788	-18,963	3,420,694	304,754	11.2	2.8	52.7	87.3	5.4	3.2
1967	4,134,544	1,111,406	1,310,811	983,708	-15,087	3,335,153	309,532	10.8	2.6	52.9	87.8	4.9	2.9
1968	4,429,594	1,314,284	1,595,037	981,100	-24,277	3,704,970	314,254	11.8	2.7	53.5	88.5	5.1	3.1
1969	4,326,159	1,390,238	1,553,486	945,907	1,396	3,544,896	319,005	11.1	2.7	52.8	88.3	5.0	3.0
1970	4,794,318	1,540,300	1,621,322	1,054,589	34,165	3,854,917	323,848	11.9	2.9	50.8	85.6	5.6	3.3
1971	4,889,162	1,595,130	1,637,600	1,035,147	-4,596	3,891,890	328,807	11.8	2.8	52.9	88.1	5.2	3.1
1972	4,680,913	1,427,900	2,114,380	1,042,945	9,792	4,334,241	333,861	13.0	3.0	53.2	87.7	5.6	3.4
1973	4,794,840	1,448,175	2,061,418	1,077,991	6,630	4,336,722	338,973	12.8	3.1	51.5	86.7	5.9	3.5
1974	4,636,988	1,459,306	1,933,919	867,671	-7,742	4,236,188	344,060	12.3	2.9	51.9	86.7	5.6	3.4
1975	4,723,982	1,481,180	1,823,432	940,097	-19,201	4,106,936	349,142	11.8	2.7	50.6	86.8	5.3	3.1
1976	5,191,147	1,656,364	2,018,369	1,053,844	-7,217	4,492,091	354,149	12.7	2.9	52.9	89.0	5.6	3.3
1977	5,359,474	1,623,555	2,010,138	1,254,102	23,543	4,515,497	359,122	12.6	2.9	52.7	88.3	5.5	3.3
1978	6,152,389	1,875,988	2,121,271	1,449,163	-1,406	4,947,102	364,077	13.6	3.2	52.3	88.2	6.0	3.6
1979	6,457,777	2,012,172	2,133,453	1,533,935	-6,973	5,038,151	369,042	13.7	3.1	51.9	88.5	6.1	3.6
1980	6,884,457	2,085,641	1,965,252	1,663,470	-12,820	5,087,777	374,038	13.6	3.3	52.7	89.1	6.2	3.7
1981	7,367,637	1,864,471	2,221,538	1,842,423	-48,288	5,833,992	379,070	15.4	3.4	52.7	91.1	6.5	3.8
1982	7,371,359	2,086,617	2,177,716	1,921,582	-13,902	5,526,974	384,136	14.4	3.3	51.9	89.6	6.3	3.6
1983	7,318,638	2,028,192	2,389,120	1,961,381	-23,173	5,695,012	389,251	14.6	3.3	52.7	90.2	6.3	3.7
1984	7,936,734	1,943,765	2,468,646	2,185,549	12,076	6,288,144	394,430	15.9	3.5	52.7	90.0	6.5	3.8
1985	8,381,412	2,152,065	2,812,460	2,562,627	-38,278	6,440,903	399,685	16.1	3.6	53.7	93.0	6.7	3.9
1986	8,663,234	1,888,266	2,908,579	3,172,705	-26,181	6,484,660	405,013	16.0	3.7	53.8	92.3	6.8	4.0
1987	9,623,812	2,121,670	3,151,619	3,357,444	-105,130	7,191,186	410,409	17.5	3.9	54.0	93.4	7.3	4.2
1988	9,634,535	1,947,191	2,954,640	3,656,972	26,762	7,011,774	415,883	16.9	3.8	53.7	93.5	7.1	4.1
1989	9,591,937	1,994,295	3,140,718	3,222,231	80,640	7,596,769	421,442	18.0	4.2	53.4	92.8	7.8	4.5
1990	9,583,843	1,796,357	2,881,886	3,228,221	62,002	7,503,094	427,087	17.6	4.1	53.5	93.6	7.6	4.4
1991	9,122,972	1,842,246	2,905,240	2,825,554	60,144	7,420,556	432,827	17.1	4.1	53.6	94.7	7.6	4.3
1992	8,782,009	1,383,532	2,869,030	2,614,788	13,915	7,666,634	438,640	17.5	4.0	54.4	95.3	7.4	4.2
1993	8,889,816	1,409,335	3,035,120	2,524,220	32,378	8,023,759	444,462	18.1	4.2	54.3	95.2	7.7	4.4
1994	8,858,250	1,578,491	3,151,080	2,462,826	17,290	7,985,303	450,207	17.7	4.1	55.1	96.6	7.5	4.3
1995	8,472,928	1,479,736	3,130,362	2,444,628	74,268	7,753,215	455,813	17.0	4.1	54.6	96.1	7.4	4.2

318 AMERICA, SOUTH	- FOOD BALANCE SHEET OF FISH AND FISHERY PRODUCTS IN LIVE WEIGHT AND FISH CONTRIBUTION TO PROTEIN SUPPLY -												FAO-FIDI
YEAR	PRODUCTION	NON-FOOD USES	IMPORTS	EXPORTS	STOCKS VARIATIONS	TOTAL FOOD SUPPLY	POPULATION	PER CAPUT SUPPLY	FISH PROTEINS	ANIMAL PROTEINS	TOTAL PROTEINS	FISH/ANIMAL PROTEINS	FISH/TOTAL PROTEINS
	(.. metric tons in live weight ....)						(thousands)	(kilograms)	(grams per caput per day)			(%)	(%)
1961	6,203,200	5,371,900	107,890	108,641	3,897	834,446	151,594	5.5	1.6	26.3	63.3	6.0	2.5
1962	8,204,600	7,247,200	113,208	120,357	13,575	963,826	155,830	6.2	1.8	26.6	63.3	6.7	2.8
1963	8,259,100	7,297,800	120,476	108,888	-4,494	968,395	160,166	6.1	1.7	26.5	63.4	6.6	2.8
1964	10,962,000	9,953,400	92,899	81,275	4,853	1,025,076	164,554	6.2	1.8	25.5	63.6	7.0	2.8
1965	8,928,900	7,889,000	82,644	107,145	1,074	1,016,473	168,959	6.0	1.7	25.8	64.3	6.7	2.7
1966	10,966,400	9,893,000	119,143	112,457	14,164	1,094,249	173,371	6.3	1.8	26.5	64.2	6.8	2.8
1967	12,016,500	10,861,342	142,077	112,344	0	1,184,891	177,799	6.7	1.9	26.3	65.2	7.2	2.9
1968	12,817,900	11,593,073	155,403	113,834	0	1,266,397	182,245	7.0	2.0	27.3	66.1	7.3	3.0
1969	11,191,600	9,995,053	180,400	137,016	0	1,239,931	186,718	6.6	1.9	27.6	65.6	6.9	2.9
1970	14,761,780	13,400,782	185,292	158,924	-4,103	1,383,263	191,226	7.2	2.0	27.4	65.0	7.4	3.1
1971	13,133,830	11,700,917	145,412	183,105	-3,358	1,391,864	195,761	7.1	2.0	25.8	64.5	7.7	3.1
1972	6,729,023	5,141,447	128,399	189,923	-3,361	1,522,691	200,326	7.6	2.1	25.6	63.1	8.2	3.3
1973	4,465,618	2,514,236	168,328	257,433	-16,039	1,846,238	204,951	9.0	2.5	26.3	62.5	9.7	4.1
1974	6,676,083	4,893,070	135,191	275,081	-22,781	1,620,342	209,678	7.7	2.2	27.3	63.7	8.0	3.4
1975	5,771,908	4,015,456	190,416	277,745	-21,151	1,647,972	214,535	7.7	2.1	28.4	64.5	7.4	3.3
1976	7,213,165	5,412,492	159,683	451,819	10,780	1,519,316	219,534	6.9	1.9	28.7	64.4	6.4	2.9
1977	5,657,847	3,528,561	169,889	646,642	22,118	1,674,651	224,660	7.5	2.0	28.8	64.4	7.0	3.1
1978	7,622,676	5,109,821	202,644	814,122	958	1,902,336	229,876	8.3	2.2	29.4	64.9	7.3	3.3
1979	8,625,715	5,729,676	273,059	874,839	-85,388	2,208,871	235,129	9.4	2.4	30.1	66.4	8.0	3.6
1980	7,783,347	4,695,405	258,690	1,195,787	-2,085	2,148,761	240,380	8.9	2.3	31.1	67.2	7.4	3.5
1981	8,290,183	5,323,861	222,460	1,212,254	17,512	1,994,041	245,613	8.1	2.1	30.6	66.7	7.0	3.2
1982	9,556,898	6,847,219	232,070	1,033,842	-666	1,907,241	250,832	7.6	2.0	29.4	65.7	6.8	3.1
1983	7,695,731	5,156,176	132,161	829,000	-14,575	1,828,141	256,047	7.1	1.9	28.6	64.4	6.7	3.0
1984	10,502,780	7,652,993	137,921	879,098	-21,372	2,087,242	261,277	8.0	2.1	29.0	65.5	7.3	3.3
1985	11,934,360	9,018,635	134,439	972,107	48,682	2,126,742	266,531	8.0	2.1	29.7	65.3	7.1	3.2
1986	14,140,110	11,017,917	247,405	993,087	29,357	2,405,863	271,812	8.9	2.3	31.4	66.3	7.4	3.5
1987	12,173,190	8,798,582	239,852	1,130,750	-17,567	2,466,139	277,104	8.9	2.3	31.7	67.2	7.3	3.5
1988	14,603,640	11,410,142	177,810	1,100,731	3,267	2,273,844	282,381	8.1	2.1	30.8	67.5	6.9	3.2
1989	16,004,040	12,476,246	299,615	1,244,839	-12,197	2,570,374	287,608	8.9	2.3	31.7	66.0	7.4	3.5
1990	14,453,000	10,870,797	337,495	1,236,465	-53,575	2,629,663	292,760	9.0	2.4	31.6	65.5	7.5	3.6
1991	15,402,760	12,037,446	314,348	1,434,529	30,379	2,275,519	297,828	7.6	2.0	32.6	67.6	6.2	3.0
1992	16,560,680	12,881,362	264,251	1,467,050	20,496	2,497,039	302,818	8.3	2.3	32.9	67.9	6.8	3.3
1993	17,835,620	13,899,672	404,930	1,611,295	8,816	2,738,393	307,743	8.9	2.4	33.3	69.0	7.1	3.4
1994	22,739,580	18,531,620	485,020	1,905,635	66,323	2,853,672	312,624	9.1	2.4	33.9	70.2	7.1	3.4
1995	19,997,310	15,353,049	641,599	2,069,485	-11,276	3,205,102	317,477	10.1	2.6	35.7	72.8	7.2	3.6

337 ASIA	- FOOD BALANCE SHEET OF FISH AND FISHERY PRODUCTS IN LIVE WEIGHT AND FISH CONTRIBUTION TO PROTEIN SUPPLY -	FAO-FIDI

YEAR	PRODUCTION	NON-FOOD USES	IMPORTS	EXPORTS	STOCKS VARIATIONS	TOTAL FOOD SUPPLY	POPULATION	PER CAPUT SUPPLY	FISH PROTEINS	ANIMAL PROTEINS	TOTAL PROTEINS	FISH/ANIMAL PROTEINS	FISH/TOTAL PROTEINS
	(.. metric tons in live weight ....)						(thousands)	(kilograms)	(grams per caput per day)			(%)	(%)
1961	14,848,650	1,477,879	624,033	754,273	-156,304	13,084,230	1,702,899	7.7	2.4	7.3	48.9	32.5	4.8
1962	15,366,050	1,462,171	532,026	830,777	32,795	13,637,920	1,739,326	7.8	2.5	7.6	49.8	33.5	5.1
1963	15,700,730	1,247,236	595,007	850,025	-47,974	14,150,500	1,777,542	8.0	2.6	8.0	49.6	32.8	5.3
1964	16,326,930	1,606,143	672,735	1,024,825	149,885	14,518,580	1,817,862	8.0	2.5	8.1	50.7	31.0	5.0
1965	17,396,240	1,563,143	711,455	1,018,566	-930	15,525,060	1,860,462	8.3	2.8	8.5	51.5	32.5	5.3
1966	18,076,640	1,623,713	896,253	1,151,274	-21,960	16,175,950	1,905,386	8.5	2.8	8.7	51.2	32.3	5.5
1967	19,254,140	1,989,669	822,257	1,088,962	-28,574	16,969,190	1,952,418	8.7	2.9	8.7	50.4	32.6	5.6
1968	20,908,150	2,557,970	889,532	1,135,442	22,773	18,127,040	2,001,141	9.1	3.0	9.0	50.9	33.5	5.9
1969	21,428,150	2,999,080	936,784	1,257,887	-21,634	18,086,330	2,050,976	8.8	2.9	8.9	50.7	32.3	5.7
1970	22,538,550	3,071,880	985,290	1,548,733	-25,855	18,877,380	2,101,430	9.0	2.9	8.9	52.1	32.7	5.6
1971	24,325,520	3,601,795	1,107,662	1,616,954	-12,653	20,201,780	2,152,466	9.4	3.0	9.2	51.8	32.9	5.9
1972	25,718,540	3,999,369	1,266,106	1,860,113	28,749	21,153,920	2,203,980	9.6	3.1	9.4	51.8	32.8	6.0
1973	27,104,950	4,328,733	1,484,178	2,104,490	633	22,156,540	2,255,340	9.8	3.1	9.5	52.6	32.8	5.9
1974	27,831,230	4,286,199	1,482,278	1,966,185	-11,462	23,049,670	2,305,782	10.0	3.2	9.6	52.1	33.0	6.1
1975	28,145,820	4,225,745	1,585,047	2,017,456	-2,563	23,485,110	2,354,810	10.0	3.2	9.6	52.2	33.1	6.1
1976	28,963,050	4,622,040	1,729,108	2,175,377	11,736	23,906,480	2,402,099	10.0	3.1	9.7	52.0	31.9	6.0
1977	30,267,350	5,465,364	1,863,431	2,403,497	6,365	24,268,280	2,447,899	9.9	3.1	9.9	52.5	31.2	5.9
1978	30,376,780	5,427,496	2,031,436	2,583,200	-5,254	24,392,270	2,493,028	9.8	3.1	10.1	54.9	30.5	5.6
1979	30,147,500	5,720,162	2,400,104	2,690,535	16,259	24,153,180	2,538,662	9.5	3.0	10.5	55.5	28.9	5.4
1980	31,047,780	5,629,163	2,076,894	2,709,143	21,320	24,807,690	2,585,668	9.6	3.0	10.7	54.6	28.4	5.6
1981	32,473,520	6,446,920	2,187,763	2,643,149	-10,593	25,560,620	2,634,198	9.7	3.1	11.0	56.4	28.1	5.5
1982	33,517,480	6,564,688	2,381,870	2,619,114	-1,421	26,714,130	2,684,031	10.0	3.2	11.4	57.3	27.9	5.5
1983	35,564,610	7,133,690	2,462,732	2,785,525	-16,096	28,092,020	2,735,190	10.3	3.3	11.8	58.8	27.9	5.6
1984	37,618,250	8,032,317	2,867,870	2,952,388	12,785	29,514,200	2,787,603	10.6	3.4	12.3	60.1	27.4	5.6
1985	38,470,360	7,399,806	3,146,761	3,119,066	4,669	31,102,910	2,841,143	11.0	3.4	12.8	59.3	26.9	5.8
1986	41,691,800	8,651,269	3,708,319	3,712,352	-6,112	33,030,390	2,895,956	11.4	3.5	13.1	60.7	26.8	5.8
1987	43,789,200	8,836,075	4,298,194	4,204,394	7,381	35,054,320	2,951,887	11.9	3.6	13.4	60.7	26.9	5.9
1988	45,502,790	9,081,743	5,040,926	4,587,365	-32,106	36,842,160	3,008,129	12.3	3.7	14.0	61.1	26.4	6.0
1989	46,391,120	8,655,861	5,153,677	4,595,176	-25,557	38,267,860	3,063,607	12.5	3.8	14.3	62.7	26.7	6.1
1990	47,122,110	8,418,886	5,367,818	4,751,232	-29,467	39,289,850	3,117,556	12.6	3.8	14.7	62.3	26.2	6.2
1991	47,715,760	8,054,414	6,221,126	5,140,718	64,613	40,805,630	3,169,611	12.9	3.9	15.2	62.9	25.9	6.2
1992	50,468,070	7,323,726	6,710,112	5,544,523	13,276	44,321,940	3,219,988	13.8	4.1	15.9	63.8	25.9	6.5
1993	54,293,120	7,161,882	6,999,832	5,856,928	31,794	48,304,760	3,269,210	14.8	4.4	16.8	64.1	26.0	6.8
1994	58,343,460	6,623,545	7,564,742	6,218,784	-22,292	53,042,340	3,318,096	16.0	4.7	17.9	65.8	26.1	7.1
1995	63,206,520	6,844,632	7,725,640	6,189,532	66,362	57,963,120	3,367,234	17.2	4.9	18.8	67.1	26.2	7.4

335 EUROPE	- FOOD BALANCE SHEET OF FISH AND FISHERY PRODUCTS IN LIVE WEIGHT AND FISH CONTRIBUTION TO PROTEIN SUPPLY -	FAO-FIDI

YEAR	PRODUCTION	NON-FOOD USES	IMPORTS	EXPORTS	STOCKS VARIATIONS	TOTAL FOOD SUPPLY	POPULATION	PER CAPUT SUPPLY	FISH PROTEINS	ANIMAL PROTEINS	TOTAL PROTEINS	FISH/ANIMAL PROTEINS	FISH/TOTAL PROTEINS
	(..		metric tons in live weight		....)		(thousands)	(kilograms)		(grams per caput per day)		(%)	(%)
1961	8,125,246	1,702,495	1,976,854	2,359,841	-121,938	5,917,826	438,322	13.5	3.8	40.4	85.6	9.5	4.5
1962	8,322,263	1,909,406	2,096,967	2,544,409	31,839	5,997,256	442,424	13.6	3.8	41.0	86.4	9.3	4.4
1963	8,690,607	1,959,817	2,243,096	2,608,090	1,314	6,367,112	446,573	14.3	4.0	41.7	86.8	9.5	4.6
1964	9,475,632	2,488,514	2,304,086	2,635,814	-41,668	6,613,720	450,578	14.7	4.1	42.3	87.2	9.8	4.7
1965	10,665,000	3,467,273	2,468,696	2,795,449	26,286	6,897,264	454,300	15.2	4.2	42.9	87.8	9.8	4.8
1966	11,350,100	4,086,733	2,412,092	2,797,308	-48,773	6,829,376	457,682	14.9	4.1	43.6	88.4	9.4	4.6
1967	11,839,440	4,486,154	2,484,658	2,727,231	-98,350	7,012,366	460,764	15.2	4.2	44.9	89.1	9.3	4.7
1968	11,701,370	4,136,638	2,605,245	2,861,096	9,656	7,318,533	463,630	15.8	4.3	45.8	89.7	9.4	4.8
1969	11,156,570	3,554,290	2,550,867	3,104,069	96,753	7,145,836	466,413	15.3	4.2	46.2	89.8	9.0	4.6
1970	11,876,720	4,048,587	2,807,630	3,282,291	100,873	7,454,348	469,207	15.9	4.3	47.8	91.1	9.0	4.7
1971	11,985,980	4,223,112	2,898,637	3,227,536	-42,325	7,391,642	472,032	15.7	4.4	48.6	91.3	9.0	4.8
1972	12,250,590	4,515,050	3,082,160	3,458,993	12,042	7,370,749	474,849	15.5	4.3	49.3	91.9	8.8	4.7
1973	12,559,340	4,825,705	3,138,419	3,501,884	60,748	7,430,926	477,627	15.6	4.4	50.0	92.8	8.7	4.7
1974	12,673,090	5,031,208	3,195,513	3,221,588	-49,841	7,565,956	480,317	15.8	4.4	50.9	93.4	8.6	4.7
1975	12,569,420	4,823,247	3,083,119	3,284,572	-4,924	7,539,798	482,880	15.6	4.4	51.6	93.6	8.5	4.7
1976	13,508,380	5,737,613	3,227,530	3,393,088	-76,407	7,528,800	485,325	15.5	4.5	52.5	94.4	8.5	4.8
1977	13,403,160	5,961,918	3,276,297	3,346,536	-89,666	7,281,330	487,662	14.9	4.3	52.8	94.9	8.2	4.6
1978	12,621,460	5,357,191	3,536,491	3,604,505	2,469	7,198,724	489,859	14.7	4.3	54.2	96.3	7.9	4.4
1979	12,445,700	5,251,668	3,899,912	3,998,025	39,653	7,135,572	491,873	14.5	4.2	55.2	97.4	7.6	4.3
1980	12,615,500	5,201,877	4,111,062	3,968,728	-89,321	7,466,653	493,682	15.1	4.4	55.9	98.0	7.9	4.5
1981	12,805,790	4,835,698	4,106,484	4,437,804	9,564	7,648,335	495,258	15.4	4.4	55.5	97.1	7.9	4.5
1982	12,432,490	4,314,594	4,314,172	4,566,231	-77,930	7,787,902	496,630	15.7	4.4	55.1	96.4	8.0	4.6
1983	12,908,320	4,541,997	4,434,902	4,736,465	109,120	8,173,892	497,909	16.4	4.6	55.8	96.8	8.2	4.7
1984	13,257,380	4,961,298	4,603,884	4,675,562	19,806	8,244,224	499,246	16.5	4.6	56.2	98.1	8.2	4.7
1985	13,080,850	4,708,155	5,233,322	5,071,662	52,891	8,587,252	500,745	17.2	4.7	57.4	99.3	8.3	4.8
1986	12,870,860	4,460,270	5,426,258	5,150,476	91,578	8,777,955	502,453	17.5	4.9	57.2	99.4	8.5	4.9
1987	12,897,820	4,686,814	6,206,411	5,274,362	-86,150	9,056,910	504,329	18.0	5.1	57.9	100.1	8.7	5.1
1988	13,218,530	5,172,556	6,718,129	5,410,393	-122,735	9,230,966	506,275	18.2	5.1	58.0	100.8	8.8	5.1
1989	12,625,250	4,656,957	7,117,478	5,857,464	2,624	9,230,940	508,149	18.2	5.2	57.7	100.4	9.0	5.2
1990	11,450,600	3,938,279	7,953,457	6,123,456	31,987	9,374,310	509,846	18.4	5.1	58.0	100.4	8.9	5.1
1991	11,454,000	3,959,969	8,120,416	6,475,828	28,413	9,167,024	511,331	17.9	5.1	57.6	99.9	8.8	5.1
1992	12,738,700	5,152,983	8,170,774	6,535,909	128,047	9,348,611	512,635	18.2	5.2	57.1	98.2	9.0	5.3
1993	12,573,760	4,761,939	8,228,611	6,898,006	93,055	9,235,480	513,804	18.0	5.1	56.3	97.5	9.0	5.2
1994	12,830,940	4,622,601	8,892,611	7,509,860	39,178	9,630,252	514,913	18.7	5.3	56.0	97.0	9.4	5.4
1995	13,508,280	4,853,686	8,830,850	7,813,652	-61,427	9,610,371	516,014	18.6	5.2	56.6	97.7	9.2	5.4

339 OCEANIA	- FOOD BALANCE SHEET OF FISH AND FISHERY PRODUCTS IN LIVE WEIGHT AND FISH CONTRIBUTION TO PROTEIN SUPPLY -	FAO-FIDI

YEAR	PRODUCTION	NON-FOOD USES	IMPORTS	EXPORTS	STOCKS VARIATIONS	TOTAL FOOD SUPPLY	POPULATION	PER CAPUT SUPPLY	FISH PROTEINS	ANIMAL PROTEINS	TOTAL PROTEINS	FISH/ANIMAL PROTEINS	FISH/TOTAL PROTEINS
	(.. metric tons in live weight ....)						(thousands)	(kilograms)	(grams per caput per day)			(%)	(%)
1961	135,530	5,300	104,004	13,706	0	220,528	16,161	13.7	3.2	52.9	84.5	6.1	3.8
1962	142,830	5,600	100,490	21,155	0	216,565	16,521	13.1	3.1	54.6	86.3	5.7	3.6
1963	148,480	5,600	64,050	19,860	0	187,070	16,884	11.1	2.8	54.3	86.8	5.2	3.3
1964	153,880	5,200	117,881	24,999	0	241,562	17,247	14.0	3.3	55.5	87.9	6.0	3.8
1965	165,580	4,700	120,723	30,410	0	251,192	17,607	14.3	3.3	55.0	86.4	6.0	3.8
1966	185,880	6,050	146,430	46,426	0	279,834	17,960	15.6	3.7	55.4	89.1	6.6	4.1
1967	193,130	5,900	142,499	50,607	0	279,122	18,306	15.3	3.6	53.1	86.1	6.8	4.2
1968	202,630	5,900	146,755	66,917	0	276,567	18,655	14.8	3.6	55.7	88.3	6.4	4.1
1969	184,230	5,550	161,201	64,478	0	275,404	19,024	14.5	3.6	56.1	88.4	6.4	4.1
1970	208,430	10,860	166,749	73,818	0	290,501	19,419	15.0	3.7	55.6	88.3	6.6	4.1
1971	244,030	11,680	210,806	114,826	0	328,330	19,849	16.5	4.0	58.3	90.3	6.9	4.4
1972	243,730	15,760	192,940	134,751	-1,000	285,159	20,306	14.0	3.5	59.6	91.3	5.8	3.8
1973	277,080	18,800	196,585	120,438	-11,000	323,427	20,765	15.6	4.0	58.8	88.6	6.8	4.5
1974	300,155	29,576	236,981	133,995	-5,950	367,615	21,192	17.4	4.4	55.2	85.6	8.0	5.2
1975	250,870	10,669	175,321	98,638	9,050	325,933	21,562	15.1	3.7	62.4	92.7	5.9	4.0
1976	286,168	12,256	198,628	119,674	-28,706	324,160	21,866	14.8	3.8	62.8	92.6	6.0	4.1
1977	289,103	16,084	205,480	149,744	12,986	341,741	22,115	15.5	3.9	61.1	91.2	6.3	4.2
1978	336,658	15,804	214,285	188,136	-1,840	345,163	22,334	15.5	3.9	63.6	93.1	6.2	4.2
1979	363,749	18,310	233,530	218,091	-1,861	359,017	22,561	15.9	4.1	60.1	90.1	6.8	4.5
1980	417,997	18,544	253,103	279,164	12,942	386,335	22,824	16.9	4.4	57.5	87.4	7.6	5.0
1981	450,277	13,112	276,800	274,040	1,188	441,113	23,129	19.1	4.9	58.2	88.5	8.4	5.5
1982	451,349	18,471	265,597	253,291	2,262	447,447	23,470	19.1	4.7	59.4	90.6	7.9	5.2
1983	506,517	19,034	258,611	302,734	-7,307	436,053	23,839	18.3	4.5	58.5	88.9	7.7	5.1
1984	520,616	15,463	281,597	308,463	-3,414	474,873	24,223	19.6	4.9	58.2	90.2	8.4	5.4
1985	515,830	32,204	304,443	306,294	5,311	487,085	24,612	19.8	5.0	58.9	91.1	8.5	5.5
1986	557,476	45,734	308,150	319,860	9,386	509,418	25,005	20.4	5.2	58.6	89.7	8.8	5.7
1987	618,437	72,097	311,387	332,009	1,274	526,992	25,405	20.7	5.3	59.0	90.1	9.0	5.9
1988	665,402	49,054	332,220	411,891	-1,071	535,607	25,808	20.8	5.2	57.9	89.2	9.0	5.8
1989	676,378	50,399	338,224	421,621	5,367	547,948	26,208	20.9	5.3	59.0	90.3	8.9	5.8
1990	746,334	51,180	329,706	432,944	-38,848	553,068	26,604	20.8	5.2	59.2	90.9	8.8	5.8
1991	829,343	65,514	316,479	552,081	398	528,625	26,815	19.7	4.9	57.4	90.2	8.6	5.5
1992	916,881	74,967	318,596	624,488	17,430	553,452	27,192	20.4	5.2	58.2	89.6	8.9	5.8
1993	887,139	58,913	331,596	597,411	18,783	581,194	27,564	21.1	5.4	59.1	91.7	9.1	5.9
1994	895,613	68,266	347,374	604,052	374	571,043	27,929	20.5	5.2	59.6	91.8	8.7	5.7
1995	1,020,058	83,465	313,874	688,447	-10,197	551,182	28,306	19.5	5.1	57.5	90.8	8.8	5.6

469 AREA OF THE FORMER USSR	- FOOD BALANCE SHEET OF FISH AND FISHERY PRODUCTS IN LIVE WEIGHT AND FISH CONTRIBUTION TO PROTEIN SUPPLY -												FAO-FIDI
YEAR	PRODUCTION	NON-FOOD USES	IMPORTS	EXPORTS	STOCKS VARIATIONS	TOTAL FOOD SUPPLY	POPULATION	PER CAPUT SUPPLY	FISH PROTEINS	ANIMAL PROTEINS	TOTAL PROTEINS	FISH/ANIMAL PROTEINS	FISH/TOTAL PROTEINS
	(.. metric tons in live weight ....)						(thousands)	(kilograms)	(grams per caput per day)			(%)	(%)
1961	3,236,000	97,800	32,037	113,697	-556	3,055,985	217,904	14.0	5.5	36.4	95.9	15.1	5.7
1962	3,601,800	97,400	64,660	134,191	-4,444	3,430,426	221,411	15.5	5.9	38.2	98.0	15.4	6.0
1963	3,963,300	128,300	79,725	188,656	6,111	3,732,180	224,794	16.6	6.4	40.2	99.7	15.9	6.4
1964	4,475,700	233,600	83,048	191,387	-3,333	4,130,429	227,984	18.1	6.6	37.4	97.3	17.5	6.7
1965	5,099,900	364,100	57,628	257,939	3,333	4,538,822	230,936	19.7	6.6	38.2	97.0	17.4	6.8
1966	5,348,800	441,100	47,799	299,025	-4,444	4,652,028	233,621	19.9	6.5	39.8	97.3	16.4	6.7
1967	5,777,200	520,300	53,235	282,909	1,667	5,028,891	236,061	21.3	6.9	42.5	99.2	16.2	6.9
1968	6,082,100	782,500	43,924	281,924	0	5,061,600	238,334	21.2	6.9	44.0	100.1	15.8	6.9
1969	6,498,400	823,300	40,423	258,325	-4,444	5,452,754	240,555	22.7	7.3	44.9	101.0	16.2	7.2
1970	7,207,929	1,058,400	45,018	314,646	1,667	5,881,566	242,806	24.2	7.7	46.2	102.4	16.6	7.5
1971	7,298,840	1,369,800	26,908	342,481	3,889	5,617,355	245,117	22.9	7.4	46.7	102.2	15.8	7.2
1972	7,860,605	1,133,200	25,199	304,593	6,667	6,454,678	247,472	26.1	8.1	47.8	102.2	16.9	7.9
1973	8,636,567	1,713,400	29,141	318,052	-556	6,633,700	249,844	26.6	8.6	47.6	102.3	18.1	8.4
1974	9,200,158	1,718,600	42,952	428,712	-1,663	7,094,134	252,187	28.1	9.1	51.0	105.2	17.8	8.6
1975	10,025,040	2,254,552	56,461	573,796	3,885	7,257,034	254,469	28.5	9.3	52.7	106.2	17.6	8.7
1976	10,167,870	2,733,650	29,867	542,686	0	6,921,398	256,687	27.0	8.0	50.3	104.1	16.0	7.7
1977	9,229,514	2,227,250	40,721	501,779	0	6,541,206	258,864	25.3	7.7	49.5	102.5	15.6	7.5
1978	9,046,691	1,877,400	85,027	541,507	0	6,712,811	261,025	25.7	8.1	50.7	103.9	16.0	7.8
1979	9,049,666	1,690,700	116,670	562,556	0	6,913,080	263,207	26.3	8.4	51.2	104.1	16.3	8.0
1980	9,515,640	2,106,100	189,426	563,584	0	7,035,382	265,436	26.5	8.5	50.4	102.7	16.8	8.3
1981	9,566,299	1,716,400	71,012	455,410	0	7,465,501	267,698	27.9	8.6	50.2	102.0	17.2	8.5
1982	9,990,642	2,199,537	55,952	369,112	0	7,477,945	269,981	27.7	8.8	50.0	102.4	17.6	8.6
1983	9,816,651	2,334,932	333,222	573,378	0	7,241,564	272,315	26.6	8.5	50.7	102.4	16.9	8.4
1984	10,592,920	2,487,536	387,869	591,598	0	7,901,658	274,737	28.8	9.2	52.4	103.9	17.5	8.8
1985	10,522,830	2,140,514	443,464	677,026	0	8,148,755	277,261	29.4	9.5	53.6	105.3	17.7	9.0
1986	11,259,960	2,330,228	467,020	861,395	0	8,535,352	279,913	30.5	9.5	54.6	106.0	17.4	9.0
1987	11,159,620	2,740,330	514,231	816,591	0	8,116,926	282,644	28.7	9.2	54.9	106.6	16.7	8.6
1988	11,332,060	3,065,035	637,764	1,034,533	0	7,870,260	285,296	27.6	8.9	55.4	106.3	16.0	8.4
1989	11,310,090	3,188,670	600,872	877,863	0	7,844,435	287,659	27.3	8.9	56.4	106.4	15.9	8.4
1990	10,389,030	2,653,265	572,880	1,250,862	0	7,057,782	289,586	24.4	8.4	55.9	106.5	15.0	7.9
1991	9,406,799	3,690,790	276,516	901,150	-120,852	4,970,524	291,021	17.1	6.4	48.8	98.8	13.1	6.5
1992	6,845,528	2,283,918	215,011	905,852	42,827	3,913,594	292,012	13.4	4.5	41.5	88.7	10.9	5.1
1993	5,410,864	1,087,960	192,950	1,508,970	-6,228	3,001,256	292,642	10.3	3.5	39.9	87.9	8.7	4.0
1994	4,529,362	915,473	577,063	1,952,468	27,246	2,265,730	293,043	7.7	2.8	40.0	86.2	7.1	3.3
1995	5,216,362	918,629	713,076	1,961,835	56,812	3,105,787	293,319	10.6	3.4	37.7	83.9	9.0	4.1

376 LOW-INCOME FOOD-DEFICIT COUNTRIES	- FOOD BALANCE SHEET OF FISH AND FISHERY PRODUCTS IN LIVE WEIGHT AND FISH CONTRIBUTION TO PROTEIN SUPPLY -	FAO-FIDI

YEAR	PRODUCTION	NON-FOOD USES	IMPORTS	EXPORTS	STOCKS VARIATIONS	TOTAL FOOD SUPPLY	POPULATION	PER CAPUT SUPPLY	FISH PROTEINS	ANIMAL PROTEINS	TOTAL PROTEINS	FISH/ANIMAL PROTEINS	FISH/TOTAL PROTEINS
	(..........	....	metric tons in live weight	....	....	..........)	(thousands)	(kilograms)		(grams per caput per day)		(%)	(%)
1961	7,982,750	458,550	715,553	379,551	-2,906	7,857,296	1,712,178	4.6	1.4	6.1	46.8	22.1	2.9
1962	8,391,950	476,400	581,744	401,897	-4,513	8,090,864	1,750,315	4.6	1.5	6.4	47.6	23.7	3.2
1963	8,586,730	429,853	678,775	406,544	-6,200	8,422,908	1,790,318	4.7	1.5	6.7	47.4	23.2	3.3
1964	9,456,730	661,155	716,176	499,484	-7,579	9,004,688	1,832,485	4.9	1.6	6.9	48.6	23.5	3.4
1965	9,833,230	584,609	656,736	445,164	-21,868	9,438,332	1,876,944	5.0	1.7	7.1	49.1	23.7	3.4
1966	10,351,520	768,693	811,364	490,691	-16,359	9,887,138	1,923,697	5.1	1.7	7.2	48.5	23.9	3.5
1967	10,466,500	714,704	728,927	483,279	-28,517	9,968,925	1,972,583	5.1	1.7	7.2	47.5	23.6	3.6
1968	11,030,520	724,800	703,540	505,494	-8,434	10,495,340	2,023,160	5.2	1.7	7.3	47.8	23.8	3.6
1969	11,469,100	883,024	729,288	560,494	-13,102	10,741,760	2,074,745	5.2	1.7	7.2	47.8	23.5	3.5
1970	11,963,550	897,942	758,009	711,781	-23,074	11,088,760	2,126,674	5.2	1.7	7.2	49.3	24.0	3.5
1971	12,941,240	1,020,226	805,434	808,676	-4,537	11,913,240	2,179,128	5.5	1.8	7.4	48.6	24.5	3.8
1972	13,893,260	1,489,246	822,379	877,396	10,297	12,359,300	2,231,962	5.5	1.9	7.5	48.2	24.6	3.8
1973	14,685,120	1,637,018	831,939	990,346	7,583	12,897,270	2,284,702	5.7	1.9	7.5	48.9	24.6	3.8
1974	15,355,940	1,482,114	926,636	973,111	4,443	13,831,800	2,336,900	5.9	1.9	7.7	48.6	25.3	4.0
1975	15,424,560	1,366,684	1,031,880	826,927	-8,722	14,254,100	2,388,212	6.0	2.0	7.6	48.7	25.7	4.0
1976	15,773,920	1,358,168	1,179,533	951,847	-11,870	14,631,570	2,438,422	6.0	1.8	7.6	48.4	23.6	3.7
1977	16,673,010	1,470,511	1,063,296	1,039,376	19,544	15,245,960	2,487,714	6.1	1.8	7.7	48.8	23.7	3.8
1978	16,861,580	1,764,439	1,314,359	1,132,142	-3,899	15,275,470	2,536,758	6.0	1.8	7.9	51.2	23.1	3.6
1979	16,643,160	1,748,906	1,444,052	1,233,965	13,433	15,117,770	2,586,465	5.8	1.8	8.4	51.8	21.6	3.5
1980	17,337,720	1,811,730	1,523,388	1,346,744	41,542	15,744,170	2,637,493	6.0	1.8	8.6	51.0	21.3	3.6
1981	17,867,560	1,779,427	1,859,461	1,376,217	3,850	16,575,230	2,690,100	6.2	1.9	8.9	52.7	21.2	3.6
1982	18,760,040	1,776,692	1,991,894	1,362,117	-11,818	17,601,310	2,744,104	6.4	2.0	9.1	53.4	21.6	3.7
1983	19,866,670	1,872,480	1,616,024	1,517,968	8,347	18,100,590	2,799,617	6.5	2.0	9.3	54.4	21.2	3.6
1984	21,689,840	2,465,223	1,524,523	1,624,450	9,489	19,134,180	2,856,719	6.7	2.0	9.6	55.3	20.9	3.6
1985	23,160,830	2,651,248	1,534,047	1,730,570	14,972	20,328,030	2,915,276	7.0	2.1	10.5	56.3	19.7	3.7
1986	25,156,370	2,713,811	1,584,067	1,971,773	921	22,055,770	2,975,723	7.4	2.2	10.9	57.6	20.0	3.8
1987	26,789,080	2,206,636	1,824,478	2,350,955	1,271	24,057,240	3,037,772	7.9	2.3	11.2	57.6	20.4	3.9
1988	28,921,540	2,477,109	1,886,394	2,527,718	-7,460	25,795,650	3,100,478	8.3	2.4	11.6	58.0	20.4	4.1
1989	30,445,860	2,724,406	2,174,040	2,534,886	-48,199	27,312,410	3,163,224	8.6	2.5	11.9	59.6	20.8	4.1
1990	31,746,110	2,603,411	2,242,438	2,774,196	-18,942	28,592,000	3,224,680	8.9	2.5	12.3	59.1	20.6	4.3
1991	33,477,240	2,384,727	1,994,746	3,058,263	40,577	30,069,580	3,284,746	9.2	2.6	12.7	59.6	20.6	4.4
1992	36,207,760	2,388,438	2,302,707	3,315,729	-2,273	32,804,020	3,347,627	9.8	2.8	13.3	60.3	20.9	4.6
1993	40,102,140	2,666,281	2,314,860	3,736,750	7,448	36,021,410	3,405,618	10.6	3.0	14.0	60.4	21.1	4.9
1994	44,502,790	2,969,163	2,273,728	4,089,498	40,182	39,758,040	3,463,736	11.5	3.2	14.9	62.1	21.3	5.1
1995	50,301,020	3,821,763	2,469,481	4,255,572	19,114	44,711,650	3,522,576	12.7	3.4	15.8	63.1	21.8	5.4

Note: for the list of countries see page 18.

SECTION 2

FISH AND FISHERY PRODUCTS

PROVISIONAL FOOD BALANCE SHEETS AND FISH CONTRIBUTION TO PROTEIN SUPPLY BY COUNTRY

FROM 1961 TO 1995

(Live Weight)

SECTION 2	- INDEX FOR FOOD BALANCE SHEETS BY COUNTRY -	FAO-FIDI

Note: countries are indicated by the FAO multilingual name used for statistical purposes. They are listed in alphabetical order and preceded by the FAOSTAT code.

2 AFGHANISTAN	- FOOD BALANCE SHEET OF FISH AND FISHERY PRODUCTS IN LIVE WEIGHT AND FISH CONTRIBUTION TO PROTEIN SUPPLY -	FAO-FIDI

YEAR	PRODUCTION	NON-FOOD USES	IMPORTS	EXPORTS	STOCKS VARIATIONS	TOTAL FOOD SUPPLY	POPULATION	PER CAPUT SUPPLY	FISH PROTEINS	ANIMAL PROTEINS	TOTAL PROTEINS	FISH/ANIMAL PROTEINS	FISH/TOTAL PROTEINS
	(..		metric tons in live weight			....)	(thousands)	(kilograms)		(grams per caput per day)		(%)	(%)
1961	300	0	...	...	0	300	11,014	0.0	0.0	10.9	69.2	0.1	0.0
1962	300	0	...	...	0	300	11,262	0.0	0.0	10.8	67.5	0.1	0.0
1963	300	0	...	...	0	300	11,521	0.0	0.0	11.4	62.7	0.1	0.0
1964	300	0	...	...	0	300	11,791	0.0	0.0	11.5	67.8	0.1	0.0
1965	300	0	...	...	0	300	12,071	0.0	0.0	12.0	68.1	0.1	0.0
1966	300	0	...	...	0	300	12,358	0.0	0.0	12.8	64.0	0.1	0.0
1967	400	0	...	...	0	400	12,652	0.0	0.0	13.4	69.7	0.1	0.0
1968	400	0	...	...	0	400	12,957	0.0	0.0	13.9	68.7	0.1	0.0
1969	400	0	...	...	0	400	13,280	0.0	0.0	13.7	69.4	0.1	0.0
1970	400	0	...	...	0	400	13,623	0.0	0.0	11.8	59.1	0.1	0.0
1971	500	0	...	...	0	500	13,984	0.0	0.0	11.0	59.2	0.1	0.0
1972	500	0	...	...	0	500	14,356	0.0	0.0	10.3	62.4	0.1	0.0
1973	500	0	...	...	0	500	14,724	0.0	0.0	11.2	64.2	0.1	0.0
1974	500	0	...	...	0	500	15,070	0.0	0.0	11.3	64.2	0.1	0.0
1975	600	0	...	...	0	600	15,378	0.0	0.0	11.5	65.1	0.1	0.0
1976	600	0	...	...	0	600	15,653	0.0	0.0	12.1	66.9	0.1	0.0
1977	600	0	...	...	0	600	15,892	0.0	0.0	11.7	60.3	0.1	0.0
1978	600	0	...	...	0	600	16,064	0.0	0.0	11.7	63.5	0.1	0.0
1979	600	0	...	...	0	600	16,128	0.0	0.0	11.8	63.1	0.1	0.0
1980	700	0	...	...	0	700	16,063	0.0	0.0	12.2	61.4	0.1	0.0
1981	700	0	...	...	0	700	15,862	0.0	0.0	12.7	61.2	0.1	0.0
1982	700	0	...	...	0	700	15,549	0.1	0.0	13.0	60.5	0.1	0.0
1983	800	0	...	...	0	800	15,177	0.1	0.0	13.4	61.1	0.1	0.0
1984	800	0	...	...	0	800	14,814	0.1	0.0	12.9	58.9	0.1	0.0
1985	800	0	...	...	0	800	14,519	0.1	0.0	11.7	56.7	0.1	0.0
1986	800	0	...	...	0	800	14,300	0.1	0.0	10.2	55.0	0.2	0.0
1987	1,000	0	...	...	0	1,000	14,165	0.1	0.0	10.5	63.1	0.2	0.0
1988	1,000	0	...	...	0	1,000	14,162	0.1	0.0	10.8	58.2	0.2	0.0
1989	1,000	0	...	...	0	1,000	14,346	0.1	0.0	11.1	56.6	0.2	0.0
1990	1,100	0	...	...	0	1,100	14,754	0.1	0.0	11.3	54.3	0.2	0.0
1991	1,100	0	...	...	0	1,100	15,403	0.1	0.0	10.9	50.5	0.2	0.0
1992	1,200	0	...	...	0	1,200	16,276	0.1	0.0	10.3	44.3	0.2	0.1
1993	1,200	0	...	...	0	1,200	17,322	0.1	0.0	9.7	44.0	0.2	0.1
1994	1,300	0	...	...	0	1,300	18,470	0.1	0.0	9.1	40.9	0.2	0.1
1995	1,300	0	...	...	0	1,300	19,661	0.1	0.0	8.6	38.5	0.2	0.1

3 ALBANIA	- FOOD BALANCE SHEET OF FISH AND FISHERY PRODUCTS IN LIVE WEIGHT AND FISH CONTRIBUTION TO PROTEIN SUPPLY -	FAO-FIDI

YEAR	PRODUCTION	NON-FOOD USES	IMPORTS	EXPORTS	STOCKS VARIATIONS	TOTAL FOOD SUPPLY	POPULATION	PER CAPUT SUPPLY	FISH PROTEINS	ANIMAL PROTEINS	TOTAL PROTEINS	FISH/ANIMAL PROTEINS	FISH/TOTAL PROTEINS
	(......................................		metric tons in live weight		....	)	(thousands)	(kilograms)		(grams per caput per day)		(%)	(%)
1961	3,200	0	...	...	0	3,200	1,660	1.9	0.6	19.0	64.1	2.9	0.9
1962	2,800	0	...	...	0	2,800	1,712	1.6	0.5	18.7	64.8	2.5	0.7
1963	3,000	0	...	...	0	3,000	1,764	1.7	0.5	18.3	62.9	2.7	0.8
1964	3,000	0	...	...	0	3,000	1,817	1.7	0.5	17.8	64.3	2.7	0.7
1965	4,000	0	...	...	0	4,000	1,870	2.1	0.6	18.1	64.1	3.4	1.0
1966	4,000	0	...	...	0	4,000	1,923	2.1	0.6	18.1	63.4	3.3	1.0
1967	5,000	0	...	...	0	5,000	1,976	2.5	0.7	19.5	64.5	3.7	1.1
1968	5,000	0	...	...	0	5,000	2,029	2.5	0.7	18.3	65.6	3.9	1.1
1969	6,000	0	...	...	0	6,000	2,083	2.9	0.8	18.3	67.1	4.5	1.2
1970	8,000	0	...	...	0	8,000	2,138	3.7	1.1	18.7	67.9	5.8	1.6
1971	8,000	0	...	...	0	8,000	2,195	3.6	1.1	19.4	66.9	5.4	1.6
1972	8,000	0	...	...	0	8,000	2,254	3.6	1.0	18.7	68.5	5.5	1.5
1973	8,000	0	...	...	0	8,000	2,312	3.5	1.0	19.3	69.0	5.2	1.4
1974	8,500	0	...	...	0	8,500	2,370	3.6	1.0	19.5	71.0	5.3	1.5
1975	8,500	0	...	...	0	8,500	2,424	3.5	1.0	19.4	69.2	5.2	1.5
1976	8,500	0	...	...	0	8,500	2,475	3.4	1.0	19.9	76.0	5.0	1.3
1977	8,500	0	...	...	0	8,500	2,524	3.4	1.0	20.3	80.2	4.8	1.2
1978	9,000	0	...	...	0	9,000	2,572	3.5	1.0	21.6	78.6	4.7	1.3
1979	9,000	0	...	...	0	9,000	2,620	3.4	1.0	22.0	75.5	4.5	1.3
1980	9,000	0	...	...	0	9,000	2,671	3.4	1.0	22.1	76.0	4.4	1.3
1981	9,376	0	...	...	0	9,376	2,724	3.4	1.0	22.9	79.6	4.3	1.2
1982	9,511	0	0	320	0	9,191	2,779	3.3	1.0	23.1	77.9	4.1	1.2
1983	8,973	0	0	140	0	8,833	2,837	3.1	0.9	23.3	82.3	3.9	1.1
1984	8,006	0	0	220	0	7,786	2,898	2.7	0.7	22.3	80.1	3.2	0.9
1985	11,399	0	0	0	0	11,399	2,962	3.9	1.1	21.9	75.6	5.1	1.5
1986	11,922	0	0	2,813	0	9,109	3,032	3.0	0.9	20.7	74.6	4.4	1.2
1987	13,134	0	0	2,583	0	10,551	3,105	3.4	1.0	21.5	74.9	4.6	1.3
1988	14,559	0	0	3,378	0	11,181	3,176	3.5	1.0	22.9	77.8	4.3	1.3
1989	11,961	0	0	3,252	0	8,709	3,239	2.7	0.8	23.3	76.9	3.3	1.0
1990	15,015	0	0	5,810	0	9,205	3,289	2.8	0.7	25.4	79.0	2.7	0.9
1991	3,871	0	0	2,170	0	1,701	3,324	0.5	0.2	28.3	71.0	0.5	0.2
1992	3,093	0	34	1,037	0	2,090	3,346	0.6	0.2	27.9	80.4	0.6	0.2
1993	3,000	0	1,154	1,307	0	2,847	3,359	0.9	0.2	30.3	83.3	0.6	0.2
1994	2,600	0	1,214	930	-8	2,876	3,369	0.9	0.2	36.6	84.1	0.5	0.2
1995	1,739	0	2,832	827	8	3,752	3,383	1.1	0.2	44.9	88.5	0.5	0.3

4 ALGERIA	- FOOD BALANCE SHEET OF FISH AND FISHERY PRODUCTS IN LIVE WEIGHT AND FISH CONTRIBUTION TO PROTEIN SUPPLY -	FAO-FIDI

YEAR	PRODUCTION	NON-FOOD USES	IMPORTS	EXPORTS	STOCKS VARIATIONS	TOTAL FOOD SUPPLY	POPULATION	PER CAPUT SUPPLY	FISH PROTEINS	ANIMAL PROTEINS	TOTAL PROTEINS	FISH/ANIMAL PROTEINS	FISH/TOTAL PROTEINS
	(.. metric tons in live weight ....)						(thousands)	(kilograms)	(grams per caput per day)			(%)	(%)
1961	30,400	0	18,647	10,804	0	38,243	11,007	3.5	1.1	11.3	48.0	9.3	2.2
1962	21,500	0	7,029	5,700	0	22,829	11,210	2.0	0.6	9.6	47.2	6.5	1.3
1963	16,900	0	3,451	3,800	0	16,551	11,421	1.5	0.5	7.9	43.4	5.8	1.1
1964	17,300	0	3,814	2,178	0	18,936	11,655	1.6	0.5	8.2	43.4	6.3	1.2
1965	18,300	0	1,014	1,900	0	17,414	11,923	1.5	0.5	8.4	45.2	5.6	1.1
1966	20,350	0	1,203	1,564	0	19,989	12,230	1.6	0.5	8.7	45.2	6.2	1.2
1967	21,150	0	403	2,700	0	18,853	12,572	1.5	0.5	8.6	46.0	5.6	1.0
1968	18,250	0	303	2,850	0	15,703	12,944	1.2	0.4	9.4	48.1	4.1	0.8
1969	23,150	0	303	2,854	0	20,599	13,338	1.5	0.5	9.6	47.8	5.1	1.0
1970	25,700	0	200	2,873	0	23,027	13,746	1.7	0.6	9.2	46.8	6.1	1.2
1971	23,831	0	0	2,900	0	20,931	14,168	1.5	0.5	9.7	47.5	5.1	1.0
1972	28,323	0	0	2,000	0	26,323	14,604	1.8	0.6	10.1	51.6	6.0	1.2
1973	31,243	0	100	4,175	0	27,168	15,057	1.8	0.6	9.9	52.1	6.0	1.1
1974	35,758	0	283	2,400	0	33,641	15,528	2.2	0.7	11.1	54.7	6.4	1.3
1975	37,693	0	492	2,033	0	36,152	16,018	2.3	0.8	11.7	57.7	6.4	1.3
1976	35,122	0	674	2,786	0	33,010	16,526	2.0	0.7	12.1	58.2	5.5	1.2
1977	43,475	0	597	3,760	0	40,312	17,050	2.4	0.8	13.1	59.5	6.0	1.3
1978	34,143	0	458	692	0	33,910	17,592	1.9	0.6	13.1	63.5	4.8	1.0
1979	38,678	0	1,182	498	0	39,363	18,155	2.2	0.7	12.8	64.6	5.6	1.1
1980	48,000	0	417	462	0	47,955	18,740	2.6	0.8	14.4	68.8	5.8	1.2
1981	56,000	0	2,723	475	0	58,248	19,348	3.0	1.0	15.8	68.3	6.2	1.4
1982	64,500	0	3,010	24	0	67,487	19,977	3.4	1.1	16.9	68.3	6.5	1.6
1983	65,000	0	5,178	69	0	70,109	20,616	3.4	1.1	18.3	70.0	6.1	1.6
1984	65,587	0	5,026	87	0	70,526	21,256	3.3	1.1	17.8	68.5	6.0	1.6
1985	66,131	45	12,058	135	0	78,009	21,887	3.6	1.2	19.0	74.2	6.1	1.6
1986	65,511	88	18,614	91	0	83,946	22,506	3.7	1.2	17.7	74.2	7.0	1.7
1987	94,368	0	9,017	132	0	103,253	23,115	4.5	1.5	18.9	74.9	8.0	2.0
1988	106,734	0	3,109	112	0	109,731	23,719	4.6	1.6	18.5	76.8	8.6	2.1
1989	99,564	0	4,712	87	0	104,189	24,324	4.3	1.4	19.5	78.2	7.3	1.8
1990	91,060	0	2,169	192	-444	92,593	24,935	3.7	1.3	19.9	78.0	6.3	1.6
1991	79,690	24	55	513	44	79,252	25,552	3.1	1.0	19.2	77.5	5.4	1.3
1992	95,274	0	138	1,077	344	94,680	26,175	3.6	1.2	19.3	79.3	6.4	1.6
1993	101,896	0	158	732	56	101,378	26,806	3.8	1.3	21.5	81.8	5.9	1.5
1994	135,402	0	924	875	0	135,451	27,450	4.9	1.7	20.3	82.8	8.2	2.0
1995	106,246	0	8,314	1,411	0	113,149	28,109	4.0	1.4	18.9	80.7	7.3	1.7

5 AMERICAN SAMOA	- FOOD BALANCE SHEET OF FISH AND FISHERY PRODUCTS IN LIVE WEIGHT AND FISH CONTRIBUTION TO PROTEIN SUPPLY -	FAO-FIDI

YEAR	PRODUCTION	NON-FOOD USES	IMPORTS	EXPORTS	STOCKS VARIATIONS	TOTAL FOOD SUPPLY	POPULATION	PER CAPUT SUPPLY	FISH PROTEINS	ANIMAL PROTEINS	TOTAL PROTEINS	FISH/ANIMAL PROTEINS	FISH/TOTAL PROTEINS
	(..	metric tons in live weight	....)				(thousands)	(kilograms)		(grams per caput per day)		(%)	(%)
1961	...	0	...	...	0	...	21	...	...	22.9	43.7	...	...
1962	...	0	...	...	0	...	21	...	...	22.2	42.6	...	...
1963	...	0	...	...	0	...	22	...	...	26.0	45.4	...	...
1964	...	0	...	...	0	...	23	...	...	22.2	42.1	...	...
1965	...	0	...	...	0	...	24	...	...	21.3	40.9	...	...
1966	...	0	...	...	0	...	25	...	...	19.3	39.3	...	...
1967	...	0	...	...	0	...	25	...	...	18.1	41.3	...	...
1968	...	0	...	...	0	...	26	...	...	17.5	41.2	...	...
1969	...	0	...	...	0	...	26	...	...	20.3	44.4	...	...
1970	...	0	...	...	0	...	27	...	...	20.2	45.3	...	
1971	...	0	...	...	0	...	28	...	...	25.9	51.7	...	...
1972	...	0	...	...	0	...	28	...	...	27.8	52.0	...	...
1973	100	0	...	...	0	100	29	3.5	1.0	34.4	55.4	2.8	1.8
1974	82	0	...	...	0	82	30	2.8	0.9	24.9	51.8	3.6	1.7
1975	136	0	...	...	0	136	30	4.5	1.5	29.7	57.3	4.9	2.6
1976	113	0	...	...	0	113	30	3.7	1.2	30.1	58.1	3.8	2.0
1977	202	0	...	...	0	202	31	6.6	2.0	35.0	62.9	5.6	3.1
1978	193	0	...	...	0	193	31	6.3	1.8	37.7	66.7	4.9	2.8
1979	183	0	...	...	0	183	31	5.9	1.7	43.9	72.1	3.9	2.4
1980	148	0	...	...	0	148	32	4.6	1.4	38.7	66.9	3.6	2.1
1981	153	0	...	...	0	153	33	4.6	1.4	31.1	64.7	4.5	2.2
1982	112	0	...	...	0	112	34	3.3	0.9	31.4	64.0	3.0	1.5
1983	112	0	...	...	0	112	36	3.1	1.0	34.3	63.7	2.8	1.5
1984	152	0	...	...	0	152	37	4.1	1.3	44.7	72.5	2.9	1.8
1985	118	0	...	...	0	118	39	3.0	0.9	42.2	69.4	2.2	1.3
1986	146	0	...	...	0	146	41	3.6	1.1	37.0	64.6	2.9	1.7
1987	67	0	...	...	0	67	42	1.6	0.5	29.3	59.4	1.8	0.9
1988	99	0	...	...	0	99	44	2.3	0.7	26.4	54.5	2.8	1.4
1989	99	0	...	...	0	99	45	2.2	0.7	30.3	58.0	2.4	1.3
1990	42	0	...	...	0	42	47	0.9	0.3	25.7	54.9	1.2	0.6
1991	45	0	...	...	0	45	48	0.9	0.3	33.8	61.2	0.9	0.5
1992	56	0	...	...	0	56	50	1.1	0.4	33.7	63.7	1.1	0.6
1993	34	0	...	...	0	34	51	0.7	0.2	32.3	61.3	0.7	0.4
1994	133	0	...	...	0	133	52	2.5	0.8	31.8	58.1	2.6	1.4
1995	175	0	...	...	0	175	54	3.2	1.1	31.3	56.8	3.5	1.9

7 ANGOLA	- FOOD BALANCE SHEET OF FISH AND FISHERY PRODUCTS IN LIVE WEIGHT AND FISH CONTRIBUTION TO PROTEIN SUPPLY -	FAO-FIDI

YEAR	PRODUCTION	NON-FOOD USES	IMPORTS	EXPORTS	STOCKS VARIATIONS	TOTAL FOOD SUPPLY	POPULATION	PER CAPUT SUPPLY	FISH PROTEINS	ANIMAL PROTEINS	TOTAL PROTEINS	FISH/ANIMAL PROTEINS	FISH/TOTAL PROTEINS
	(.. metric tons in live weight ....)						(thousands)	(kilograms)	(grams per caput per day)			(%)	(%)
1961	246,800	169,500	6,941	51,150	0	33,090	4,890	6.8	2.1	7.6	37.9	28.0	5.6
1962	274,800	178,700	8,988	59,714	0	45,374	4,962	9.1	2.4	8.0	38.6	30.3	6.2
1963	245,300	148,900	10,467	63,746	0	43,121	5,034	8.6	2.2	7.9	39.0	27.2	5.5
1964	362,000	261,900	9,082	47,672	0	61,510	5,106	12.1	3.3	9.5	41.9	35.1	7.9
1965	263,300	179,700	7,858	35,157	0	56,301	5,180	10.9	3.5	9.9	41.0	35.3	8.5
1966	335,300	241,300	9,132	43,592	0	59,539	5,256	11.3	3.6	10.0	41.0	35.9	8.8
1967	299,300	194,900	11,700	36,363	0	79,737	5,335	15.0	5.1	12.1	44.1	42.6	11.6
1968	303,100	184,000	13,705	45,608	0	87,198	5,416	16.1	5.9	12.7	42.7	46.4	13.8
1969	425,200	328,300	13,587	30,991	0	79,497	5,501	14.5	5.5	12.8	44.4	43.3	12.5
1970	374,501	256,700	13,069	40,357	0	90,513	5,588	16.2	5.6	13.3	44.9	42.1	12.5
1971	324,322	210,800	14,527	46,080	0	81,969	5,678	14.4	5.7	13.4	46.3	42.3	12.3
1972	606,656	500,700	7,101	53,559	0	59,497	5,771	10.3	5.2	12.8	45.2	40.4	11.4
1973	478,634	385,500	9,592	58,266	0	44,460	5,873	7.6	3.6	11.5	42.9	30.9	8.3
1974	399,759	312,409	9,523	42,818	0	54,054	5,989	9.0	4.1	12.0	44.4	34.3	9.3
1975	160,581	112,288	6,674	20,000	0	34,967	6,123	5.7	2.1	10.6	45.3	20.2	4.7
1976	81,542	27,000	...	...	0	54,542	6,277	8.7	3.6	11.8	45.5	30.6	7.9
1977	120,407	17,500	...	...	0	102,907	6,450	16.0	6.3	16.5	50.3	38.5	12.6
1978	125,630	25,000	...	...	0	100,630	6,636	15.2	6.1	16.6	51.3	36.6	11.9
1979	113,072	20,000	...	...	0	93,072	6,828	13.6	5.5	16.4	50.8	33.4	10.8
1980	85,085	20,000	...	...	0	65,085	7,019	9.3	4.3	15.3	50.8	27.8	8.4
1981	130,957	26,000	...	...	0	104,957	7,209	14.6	5.9	18.5	52.0	32.0	11.4
1982	111,488	26,000	97,292	...	0	182,780	7,399	24.7	9.4	20.4	51.0	45.9	18.4
1983	110,376	19,000	96,926	...	0	188,302	7,592	24.8	8.9	20.0	50.1	44.8	17.8
1984	72,200	11,500	132,353	...	0	193,053	7,793	24.8	8.7	20.3	48.5	42.9	18.0
1985	73,997	8,200	148,122	...	0	213,919	8,003	26.7	9.7	24.0	51.7	40.5	18.8
1986	57,436	7,300	162,928	...	0	213,064	8,225	25.9	9.4	19.6	46.5	47.8	20.2
1987	82,936	4,400	135,301	...	0	213,837	8,457	25.3	8.8	20.3	46.0	43.5	19.1
1988	101,783	2,700	124,661	3,105	0	220,639	8,700	25.4	9.2	20.3	45.0	45.1	20.3
1989	111,302	2,500	97,847	1,896	0	204,753	8,958	22.9	8.2	19.7	42.8	41.8	19.2
1990	106,945	1,140	69,674	947	0	174,532	9,229	18.9	6.8	17.0	42.7	39.8	15.9
1991	74,652	0	12,225	962	0	85,915	9,515	9.0	3.4	12.3	37.2	27.5	9.1
1992	75,430	0	10,373	1,456	-300	84,047	9,816	8.6	3.1	11.5	36.9	27.4	8.5
1993	80,758	0	8,752	2,279	-200	87,031	10,133	8.6	3.2	10.5	36.3	30.0	8.7
1994	77,924	0	8,281	2,221	-100	83,884	10,466	8.0	2.9	9.7	41.7	30.2	7.0
1995	93,847	0	9,538	4,318	600	99,667	10,816	9.2	3.3	10.0	42.4	32.8	7.8

Note: from 1976 to 1981 apparent consumption data not fully comparable as trade statistics not available.

258 ANGUILLA	- FOOD BALANCE SHEET OF FISH AND FISHERY PRODUCTS IN LIVE WEIGHT AND FISH CONTRIBUTION TO PROTEIN SUPPLY -												FAO-FIDI
YEAR	PRODUCTION	NON-FOOD USES	IMPORTS	EXPORTS	STOCKS VARIATIONS	TOTAL FOOD SUPPLY	POPULATION	PER CAPUT SUPPLY	FISH PROTEINS	ANIMAL PROTEINS	TOTAL PROTEINS	FISH/ANIMAL PROTEINS	FISH/TOTAL PROTEINS
	(..		metric tons in live weight		....)		(thousands)	(kilograms)	(grams per caput per day)			(%)	(%)
1961	...	0	0	0	0	...	6	...	...	...	...	...	...
1962	...	0	0	0	0	...	6	...	...	...	...	...	...
1963	...	0	0	0	0	...	6	...	...	...	...	...	...
1964	...	0	0	0	0	...	6	...	...	...	...	...	...
1965	...	0	0	0	0	...	6	...	...	...	...	...	...
1966	...	0	0	0	0	...	6	...	...	...	...	...	...
1967	...	0	0	0	0	...	6	...	...	...	...	...	...
1968	...	0	0	0	0	...	6	...	...	...	...	...	...
1969	...	0	0	0	0	...	6	...	...	...	...	...	...
1970	100	0	0	0	0	100	6	16.7	4.7	...	...	...	...
1971	100	0	0	0	0	100	6	16.7	4.7	...	...	...	...
1972	100	0	0	0	0	100	6	16.9	4.8	...	...	...	...
1973	100	0	0	0	0	100	6	16.9	4.8	...	...	...	...
1974	175	0	0	0	0	175	6	29.6	7.9	...	...	...	...
1975	175	0	0	0	0	175	6	29.2	7.8	...	...	...	...
1976	175	0	0	0	0	175	6	28.4	7.6	...	...	...	...
1977	225	0	0	0	0	225	6	35.3	9.5	...	...	...	...
1978	225	0	0	0	0	225	7	34.0	9.1	...	...	...	...
1979	225	0	0	0	0	225	7	32.9	8.8	...	...	...	...
1980	280	0	0	0	0	280	7	40.0	10.6	...	...	...	...
1981	280	0	0	0	0	280	7	39.5	10.5	...	...	...	...
1982	280	0	0	0	0	280	7	39.4	10.5	...	...	...	...
1983	310	0	0	0	0	310	7	43.8	11.7	...	...	...	...
1984	330	0	0	0	0	330	7	46.9	12.5	...	...	...	...
1985	330	0	0	0	0	330	7	47.1	12.6	...	...	...	...
1986	442	0	0	0	0	442	7	63.5	16.7	...	...	...	...
1987	419	0	0	0	0	419	7	60.6	16.0	...	...	...	...
1988	397	0	0	0	0	397	7	57.6	15.2	...	...	...	...
1989	374	0	0	0	0	374	7	54.1	14.4	...	...	...	...
1990	350	0	0	0	0	350	7	50.0	13.3	...	...	...	...
1991	351	0	0	0	0	351	7	49.0	13.5	...	...	...	...
1992	386	0	0	0	0	386	7	52.3	13.9	...	...	...	...
1993	330	0	0	0	0	330	8	43.2	11.6	...	...	...	...
1994	333	0	0	0	0	333	8	42.4	11.4	...	...	...	...
1995	350	0	0	0	0	350	8	43.8	11.7	...	...	...	...

8 ANTIGUA and BARBUDA	- FOOD BALANCE SHEET OF FISH AND FISHERY PRODUCTS IN LIVE WEIGHT AND FISH CONTRIBUTION TO PROTEIN SUPPLY -	FAO-FIDI

YEAR	PRODUCTION	NON-FOOD USES	IMPORTS	EXPORTS	STOCKS VARIATIONS	TOTAL FOOD SUPPLY	POPULATION	PER CAPUT SUPPLY	FISH PROTEINS	ANIMAL PROTEINS	TOTAL PROTEINS	FISH/ANIMAL PROTEINS	FISH/TOTAL PROTEINS
	(..		metric tons in live weight	....)			(thousands)	(kilograms)	(grams per caput per day)			(%)	(%)
1961	800	0	1,225	0	0	2,025	55	36.6	11.9	31.8	58.5	37.4	20.3
1962	800	0	1,250	0	0	2,050	56	36.9	12.0	32.7	58.3	36.7	20.6
1963	800	0	1,250	0	0	2,050	56	36.8	12.0	31.3	56.9	38.2	21.0
1964	700	0	1,425	0	0	2,125	56	38.1	12.5	31.9	60.8	39.3	20.6
1965	800	0	1,425	0	0	2,225	56	39.7	13.0	31.9	61.9	40.7	21.0
1966	800	0	1,425	0	0	2,225	56	39.6	13.0	34.2	61.4	37.9	21.1
1967	800	0	1,175	0	0	1,975	56	35.1	11.4	32.2	52.4	35.4	21.7
1968	800	0	925	100	0	1,625	57	28.8	9.3	29.9	52.0	31.0	17.9
1969	800	0	1,425	0	0	2,225	57	39.2	12.8	36.9	59.9	34.8	21.4
1970	900	0	1,175	100	0	1,975	57	34.7	11.2	38.8	63.9	29.0	17.6
1971	900	0	1,175	100	0	1,975	57	34.5	11.2	38.3	63.3	29.2	17.7
1972	1,200	0	1,175	100	0	2,275	58	39.4	12.6	38.5	63.6	32.6	19.8
1973	1,500	0	1,175	100	0	2,575	58	44.3	13.9	36.5	59.9	38.2	23.3
1974	1,590	0	925	90	0	2,425	59	41.4	12.8	35.1	59.9	36.5	21.4
1975	1,601	0	1,332	86	0	2,847	59	48.3	15.1	36.4	61.4	41.6	24.6
1976	1,613	0	1,525	86	0	3,052	59	51.3	15.7	35.7	60.5	44.1	26.0
1977	1,821	4	1,291	92	0	3,016	60	50.4	15.2	37.2	60.4	40.9	25.1
1978	1,620	1	1,195	187	0	2,628	60	43.6	13.1	39.5	60.3	33.0	21.6
1979	1,219	0	1,719	50	0	2,888	61	47.6	14.6	41.9	62.9	34.9	23.3
1980	1,171	0	2,369	533	0	3,007	61	49.3	15.7	48.2	71.2	32.5	22.0
1981	1,157	0	1,692	214	0	2,635	61	43.0	13.4	49.3	69.9	27.2	19.2
1982	950	0	1,348	216	0	2,082	61	33.9	11.1	48.2	70.1	23.0	15.8
1983	1,063	0	1,263	200	0	2,126	62	34.5	11.2	48.7	72.4	23.0	15.4
1984	1,465	0	1,350	85	0	2,730	62	44.2	13.3	56.1	85.8	23.8	15.5
1985	2,407	0	1,564	549	0	3,422	62	55.2	17.2	57.2	82.4	30.1	20.9
1986	2,400	0	1,461	300	0	3,561	62	57.1	17.7	57.6	82.7	30.7	21.4
1987	2,000	0	1,213	139	0	3,074	63	49.0	14.6	56.4	79.9	25.8	18.2
1988	1,500	0	1,565	194	0	2,871	63	45.5	14.1	58.4	82.6	24.2	17.1
1989	1,000	0	2,217	204	0	3,013	64	47.4	15.3	60.7	85.7	25.2	17.9
1990	885	0	1,942	205	0	2,622	64	41.0	13.2	61.5	87.6	21.4	15.0
1991	1,498	0	1,967	172	0	3,293	64	51.1	15.8	62.1	88.8	25.5	17.8
1992	1,712	0	2,081	36	0	3,757	65	58.0	17.7	61.9	88.9	28.6	19.9
1993	580	0	1,485	99	0	1,966	65	30.2	9.7	53.6	80.8	18.2	12.1
1994	629	0	1,552	100	0	2,081	66	31.7	10.2	53.8	80.9	19.0	12.6
1995	470	0	1,157	130	0	1,497	66	22.7	7.0	49.2	75.4	14.3	9.3

9 ARGENTINA	- FOOD BALANCE SHEET OF FISH AND FISHERY PRODUCTS IN LIVE WEIGHT AND FISH CONTRIBUTION TO PROTEIN SUPPLY -												FAO-FIDI
YEAR	PRODUCTION	NON-FOOD USES	IMPORTS	EXPORTS	STOCKS VARIATIONS	TOTAL FOOD SUPPLY	POPULATION	PER CAPUT SUPPLY	FISH PROTEINS	ANIMAL PROTEINS	TOTAL PROTEINS	FISH/ANIMAL PROTEINS	FISH/TOTAL PROTEINS
	(........ metric tons in live weight ....)						(thousands)	(kilograms)	(grams per caput per day)			(%)	(%)
1961	91,800	9,500	5,496	400	0	87,396	20,951	4.2	1.1	65.4	105.8	1.7	1.0
1962	92,300	13,300	4 543	1,053	0	82,491	21,286	3.9	1.0	66.1	103.8	1.5	1.0
1963	122,300	23,700	3,788	3,866	0	98,522	21,621	4.6	1.2	66.3	104.4	1.9	1.2
1964	158,300	29,900	5,493	1,570	0	132,323	21,954	6.0	1.7	56.8	95.6	3.0	1.8
1965	185,100	54,300	6,525	8,548	0	128,777	22,283	5.8	1.7	59.0	100.1	2.8	1.7
1966	221,200	101,100	9,800	5,366	0	124,534	22,607	5.5	1.5	65.1	100.5	2.3	1.5
1967	208,100	104,042	6,100	6,254	0	103,904	22,929	4.5	1.2	65.4	101.8	1.8	1.2
1968	198,300	77,973	8,896	12,814	0	116,409	23,255	5.0	1.3	68.9	105.9	1.9	1.2
1969	180,700	68,953	8,795	10,241	0	110,302	23,598	4.7	1.1	71.5	108.2	1.6	1.0
1970	191,200	58,682	10,464	23,149	0	119,833	23,962	5.0	1.2	67.5	105.9	1.7	1.1
1971	207,006	44,417	8,930	25,359	0	146,160	24,353	6.0	1.5	60.1	97.2	2.4	1.5
1972	217,400	37,747	6,100	45,759	0	139,994	24,768	5.7	1.5	57.5	94.2	2.6	1.6
1973	277,687	54,716	4,000	90,948	-6,067	129,957	25,196	5.2	1.3	59.5	94.1	2.2	1.4
1974	276,565	54,883	6,700	90,335	5,976	144,023	25,626	5.6	1.6	65.3	101.5	2.5	1.6
1975	214,136	50,273	8,220	72,256	1,983	101,810	26,049	3.9	1.2	70.4	105.6	1.7	1.2
1976	266,052	22,260	900	156,439	0	88,253	26,462	3.3	1.0	70.2	104.9	1.4	1.0
1977	379,867	55,700	9,918	211,856	0	122,229	26,866	4.6	1.2	69.9	104.8	1.7	1.1
1978	518,731	75,700	10,567	262,141	0	191,458	27,268	7.0	1.9	72.3	107.0	2.6	1.7
1979	567,676	68,400	20,063	304,600	0	214,739	27,676	7.8	2.1	71.9	106.6	3.0	2.0
1980	385,272	16,100	19,223	222,491	-180	165,724	28,094	5.9	1.6	72.7	106.8	2.2	1.5
1981	361,546	10,200	18,920	199,249	-200	170,817	28,524	6.0	1.6	71.8	106.7	2.2	1.5
1982	475,043	15,000	16,589	275,698	-24,816	176,119	28,963	6.1	1.6	64.1	99.5	2.5	1.6
1983	416,365	14,100	11,465	227,653	-8,545	177,532	29,409	6.0	1.5	61.7	96.8	2.5	1.6
1984	315,170	11,168	15,829	167,101	0	152,731	29,858	5.1	1.4	63.7	99.3	2.1	1.4
1985	406,817	16,200	11,972	199,921	8,750	211,418	30,305	7.0	1.9	69.9	104.5	2.7	1.8
1986	420,766	9,201	14,596	272,455	25,006	178,712	30,750	5.8	1.6	65.2	100.2	2.4	1.6
1987	559,780	11,390	29,473	328,730	-16,966	232,166	31,194	7.4	2.1	66.8	101.3	3.1	2.0
1988	493,517	12,500	18,652	288,058	-7,668	203,943	31,638	6.5	1.7	61.5	97.1	2.8	1.8
1989	486,632	25,000	15,317	308,416	15,197	183,729	32,082	5.7	1.5	61.3	95.8	2.4	1.6
1990	555,572	12,500	5,984	356,457	-300	192,299	32,527	5.9	1.5	58.8	91.8	2.6	1.7
1991	640,495	25,003	20,394	420,983	-1,381	213,523	32,974	6.5	1.6	60.6	95.4	2.6	1.7
1992	704,827	52,001	32,448	432,966	4,333	256,642	33,421	7.7	1.9	63.5	96.8	3.0	2.0
1993	932,258	75,000	28,396	616,445	6,283	275,492	33,869	8.1	2.0	63.8	98.2	3.2	2.1
1994	949,508	50,000	40,969	671,967	3,897	272,407	34,318	7.9	2.1	63.2	98.4	3.3	2.1
1995	1,148,789	90,000	34,887	751,721	166	342,121	34,768	9.8	2.6	60.7	96.2	4.2	2.7

1 ARMENIA	- FOOD BALANCE SHEET OF FISH AND FISHERY PRODUCTS IN LIVE WEIGHT AND FISH CONTRIBUTION TO PROTEIN SUPPLY -											FAO-FIDI
YEAR	PRODUCTION	NON-FOOD USES	IMPORTS	EXPORTS	STOCKS VARIATIONS	TOTAL FOOD SUPPLY	POPULATION	PER CAPUT SUPPLY	FISH PROTEINS	ANIMAL PROTEINS	TOTAL PROTEINS	FISH/ANIMAL PROTEINS	FISH/TOTAL PROTEINS
	(............ metric tons in live weight ....)						(thousands)	(kilograms)	(grams per caput per day)			(%)	(%)
1961	0	0	0	0	0	0	1,946	0.0	0.0	0.0	0.0	0.0	0.0
1962	0	0	0	0	0	0	2,009	0.0	0.0	0.0	0.0	0.0	0.0
1963	0	0	0	0	0	0	2,068	0.0	0.0	0.0	0.0	0.0	0.0
1964	0	0	0	0	0	0	2,138	0.0	0.0	0.0	0.0	0.0	0.0
1965	0	0	0	0	0	0	2,213	0.0	0.0	0.0	0.0	0.0	0.0
1966	0	0	0	0	0	0	2,282	0.0	0.0	0.0	0.0	0.0	0.0
1967	0	0	0	0	0	0	2,346	0.0	0.0	0.0	0.0	0.0	0.0
1968	0	0	0	0	0	0	2,411	0.0	0.0	0.0	0.0	0.0	0.0
1969	0	0	0	0	0	0	2,473	0.0	0.0	0.0	0.0	0.0	0.0
1970	0	0	0	0	0	0	2,520	0.0	0.0	0.0	0.0	0.0	0.0
1971	0	0	0	0	0	0	2,580	0.0	0.0	0.0	0.0	0.0	0.0
1972	0	0	0	0	0	0	2,641	0.0	0.0	0.0	0.0	0.0	0.0
1973	0	0	0	0	0	0	2,708	0.0	0.0	0.0	0.0	0.0	0.0
1974	0	0	0	0	0	0	2,770	0.0	0.0	0.0	0.0	0.0	0.0
1975	0	0	0	0	0	0	2,826	0.0	0.0	0.0	0.0	0.0	0.0
1976	0	0	0	0	0	0	2,883	0.0	0.0	0.0	0.0	0.0	0.0
1977	0	0	0	0	0	0	2,934	0.0	0.0	0.0	0.0	0.0	0.0
1978	0	0	0	0	0	0	2,984	0.0	0.0	0.0	0.0	0.0	0.0
1979	0	0	0	0	0	0	3,030	0.0	0.0	0.0	0.0	0.0	0.0
1980	0	0	0	0	0	0	3,072	0.0	0.0	0.0	0.0	0.0	0.0
1981	0	0	0	0	0	0	3,111	0.0	0.0	0.0	0.0	0.0	0.0
1982	0	0	0	0	0	0	3,146	0.0	0.0	0.0	0.0	0.0	0.0
1983	0	0	0	0	0	0	3,177	0.0	0.0	0.0	0.0	0.0	0.0
1984	0	0	0	0	0	0	3,206	0.0	0.0	0.0	0.0	0.0	0.0
1985	0	0	0	0	0	0	3,232	0.0	0.0	0.0	0.0	0.0	0.0
1986	0	0	0	0	0	0	3,254	0.0	0.0	0.0	0.0	0.0	0.0
1987	0	0	0	0	0	0	3,274	0.0	0.0	0.0	0.0	0.0	0.0
1988	0	0	0	0	0	0	3,294	0.0	0.0	0.0	0.0	0.0	0.0
1989	0	0	0	0	0	0	3,512	0.0	0.0	0.0	0.0	0.0	0.0
1990	0	0	0	0	0	0	3,545	0.0	0.0	0.0	0.0	0.0	0.0
1991	4,540	0	...	...	0	4,540	3,571	1.3	0.4	32.0	76.8	1.2	0.5
1992	4,750	0	...	...	0	4,750	3,593	1.3	0.4	18.7	58.7	2.1	0.7
1993	4,550	0	...	...	0	4,550	3,609	1.3	0.4	13.7	52.7	2.8	0.7
1994	4,350	0	...	...	0	4,350	3,622	1.2	0.4	19.6	58.2	1.8	0.6
1995	4,000	0	...	...	0	4,000	3,632	1.1	0.3	21.2	63.9	1.6	0.5

Note: for years before 1991 see USSR.

22 ARUBA	- FOOD BALANCE SHEET OF FISH AND FISHERY PRODUCTS IN LIVE WEIGHT AND FISH CONTRIBUTION TO PROTEIN SUPPLY -	FAO-FIDI

YEAR	PRODUCTION	NON-FOOD USES	IMPORTS	EXPORTS	STOCKS VARIATIONS	TOTAL FOOD SUPPLY	POPULATION	PER CAPUT SUPPLY	FISH PROTEINS	ANIMAL PROTEINS	TOTAL PROTEINS	FISH/ANIMAL PROTEINS	FISH/TOTAL PROTEINS
	(..		metric tons in live weight		....)		(thousands)	(kilograms)	(grams per caput per day)			(%)	(%)
1961	200	0	...	...	0	200	59	3.4	1.0	...	...	...	...
1962	300	0	...	...	0	300	59	5.1	1.4	...	...	...	...
1963	300	0	...	...	0	300	60	5.0	1.4	...	...	...	...
1964	300	0	...	...	0	300	60	5.0	1.4	...	...	...	...
1965	300	0	...	...	0	300	60	5.0	1.4	...	...	...	...
1966	300	0	...	...	0	300	60	5.0	1.4	...	...	...	...
1967	400	0	...	...	0	400	60	6.6	2.0	...	...	...	...
1968	400	0	...	...	0	400	61	6.6	2.0	...	...	...	...
1969	400	0	...	...	0	400	61	6.6	2.0	...	...	...	...
1970	400	0	...	...	0	400	61	6.6	2.0	...	...	...	...
1971	400	0	...	...	0	400	61	6.5	1.9	...	...	...	...
1972	400	0	...	...	0	400	62	6.5	1.9	...	...	...	...
1973	500	0	...	...	0	500	62	8.1	2.4	...	...	...	...
1974	550	0	...	...	0	550	62	8.9	2.6	...	...	...	...
1975	600	0	...	...	0	600	62	9.7	2.8	...	...	...	...
1976	650	0	...	...	0	650	62	10.5	3.1	...	...	...	...
1977	700	0	...	...	0	700	61	11.5	3.4	...	...	...	...
1978	770	0	...	...	0	770	61	12.7	3.7	...	...	...	...
1979	770	0	...	...	0	770	60	12.8	3.8	...	...	...	...
1980	770	0	...	...	0	770	60	12.8	3.8	...	...	...	...
1981	770	0	...	...	0	770	60	12.8	3.8	...	...	...	...
1982	770	0	...	...	0	770	61	12.7	3.7	...	...	...	...
1983	770	0	...	...	0	770	61	12.6	3.7	...	...	...	...
1984	770	0	1,474	14	0	2,230	62	35.9	10.0	...	...	...	...
1985	770	0	1,998	132	0	2,636	63	41.8	11.4	...	...	...	...
1986	700	0	2,985	296	0	3,390	64	53.1	14.2	...	...	...	...
1987	620	0	2,082	751	0	1,951	65	30.2	8.5	...	...	...	...
1988	550	0	1,778	24	0	2,304	65	35.2	9.8	...	...	...	...
1989	480	0	2,132	13	0	2,600	66	39.2	10.3	...	...	...	...
1990	420	0	1,781	23	0	2,179	67	32.5	9.2	...	...	...	...
1991	350	0	1,859	6	0	2,202	68	32.5	9.2	...	...	...	...
1992	300	0	359	8	0	651	68	9.5	2.8	...	...	...	...
1993	250	0	368	7	0	611	69	8.9	2.6	...	...	...	...
1994	250	0	402	9	0	643	69	9.3	2.7	...	...	...	...
1995	130	0	501	0	0	631	70	9.0	2.5	...	...	...	...

Note: up to 1983 trade included with that of the Netherlands Antilles.

10 AUSTRALIA	- FOOD BALANCE SHEET OF FISH AND FISHERY PRODUCTS IN LIVE WEIGHT AND FISH CONTRIBUTION TO PROTEIN SUPPLY -	FAO-FIDI

YEAR	PRODUCTION	NON-FOOD USES	IMPORTS	EXPORTS	STOCKS VARIATIONS	TOTAL FOOD SUPPLY	POPULATION	PER CAPUT SUPPLY	FISH PROTEINS	ANIMAL PROTEINS	TOTAL PROTEINS	FISH/ANIMAL PROTEINS	FISH/TOTAL PROTEINS
	(........................ metric tons in live weight					........................)	(thousands)	(kilograms)	(grams per caput per day)			(%)	(%)
1961	60,900	2,200	82,736	6,906	0	134,530	10,493	12.8	3.0	64.6	97.9	4.7	3.1
1962	66,100	1,500	74,815	13,818	0	125,598	10,717	11.7	2.7	66.6	100.4	4.0	2.7
1963	69,900	1,500	41,893	14,041	0	96,252	10,941	8.8	2.3	65.8	100.4	3.4	2.3
1964	72,000	1,400	85,864	15,849	0	140,615	11,165	12.6	2.9	68.0	102.0	4.3	2.9
1965	78,900	1,100	86,543	18,644	0	145,699	11,388	12.8	2.9	67.2	99.7	4.3	2.9
1966	88,500	1,100	103,405	25,162	0	165,642	11,607	14.3	3.3	67.6	103.8	4.9	3.2
1967	91,800	1,000	95,535	27,309	0	159,026	11,824	13.5	3.0	64.3	100.0	4.7	3.0
1968	101,100	1,100	98,384	39,462	0	158,922	12,044	13.2	3.1	67.5	102.6	4.6	3.0
1969	91,500	800	106,993	33,361	0	164,332	12,278	13.4	3.2	68.2	102.4	4.7	3.1
1970	102,400	1,000	101,977	36,044	0	167,333	12,530	13.4	3.2	66.7	101.8	4.7	3.1
1971	111,200	4,400	128,449	52,088	0	183,162	12,807	14.3	3.2	70.2	104.2	4.6	3.1
1972	117,700	6,200	108,082	56,316	0	163,266	13,103	12.5	2.9	72.0	105.5	4.0	2.7
1973	125,300	8,500	110,338	51,715	0	175,423	13,399	13.1	3.3	69.8	100.3	4.8	3.3
1974	135,801	6,600	136,078	47,377	0	217,901	13,669	15.9	4.0	64.2	95.7	6.2	4.2
1975	108,393	2,000	124,601	43,796	0	187,197	13,895	13.5	3.2	75.0	106.1	4.3	3.1
1976	110,610	4,500	114,160	41,587	-2,265	176,418	14,069	12.5	3.1	75.8	106.5	4.1	2.9
1977	128,994	6,000	133,284	51,998	-2,226	202,055	14,200	14.2	3.5	73.5	104.1	4.8	3.4
1978	123,208	5,200	128,073	51,484	-2,527	192,070	14,309	13.4	3.3	77.7	107.3	4.2	3.1
1979	127,633	4,700	122,962	55,179	-6,248	184,468	14,423	12.8	3.1	72.5	102.8	4.3	3.1
1980	131,957	2,200	143,219	80,394	11,770	204,353	14,564	14.0	3.5	68.9	98.4	5.1	3.6
1981	146,485	1,600	159,935	64,789	0	240,031	14,737	16.3	4.0	69.8	100.7	5.7	4.0
1982	166,408	...	166,159	86,451	-571	245,544	14,937	16.4	3.9	71.6	103.4	5.4	3.7
1983	169,241	...	151,997	95,398	-152	225,689	15,159	14.9	3.5	70.3	101.3	5.0	3.4
1984	169,373	...	180,025	87,328	-152	261,918	15,395	17.0	4.2	71.0	104.6	5.9	4.0
1985	161,367	6,000	195,804	78,354	-606	272,211	15,637	17.4	4.2	70.7	104.0	5.9	4.0
1986	180,338	13,277	187,372	74,637	0	279,796	15,886	17.6	4.3	70.9	102.5	6.1	4.2
1987	204,794	28,316	195,277	79,093	1,195	293,857	16,143	18.2	4.4	71.0	102.9	6.2	4.3
1988	213,326	16,594	200,932	87,079	-8,808	301,776	16,400	18.4	4.4	71.0	103.2	6.2	4.2
1989	181,783	16,592	206,292	74,945	9,092	305,629	16,650	18.4	4.4	72.4	104.4	6.0	4.2
1990	220,467	23,382	206,004	92,383	-8	310,697	16,884	18.4	4.4	72.3	104.8	6.0	4.2
1991	237,421	26,088	206,398	101,249	0	316,482	17,100	18.5	4.3	70.1	103.7	6.1	4.1
1992	248,672	32,093	223,679	101,232	0	339,026	17,301	19.6	4.5	70.8	102.2	6.4	4.4
1993	246,372	21,000	223,201	106,247	-158	342,168	17,490	19.6	4.6	71.4	104.2	6.5	4.5
1994	222,280	26,000	251,939	99,319	0	348,900	17,675	19.7	4.7	72.8	104.5	6.5	4.5
1995	219,499	12,000	225,031	104,408	167	328,289	17,862	18.4	4.4	69.5	104.4	6.3	4.2

11 AUSTRIA	- FOOD BALANCE SHEET OF FISH AND FISHERY PRODUCTS IN LIVE WEIGHT AND FISH CONTRIBUTION TO PROTEIN SUPPLY -											FAO-FIDI
YEAR	PRODUCTION	NON-FOOD USES	IMPORTS	EXPORTS	STOCKS VARIATIONS	TOTAL FOOD SUPPLY	POPULATION	PER CAPUT SUPPLY	FISH PROTEINS	ANIMAL PROTEINS	TOTAL PROTEINS	FISH/ANIMAL PROTEINS	FISH/TOTAL PROTEINS
	(.. metric tons in live weight ....)						(thousands)	(kilograms)	(grams per caput per day)			(%)	(%)
1961	4,200	0	45,989	300	0	49,889	7,086	7.0	1.6	48.7	88.7	3.4	1.8
1962	4,500	0	48,014	400	0	52,114	7,129	7.3	1.7	50.5	89.6	3.3	1.9
1963	4,600	0	47,418	400	0	51,618	7,176	7.2	1.6	50.4	90.3	3.3	1.8
1964	4,900	0	48,559	500	0	52,959	7,224	7.3	1.7	51.6	91.3	3.3	1.9
1965	5,100	0	51,598	500	0	56,198	7,271	7.7	1.8	50.6	89.8	3.6	2.0
1966	4,900	0	48,564	500	0	52,964	7,315	7.2	1.7	50.2	89.2	3.4	1.9
1967	4,200	0	51,796	600	0	55,396	7,357	7.5	1.8	51.2	89.0	3.5	2.0
1968	4,000	0	52,575	600	0	55,975	7,396	7.6	1.8	51.9	89.9	3.4	2.0
1969	4,000	0	52,765	600	0	56,165	7,433	7.6	1.8	52.0	89.3	3.4	2.0
1970	3,200	0	54,570	700	0	57,070	7,467	7.6	1.8	52.3	89.7	3.4	2.0
1971	2,900	0	55,945	600	0	58,245	7,499	7.8	1.8	53.5	90.2	3.4	2.0
1972	2,600	0	54,579	600	0	56,579	7,527	7.5	1.8	54.7	90.9	3.2	2.0
1973	2,200	0	57,921	500	0	59,621	7,551	7.9	2.0	55.2	91.3	3.5	2.1
1974	2,070	0	53,044	600	0	54,514	7,569	7.2	1.8	55.5	89.7	3.2	2.0
1975	2,040	0	49,505	463	0	51,082	7,579	6.7	1.7	56.0	90.8	2.9	1.8
1976	2,360	0	43,530	802	0	45,088	7,581	6.0	1.9	58.7	92.6	3.3	2.1
1977	2,470	0	40,490	547	0	42,413	7,577	5.6	1.8	58.0	91.6	3.1	2.0
1978	3,700	0	35,450	442	0	38,708	7,568	5.1	1.8	59.7	93.1	3.0	1.9
1979	4,100	0	40,298	660	0	43,738	7,558	5.8	2.0	60.7	94.6	3.3	2.1
1980	4,300	0	39,541	442	0	43,400	7,549	5.8	2.0	61.1	94.8	3.2	2.1
1981	4,400	0	39,992	942	0	43,450	7,544	5.8	2.0	63.0	96.9	3.1	2.0
1982	4,500	0	40,225	1,394	0	43,331	7,541	5.8	2.0	61.7	95.6	3.2	2.1
1983	4,700	0	43,769	1,670	0	46,799	7,542	6.2	2.1	62.9	95.7	3.3	2.2
1984	4,700	8	41,782	1,217	0	45,257	7,547	6.0	2.0	62.4	95.9	3.2	2.1
1985	4,800	12	42,622	804	0	46,606	7,558	6.2	2.1	62.7	96.0	3.3	2.2
1986	4,750	17	45,552	1,128	0	49,157	7,573	6.5	2.2	63.8	97.6	3.4	2.2
1987	4,600	24	48,748	647	0	52,677	7,593	6.9	2.3	64.0	98.5	3.7	2.4
1988	5,100	41	56,455	677	0	60,837	7,621	8.0	2.3	63.5	97.7	3.7	2.4
1989	5,000	37	64,776	887	0	68,852	7,658	9.0	2.6	64.8	100.4	4.0	2.6
1990	4,800	198	67,377	995	0	70,984	7,705	9.2	2.7	65.5	102.0	4.1	2.6
1991	4,505	130	68,866	1,145	0	72,096	7,764	9.3	2.7	66.3	102.8	4.1	2.7
1992	4,105	178	78,225	2,180	0	79,972	7,832	10.2	3.0	67.6	103.3	4.4	2.9
1993	4,605	73	75,356	2,139	0	77,749	7,905	9.8	2.9	66.3	102.0	4.4	2.9
1994	4,548	131	93,352	2,609	0	95,161	7,978	11.9	3.5	65.1	99.5	5.4	3.5
1995	4,458	335	78,940	7,298	0	75,765	8,045	9.4	2.4	65.3	100.8	3.7	2.4

52 AZERBAIJAN	- FOOD BALANCE SHEET OF FISH AND FISHERY PRODUCTS IN LIVE WEIGHT AND FISH CONTRIBUTION TO PROTEIN SUPPLY -												FAO-FIDI
YEAR	PRODUCTION	NON-FOOD USES	IMPORTS	EXPORTS	STOCKS VARIATIONS	TOTAL FOOD SUPPLY	POPULATION	PER CAPUT SUPPLY	FISH PROTEINS	ANIMAL PROTEINS	TOTAL PROTEINS	FISH/ANIMAL PROTEINS	FISH/TOTAL PROTEINS
	(.. metric tons in live weight ....)						(thousands)	(kilograms)	(grams per caput per day)			(%)	(%)
1961	0	0	0	0	0	0	4,042	0.0	0.0	0.0	0.0	0.0	0.0
1962	0	0	0	0	0	0	4,164	0.0	0.0	0.0	0.0	0.0	0.0
1963	0	0	0	0	0	0	4,289	0.0	0.0	0.0	0.0	0.0	0.0
1964	0	0	0	0	0	0	4,435	0.0	0.0	0.0	0.0	0.0	0.0
1965	0	0	0	0	0	0	4,570	0.0	0.0	0.0	0.0	0.0	0.0
1966	0	0	0	0	0	0	4,704	0.0	0.0	0.0	0.0	0.0	0.0
1967	0	0	0	0	0	0	4,828	0.0	0.0	0.0	0.0	0.0	0.0
1968	0	0	0	0	0	0	4,944	0.0	0.0	0.0	0.0	0.0	0.0
1969	0	0	0	0	0	0	5,058	0.0	0.0	0.0	0.0	0.0	0.0
1970	0	0	0	0	0	0	5,172	0.0	0.0	0.0	0.0	0.0	0.0
1971	0	0	0	0	0	0	5,283	0.0	0.0	0.0	0.0	0.0	0.0
1972	0	0	0	0	0	0	5,392	0.0	0.0	0.0	0.0	0.0	0.0
1973	0	0	0	0	0	0	5,494	0.0	0.0	0.0	0.0	0.0	0.0
1974	0	0	0	0	0	0	5,594	0.0	0.0	0.0	0.0	0.0	0.0
1975	0	0	0	0	0	0	5,689	0.0	0.0	0.0	0.0	0.0	0.0
1976	0	0	0	0	0	0	5,781	0.0	0.0	0.0	0.0	0.0	0.0
1977	0	0	0	0	0	0	5,876	0.0	0.0	0.0	0.0	0.0	0.0
1978	0	0	0	0	0	0	5,976	0.0	0.0	0.0	0.0	0.0	0.0
1979	0	0	0	0	0	0	6,070	0.0	0.0	0.0	0.0	0.0	0.0
1980	0	0	0	0	0	0	6,157	0.0	0.0	0.0	0.0	0.0	0.0
1981	0	0	0	0	0	0	6,252	0.0	0.0	0.0	0.0	0.0	0.0
1982	0	0	0	0	0	0	6,350	0.0	0.0	0.0	0.0	0.0	0.0
1983	0	0	0	0	0	0	6,452	0.0	0.0	0.0	0.0	0.0	0.0
1984	0	0	0	0	0	0	6,559	0.0	0.0	0.0	0.0	0.0	0.0
1985	0	0	0	0	0	0	6,659	0.0	0.0	0.0	0.0	0.0	0.0
1986	0	0	0	0	0	0	6,757	0.0	0.0	0.0	0.0	0.0	0.0
1987	0	0	0	0	0	0	6,868	0.0	0.0	0.0	0.0	0.0	0.0
1988	0	0	0	0	0	0	6,982	0.0	0.0	0.0	0.0	0.0	0.0
1989	0	0	0	0	0	0	7,069	0.0	0.0	0.0	0.0	0.0	0.0
1990	0	0	0	0	0	0	7,159	0.0	0.0	0.0	0.0	0.0	0.0
1991	39,663	0	...	...	0	39,663	7,243	5.5	1.6	35.3	78.4	4.6	2.1
1992	31,200	0	...	...	0	31,200	7,321	4.3	1.3	21.8	69.3	5.9	1.8
1993	22,200	0	...	...	0	22,200	7,395	3.0	0.9	18.9	67.5	4.8	1.3
1994	19,200	0	1,179	3,230	0	17,149	7,464	2.3	0.7	16.4	61.6	4.3	1.1
1995	9,800	0	200	340	0	9,660	7,531	1.3	0.4	16.4	61.4	2.4	0.6

Note: for years before 1991 see USSR.

12 BAHAMAS	- FOOD BALANCE SHEET OF FISH AND FISHERY PRODUCTS IN LIVE WEIGHT AND FISH CONTRIBUTION TO PROTEIN SUPPLY -	FAO-FIDI

YEAR	PRODUCTION	NON-FOOD USES	IMPORTS	EXPORTS	STOCKS VARIATIONS	TOTAL FOOD SUPPLY	POPULATION	PER CAPUT SUPPLY	FISH PROTEINS	ANIMAL PROTEINS	TOTAL PROTEINS	FISH/ANIMAL PROTEINS	FISH/TOTAL PROTEINS
	(............		metric tons in live weight			............)	(thousands)	(kilograms)	(grams per caput per day)			(%)	(%)
1961	1,500	0	235	600	0	1,135	114	9.9	2.9	38.6	73.5	7.6	4.0
1962	1,600	0	235	600	0	1,235	121	10.2	3.0	38.1	71.8	7.9	4.2
1963	1,700	0	235	300	0	1,635	128	12.8	3.7	37.9	70.8	9.8	5.2
1964	1,900	0	343	900	0	1,343	134	10.0	3.0	40.5	72.9	7.4	4.1
1965	2,300	0	333	500	0	2,133	140	15.2	4.3	41.2	72.9	10.5	5.9
1966	2,500	0	423	200	0	2,723	147	18.5	5.2	44.3	77.0	11.7	6.8
1967	2,300	0	501	200	0	2,601	153	17.0	4.8	44.1	76.3	10.8	6.3
1968	2,100	0	521	200	0	2,421	159	15.2	4.3	47.2	78.0	9.1	5.5
1969	1,700	0	816	100	0	2,416	164	14.7	4.3	45.4	76.1	9.5	5.7
1970	2,100	0	1,252	400	0	2,952	170	17.4	4.7	48.1	78.3	9.8	6.1
1971	2,300	0	1,262	963	0	2,599	174	14.9	4.1	46.2	76.8	8.8	5.3
1972	2,000	0	1,039	954	0	2,085	178	11.7	3.0	44.3	73.0	6.8	4.1
1973	3,100	0	1,318	1,346	0	3,072	183	16.8	4.5	44.0	71.8	10.2	6.3
1974	2,992	0	1,176	1,529	0	2,639	185	14.3	3.8	42.4	70.1	8.9	5.4
1975	2,801	0	1,227	1,200	0	2,828	190	14.9	3.9	41.2	68.4	9.6	5.8
1976	2,917	0	1,381	1,730	0	2,568	193	13.3	3.3	40.8	69.1	8.0	4.7
1977	3,741	0	827	2,051	0	2,518	197	12.8	3.2	36.2	62.4	8.8	5.1
1978	3,762	0	827	1,650	0	2,939	201	14.6	3.6	37.1	62.1	9.8	5.8
1979	4,029	0	380	1,834	0	2,576	206	12.5	3.0	39.0	64.6	7.8	4.7
1980	5,026	0	1,957	2,177	-1,000	3,806	210	18.1	4.4	47.5	76.3	9.2	5.7
1981	4,372	0	1,196	2,227	1,000	4,341	215	20.2	5.6	45.6	71.7	12.3	7.8
1982	4,686	0	1,622	2,288	0	4,020	220	18.3	4.8	49.3	81.0	9.7	5.9
1983	5,211	0	1,481	2,400	0	4,292	223	19.3	4.9	48.0	79.7	10.1	6.1
1984	5,351	0	1,636	2,697	0	4,290	227	18.9	4.9	50.1	80.1	9.8	6.1
1985	7,671	0	2,088	3,310	-1,316	5,134	232	22.1	5.3	50.0	81.1	10.5	6.5
1986	5,935	0	1,978	3,853	1,316	5,376	236	22.8	6.5	52.5	84.2	12.4	7.8
1987	7,106	0	2,341	3,222	-1,000	5,225	241	21.7	5.1	51.5	81.8	9.8	6.2
1988	7,251	53	1,971	4,502	1,000	5,667	246	23.0	6.4	52.7	84.2	12.1	7.6
1989	8,151	0	2,420	3,274	-1,053	6,244	250	25.0	6.0	50.7	82.5	11.8	7.3
1990	7,547	0	2,599	5,220	1,053	5,978	256	23.4	6.9	52.6	82.9	13.1	8.4
1991	9,205	29	3,039	6,017	0	6,198	260	23.8	6.7	51.1	79.2	13.1	8.4
1992	9,856	26	3,071	6,121	0	6,781	264	25.7	7.1	47.5	76.4	14.9	9.3
1993	10,081	28	3,413	6,802	0	6,664	268	24.9	7.2	47.8	74.8	15.1	9.6
1994	9,707	24	3,444	7,022	0	6,105	274	22.3	6.6	52.1	80.5	12.7	8.2
1995	9,638	22	3,226	6,086	0	6,756	279	24.2	6.8	49.2	78.1	13.7	8.6

13 BAHRAIN	- FOOD BALANCE SHEET OF FISH AND FISHERY PRODUCTS IN LIVE WEIGHT AND FISH CONTRIBUTION TO PROTEIN SUPPLY -													FAO-FIDI
YEAR	PRODUCTION	NON-FOOD USES	IMPORTS	EXPORTS	STOCKS VARIATIONS	TOTAL FOOD SUPPLY	POPULATION	PER CAPUT SUPPLY	FISH PROTEINS	ANIMAL PROTEINS	TOTAL PROTEINS	FISH/ANIMAL PROTEINS	FISH/TOTAL PROTEINS
	(.......................... metric tons in live weight ....)						(thousands)	(kilograms)	(grams per caput per day)			(%)	(%)
1961	1,500	0	0	0	0	1,500	163	9.2	2.5	15.4	47.4	16.2	5.3
1962	1,500	0	0	0	0	1,500	170	8.8	2.4	14.9	47.0	16.0	5.1
1963	1,800	0	0	0	0	1,800	177	10.2	2.7	14.9	46.1	18.4	5.9
1964	1,800	0	0	0	0	1,800	184	9.8	2.6	15.2	48.8	17.2	5.4
1965	2,000	0	0	0	0	2,000	191	10.5	2.8	14.8	51.3	18.9	5.5
1966	2,500	0	0	36	0	2,464	197	12.5	3.4	16.5	50.5	20.5	6.7
1967	2,800	0	0	29	0	2,771	202	13.7	3.7	18.3	56.3	20.2	6.6
1968	3,000	0	0	566	0	2,434	207	11.8	3.2	17.5	56.4	18.2	5.7
1969	3,300	0	0	802	0	2,498	213	11.8	3.2	18.2	51.9	17.8	6.2
1970	3,500	0	0	380	0	3,120	220	14.2	3.7	20.5	59.7	17.9	6.1
1971	3,600	0	712	640	0	3,672	228	16.1	4.3	26.4	66.0	16.3	6.5
1972	3,700	0	1,925	1,900	0	3,725	237	15.7	4.2	23.0	59.1	18.3	7.1
1973	3,800	0	1,500	1,150	0	4,150	248	16.8	4.6	25.6	61.8	17.9	7.4
1974	3,900	0	2,576	1,675	-400	4,401	259	17.0	4.7	29.3	52.6	16.1	8.9
1975	4,000	0	1,545	1,283	0	4,262	272	15.7	4.2	27.1	63.8	15.5	6.6
1976	4,084	0	1,381	1,631	400	4,234	286	14.8	4.0	62.0	101.5	6.5	3.9
1977	4,837	0	2,318	1,208	0	5,948	301	19.8	5.5	38.8	85.4	14.1	6.4
1978	4,000	0	1,333	558	0	4,775	317	15.1	4.1	36.5	85.6	11.3	4.8
1979	3,801	0	587	0	0	4,388	332	13.2	3.7	38.3	93.0	9.5	3.9
1980	5,115	0	1,320	28	0	6,407	347	18.5	5.0	41.9	96.5	11.9	5.2
1981	5,747	0	2,267	8	0	8,006	361	22.2	6.3	39.8	99.7	15.8	6.3
1982	5,594	0	1,473	3	0	7,064	374	18.9	5.2	45.6	96.6	11.3	5.3
1983	4,812	0	2,231	0	0	7,043	387	18.2	5.1	48.1	90.5	10.6	5.6
1984	5,599	0	2,402	0	0	8,001	400	20.0	5.6	48.5	103.5	11.5	5.4
1985	7,763	0	2,738	0	-1,111	9,390	414	22.7	6.2	50.3	96.1	12.3	6.5
1986	8,057	0	2,385	381	-556	9,505	428	22.2	6.0	48.7	90.4	12.3	6.6
1987	7,842	0	2,907	821	0	9,928	444	22.4	6.1	47.4	100.7	13.0	6.1
1988	6,736	0	2,177	603	0	8,310	459	18.1	5.0	47.3	92.2	10.5	5.4
1989	9,208	0	2,540	1,034	0	10,714	475	22.6	6.0	49.5	113.1	12.2	5.3
1990	8,105	0	2,047	1,519	0	8,633	490	17.6	4.7	48.8	89.4	9.5	5.2
1991	7,553	0	3,040	2,535	1,111	9,169	504	18.2	5.0	43.2	88.1	11.5	5.6
1992	7,983	0	2,614	2,067	556	9,085	518	17.5	4.7	50.3	129.5	9.3	3.6
1993	8,958	0	3,318	2,673	0	9,603	531	18.1	4.8	49.9	120.6	9.6	4.0
1994	7,629	0	3,150	2,872	0	7,907	544	14.5	3.8	49.2	108.6	7.7	3.5
1995	9,388	0	4,103	3,646	0	9,844	557	17.7	4.6	48.3	109.2	9.6	4.2

16 BANGLADESH	- FOOD BALANCE SHEET OF FISH AND FISHERY PRODUCTS IN LIVE WEIGHT AND FISH CONTRIBUTION TO PROTEIN SUPPLY -												FAO-FIDI
YEAR	PRODUCTION	NON-FOOD USES	IMPORTS	EXPORTS	STOCKS VARIATIONS	TOTAL FOOD SUPPLY	POPULATION	PER CAPUT SUPPLY	FISH PROTEINS	ANIMAL PROTEINS	TOTAL PROTEINS	FISH/ANIMAL PROTEINS	FISH/TOTAL PROTEINS
	(....... metric tons in live weight ......)						(thousands)	(kilograms)	(grams per caput per day)			(%)	(%)
1961	450,000	0	0	16,000	0	434,000	52,724	8.2	2.5	5.3	44.0	46.5	5.6
1962	500,000	0	0	27,505	0	472,495	54,053	8.7	2.6	5.5	43.5	47.6	6.0
1963	500,000	0	0	27,497	0	472,503	55,417	8.5	2.5	5.5	44.6	46.1	5.7
1964	550,000	0	0	23,199	0	526,801	56,832	9.3	2.8	5.9	44.8	47.4	6.2
1965	600,000	0	0	7,189	0	592,811	58,312	10.2	3.0	6.1	43.9	49.7	6.9
1966	600,000	0	0	1,293	0	598,707	59,859	10.0	3.0	5.9	43.5	50.6	6.9
1967	600,000	0	0	1,703	0	598,297	61,469	9.7	2.9	5.9	41.1	49.6	7.1
1968	650,000	0	0	2,708	0	647,292	63,143	10.3	3.1	6.0	43.2	50.9	7.1
1969	650,000	0	0	3,583	0	646,417	64,877	10.0	3.0	6.0	45.0	49.8	6.6
1970	690,100	0	0	1,900	0	688,200	66,671	10.3	3.1	6.2	46.7	49.9	6.6
1971	740,100	0	0	1,300	0	738,800	68,517	10.8	3.2	6.2	43.5	52.0	7.4
1972	818,100	0	0	700	0	817,400	70,419	11.6	3.5	6.3	41.0	55.1	8.5
1973	820,200	0	400	3,400	0	817,200	72,387	11.3	3.4	6.3	44.0	53.6	7.7
1974	822,082	0	80	5,203	0	816,959	74,438	11.0	3.3	6.1	44.4	53.7	7.4
1975	640,070	0	0	2,029	0	638,041	76,582	8.3	2.5	5.3	41.7	47.1	6.0
1976	641,605	0	0	4,018	0	637,588	78,821	8.1	2.4	5.4	43.7	45.3	5.6
1977	643,769	0	400	3,737	0	640,432	81,143	7.9	2.4	5.2	40.7	45.7	5.8
1978	646,895	0	40	3,123	0	643,812	83,512	7.7	2.3	5.6	43.8	41.5	5.3
1979	647,128	0	225	5,246	0	642,107	85,883	7.5	2.2	5.4	43.6	41.8	5.1
1980	646,970	0	101	10,324	0	636,747	88,221	7.2	2.2	4.5	42.3	47.6	5.1
1981	651,256	0	33	8,696	0	642,593	90,508	7.1	2.1	4.5	40.9	46.9	5.2
1982	689,498	0	0	10,299	0	679,199	92,746	7.3	2.2	4.5	41.6	48.9	5.3
1983	726,587	0	0	15,658	0	710,929	94,947	7.5	2.2	4.6	42.4	48.4	5.3
1984	756,013	0	0	22,093	0	733,920	97,131	7.6	2.3	4.7	43.1	48.5	5.3
1985	775,631	0	0	27,850	0	747,781	99,310	7.5	2.3	5.0	43.6	44.8	5.2
1986	796,910	0	0	26,616	0	770,294	101,494	7.6	2.3	5.0	42.2	45.4	5.4
1987	817,003	0	0	28,535	0	788,468	103,670	7.6	2.3	5.1	43.8	45.2	5.2
1988	829,929	0	0	29,216	0	800,713	105,803	7.6	2.3	5.0	43.9	45.5	5.1
1989	843,611	0	22	27,743	0	815,890	107,844	7.6	2.3	5.0	43.6	45.3	5.2
1990	847,830	0	2	34,068	0	813,764	109,765	7.4	2.2	5.0	44.8	44.4	4.9
1991	892,700	6,250	11	31,671	0	854,791	111,550	7.7	2.3	5.1	44.7	45.1	5.1
1992	966,727	6,500	100	39,092	0	921,235	113,223	8.1	2.4	5.3	44.7	46.0	5.4
1993	1,047,170	2,890	76	38,796	0	1,005,561	114,844	8.8	2.6	5.6	43.1	46.7	6.0
1994	1,090,610	3,500	76	40,527	0	1,046,659	116,493	9.0	2.7	5.6	45.1	47.2	5.9
1995	1,170,365	5,000	30	37,619	0	1,127,776	118,229	9.5	2.8	5.9	43.6	48.0	6.5

14 BARBADOS	- FOOD BALANCE SHEET OF FISH AND FISHERY PRODUCTS IN LIVE WEIGHT AND FISH CONTRIBUTION TO PROTEIN SUPPLY -												FAO-FIDI
YEAR	PRODUCTION	NON-FOOD USES	IMPORTS	EXPORTS	STOCKS VARIATIONS	TOTAL FOOD SUPPLY	POPULATION	PER CAPUT SUPPLY	FISH PROTEINS	ANIMAL PROTEINS	TOTAL PROTEINS	FISH/ANIMAL PROTEINS	FISH/TOTAL PROTEINS
	(.......... metric tons in live weight ....)						(thousands)	(kilograms)	(grams per caput per day)			(%)	(%)
1961	4,600	0	1,127	0	0	5,727	231	24.8	8.3	29.2	62.4	28.4	13.3
1962	4,700	0	952	0	0	5,652	232	24.3	8.1	29.1	62.5	28.0	13.0
1963	2,800	0	2,652	0	0	5,452	233	23.4	7.9	32.4	66.4	24.5	12.0
1964	2,100	0	3,302	800	0	4,602	234	19.6	6.8	33.0	66.1	20.5	10.2
1965	2,600	0	3,477	900	0	5,177	235	22.0	7.4	36.8	68.4	20.2	10.9
1966	4,800	0	3,403	1,300	0	6,903	236	29.3	9.9	36.2	68.3	27.4	14.6
1967	4,600	0	3,153	1,200	0	6,553	237	27.7	9.3	37.7	74.7	24.8	12.5
1968	3,500	0	3,227	1,200	0	5,527	237	23.3	7.9	36.4	70.7	21.7	11.2
1969	3,600	0	3,303	1,200	0	5,703	238	24.0	8.1	41.3	74.4	19.7	10.9
1970	2,300	0	3,429	250	0	5,479	239	23.0	7.8	41.8	76.6	18.8	10.2
1971	2,300	0	2,928	0	0	5,228	240	21.8	7.4	46.6	82.0	15.9	9.0
1972	2,300	0	3,003	175	0	5,127	241	21.2	7.2	44.5	77.3	16.2	9.3
1973	2,600	0	2,728	0	0	5,328	243	21.9	7.4	44.8	78.5	16.4	9.4
1974	2,685	0	3,611	0	-183	6,113	244	25.0	8.4	48.7	79.4	17.2	10.6
1975	4,412	0	2,811	0	-183	7,039	246	28.7	9.6	44.2	76.3	21.7	12.6
1976	5,240	0	2,758	147	83	7,934	247	32.2	10.4	49.6	82.2	20.9	12.6
1977	3,352	0	3,105	418	283	6,322	247	25.6	8.2	47.7	80.8	17.1	10.1
1978	3,812	0	4,031	50	0	7,793	248	31.4	9.9	52.3	86.7	19.0	11.5
1979	4,465	0	3,555	262	0	7,758	248	31.2	10.4	48.6	82.9	21.5	12.6
1980	3,735	0	3,044	40	0	6,740	249	27.1	8.7	48.3	86.3	18.1	10.1
1981	3,411	0	4,943	36	0	8,318	250	33.3	10.5	53.6	89.0	19.7	11.8
1982	3,480	0	4,164	14	-56	7,574	250	30.3	10.3	56.0	87.1	18.3	11.8
1983	6,522	0	3,010	35	56	9,553	251	38.0	13.0	57.4	91.4	22.7	14.3
1984	5,787	0	3,527	26	0	9,289	252	36.9	12.6	58.1	92.7	21.7	13.6
1985	3,915	0	3,993	22	-2	7,885	253	31.2	10.6	57.8	93.3	18.3	11.3
1986	4,227	0	3,744	31	2	7,943	254	31.3	10.7	57.9	94.5	18.4	11.3
1987	3,702	0	3,765	28	0	7,439	255	29.2	10.0	59.3	97.8	16.8	10.2
1988	9,097	4,000	3,461	182	0	8,375	255	32.8	11.2	59.0	95.0	19.0	11.8
1989	2,547	7	4,649	69	-2	7,118	256	27.8	8.9	63.0	100.1	14.1	8.9
1990	2,967	3	4,356	58	2	7,264	257	28.3	9.1	60.8	98.9	15.0	9.2
1991	2,074	38	4,480	52	-11	6,453	258	25.0	7.9	55.5	94.9	14.2	8.3
1992	3,342	1	3,541	96	0	6,785	259	26.2	8.5	47.6	84.6	17.8	10.0
1993	2,852	2	4,011	105	11	6,767	259	26.1	8.2	47.1	83.3	17.5	9.9
1994	2,585	2	4,196	157	-89	6,534	260	25.1	7.9	48.6	86.8	16.3	9.1
1995	3,284	1	4,767	324	89	7,815	261	30.0	9.4	52.8	91.4	17.8	10.3

57 BELARUS	- FOOD BALANCE SHEET OF FISH AND FISHERY PRODUCTS IN LIVE WEIGHT AND FISH CONTRIBUTION TO PROTEIN SUPPLY -	FAO-FIDI

YEAR	PRODUCTION	NON-FOOD USES	IMPORTS	EXPORTS	STOCKS VARIATIONS	TOTAL FOOD SUPPLY	POPULATION	PER CAPUT SUPPLY	FISH PROTEINS	ANIMAL PROTEINS	TOTAL PROTEINS	FISH/ANIMAL PROTEINS	FISH/TOTAL PROTEINS
	(.. metric tons in live weight ....)						(thousands)	(kilograms)	(grams per caput per day)			(%)	(%)
1961	0	0	0	0	0	0	8,292	0.0	0.0	0.0	0.0	0.0	0.0
1962	0	0	0	0	0	0	8,393	0.0	0.0	0.0	0.0	0.0	0.0
1963	0	0	0	0	0	0	8,465	0.0	0.0	0.0	0.0	0.0	0.0
1964	0	0	0	0	0	0	8,527	0.0	0.0	0.0	0.0	0.0	0.0
1965	0	0	0	0	0	0	8,610	0.0	0.0	0.0	0.0	0.0	0.0
1966	0	0	0	0	0	0	8,712	0.0	0.0	0.0	0.0	0.0	0.0
1967	0	0	0	0	0	0	8,803	0.0	0.0	0.0	0.0	0.0	0.0
1968	0	0	0	0	0	0	8,880	0.0	0.0	0.0	0.0	0.0	0.0
1969	0	0	0	0	0	0	8,962	0.0	0.0	0.0	0.0	0.0	0.0
1970	0	0	0	0	0	0	9,040	0.0	0.0	0.0	0.0	0.0	0.0
1971	0	0	0	0	0	0	9,112	0.0	0.0	0.0	0.0	0.0	0.0
1972	0	0	0	0	0	0	9,179	0.0	0.0	0.0	0.0	0.0	0.0
1973	0	0	0	0	0	0	9,245	0.0	0.0	0.0	0.0	0.0	0.0
1974	0	0	0	0	0	0	9,312	0.0	0.0	0.0	0.0	0.0	0.0
1975	0	0	0	0	0	0	9,367	0.0	0.0	0.0	0.0	0.0	0.0
1976	0	0	0	0	0	0	9,412	0.0	0.0	0.0	0.0	0.0	0.0
1977	0	0	0	0	0	0	9,463	0.0	0.0	0.0	0.0	0.0	0.0
1978	0	0	0	0	0	0	9,526	0.0	0.0	0.0	0.0	0.0	0.0
1979	0	0	0	0	0	0	9,586	0.0	0.0	0.0	0.0	0.0	0.0
1980	0	0	0	0	0	0	9,627	0.0	0.0	0.0	0.0	0.0	0.0
1981	0	0	0	0	0	0	9,689	0.0	0.0	0.0	0.0	0.0	0.0
1982	0	0	0	0	0	0	9,754	0.0	0.0	0.0	0.0	0.0	0.0
1983	0	0	0	0	0	0	9,823	0.0	0.0	0.0	0.0	0.0	0.0
1984	0	0	0	0	0	0	9,891	0.0	0.0	0.0	0.0	0.0	0.0
1985	0	0	0	0	0	0	9,958	0.0	0.0	0.0	0.0	0.0	0.0
1986	0	0	0	0	0	0	10,023	0.0	0.0	0.0	0.0	0.0	0.0
1987	0	0	0	0	0	0	10,086	0.0	0.0	0.0	0.0	0.0	0.0
1988	0	0	0	0	0	0	10,142	0.0	0.0	0.0	0.0	0.0	0.0
1989	0	0	0	0	0	0	10,219	0.0	0.0	0.0	0.0	0.0	0.0
1990	0	0	0	0	0	0	10,260	0.0	0.0	0.0	0.0	0.0	0.0
1991	15,543	0	...	...	0	15,543	10,293	1.5	0.5	46.3	95.9	1.0	0.5
1992	9,900	0	...	...	0	9,900	10,318	1.0	0.3	51.9	95.8	0.6	0.3
1993	10,000	0	1,833	...	0	11,833	10,337	1.1	0.3	52.7	97.7	0.7	0.4
1994	7,600	0	9,981	3,279	0	14,303	10,348	1.4	0.4	51.7	95.7	0.8	0.4
1995	5,100	10	7,238	2,679	0	9,649	10,352	0.9	0.3	49.2	93.8	0.6	0.3

Note: for years before 1991 see USSR.

15 BELGIUM-LUXEMBOURG	- FOOD BALANCE SHEET OF FISH AND FISHERY PRODUCTS IN LIVE WEIGHT AND FISH CONTRIBUTION TO PROTEIN SUPPLY -	FAO-FIDI

YEAR	PRODUCTION	NON-FOOD USES	IMPORTS	EXPORTS	STOCKS VARIATIONS	TOTAL FOOD SUPPLY	POPULATION	PER CAPUT SUPPLY	FISH PROTEINS	ANIMAL PROTEINS	TOTAL PROTEINS	FISH/ANIMAL PROTEINS	FISH/TOTAL PROTEINS
	(........ metric tons in live weight					........)	(thousands)	(kilograms)	(grams per caput per day)			(%)	(%)
1961	61,600	1,500	129,527	22,115	0	167,512	9,534	17.6	5.0	49.3	90.1	10.2	5.6
1962	59,900	2,000	131,144	32,072	0	156,972	9,604	16.3	4.5	48.2	89.0	9.4	5.1
1963	62,000	2,000	126,098	25,583	0	160,515	9,673	16.6	4.7	49.0	89.6	9.6	5.3
1964	59,500	2,000	137,774	28,142	0	167,133	9,738	17.2	4.8	48.8	91.1	9.7	5.2
1965	59,700	2,200	143,056	32,985	0	167,571	9,795	17.1	4.7	48.4	88.7	9.7	5.3
1966	62,800	1,500	154,141	43,964	0	171,477	9,846	17.4	4.8	50.6	89.9	9.4	5.3
1967	63,800	1,500	146,340	43,841	0	164,799	9,888	16.7	4.6	50.7	90.3	9.1	5.1
1968	68,200	500	152,264	53,324	0	166,640	9,925	16.8	4.5	50.7	89.8	8.9	5.0
1969	58,700	500	154,099	49,469	0	162,830	9,960	16.4	4.4	52.7	93.2	8.3	4.7
1970	53,000	1,000	166,485	52,848	0	165,638	9,995	16.6	4.7	54.2	92.2	8.6	5.1
1971	60,200	1,600	166,162	53,063	0	171,699	10,031	17.1	4.7	56.2	94.8	8.3	4.9
1972	59,000	1,300	160,382	51,660	0	166,422	10,067	16.5	4.4	55.4	93.5	8.0	4.7
1973	53,300	700	156,972	46,401	0	163,172	10,102	16.2	4.2	56.3	94.4	7.4	4.4
1974	46,820	900	165,344	39,761	0	171,503	10,133	16.9	4.3	58.3	96.1	7.4	4.5
1975	49,027	2,500	151,280	35,724	0	162,083	10,158	16.0	3.9	58.2	95.0	6.7	4.1
1976	44,496	11,701	176,423	33,437	0	175,779	10,178	17.3	4.4	57.7	95.0	7.7	4.7
1977	45,360	13,025	176,632	35,960	0	173,002	10,194	17.0	4.1	59.3	97.3	7.0	4.2
1978	50,577	12,630	175,342	38,459	0	174,827	10,204	17.1	4.3	61.0	98.9	7.0	4.3
1979	47,125	11,653	183,660	34,751	-400	183,981	10,212	18.0	4.4	61.8	98.3	7.2	4.5
1980	45,615	13,056	188,829	33,732	0	187,656	10,216	18.4	4.6	63.0	100.5	7.2	4.5
1981	49,279	13,272	187,900	39,543	-200	184,159	10,217	18.0	4.5	62.2	100.7	7.2	4.4
1982	47,841	12,481	190,317	40,541	-200	184,935	10,216	18.1	4.7	63.7	102.2	7.3	4.6
1983	48,580	11,536	197,522	44,616	200	190,150	10,216	18.6	4.8	63.1	100.7	7.6	4.8
1984	48,181	13,134	187,241	44,450	200	178,038	10,217	17.4	4.6	62.7	101.2	7.4	4.6
1985	45,555	12,226	194,858	47,317	-200	180,661	10,225	17.7	4.7	63.6	101.1	7.3	4.6
1986	39,996	11,254	207,392	52,080	-198	183,857	10,235	18.0	4.8	63.7	101.2	7.5	4.7
1987	40,831	11,123	215,918	60,785	-198	184,643	10,252	18.0	4.9	63.7	101.7	7.7	4.8
1988	42,241	12,629	220,666	65,780	-444	184,054	10,273	17.9	4.8	63.0	101.9	7.7	4.7
1989	40,368	29,642	238,560	68,910	154	180,530	10,300	17.5	5.0	63.0	101.5	8.0	4.9
1990	42,135	37,765	269,292	77,750	77	195,989	10,331	19.0	5.4	63.6	102.8	8.4	5.2
1991	40,737	32,858	261,127	75,071	131	194,065	10,368	18.7	5.2	65.6	105.0	8.0	5.0
1992	37,966	38,060	273,344	70,393	71	202,929	10,409	19.5	5.4	65.8	107.5	8.2	5.0
1993	36,944	32,743	261,526	79,210	707	187,224	10,452	17.9	5.1	67.2	106.8	7.6	4.8
1994	35,102	38,208	309,786	105,005	300	201,975	10,495	19.2	5.5	60.2	100.8	9.1	5.4
1995	36,445	35,332	325,949	112,032	0	215,031	10,534	20.4	5.7	59.1	101.7	9.6	5.6

23 BELIZE	- FOOD BALANCE SHEET OF FISH AND FISHERY PRODUCTS IN LIVE WEIGHT AND FISH CONTRIBUTION TO PROTEIN SUPPLY -	FAO-FIDI

YEAR	PRODUCTION	NON-FOOD USES	IMPORTS	EXPORTS	STOCKS VARIATIONS	TOTAL FOOD SUPPLY	POPULATION	PER CAPUT SUPPLY	FISH PROTEINS	ANIMAL PROTEINS	TOTAL PROTEINS	FISH/ANIMAL PROTEINS	FISH/TOTAL PROTEINS
	(..		metric tons in live weight			....)	(thousands)	(kilograms)		(grams per caput per day)		(%)	(%)
1961	600	0	175	200	0	575	96	6.0	1.4	25.4	55.6	5.7	2.6
1962	800	0	175	300	0	675	98	6.9	1.6	29.1	61.7	5.3	2.5
1963	900	0	175	300	0	775	101	7.7	1.8	22.6	57.8	7.8	3.0
1964	900	0	175	200	0	875	104	8.4	2.0	23.9	57.9	8.3	3.4
1965	800	0	175	400	0	575	107	5.4	1.4	22.5	56.1	6.1	2.4
1966	900	0	175	400	0	675	110	6.1	1.4	22.6	55.3	6.1	2.5
1967	1,300	0	175	600	0	875	113	7.7	1.3	23.2	56.3	5.5	2.3
1968	1,300	0	175	500	0	975	117	8.4	1.7	24.4	57.0	6.8	2.9
1969	1,600	0	175	700	0	1,075	120	9.0	1.2	21.3	54.1	5.4	2.1
1970	1,500	0	175	700	0	975	123	8.0	1.1	25.3	57.3	4.3	1.9
1971	1,800	0	175	800	0	1,175	125	9.4	1.2	26.0	59.1	4.5	2.0
1972	1,900	0	175	1,263	0	812	128	6.4	0.8	25.8	58.7	3.1	1.4
1973	1,600	0	175	750	0	1,025	130	7.9	1.3	24.2	56.9	5.2	2.2
1974	1,618	0	175	762	0	1,031	132	7.8	1.5	27.0	60.9	5.6	2.5
1975	1,603	0	209	767	-88	957	134	7.2	1.3	29.4	62.4	4.5	2.1
1976	1,613	0	295	794	0	1,114	136	8.2	1.6	29.3	62.0	5.6	2.7
1977	1,554	0	345	854	35	1,080	138	7.8	1.7	30.4	63.4	5.5	2.6
1978	1,522	0	346	1,005	-60	803	140	5.7	1.2	30.7	64.2	4.0	1.9
1979	1,440	0	507	1,002	53	997	143	7.0	1.9	32.1	66.2	6.0	2.9
1980	1,349	0	601	768	0	1,182	146	8.1	2.2	33.8	69.0	6.5	3.2
1981	1,365	0	324	785	60	964	149	6.5	1.9	34.0	60.2	5.5	2.7
1982	1,358	0	193	690	0	861	153	5.6	1.7	32.6	68.3	5.3	2.5
1983	1,519	0	114	791	0	842	157	5.4	1.6	34.4	67.1	4.5	2.3
1984	1,304	0	354	704	0	954	162	5.9	1.8	33.9	66.6	5.2	2.6
1985	1,391	0	246	715	0	922	166	5.6	1.6	33.7	66.4	4.7	2.4
1986	1,459	0	432	766	0	1,125	170	6.6	1.8	33.0	66.5	5.6	2.8
1987	1,506	0	655	725	0	1,436	174	8.2	2.4	34.2	68.3	6.9	3.5
1988	1,498	0	456	753	0	1,201	178	6.7	1.9	33.6	67.2	5.5	2.8
1989	1,775	0	904	771	0	1,908	183	10.4	2.8	31.4	67.0	8.8	4.1
1990	1,513	0	365	773	0	1,105	187	5.9	1.6	33.0	69.0	5.0	2.4
1991	1,639	0	455	875	0	1,219	192	6.3	1.8	28.8	67.0	6.2	2.7
1992	2,196	0	527	1,078	0	1,645	197	8.3	2.4	32.2	69.5	7.5	3.5
1993	2,129	0	554	1,199	0	1,483	203	7.3	2.0	30.2	66.5	6.7	3.0
1994	2,034	0	313	958	-40	1,349	208	6.5	2.1	30.4	65.2	6.8	3.2
1995	2,179	0	434	1,215	40	1,437	213	6.7	1.8	28.5	63.0	6.4	2.9

53 BENIN - FOOD BALANCE SHEET OF FISH AND FISHERY PRODUCTS IN LIVE WEIGHT AND FISH CONTRIBUTION TO PROTEIN SUPPLY - FAO-FIDI

YEAR	PRODUCTION	NON-FOOD USES	IMPORTS	EXPORTS	STOCKS VARIATIONS	TOTAL FOOD SUPPLY	POPULATION	PER CAPUT SUPPLY	FISH PROTEINS	ANIMAL PROTEINS	TOTAL PROTEINS	FISH/ANIMAL PROTEINS	FISH/TOTAL PROTEINS
	(..		metric tons in live weight			....)	(thousands)	(kilograms)		(grams per caput per day)		(%)	(%)
1961	28,000	0	785	7,822	0	20,963	2,270	9.2	2.7	7.1	45.5	37.6	5.9
1962	30,000	0	1,433	4,529	0	26,903	2,305	11.7	3.4	7.8	46.9	43.2	7.2
1963	25,000	0	1,004	1,211	0	24,794	2,344	10.6	3.1	7.8	44.5	39.4	6.9
1964	26,000	0	2,802	300	0	28,502	2,385	12.0	3.5	8.3	43.8	42.0	8.0
1965	20,000	0	6,402	300	0	26,102	2,430	10.7	3.1	8.2	46.9	38.3	6.7
1966	18,800	0	3,202	300	0	21,702	2,479	8.8	2.6	7.6	44.7	33.6	5.7
1967	35,600	0	3,302	1,204	0	37,698	2,531	14.9	4.3	9.5	48.0	45.3	9.0
1968	29,000	0	5,204	802	0	33,402	2,586	12.9	3.6	8.9	49.9	41.2	7.3
1969	29,700	0	2,502	602	0	31,600	2,644	12.0	3.5	8.8	48.1	39.5	7.2
1970	43,000	0	3,900	602	0	46,298	2,705	17.1	4.9	10.2	48.3	48.4	10.2
1971	43,300	0	5,178	879	0	47,598	2,768	17.2	4.9	10.2	48.1	48.1	10.2
1972	42,500	0	4,553	1,179	0	45,873	2,834	16.2	4.6	9.9	45.8	46.2	10.0
1973	40,700	0	7,305	1,560	0	46,445	2,902	16.0	4.6	9.9	46.3	46.5	9.9
1974	39,402	0	8,779	1,643	0	46,538	2,973	15.7	4.5	10.0	44.8	45.1	10.0
1975	37,533	0	11,234	1,116	0	47,651	3,046	15.6	4.5	10.1	43.3	44.5	10.4
1976	37,104	0	6,247	570	0	42,781	3,121	13.7	4.0	9.8	46.3	40.9	8.6
1977	36,528	0	6,216	570	0	42,174	3,198	13.2	3.8	9.5	47.1	40.3	8.2
1978	35,879	0	7,812	125	0	43,566	3,279	13.3	3.9	9.7	47.9	40.1	8.1
1979	35,700	0	7,953	133	0	43,520	3,365	12.9	3.8	9.7	50.9	39.4	7.5
1980	37,832	0	4,266	4	0	42,094	3,459	12.2	3.6	9.7	50.9	36.6	7.0
1981	37,768	0	5,085	17	0	42,836	3,560	12.0	3.5	9.8	47.3	36.2	7.5
1982	37,521	0	6,345	6	0	43,860	3,668	12.0	3.5	10.1	45.7	34.9	7.7
1983	34,628	0	4,590	0	0	39,218	3,783	10.4	3.1	10.4	44.1	29.4	6.9
1984	35,290	0	5,998	66	0	41,222	3,902	10.6	3.1	10.1	48.1	30.9	6.5
1985	36,371	0	7,338	139	0	43,570	4,025	10.8	3.2	10.2	51.2	31.3	6.3
1986	38,753	0	8,323	165	0	46,911	4,151	11.3	3.4	10.1	51.8	33.1	6.5
1987	41,903	0	6,098	115	0	47,886	4,280	11.2	3.3	10.3	50.6	32.4	6.6
1988	37,267	0	5,961	176	0	43,052	4,412	9.8	2.9	9.5	48.9	30.5	5.9
1989	41,860	0	2,828	64	0	44,624	4,547	9.8	2.9	9.3	54.8	31.0	5.3
1990	38,234	0	3,366	41	0	41,559	4,684	8.9	2.6	9.4	54.5	27.5	4.7
1991	35,083	0	10,866	44	0	45,905	4,824	9.5	2.8	9.3	55.1	30.2	5.1
1992	32,488	0	14,062	32	0	46,518	4,966	9.4	2.8	9.6	56.8	29.1	4.9
1993	39,221	0	19,203	376	0	58,048	5,111	11.4	3.4	10.5	57.2	32.2	5.9
1994	39,924	0	11,945	516	0	51,353	5,259	9.8	2.9	9.5	55.1	30.5	5.3
1995	44,449	0	14,190	446	0	58,193	5,409	10.8	3.2	10.0	57.9	31.9	5.5

17 BERMUDA	- FOOD BALANCE SHEET OF FISH AND FISHERY PRODUCTS IN LIVE WEIGHT AND FISH CONTRIBUTION TO PROTEIN SUPPLY -	FAO-FIDI

YEAR	PRODUCTION	NON-FOOD USES	IMPORTS	EXPORTS	STOCKS VARIATIONS	TOTAL FOOD SUPPLY	POPULATION	PER CAPUT SUPPLY	FISH PROTEINS	ANIMAL PROTEINS	TOTAL PROTEINS	FISH/ANIMAL PROTEINS	FISH/TOTAL PROTEINS
	(.................................... metric tons in live weight					....................................)	(thousands)	(kilograms)	(grams per caput per day)			(%)	(%)
1961	500	0	1,051	0	0	1,551	46	33.8	10.7	63.3	88.6	16.9	12.1
1962	400	0	1,051	0	0	1,451	47	30.9	9.9	63.9	89.3	15.4	11.0
1963	400	0	1,051	0	0	1,451	48	30.2	9.6	63.2	87.3	15.2	11.0
1964	450	0	1,051	0	0	1,501	49	30.6	9.7	62.4	87.6	15.6	11.1
1965	450	0	951	0	0	1,401	50	28.0	9.1	62.9	87.5	14.4	10.3
1966	500	0	1,402	0	0	1,902	51	37.4	12.1	68.3	91.9	17.8	13.2
1967	500	0	1,426	0	0	1,926	52	37.4	11.8	65.1	89.7	18.1	13.1
1968	550	0	1,402	0	0	1,952	52	37.4	12.1	68.1	92.4	17.7	13.0
1969	550	0	1,226	0	0	1,776	53	33.7	10.8	66.8	91.6	16.1	11.8
1970	900	0	1,326	0	0	2,226	53	42.0	13.1	66.7	93.5	19.6	14.0
1971	1,000	0	1,051	0	0	2,051	53	38.6	11.9	65.5	91.0	18.2	13.1
1972	1,000	0	1,151	0	0	2,151	53	40.5	12.4	64.9	91.7	19.1	13.5
1973	700	0	1,151	0	0	1,851	53	34.9	10.8	65.0	92.7	16.7	11.7
1974	699	0	1,326	0	0	2,025	53	38.2	12.1	69.1	96.5	17.4	12.5
1975	4,582	4,000	1,650	0	0	2,232	53	42.1	13.2	71.1	98.4	18.6	13.4
1976	7,218	6,700	1,782	0	0	2,300	53	43.3	13.5	75.2	103.6	18.0	13.1
1977	7,635	6,800	1,375	0	0	2,210	53	41.5	13.2	71.0	99.8	18.6	13.3
1978	8,723	8,000	1,509	0	0	2,232	53	41.7	13.3	70.3	100.8	19.0	13.2
1979	6,327	5,900	1,843	0	0	2,270	54	42.2	13.6	77.3	107.8	17.6	12.7
1980	525	0	1,731	0	0	2,256	54	41.8	13.2	77.4	108.3	17.0	12.2
1981	424	0	1,686	0	0	2,110	54	38.9	12.3	75.1	106.8	16.3	11.5
1982	418	0	1,844	0	0	2,262	55	41.5	13.1	74.6	108.1	17.6	12.1
1983	517	0	1,950	0	0	2,467	55	45.0	14.3	75.8	110.4	18.8	12.9
1984	538	0	2,057	0	0	2,595	55	46.9	14.8	75.0	111.6	19.7	13.2
1985	670	0	1,995	0	0	2,665	56	47.6	15.4	73.3	109.0	21.0	14.1
1986	820	0	1,956	0	0	2,776	57	48.8	15.7	72.6	109.0	21.7	14.4
1987	821	0	1,891	0	0	2,712	58	46.8	14.1	71.0	107.5	19.8	13.1
1988	773	0	2,164	0	0	2,937	59	49.7	14.9	74.6	108.3	20.0	13.7
1989	774	0	1,942	0	0	2,716	60	45.2	13.4	68.4	100.0	19.5	13.4
1990	463	0	2,499	0	0	2,962	61	48.6	15.9	71.5	100.0	22.2	15.9
1991	428	0	2,171	0	0	2,599	62	42.2	13.4	65.2	93.8	20.6	14.3
1992	432	0	2,268	0	0	2,700	62	43.5	13.8	66.1	100.6	20.9	13.7
1993	393	0	2,014	0	0	2,407	62	38.6	12.3	65.6	102.2	18.7	12.0
1994	386	0	1,989	0	0	2,375	63	37.9	12.0	63.9	100.1	18.8	12.0
1995	444	0	2,079	0	0	2,523	63	40.1	10.8	57.2	95.4	18.8	11.3

Note: 1975-1979 production includes substantial amounts of fish caught by foreign vessels flying national flag and landing abroad.

18 BHUTAN	- FOOD BALANCE SHEET OF FISH AND FISHERY PRODUCTS IN LIVE WEIGHT AND FISH CONTRIBUTION TO PROTEIN SUPPLY -	FAO-FIDI

YEAR	PRODUCTION	NON-FOOD USES	IMPORTS	EXPORTS	STOCKS VARIATIONS	TOTAL FOOD SUPPLY	POPULATION	PER CAPUT SUPPLY	FISH PROTEINS	ANIMAL PROTEINS	TOTAL PROTEINS	FISH/ANIMAL PROTEINS	FISH/TOTAL PROTEINS
	(........ metric tons in live weight ....)						(thousands)	(kilograms)	(grams per caput per day)			(%)	(%)
1961	100	0	0	0	0	100	883	0.1	0.0	3.2	26.9	1.1	0.1
1962	100	0	0	0	0	100	899	0.1	0.0	3.4	27.3	1.0	0.1
1963	150	0	0	0	0	150	916	0.2	0.1	3.5	27.7	1.4	0.2
1964	150	0	0	0	0	150	933	0.2	0.1	3.5	27.9	1.4	0.2
1965	150	0	0	0	0	150	951	0.2	0.1	3.5	28.1	1.3	0.2
1966	150	0	0	0	0	150	969	0.2	0.1	3.5	28.3	1.3	0.2
1967	150	0	0	0	0	150	987	0.2	0.1	3.5	28.5	1.3	0.2
1968	150	0	0	0	0	150	1,007	0.2	0.0	3.5	28.8	1.3	0.2
1969	150	0	0	0	0	150	1,026	0.2	0.0	3.6	29.0	1.2	0.2
1970	200	0	0	0	0	200	1,047	0.2	0.1	3.6	29.1	1.6	0.2
1971	200	0	0	0	0	200	1,068	0.2	0.1	3.6	29.3	1.6	0.2
1972	200	0	0	0	0	200	1,089	0.2	0.1	3.6	25.7	1.5	0.2
1973	200	0	0	0	0	200	1,112	0.2	0.1	3.6	25.7	1.5	0.2
1974	200	0	0	0	0	200	1,135	0.2	0.1	3.7	25.8	1.4	0.2
1975	200	0	0	0	0	200	1,159	0.2	0.1	3.7	25.9	1.4	0.2
1976	200	0	0	0	0	200	1,184	0.2	0.1	3.7	25.9	1.4	0.2
1977	200	0	0	0	0	200	1,209	0.2	0.1	3.7	26.0	1.3	0.2
1978	250	0	0	0	0	250	1,236	0.2	0.1	3.8	26.0	1.6	0.2
1979	250	0	0	0	0	250	1,263	0.2	0.1	3.7	26.1	1.6	0.2
1980	250	0	0	0	0	250	1,292	0.2	0.1	3.8	27.1	1.5	0.2
1981	250	0	0	0	0	250	1,321	0.2	0.1	3.8	28.3	1.5	0.2
1982	250	0	0	0	0	250	1,351	0.2	0.1	3.8	28.6	1.5	0.2
1983	250	0	0	0	0	250	1,382	0.2	0.1	3.9	29.4	1.4	0.2
1984	250	0	0	0	0	250	1,415	0.2	0.1	3.9	30.7	1.4	0.2
1985	280	0	0	0	0	280	1,451	0.2	0.1	3.9	28.1	1.5	0.2
1986	280	0	0	0	0	280	1,490	0.2	0.1	3.8	28.2	1.5	0.2
1987	300	0	0	0	0	300	1,531	0.2	0.1	3.8	25.8	1.6	0.2
1988	315	0	0	0	0	315	1,573	0.2	0.1	3.7	18.5	1.6	0.3
1989	312	0	0	0	0	312	1,612	0.2	0.1	3.7	17.8	1.6	0.3
1990	331	0	0	0	0	331	1,645	0.2	0.1	3.9	22.5	1.5	0.3
1991	330	0	0	0	0	330	1,672	0.2	0.1	3.8	21.5	1.6	0.3
1992	350	0	0	0	0	350	1,693	0.2	0.1	3.9	20.4	1.6	0.3
1993	350	0	0	0	0	350	1,713	0.2	0.1	3.9	20.2	1.6	0.3
1994	340	0	0	0	0	340	1,738	0.2	0.1	3.9	21.3	1.5	0.3
1995	340	0	0	0	0	340	1,770	0.2	0.1	3.7	22.1	1.5	0.3

19 BOLIVIA	- FOOD BALANCE SHEET OF FISH AND FISHERY PRODUCTS IN LIVE WEIGHT AND FISH CONTRIBUTION TO PROTEIN SUPPLY -												FAO-FIDI
YEAR	PRODUCTION	NON-FOOD USES	IMPORTS	EXPORTS	STOCKS VARIATIONS	TOTAL FOOD SUPPLY	POPULATION	PER CAPUT SUPPLY	FISH PROTEINS	ANIMAL PROTEINS	TOTAL PROTEINS	FISH/ANIMAL PROTEINS	FISH/TOTAL PROTEINS
	(.......... metric tons in live weight ....)						(thousands)	(kilograms)	(grams per caput per day)			(%)	(%)
1961	500	0	200	0	0	700	3,426	0.2	0.1	14.0	46.7	0.4	0.1
1962	1,300	0	3,200	0	0	4,500	3,503	1.3	0.4	14.4	47.5	2.6	0.8
1963	1,400	0	3,200	0	0	4,600	3,582	1.3	0.4	15.0	49.4	2.5	0.8
1964	1,900	0	3,500	0	0	5,400	3,664	1.5	0.4	14.4	46.7	3.0	0.9
1965	1,800	0	3,000	0	0	4,800	3,748	1.3	0.4	14.5	49.0	2.6	0.8
1966	1,700	0	3,850	0	0	5,550	3,835	1.5	0.4	14.5	48.2	2.9	0.9
1967	1,500	0	4,250	0	0	5,750	3,924	1.5	0.4	14.9	48.7	2.9	0.9
1968	1,300	0	2,450	0	0	3,750	4,016	0.9	0.3	14.8	49.4	1.9	0.6
1969	1,200	0	4,400	0	0	5,600	4,112	1.4	0.4	15.3	48.7	2.6	0.8
1970	1,100	0	3,250	0	0	4,350	4,212	1.0	0.3	15.7	49.7	1.9	0.6
1971	1,300	0	6,000	0	0	7,300	4,315	1.7	0.5	15.5	50.7	3.1	1.0
1972	1,500	0	4,600	0	0	6,100	4,421	1.4	0.4	16.6	51.3	2.4	0.8
1973	1,500	0	6,800	0	0	8,300	4,531	1.8	0.5	16.6	50.1	3.2	1.1
1974	1,050	0	6,800	0	0	7,850	4,643	1.7	0.5	17.1	53.8	2.8	0.9
1975	1,050	0	7,012	0	0	8,062	4,759	1.7	0.5	17.7	55.6	2.7	0.9
1976	1,250	0	4,853	0	0	6,103	4,878	1.3	0.4	18.0	54.9	2.3	0.8
1977	1,550	0	7,185	0	0	8,735	4,999	1.8	0.6	19.0	52.8	3.2	1.1
1978	1,550	0	7,659	0	0	9,209	5,121	1.8	0.6	19.8	54.2	3.1	1.1
1979	3,650	0	9,707	0	0	13,357	5,240	2.6	0.9	19.8	53.1	4.4	1.6
1980	4,379	0	12,456	0	0	16,835	5,355	3.1	1.1	22.8	55.5	4.7	1.9
1981	5,617	0	9,708	0	0	15,325	5,465	2.8	0.9	21.7	56.6	4.3	1.7
1982	4,105	0	1,614	0	0	5,719	5,570	1.0	0.3	19.9	53.6	1.6	0.6
1983	4,105	0	2,286	20	0	6,371	5,674	1.1	0.4	20.4	50.9	1.8	0.7
1984	4,105	0	3,922	50	0	7,977	5,781	1.4	0.5	19.3	54.0	2.4	0.8
1985	4,170	0	3,998	20	0	8,148	5,895	1.4	0.5	20.0	57.7	2.3	0.8
1986	3,871	0	2,357	122	0	6,106	6,017	1.0	0.3	18.7	53.4	1.8	0.6
1987	4,271	0	2,792	185	0	6,878	6,145	1.1	0.4	20.1	56.2	1.8	0.6
1988	4,427	0	511	134	0	4,804	6,281	0.8	0.2	20.1	55.3	1.2	0.4
1989	6,024	0	1,544	164	0	7,404	6,423	1.2	0.4	21.0	55.3	1.7	0.7
1990	7,424	0	387	192	0	7,619	6,573	1.2	0.4	20.8	53.3	1.8	0.7
1991	5,367	0	283	476	0	5,174	6,730	0.8	0.2	21.0	57.3	1.1	0.4
1992	5,171	0	922	455	0	5,638	6,893	0.8	0.3	20.4	55.8	1.2	0.4
1993	6,167	0	5,261	477	-9	10,943	7,063	1.6	0.4	21.2	58.2	2.1	0.8
1994	5,970	0	6,303	127	-4	12,142	7,237	1.7	0.5	22.0	56.6	2.2	0.9
1995	6,308	0	3,586	55	13	9,853	7,414	1.3	0.4	22.8	57.6	1.7	0.7

80 BOSNIA and HERZEGOVINA	- FOOD BALANCE SHEET OF FISH AND FISHERY PRODUCTS IN LIVE WEIGHT AND FISH CONTRIBUTION TO PROTEIN SUPPLY -	FAO-FIDI

YEAR	PRODUCTION	NON-FOOD USES	IMPORTS	EXPORTS	STOCKS VARIATIONS	TOTAL FOOD SUPPLY	POPULATION	PER CAPUT SUPPLY	FISH PROTEINS	ANIMAL PROTEINS	TOTAL PROTEINS	FISH/ANIMAL PROTEINS	FISH/TOTAL PROTEINS
	(............		metric tons in live weight			............)	(thousands)	(kilograms)		(grams per caput per day)		(%)	(%)
1961	0	0	0	0	0	0	3,225	0.0	0.0	0.0	0.0	0.0	0.0
1962	0	0	0	0	0	0	3,271	0.0	0.0	0.0	0.0	0.0	0.0
1963	0	0	0	0	0	0	3,315	0.0	0.0	0.0	0.0	0.0	0.0
1964	0	0	0	0	0	0	3,357	0.0	0.0	0.0	0.0	0.0	0.0
1965	0	0	0	0	0	0	3,396	0.0	0.0	0.0	0.0	0.0	0.0
1966	0	0	0	0	0	0	3,433	0.0	0.0	0.0	0.0	0.0	0.0
1967	0	0	0	0	0	0	3,466	0.0	0.0	0.0	0.0	0.0	0.0
1968	0	0	0	0	0	0	3,498	0.0	0.0	0.0	0.0	0.0	0.0
1969	0	0	0	0	0	0	3,531	0.0	0.0	0.0	0.0	0.0	0.0
1970	0	0	0	0	0	0	3,564	0.0	0.0	0.0	0.0	0.0	0.0
1971	0	0	0	0	0	0	3,600	0.0	0.0	0.0	0.0	0.0	0.0
1972	0	0	0	0	0	0	3,637	0.0	0.0	0.0	0.0	0.0	0.0
1973	0	0	0	0	0	0	3,675	0.0	0.0	0.0	0.0	0.0	0.0
1974	0	0	0	0	0	0	3,712	0.0	0.0	0.0	0.0	0.0	0.0
1975	0	0	0	0	0	0	3,747	0.0	0.0	0.0	0.0	0.0	0.0
1976	0	0	0	0	0	0	3,781	0.0	0.0	0.0	0.0	0.0	0.0
1977	0	0	0	0	0	0	3,813	0.0	0.0	0.0	0.0	0.0	0.0
1978	0	0	0	0	0	0	3,844	0.0	0.0	0.0	0.0	0.0	0.0
1979	0	0	0	0	0	0	3,878	0.0	0.0	0.0	0.0	0.0	0.0
1980	0	0	0	0	0	0	3,914	0.0	0.0	0.0	0.0	0.0	0.0
1981	0	0	0	0	0	0	3,950	0.0	0.0	0.0	0.0	0.0	0.0
1982	0	0	0	0	0	0	3,985	0.0	0.0	0.0	0.0	0.0	0.0
1983	0	0	0	0	0	0	4,023	0.0	0.0	0.0	0.0	0.0	0.0
1984	0	0	0	0	0	0	4,068	0.0	0.0	0.0	0.0	0.0	0.0
1985	0	0	0	0	0	0	4,122	0.0	0.0	0.0	0.0	0.0	0.0
1986	0	0	0	0	0	0	4,192	0.0	0.0	0.0	0.0	0.0	0.0
1987	0	0	0	0	0	0	4,275	0.0	0.0	0.0	0.0	0.0	0.0
1988	0	0	0	0	0	0	4,344	0.0	0.0	0.0	0.0	0.0	0.0
1989	0	0	0	0	0	0	4,353	0.0	0.0	0.0	0.0	0.0	0.0
1990	0	0	0	0	0	0	4,308	0.0	0.0	0.0	0.0	0.0	0.0
1991	0	0	0	0	0	0	4,183	0.0	0.0	0.0	0.0	0.0	0.0
1992	3,000	0	436	0	0	3,436	3,992	0.9	0.3	22.9	68.8	1.1	0.4
1993	2,500	0	3,138	0	0	5,638	3,785	1.5	0.5	22.3	70.1	2.2	0.7
1994	2,400	0	535	0	0	2,935	3,628	0.8	0.2	18.3	67.8	1.3	0.4
1995	2,500	0	2,585	0	0	5,085	3,569	1.4	0.4	15.9	66.2	2.6	0.6

Note: for years before 1992 see Yugoslavia, S.F.R.

20 BOTSWANA	- FOOD BALANCE SHEET OF FISH AND FISHERY PRODUCTS IN LIVE WEIGHT AND FISH CONTRIBUTION TO PROTEIN SUPPLY -	FAO-FIDI

YEAR	PRODUCTION	NON-FOOD USES	IMPORTS	EXPORTS	STOCKS VARIATIONS	TOTAL FOOD SUPPLY	POPULATION	PER CAPUT SUPPLY	FISH PROTEINS	ANIMAL PROTEINS	TOTAL PROTEINS	FISH/ANIMAL PROTEINS	FISH/TOTAL PROTEINS
	(.......... metric tons in live weight					..........)	(thousands)	(kilograms)	(grams per caput per day)			(%)	(%)
1961	400	0	...	0	0	400	492	0.8	0.2	23.5	74.8	1.0	0.3
1962	500	0	...	0	0	500	505	1.0	0.3	24.2	70.7	1.2	0.4
1963	500	0	...	0	0	500	519	1.0	0.3	26.4	74.0	1.1	0.4
1964	500	0	...	0	0	500	534	0.9	0.3	26.1	73.8	1.1	0.4
1965	500	0	...	0	0	500	549	0.9	0.3	28.1	76.1	1.0	0.4
1966	600	0	...	0	0	600	565	1.1	0.3	26.1	74.0	1.2	0.4
1967	600	0	...	0	0	600	581	1.0	0.3	25.3	75.0	1.2	0.4
1968	800	0	...	0	0	800	598	1.3	0.4	25.2	74.8	1.6	0.5
1969	800	0	...	0	0	800	617	1.3	0.4	26.1	75.1	1.5	0.5
1970	800	0	...	0	0	800	637	1.3	0.4	27.5	75.2	1.4	0.5
1971	1,100	0	...	0	0	1,100	659	1.7	0.5	27.0	76.8	1.9	0.7
1972	1,200	0	...	0	0	1,200	682	1.8	0.5	27.6	77.0	1.9	0.7
1973	1,200	0	...	0	0	1,200	707	1.7	0.5	27.4	76.6	1.9	0.7
1974	1,200	0	...	0	0	1,200	732	1.6	0.5	24.2	69.7	2.0	0.7
1975	1,200	0	...	0	0	1,200	759	1.6	0.5	22.6	71.1	2.1	0.7
1976	1,600	0	...	0	0	1,600	787	2.0	0.6	22.3	70.0	2.7	0.9
1977	1,500	0	565	25	0	2,040	815	2.5	0.8	22.3	73.6	3.5	1.1
1978	1,000	0	1,445	9	0	2,436	844	2.9	0.9	21.6	67.7	4.3	1.4
1979	1,000	0	1,096	31	0	2,065	875	2.4	0.8	25.6	73.7	2.9	1.0
1980	1,250	0	1,828	14	0	3,064	906	3.4	1.1	21.9	67.3	5.0	1.6
1981	1,450	0	2,462	58	0	3,855	938	4.1	1.3	21.2	66.1	6.3	2.0
1982	1,400	0	1,394	33	0	2,761	971	2.8	0.9	20.9	66.1	4.4	1.4
1983	1,250	0	1,563	0	0	2,813	1,005	2.8	0.9	20.4	68.2	4.4	1.3
1984	1,500	0	1,874	15	0	3,359	1,041	3.2	1.1	28.8	76.9	3.6	1.4
1985	1,500	0	1,606	79	0	3,027	1,077	2.8	0.9	22.2	73.0	4.1	1.3
1986	1,700	0	2,220	283	0	3,637	1,115	3.3	1.1	21.2	71.8	5.1	1.5
1987	1,900	0	3,041	420	0	4,521	1,154	3.9	1.3	21.5	69.6	5.9	1.8
1988	1,900	0	4,293	438	0	5,755	1,194	4.8	1.5	23.7	71.9	6.5	2.2
1989	1,900	0	5,232	992	0	6,140	1,234	5.0	1.4	24.5	70.3	5.8	2.0
1990	1,900	0	10,325	363	0	11,862	1,272	9.3	2.5	29.7	73.2	8.5	3.4
1991	1,900	0	12,900	168	0	14,633	1,310	11.2	3.1	31.9	72.3	9.8	4.3
1992	1,900	0	7,713	82	0	9,530	1,346	7.1	2.0	32.6	70.0	6.3	2.9
1993	2,000	0	6,560	65	-5	8,490	1,381	6.2	1.8	27.7	66.8	6.5	2.7
1994	2,000	0	6,151	12	0	8,139	1,416	5.8	1.7	24.1	70.1	7.0	2.4
1995	2,000	0	5,272	57	5	7,220	1,450	5.0	1.6	28.2	71.2	5.5	2.2

21 BRAZIL	- FOOD BALANCE SHEET OF FISH AND FISHERY PRODUCTS IN LIVE WEIGHT AND FISH CONTRIBUTION TO PROTEIN SUPPLY -	FAO-FIDI

YEAR	PRODUCTION	NON-FOOD USES	IMPORTS	EXPORTS	STOCKS VARIATIONS	TOTAL FOOD SUPPLY	POPULATION	PER CAPUT SUPPLY	FISH PROTEINS	ANIMAL PROTEINS	TOTAL PROTEINS	FISH/ANIMAL PROTEINS	FISH/TOTAL PROTEINS
	(.. metric tons in live weight ....)						(thousands)	(kilograms)	(grams per caput per day)			(%)	(%)
1961	274,800	5,000	77,905	3,252	0	344,454	74,967	4.6	1.4	17.8	56.3	7.8	2.5
1962	379,300	5,000	81,500	3,780	0	452,020	77,264	5.9	1.7	18.1	57.2	9.6	3.0
1963	335,400	5,000	90,300	3,240	0	417,460	79,618	5.2	1.6	18.0	57.7	8.7	2.7
1964	368,800	5,000	63,261	3,173	0	423,888	81,990	5.2	1.5	18.7	59.4	8.2	2.6
1965	388,700	5,000	49,861	4,010	0	429,551	84,351	5.1	1.6	19.1	60.5	8.2	2.6
1966	393,000	5,000	84,583	4,271	0	468,313	86,691	5.4	1.7	19.1	60.6	8.8	2.8
1967	419,600	5,000	112,468	4,784	0	522,284	89,020	5.9	1.8	19.1	62.3	9.6	2.9
1968	495,000	5,000	126,327	8,671	0	607,656	91,342	6.7	2.1	20.0	63.2	10.4	3.3
1969	492,100	5,000	141,867	13,959	0	615,008	93,673	6.6	2.0	19.8	61.0	10.3	3.3
1970	572,940	5,000	149,642	14,911	-100	702,571	96,021	7.3	2.2	20.1	60.4	11.1	3.7
1971	613,401	5,000	106,071	17,945	0	696,527	98,384	7.1	2.1	19.3	61.8	11.1	3.4
1972	620,263	3,900	99,423	26,725	0	689,061	100,761	6.8	2.1	19.7	60.0	10.5	3.4
1973	759,861	3,900	134,354	40,202	-100	850,013	103,170	8.2	2.5	20.6	59.4	12.2	4.2
1974	715,042	3,900	95,102	28,149	-19,882	758,213	105,634	7.2	2.2	21.1	59.4	10.6	3.8
1975	752,513	6,500	157,365	32,989	0	870,388	108,167	8.1	2.3	22.1	59.8	10.6	3.9
1976	660,099	54,500	126,254	39,510	21,775	714,118	110,776	6.5	1.8	22.6	59.2	8.1	3.1
1977	731,486	90,400	108,457	30,697	100	718,946	113,451	6.3	1.8	22.1	59.2	8.2	3.1
1978	754,413	100,300	122,650	39,195	175	737,743	116,174	6.4	1.8	22.7	59.9	7.8	3.0
1979	783,277	37,000	172,680	40,014	25	878,969	118,922	7.4	2.0	23.9	62.3	8.5	3.3
1980	808,330	50,600	128,566	61,698	0	824,598	121,672	6.8	1.9	25.7	62.8	7.5	3.1
1981	811,260	81,800	94,235	70,136	-11,054	742,505	124,417	6.0	1.7	24.4	61.6	6.9	2.7
1982	831,546	100,700	115,298	70,893	11,116	786,367	127,154	6.2	1.7	24.2	62.1	7.0	2.7
1983	880,904	89,400	79,324	64,981	0	805,847	129,878	6.2	1.7	24.1	62.3	7.0	2.7
1984	960,569	84,634	70,644	60,453	0	886,125	132,582	6.7	1.9	24.3	62.2	7.7	3.0
1985	967,567	69,700	79,068	80,100	0	896,836	135,260	6.6	1.8	24.7	62.4	7.4	2.9
1986	957,550	90,200	186,194	65,256	0	988,288	137,910	7.2	2.0	28.8	65.2	6.9	3.0
1987	947,992	75,000	163,751	64,236	0	972,507	140,524	6.9	1.9	28.5	65.5	6.6	2.9
1988	830,122	60,000	117,221	67,508	0	819,836	143,088	5.7	1.6	27.3	65.9	5.8	2.4
1989	850,030	60,000	241,614	68,094	-250	963,299	145,585	6.6	1.8	29.8	65.8	6.0	2.7
1990	802,940	50,027	275,379	51,164	-317	976,812	148,002	6.6	1.8	30.0	65.3	5.9	2.7
1991	800,080	50,004	224,890	67,863	-550	906,553	150,335	6.0	1.7	31.4	68.0	5.3	2.4
1992	790,110	60,006	145,729	68,858	533	807,509	152,588	5.3	1.4	30.6	67.3	4.7	2.2
1993	780,130	50,007	265,453	74,441	583	921,718	154,774	6.0	1.6	30.8	68.2	5.2	2.4
1994	820,150	90,007	297,041	61,183	0	966,001	156,912	6.2	1.6	31.8	69.7	5.2	2.4
1995	805,160	60,001	409,122	40,305	0	1,113,976	159,015	7.0	1.8	35.0	74.1	5.3	2.5

239 BRITISH VIRGIN ISLANDS - FOOD BALANCE SHEET OF FISH AND FISHERY PRODUCTS IN LIVE WEIGHT AND FISH CONTRIBUTION TO PROTEIN SUPPLY - FAO-FIDI

YEAR	PRODUCTION	NON-FOOD USES	IMPORTS	EXPORTS	STOCKS VARIATIONS	TOTAL FOOD SUPPLY	POPULATION	PER CAPUT SUPPLY	FISH PROTEINS	ANIMAL PROTEINS	TOTAL PROTEINS	FISH/ANIMAL PROTEINS	FISH/TOTAL PROTEINS
	(.........		metric tons in live weight			.........)	(thousands)	(kilograms)		(grams per caput per day)		(%)	(%)
1961	400	0	...	...	0	400	7	55.0	15.1	30.7	47.1	49.3	32.1
1962	400	0	...	...	0	400	8	52.1	14.3	30.9	45.5	46.4	31.5
1963	400	0	...	...	0	400	8	49.1	13.5	27.4	46.1	49.3	29.3
1964	400	0	...	...	0	400	9	46.4	12.8	29.7	43.5	43.1	29.4
1965	400	0	...	...	0	400	9	44.4	12.2	29.0	45.4	42.2	26.9
1966	400	0	...	...	0	400	9	43.1	11.9	28.4	44.7	41.8	26.5
1967	300	0	...	...	0	300	10	31.6	8.9	23.7	40.2	37.5	22.2
1968	300	0	...	...	0	300	10	31.0	8.8	22.1	39.2	39.6	22.3
1969	300	0	...	...	0	300	10	30.5	8.6	25.0	43.8	34.4	19.7
1970	300	0	...	...	0	300	10	30.0	8.5	25.3	43.1	33.5	19.6
1971	300	0	...	...	0	300	10	29.4	8.3	24.2	41.7	34.3	19.9
1972	300	0	...	...	0	300	10	28.9	8.1	24.7	43.7	33.0	18.6
1973	300	0	...	...	0	300	11	28.3	8.0	25.4	44.0	31.5	18.2
1974	306	0	...	...	0	306	11	28.3	7.9	20.7	37.3	38.3	21.2
1975	318	0	...	...	0	318	11	28.9	8.1	19.5	36.9	41.6	22.0
1976	380	0	...	...	0	380	11	34.0	9.5	22.4	37.1	42.6	25.7
1977	440	0	...	...	0	440	11	38.8	10.9	23.7	36.3	45.9	29.9
1978	550	0	...	...	0	550	12	47.8	13.4	26.6	39.0	50.2	34.3
1979	610	0	...	...	0	610	12	52.0	14.5	27.0	37.5	53.9	38.8
1980	670	0	...	...	0	670	12	55.8	15.6	27.8	38.4	56.1	40.6
1981	730	0	...	...	0	730	12	59.2	16.5	28.9	39.5	57.3	41.9
1982	790	0	...	...	0	790	13	62.1	17.3	29.5	39.9	58.7	43.4
1983	898	0	...	...	0	898	13	68.3	19.0	29.8	39.8	64.0	47.9
1984	1,038	0	...	...	0	1,038	14	76.4	21.3	32.0	41.6	66.6	51.2
1985	1,148	0	...	...	0	1,148	14	82.0	22.9	33.4	43.1	68.5	53.1
1986	1,198	0	...	...	0	1,198	14	83.3	23.2	35.2	45.4	65.9	51.2
1987	1,248	0	...	...	0	1,248	15	84.6	23.6	34.2	44.3	69.1	53.4
1988	1,284	0	...	...	0	1,284	15	85.0	23.7	34.6	44.4	68.4	53.4
1989	1,357	0	...	...	0	1,357	16	87.4	24.4	34.2	43.0	71.4	56.7
1990	1,377	0	...	...	0	1,377	16	86.1	24.0	41.4	54.6	58.0	44.0
1991	1,400	0	...	...	0	1,400	17	84.5	23.5	32.5	44.0	72.4	53.4
1992	1,000	0	...	...	0	1,000	17	58.2	16.2	24.4	34.1	66.4	47.4
1993	757	0	...	...	0	757	18	42.4	11.8	20.0	29.5	59.0	40.0
1994	900	0	...	...	0	900	18	48.8	13.6	21.6	30.2	62.9	44.9
1995	1,000	0	...	...	0	1,000	19	52.6	14.6	22.4	36.2	65.3	40.4

26 BRUNEI DARUSSALAM	- FOOD BALANCE SHEET OF FISH AND FISHERY PRODUCTS IN LIVE WEIGHT AND FISH CONTRIBUTION TO PROTEIN SUPPLY -	FAO-FIDI

YEAR	PRODUCTION	NON-FOOD USES	IMPORTS	EXPORTS	STOCKS VARIATIONS	TOTAL FOOD SUPPLY	POPULATION	PER CAPUT SUPPLY	FISH PROTEINS	ANIMAL PROTEINS	TOTAL PROTEINS	FISH/ANIMAL PROTEINS	FISH/TOTAL PROTEINS
	(..		metric tons in live weight		....)		(thousands)	(kilograms)	(grams per caput per day)			(%)	(%)
1961	900	0	1,000	0	0	1,900	86	22.2	6.6	15.8	44.7	41.5	14.7
1962	900	0	1,050	0	0	1,950	90	21.8	6.4	16.3	45.1	39.2	14.2
1963	1,000	0	1,100	0	0	2,100	94	22.4	6.6	18.2	47.5	36.0	13.8
1964	1,000	0	1,050	0	0	2,050	98	20.9	5.3	16.8	45.1	31.8	11.8
1965	1,100	0	950	0	0	2,050	103	20.0	5.1	17.3	47.2	29.6	10.9
1966	1,200	0	1,600	0	0	2,800	107	26.1	6.3	19.3	52.0	32.7	12.1
1967	1,200	0	1,800	0	0	3,000	113	26.6	7.2	19.7	53.5	36.6	13.5
1968	1,300	0	1,400	0	0	2,700	118	22.9	6.1	19.8	53.4	31.1	11.5
1969	1,300	0	1,900	0	0	3,200	124	25.8	6.5	20.3	54.1	32.2	12.1
1970	1,500	0	1,900	0	0	3,400	130	26.2	6.6	20.6	54.2	32.3	12.3
1971	1,500	0	2,300	100	0	3,700	136	27.3	7.0	21.6	54.2	32.3	12.9
1972	1,500	0	2,700	100	0	4,100	142	28.9	7.5	21.5	54.3	34.7	13.8
1973	1,500	0	2,150	100	0	3,550	148	24.0	6.6	21.0	56.2	31.6	11.8
1974	1,560	0	2,350	100	0	3,810	154	24.7	6.2	19.8	54.2	31.4	11.5
1975	1,570	0	2,697	71	0	4,197	161	26.1	6.5	22.9	56.3	28.2	11.5
1976	1,644	0	5,156	195	0	6,605	167	39.5	9.3	28.2	64.0	33.0	14.5
1977	2,193	0	4,188	182	0	6,199	174	35.7	8.1	27.9	62.7	29.2	13.0
1978	2,704	0	4,264	108	0	6,860	180	38.0	9.0	29.3	64.8	30.7	13.9
1979	2,792	0	3,883	100	0	6,575	187	35.2	8.5	31.8	69.2	26.6	12.2
1980	2,225	0	5,228	86	0	7,367	193	38.2	8.7	35.6	72.9	24.4	11.9
1981	2,367	0	6,397	155	0	8,608	199	43.2	10.1	37.5	75.7	26.9	13.3
1982	2,357	0	6,461	82	0	8,735	205	42.6	9.8	38.3	77.3	25.6	12.7
1983	3,055	0	7,241	99	0	10,197	211	48.4	11.0	39.5	77.5	27.9	14.2
1984	3,425	0	7,221	222	0	10,424	217	48.1	11.3	39.2	79.8	28.8	14.2
1985	3,967	0	7,255	279	-133	10,810	223	48.5	11.4	40.0	83.8	28.4	13.6
1986	4,106	0	5,694	185	0	9,616	229	41.9	11.0	42.0	87.5	26.2	12.6
1987	3,916	0	4,202	269	133	7,982	236	33.8	8.9	48.0	90.5	18.5	9.8
1988	2,042	0	4,183	200	0	6,025	243	24.8	6.7	47.2	88.1	14.1	7.6
1989	2,318	0	8,060	240	0	10,138	250	40.6	10.1	46.9	90.0	21.6	11.3
1990	2,350	0	5,630	260	0	7,720	257	30.0	7.7	42.5	84.6	18.2	9.1
1991	1,652	0	5,048	300	0	6,400	264	24.2	6.1	49.1	89.7	12.4	6.8
1992	1,709	0	4,006	300	0	5,415	272	19.9	4.7	45.2	84.8	10.5	5.6
1993	1,764	0	3,987	340	0	5,411	279	19.4	4.6	45.0	85.1	10.3	5.4
1994	4,517	0	5,097	340	0	9,274	286	32.4	7.9	51.5	91.8	15.4	8.6
1995	4,822	0	3,586	368	0	8,040	293	27.4	7.4	47.1	88.1	15.8	8.4

27 BULGARIA - FOOD BALANCE SHEET OF FISH AND FISHERY PRODUCTS IN LIVE WEIGHT AND FISH CONTRIBUTION TO PROTEIN SUPPLY - FAO-FIDI

YEAR	PRODUCTION	NON-FOOD USES	IMPORTS	EXPORTS	STOCKS VARIATIONS	TOTAL FOOD SUPPLY	POPULATION	PER CAPUT SUPPLY	FISH PROTEINS	ANIMAL PROTEINS	TOTAL PROTEINS	FISH/ANIMAL PROTEINS	FISH/TOTAL PROTEINS
	(..		metric tons in live weight		....)		(thousands)	(kilograms)	(grams per caput per day)			(%)	(%)
1961	8,100	0	13,650	0	0	21,750	7,938	2.7	0.8	23.4	91.5	3.5	0.9
1962	9,600	0	9,000	0	0	18,600	8,007	2.3	0.7	23.4	91.5	2.9	0.7
1963	7,500	0	28,900	200	0	36,200	8,073	4.5	1.3	23.7	92.5	5.3	1.4
1964	13,700	0	22,750	0	0	36,450	8,137	4.5	1.3	24.9	93.7	5.1	1.4
1965	21,600	0	15,100	600	0	36,100	8,201	4.4	1.2	26.2	96.4	4.7	1.3
1966	27,300	0	8,650	5,392	0	30,558	8,262	3.7	1.0	28.1	99.7	3.4	1.0
1967	44,900	0	8,609	9,462	0	44,047	8,322	5.3	1.4	29.2	98.1	4.8	1.4
1968	59,900	0	10,500	15,956	0	54,444	8,380	6.5	1.7	28.3	95.3	6.1	1.8
1969	82,800	0	6,400	23,529	0	65,671	8,436	7.8	2.3	28.9	95.8	7.8	2.4
1970	93,700	0	7,300	35,270	0	65,729	8,490	7.7	2.3	29.3	94.8	7.8	2.4
1971	101,100	0	9,700	37,662	0	73,139	8,542	8.6	2.6	30.8	97.4	8.3	2.6
1972	111,000	0	7,715	46,401	0	72,314	8,592	8.4	2.5	33.5	98.2	7.4	2.5
1973	102,028	0	3,668	55,680	0	50,016	8,639	5.8	1.7	33.5	96.7	5.1	1.8
1974	120,646	0	10,408	48,066	0	82,989	8,683	9.6	2.8	35.8	98.0	7.8	2.9
1975	158,121	0	1,895	37,241	-21,276	101,499	8,722	11.6	3.6	39.4	100.5	9.2	3.6
1976	167,097	0	4,139	62,312	-21,261	87,663	8,756	10.0	3.0	40.9	103.3	7.2	2.9
1977	137,996	0	3,896	57,861	-10,753	73,278	8,784	8.3	2.6	40.7	100.7	6.5	2.6
1978	101,116	0	5,568	65,847	26,604	67,440	8,809	7.7	2.4	41.8	102.6	5.7	2.3
1979	89,515	0	5,872	64,797	26,582	57,172	8,835	6.5	2.0	42.5	103.1	4.6	1.9
1980	126,412	0	5,062	71,644	0	59,830	8,862	6.8	2.1	43.1	103.8	4.8	2.0
1981	106,673	0	2,253	53,746	0	55,180	8,892	6.2	1.9	44.9	105.7	4.3	1.8
1982	115,607	0	1,378	57,602	0	59,383	8,924	6.7	2.0	46.1	105.9	4.4	1.9
1983	121,112	0	1,484	35,714	0	86,882	8,951	9.7	3.0	48.4	106.0	6.2	2.8
1984	115,567	0	2,142	30,765	0	86,944	8,965	9.7	2.9	49.7	107.1	5.9	2.7
1985	100,223	0	2,228	32,518	0	69,933	8,960	7.8	2.3	50.2	106.7	4.6	2.2
1986	109,685	0	3,752	35,524	0	77,913	8,934	8.7	2.5	51.0	107.6	4.9	2.3
1987	111,652	0	1,448	50,681	0	62,419	8,889	7.0	2.0	51.3	110.2	4.0	1.8
1988	118,018	0	2	60,881	0	57,139	8,833	6.5	1.9	51.3	111.0	3.7	1.7
1989	102,966	0	280	46,681	0	56,566	8,773	6.5	2.0	52.3	109.9	3.8	1.8
1990	57,103	0	1,091	25,509	-400	32,285	8,718	3.7	1.3	51.9	108.6	2.4	1.2
1991	57,854	0	1,099	33,880	0	25,073	8,670	2.9	1.0	43.2	94.9	2.3	1.0
1992	32,118	0	6,611	11,724	-4,744	22,260	8,627	2.6	0.9	42.1	89.7	2.2	1.0
1993	21,585	0	5,603	8,218	933	19,904	8,587	2.3	0.7	41.2	87.2	1.6	0.8
1994	16,400	0	6,237	8,469	2,267	16,435	8,549	1.9	0.6	36.3	82.1	1.6	0.7
1995	17,352	0	20,712	4,869	1,944	35,139	8,509	4.1	1.3	38.6	85.3	3.3	1.5

233 BURKINA FASO	- FOOD BALANCE SHEET OF FISH AND FISHERY PRODUCTS IN LIVE WEIGHT AND FISH CONTRIBUTION TO PROTEIN SUPPLY -	FAO-FIDI

YEAR	PRODUCTION	NON-FOOD USES	IMPORTS	EXPORTS	STOCKS VARIATIONS	TOTAL FOOD SUPPLY	POPULATION	PER CAPUT SUPPLY	FISH PROTEINS	ANIMAL PROTEINS	TOTAL PROTEINS	FISH/ANIMAL PROTEINS	FISH/TOTAL PROTEINS
	(.................................... metric tons in live weight					....................................)	(thousands)	(kilograms)	(grams per caput per day)			(%)	(%)
1961	2,500	0	400	0	0	2,900	4,535	0.6	0.2	5.3	48.6	3.4	0.4
1962	2,500	0	2,486	0	0	4,986	4,615	1.1	0.3	5.5	47.9	5.9	0.7
1963	3,000	0	6,200	0	0	9,200	4,695	2.0	0.6	5.9	53.5	9.9	1.1
1964	3,500	0	5,343	0	0	8,843	4,779	1.9	0.6	6.1	56.4	9.1	1.0
1965	3,500	0	3,429	0	0	6,929	4,870	1.4	0.4	6.3	57.9	6.8	0.7
1966	4,000	0	4,286	0	0	8,286	4,968	1.7	0.5	6.6	56.3	7.6	0.9
1967	4,500	0	3,750	0	0	8,250	5,073	1.6	0.5	6.4	57.2	7.1	0.8
1968	4,500	0	3,438	0	0	7,938	5,185	1.5	0.4	6.1	56.8	7.0	0.8
1969	4,500	0	3,638	0	0	8,138	5,302	1.5	0.4	6.7	56.4	6.4	0.8
1970	5,000	0	2,450	0	0	7,450	5,424	1.4	0.4	6.9	54.7	6.1	0.8
1971	5,000	0	1,750	0	0	6,750	5,551	1.2	0.4	6.8	53.2	5.5	0.7
1972	4,000	0	2,950	0	0	6,950	5,683	1.2	0.4	6.1	46.2	6.4	0.9
1973	5,000	0	1,450	0	0	6,450	5,820	1.1	0.3	4.8	43.6	7.0	0.8
1974	5,000	0	3,581	0	0	8,581	5,962	1.4	0.5	4.9	46.5	9.2	1.0
1975	5,000	0	2,059	0	0	7,059	6,108	1.2	0.4	5.0	50.5	7.0	0.7
1976	6,000	0	1,969	0	0	7,969	6,258	1.3	0.4	5.5	52.2	7.2	0.8
1977	6,000	0	2,582	25	0	8,557	6,412	1.3	0.4	5.8	46.1	7.3	0.9
1978	7,000	0	2,366	25	0	9,341	6,571	1.4	0.4	6.5	52.1	6.9	0.9
1979	7,000	0	2,441	10	0	9,431	6,736	1.4	0.4	6.8	53.8	6.4	0.8
1980	6,500	0	3,317	0	0	9,817	6,909	1.4	0.5	6.8	50.2	6.6	0.9
1981	7,500	0	3,423	28	0	10,896	7,087	1.5	0.5	6.7	47.9	7.2	1.0
1982	7,000	0	3,614	78	0	10,536	7,273	1.5	0.4	6.6	52.0	6.7	0.9
1983	7,100	0	4,085	5	0	11,180	7,467	1.5	0.5	7.0	49.7	6.7	0.9
1984	7,374	0	3,152	0	0	10,526	7,670	1.4	0.4	6.8	48.6	6.3	0.9
1985	7,443	0	4,583	0	0	12,026	7,882	1.5	0.5	7.3	53.0	6.5	0.9
1986	7,640	0	6,762	13	0	14,389	8,104	1.8	0.6	7.8	65.0	7.2	0.9
1987	7,836	0	8,705	0	0	16,541	8,335	2.0	0.6	7.8	65.3	7.8	0.9
1988	7,916	0	10,851	0	0	18,767	8,576	2.2	0.7	7.7	63.7	8.5	1.0
1989	8,015	0	9,518	0	0	17,533	8,825	2.0	0.6	7.1	67.4	8.3	0.9
1990	7,015	0	8,007	0	0	15,022	9,082	1.7	0.5	7.2	60.6	7.0	0.8
1991	7,012	0	6,712	0	0	13,724	9,347	1.5	0.4	7.5	55.8	5.9	0.8
1992	7,508	0	6,009	0	0	13,517	9,619	1.4	0.4	7.0	72.1	6.0	0.6
1993	7,000	0	5,324	0	0	12,324	9,899	1.3	0.4	7.0	73.0	5.2	0.5
1994	8,000	0	5,781	0	0	13,781	10,186	1.4	0.4	6.7	70.9	6.3	0.6
1995	8,000	0	6,192	3	0	14,189	10,479	1.4	0.4	6.7	66.8	6.2	0.6

29 BURUNDI	- FOOD BALANCE SHEET OF FISH AND FISHERY PRODUCTS IN LIVE WEIGHT AND FISH CONTRIBUTION TO PROTEIN SUPPLY -	FAO-FIDI

YEAR	PRODUCTION	NON-FOOD USES	IMPORTS	EXPORTS	STOCKS VARIATIONS	TOTAL FOOD SUPPLY	POPULATION	PER CAPUT SUPPLY	FISH PROTEINS	ANIMAL PROTEINS	TOTAL PROTEINS	FISH/ANIMAL PROTEINS	FISH/TOTAL PROTEINS
	(.. metric tons in live weight ....)						(thousands)	(kilograms)	(grams per caput per day)			(%)	(%)
1961	5,300	0	...	0	0	5,300	2,993	1.8	0.5	2.8	70.6	18.0	0.7
1962	7,200	0	...	0	0	7,200	3,045	2.4	0.7	3.1	69.2	22.2	1.0
1963	10,600	0	...	0	0	10,600	3,099	3.4	1.0	3.4	67.4	30.1	1.5
1964	9,700	0	...	0	0	9,700	3,154	3.1	0.9	3.4	73.2	26.7	1.2
1965	20,200	0	...	0	0	20,200	3,213	6.3	1.9	4.5	72.5	41.4	2.6
1966	16,600	0	...	0	0	16,600	3,275	5.1	1.5	4.2	71.1	35.6	2.1
1967	12,300	0	...	0	0	12,300	3,341	3.7	1.1	3.9	73.1	27.5	1.5
1968	15,000	0	...	0	0	15,000	3,407	4.4	1.3	4.2	74.9	30.3	1.7
1969	15,600	0	...	0	0	15,600	3,465	4.5	1.3	4.5	70.0	28.6	1.9
1970	13,300	0	1,000	0	0	14,300	3,514	4.1	1.2	4.4	73.1	27.1	1.6
1971	16,900	0	1,200	0	0	18,100	3,550	5.1	1.5	4.8	73.7	31.2	2.0
1972	7,500	0	590	0	0	8,090	3,576	2.3	0.7	4.2	66.9	15.8	1.0
1973	8,500	0	1,700	0	0	10,200	3,601	2.8	0.8	4.5	72.8	18.7	1.2
1974	10,982	0	1,000	0	0	11,982	3,633	3.3	1.0	4.8	62.8	20.2	1.5
1975	14,547	0	750	0	0	15,297	3,680	4.2	1.2	5.1	74.3	23.5	1.6
1976	20,333	0	108	30	0	20,411	3,745	5.5	1.6	5.6	74.9	29.4	2.2
1977	18,900	0	365	11	0	19,254	3,826	5.0	1.5	5.4	75.3	28.1	2.0
1978	15,682	0	95	16	0	15,761	3,920	4.0	1.2	5.2	69.3	23.3	1.7
1979	11,250	0	420	4	0	11,667	4,022	2.9	0.9	4.9	67.0	17.9	1.3
1980	14,767	0	765	0	0	15,532	4,130	3.8	1.1	5.0	67.8	22.7	1.7
1981	11,880	0	925	8	0	12,798	4,242	3.0	0.9	4.2	67.5	21.9	1.4
1982	12,131	0	1,055	0	0	13,186	4,358	3.0	0.9	4.4	64.1	20.9	1.4
1983	11,366	0	1,153	0	0	12,519	4,481	2.8	0.9	4.1	62.0	20.5	1.4
1984	11,363	0	1,070	24	0	12,410	4,610	2.7	0.8	4.5	55.7	18.2	1.5
1985	11,371	0	485	0	0	11,856	4,748	2.5	0.8	4.1	64.0	18.4	1.2
1986	11,861	0	10	1	0	11,870	4,894	2.4	0.7	4.1	64.3	17.9	1.1
1987	12,034	0	35	2	0	12,067	5,048	2.4	0.7	3.7	63.2	19.3	1.1
1988	11,701	0	93	0	0	11,794	5,203	2.3	0.7	3.9	61.6	17.3	1.1
1989	10,787	0	457	0	0	11,244	5,351	2.1	0.6	3.6	53.2	17.7	1.2
1990	17,425	0	496	0	0	17,921	5,487	3.3	1.0	4.1	61.6	23.7	1.6
1991	21,044	0	382	0	0	21,426	5,606	3.8	1.1	4.2	61.4	27.2	1.9
1992	23,083	0	315	0	0	23,398	5,713	4.1	1.2	4.3	60.7	28.4	2.0
1993	22,055	0	698	0	0	22,753	5,817	3.9	1.2	4.2	57.8	28.0	2.0
1994	20,705	0	1,088	0	0	21,793	5,930	3.7	1.1	3.8	53.1	28.8	2.1
1995	21,101	0	944	9	0	22,036	6,064	3.6	1.1	3.6	56.2	29.9	1.9

115 CAMBODIA	- FOOD BALANCE SHEET OF FISH AND FISHERY PRODUCTS IN LIVE WEIGHT AND FISH CONTRIBUTION TO PROTEIN SUPPLY -	FAO-FIDI

YEAR	PRODUCTION	NON-FOOD USES	IMPORTS	EXPORTS	STOCKS VARIATIONS	TOTAL FOOD SUPPLY	POPULATION	PER CAPUT SUPPLY	FISH PROTEINS	ANIMAL PROTEINS	TOTAL PROTEINS	FISH/ANIMAL PROTEINS	FISH/TOTAL PROTEINS
	(........................ metric tons in live weight					........................)	(thousands)	(kilograms)	(grams per caput per day)			(%)	(%)
1961	45,000	0	0	15,500	0	29,500	5,564	5.3	1.5	3.8	43.6	40.6	3.5
1962	45,000	0	726	14,900	0	30,826	5,698	5.4	1.6	4.0	43.0	39.1	3.7
1963	40,000	0	702	7,700	0	33,002	5,836	5.7	1.8	4.6	45.0	39.2	4.0
1964	40,000	0	100	2,300	0	37,800	5,984	6.3	1.9	4.8	46.3	39.9	4.1
1965	45,000	0	200	900	0	44,300	6,141	7.2	2.2	5.2	44.7	42.6	4.9
1966	45,000	0	0	3,600	0	41,400	6,310	6.6	2.0	5.1	44.9	39.9	4.5
1967	50,000	0	600	7,200	0	43,400	6,486	6.7	2.1	5.3	44.7	38.6	4.6
1968	50,000	0	0	1,600	0	48,400	6,657	7.3	2.2	5.6	44.8	39.5	5.0
1969	52,000	0	0	1,000	0	51,000	6,812	7.5	2.3	5.9	45.2	38.8	5.0
1970	52,200	0	2,807	200	0	54,807	6,938	7.9	2.4	6.4	46.9	37.9	5.1
1971	77,200	0	877	200	0	77,877	7,038	11.1	3.3	7.4	48.7	44.1	6.7
1972	87,200	0	0	303	0	86,897	7,114	12.2	3.7	7.5	49.3	49.4	7.6
1973	84,700	0	175	400	0	84,475	7,156	11.8	3.6	7.4	48.4	48.7	7.4
1974	84,700	0	351	100	0	84,951	7,152	11.9	3.6	7.1	46.8	51.1	7.8
1975	84,700	0	...	...	0	84,700	7,098	11.9	3.7	6.6	39.6	55.5	9.2
1976	84,700	0	...	...	0	84,700	6,984	12.1	3.6	6.0	35.2	60.0	10.2
1977	84,700	0	...	...	0	84,700	6,824	12.4	3.7	5.9	36.4	62.2	10.1
1978	40,900	0	...	...	0	40,900	6,657	6.1	1.8	3.9	35.0	46.2	5.2
1979	30,000	0	...	...	0	30,000	6,535	4.6	1.4	4.0	35.9	33.6	3.8
1980	19,600	0	...	...	0	19,600	6,498	3.0	0.9	3.3	36.9	27.3	2.4
1981	51,594	0	...	...	0	51,594	6,561	7.9	2.3	4.8	40.9	48.8	5.7
1982	68,715	0	...	...	0	68,715	6,711	10.2	3.1	6.1	42.5	51.4	7.4
1983	68,161	0	...	...	0	68,161	6,927	9.8	3.3	6.5	43.5	51.6	7.7
1984	64,424	0	...	10	0	64,414	7,173	9.0	3.0	6.4	42.4	46.1	7.0
1985	70,578	0	...	81	0	70,498	7,422	9.5	3.0	7.1	42.0	42.7	7.2
1986	73,651	0	...	196	0	73,455	7,669	9.6	3.0	7.0	41.6	42.7	7.2
1987	82,074	0	...	113	0	81,962	7,919	10.4	3.3	7.3	44.0	44.7	7.4
1988	86,800	0	...	448	0	86,353	8,172	10.6	3.4	7.7	43.3	43.8	7.8
1989	82,088	0	...	1,151	0	80,937	8,430	9.6	2.9	7.9	44.9	37.5	6.6
1990	111,400	0	...	1,215	0	110,185	8,695	12.7	3.9	8.4	46.0	46.3	8.5
1991	117,800	0	...	1,080	0	116,720	8,965	13.0	4.0	8.5	44.8	46.9	8.9
1992	111,150	0	...	23,854	0	87,296	9,235	9.5	2.8	8.0	43.7	35.4	6.5
1993	108,900	0	...	27,295	0	81,605	9,503	8.6	2.6	7.9	42.7	33.3	6.2
1994	103,200	0	...	23,445	0	79,756	9,767	8.2	2.5	7.6	42.9	33.1	5.9
1995	112,510	0	...	22,816	0	89,694	10,024	9.0	2.8	7.8	46.0	35.3	6.0

32 CAMEROON	- FOOD BALANCE SHEET OF FISH AND FISHERY PRODUCTS IN LIVE WEIGHT AND FISH CONTRIBUTION TO PROTEIN SUPPLY -	FAO-FIDI

YEAR	PRODUCTION	NON-FOOD USES	IMPORTS	EXPORTS	STOCKS VARIATIONS	TOTAL FOOD SUPPLY	POPULATION	PER CAPUT SUPPLY	FISH PROTEINS	ANIMAL PROTEINS	TOTAL PROTEINS	FISH/ANIMAL PROTEINS	FISH/TOTAL PROTEINS
	(.................................... metric tons in live weight					....................................)	(thousands)	(kilograms)	(grams per caput per day)			(%)	(%)
1961	56,500	0	9,800	1,800	0	64,500	5,401	11.9	3.6	9.8	55.8	36.1	6.4
1962	58,100	0	8,533	5,100	0	61,533	5,513	11.2	3.3	9.6	54.6	34.5	6.1
1963	59,400	0	6,333	6,300	0	59,433	5,630	10.6	3.1	9.6	55.6	32.7	5.6
1964	59,800	0	6,433	7,500	0	58,733	5,753	10.2	3.0	9.0	53.8	33.4	5.6
1965	61,100	0	7,233	8,700	0	59,633	5,881	10.1	3.0	9.2	54.2	32.4	5.5
1966	62,600	0	2,833	3,900	0	61,533	6,014	10.2	3.0	9.5	55.6	31.6	5.4
1967	63,400	0	6,820	1,800	0	68,420	6,154	11.1	3.3	10.0	53.2	32.7	6.1
1968	65,100	0	8,620	1,800	0	71,920	6,300	11.4	3.4	10.4	54.8	32.5	6.2
1969	68,100	0	9,820	1,000	0	76,920	6,452	11.9	3.5	10.6	59.1	33.2	6.0
1970	68,800	0	12,000	2,689	0	78,111	6,612	11.8	3.5	11.3	60.3	30.7	5.7
1971	64,700	0	11,433	4,721	0	71,412	6,779	10.5	3.1	10.9	61.7	28.5	5.0
1972	60,400	0	14,212	2,800	0	71,812	6,953	10.3	3.1	10.6	61.8	29.4	5.0
1973	57,600	0	14,323	4,600	0	67,323	7,135	9.4	2.9	10.2	59.6	27.9	4.8
1974	55,736	0	14,094	7,500	0	62,330	7,326	8.5	2.6	10.1	61.6	25.7	4.2
1975	66,280	0	9,916	7,542	0	68,654	7,526	9.1	2.8	10.1	62.7	27.9	4.5
1976	70,397	0	12,370	7,575	0	75,192	7,734	9.7	2.9	10.6	62.3	27.3	4.6
1977	70,167	0	18,620	7,278	0	81,508	7,952	10.3	3.0	10.9	60.5	27.8	5.0
1978	68,867	0	16,940	11,786	0	74,021	8,179	9.1	2.7	10.6	58.0	25.1	4.6
1979	81,214	0	21,328	4,619	0	97,923	8,413	11.6	3.4	11.6	59.1	29.8	5.8
1980	81,045	0	24,736	3,184	0	102,597	8,655	11.9	3.5	11.7	59.0	29.7	5.9
1981	79,761	0	33,487	3,972	0	109,276	8,904	12.3	3.6	11.4	57.7	31.3	6.2
1982	83,012	0	39,884	4,890	0	118,005	9,161	12.9	3.7	11.3	57.8	32.9	6.4
1983	77,277	0	44,510	5,578	0	116,209	9,426	12.3	3.6	11.2	54.2	32.3	6.7
1984	75,581	0	54,648	10,847	0	119,382	9,698	12.3	3.6	11.8	51.5	30.4	6.9
1985	74,085	0	77,968	20,616	0	131,437	9,977	13.2	3.8	12.4	51.2	30.8	7.5
1986	72,104	0	103,412	18,516	0	157,000	10,263	15.3	4.4	14.0	54.0	31.5	8.2
1987	70,666	0	77,446	20,671	-500	126,941	10,557	12.0	3.4	13.1	51.4	26.2	6.7
1988	70,696	0	79,125	8,196	-800	140,825	10,858	13.0	3.7	12.9	52.5	28.5	7.0
1989	69,782	0	75,126	1,716	0	143,192	11,167	12.8	3.7	13.2	52.5	27.7	7.0
1990	70,750	0	69,141	4,523	1,300	136,669	11,484	11.9	3.6	12.8	51.9	27.8	6.9
1991	69,424	0	29,270	561	0	98,133	11,808	8.3	2.5	11.4	50.8	22.3	5.0
1992	72,032	0	28,026	530	0	99,528	12,142	8.2	2.5	11.3	50.6	22.1	4.9
1993	65,301	0	53,262	493	0	118,070	12,483	9.5	2.8	11.5	50.7	24.1	5.5
1994	66,048	0	44,368	385	0	110,032	12,833	8.6	2.6	11.1	52.0	23.0	4.9
1995	63,947	0	53,477	1,009	0	116,416	13,192	8.8	2.6	10.9	52.2	23.6	4.9

33 CANADA	- FOOD BALANCE SHEET OF FISH AND FISHERY PRODUCTS IN LIVE WEIGHT AND FISH CONTRIBUTION TO PROTEIN SUPPLY -	FAO-FIDI

YEAR	PRODUCTION	NON-FOOD USES	IMPORTS	EXPORTS	STOCKS VARIATIONS	TOTAL FOOD SUPPLY	POPULATION	PER CAPUT SUPPLY	FISH PROTEINS	ANIMAL PROTEINS	TOTAL PROTEINS	FISH/ANIMAL PROTEINS	FISH/TOTAL PROTEINS
	(.....................		metric tons in live weight			.....................)	(thousands)	(kilograms)		(grams per caput per day)		(%)	(%)
1961	1,018,600	254,000	43,778	529,442	0	278,935	18,296	15.3	3.7	58.4	90.4	6.3	4.1
1962	1,100,000	263,300	48,200	591,875	0	293,025	18,662	15.7	3.6	57.3	89.5	6.3	4.0
1963	1,190,700	347,600	49,374	594,041	0	298,433	19,009	15.7	3.3	57.8	91.9	5.7	3.6
1964	1,194,200	297,500	53,385	671,405	0	278,681	19,346	14.4	3.3	59.1	90.5	5.5	3.6
1965	1,217,700	288,400	58,574	724,541	0	263,333	19,678	13.4	3.1	58.6	94.7	5.2	3.2
1966	1,327,900	321,100	63,196	744,780	-17,613	307,603	20,003	15.4	3.5	60.1	92.9	5.7	3.7
1967	1,251,600	282,400	67,765	755,272	-16,233	265,459	20,322	13.1	3.2	60.9	93.6	5.2	3.4
1968	1,451,800	409,600	72,582	780,654	-16,606	317,523	20,643	15.4	3.4	61.2	94.4	5.5	3.6
1969	1,351,200	416,000	80,847	712,108	-7,492	296,446	20,975	14.1	3.0	62.0	94.8	4.8	3.1
1970	1,342,898	372,900	95,660	790,182	38,539	314,015	21,324	14.7	3.2	61.5	94.7	5.3	3.4
1971	1,277,146	277,700	97,654	749,467	38	347,672	21,696	16.0	3.8	61.4	94.3	6.2	4.0
1972	1,141,394	189,800	115,705	747,757	12,702	332,243	22,086	15.0	3.3	60.2	93.6	5.5	3.5
1973	1,128,801	138,400	141,379	725,056	2,261	408,985	22,481	18.2	3.9	60.9	94.4	6.4	4.1
1974	979,313	128,505	130,815	564,638	3,215	420,200	22,860	18.4	3.9	61.7	95.0	6.4	4.1
1975	991,329	133,942	133,170	612,551	-10,773	367,233	23,209	15.8	3.3	59.1	92.5	5.6	3.6
1976	1,103,567	91,647	159,459	706,518	-11,111	453,750	23,525	19.3	4.1	62.4	97.2	6.6	4.2
1977	1,236,259	105,137	180,352	825,316	28,658	514,816	23,814	21.6	4.6	62.8	96.6	7.4	4.8
1978	1,371,593	119,622	172,384	929,191	4,762	499,927	24,081	20.8	4.8	58.9	93.1	8.1	5.1
1979	1,414,965	132,100	170,169	910,142	-10,252	532,639	24,338	21.9	4.6	58.4	92.2	7.9	5.0
1980	1,347,029	60,165	190,595	950,062	-34,424	492,972	24,593	20.1	4.4	58.2	92.1	7.5	4.7
1981	1,416,735	52,362	187,522	1,024,353	11,175	538,717	24,846	21.7	4.6	58.5	90.8	7.9	5.1
1982	1,403,325	39,000	169,909	1,035,909	-12,515	485,810	25,096	19.4	4.2	59.2	94.6	7.1	4.4
1983	1,349,357	53,209	187,071	941,020	-25,599	516,599	25,355	20.4	4.8	58.2	93.5	8.3	5.2
1984	1,283,979	72,100	236,202	987,780	42,404	502,704	25,634	19.6	4.0	57.2	91.4	7.0	4.4
1985	1,453,299	97,119	224,750	1,006,579	-28,927	545,424	25,942	21.0	4.9	58.5	94.5	8.3	5.2
1986	1,512,854	95,395	248,499	1,089,458	-22,267	554,234	26,282	21.1	5.0	58.0	96.0	8.6	5.2
1987	1,567,932	144,011	284,273	1,040,031	-87,533	580,630	26,651	21.8	5.2	58.7	97.2	8.8	5.3
1988	1,679,131	215,281	298,017	1,108,708	-45,647	607,512	27,036	22.5	5.4	59.0	97.1	9.1	5.5
1989	1,636,792	284,513	312,939	1,081,003	60,678	644,892	27,421	23.5	5.6	60.2	96.4	9.4	5.9
1990	1,682,105	269,964	287,674	1,109,491	78,408	668,731	27,791	24.1	6.1	58.7	95.3	10.4	6.4
1991	1,564,727	260,301	298,530	969,905	58,064	691,115	28,144	24.6	6.1	58.2	94.6	10.5	6.5
1992	1,371,814	168,503	335,279	889,141	19,433	668,882	28,481	23.5	5.4	57.0	95.4	9.4	5.6
1993	1,211,618	94,364	413,473	817,665	-12,935	700,127	28,802	24.3	6.1	56.9	94.9	10.7	6.4
1994	1,089,276	94,483	436,902	743,252	1,471	689,913	29,109	23.7	6.2	57.9	97.9	10.6	6.3
1995	901,225	66,575	509,566	724,476	14,950	634,689	29,402	21.6	5.8	57.1	98.3	10.2	5.9

35 CAPE VERDE	- FOOD BALANCE SHEET OF FISH AND FISHERY PRODUCTS IN LIVE WEIGHT AND FISH CONTRIBUTION TO PROTEIN SUPPLY -	FAO-FIDI

YEAR	PRODUCTION	NON-FOOD USES	IMPORTS	EXPORTS	STOCKS VARIATIONS	TOTAL FOOD SUPPLY	POPULATION	PER CAPUT SUPPLY	FISH PROTEINS	ANIMAL PROTEINS	TOTAL PROTEINS	FISH/ANIMAL PROTEINS	FISH/TOTAL PROTEINS
	(............ metric tons in live weight ....)						(thousands)	(kilograms)	(grams per caput per day)			(%)	(%)
1961	1,600	0	0	590	0	1,010	202	5.0	1.4	4.1	45.3	34.3	3.1
1962	1,500	0	0	400	0	1,100	208	5.3	1.5	4.1	42.1	36.1	3.5
1963	2,000	0	0	820	0	1,180	215	5.5	1.6	4.4	42.2	35.5	3.7
1964	2,500	0	0	1,170	0	1,330	222	6.0	1.7	4.6	44.7	37.2	3.8
1965	3,500	0	0	1,950	0	1,550	229	6.8	1.9	4.9	49.0	39.4	3.9
1966	4,000	0	0	2,200	0	1,800	237	7.6	2.1	5.1	42.1	42.1	5.1
1967	5,900	0	0	3,200	0	2,700	245	11.0	3.1	6.1	36.8	51.0	8.4
1968	4,900	0	0	2,310	0	2,590	254	10.2	2.9	5.9	30.2	49.7	9.7
1969	4,000	0	0	2,150	0	1,850	261	7.1	2.1	5.5	35.2	38.7	6.0
1970	5,181	0	0	2,178	0	3,003	267	11.3	3.4	7.0	35.3	49.1	9.8
1971	4,153	0	0	1,700	0	2,453	271	9.1	2.8	7.9	42.6	34.8	6.5
1972	4,078	0	200	1,700	0	2,578	274	9.4	3.0	7.8	48.1	38.4	6.2
1973	8,333	0	0	5,250	0	3,083	275	11.2	3.5	7.4	49.6	46.8	7.0
1974	3,428	0	200	1,681	-300	1,647	276	6.0	1.9	5.8	46.8	33.5	4.1
1975	3,900	0	200	1,125	0	2,975	278	10.7	3.3	9.9	51.8	33.1	6.4
1976	3,800	0	441	1,481	125	2,886	280	10.3	3.1	9.3	47.9	33.0	6.4
1977	6,000	0	16	981	125	5,160	282	18.3	5.2	13.9	59.5	37.4	8.7
1978	7,000	0	10	598	-99	6,313	284	22.2	6.6	22.9	71.7	28.9	9.2
1979	7,476	0	7	1,149	0	6,334	286	22.1	6.3	12.9	58.3	49.0	10.8
1980	8,837	0	5	1,348	-899	6,595	289	22.8	6.6	13.7	70.0	48.5	9.5
1981	14,730	0	3	278	-1,500	12,955	293	44.3	14.1	24.6	71.4	57.2	19.7
1982	12,453	0	1	3,703	1,801	10,552	296	35.6	11.1	21.9	68.5	50.5	16.2
1983	11,863	0	78	2,487	500	9,954	300	33.1	10.5	21.6	65.9	48.6	16.0
1984	10,730	0	73	2,182	198	8,819	305	28.9	8.9	20.9	70.7	42.7	12.6
1985	10,190	0	34	1,847	0	8,377	310	27.0	8.1	17.1	71.2	47.3	11.4
1986	6,386	0	833	2,255	0	4,964	315	15.8	4.8	15.8	74.7	30.3	6.4
1987	6,931	0	29	3,621	0	3,339	321	10.4	3.2	14.7	80.8	21.6	4.0
1988	6,276	0	48	1,172	0	5,151	327	15.8	5.1	17.8	78.4	28.4	6.5
1989	8,624	0	70	2,926	0	5,769	333	17.3	5.2	16.3	74.7	31.9	7.0
1990	7,162	0	88	1,657	0	5,593	341	16.4	4.9	17.6	66.8	27.8	7.3
1991	7,556	0	78	244	0	7,390	349	21.2	6.6	18.8	68.3	35.2	9.7
1992	6,555	0	58	2,277	0	4,336	358	12.1	3.7	21.4	75.0	17.1	4.9
1993	7,188	0	88	994	0	6,282	367	17.1	5.2	22.7	78.1	22.7	6.6
1994	7,188	0	48	828	0	6,408	377	17.0	5.2	19.5	69.8	26.8	7.5
1995	7,081	0	201	265	0	7,017	386	18.2	5.6	24.5	67.9	22.9	8.3

36 CAYMAN ISLANDS	- FOOD BALANCE SHEET OF FISH AND FISHERY PRODUCTS IN LIVE WEIGHT AND FISH CONTRIBUTION TO PROTEIN SUPPLY -												FAO-FIDI
YEAR	PRODUCTION	NON-FOOD USES	IMPORTS	EXPORTS	STOCKS VARIATIONS	TOTAL FOOD SUPPLY	POPULATION	PER CAPUT SUPPLY	FISH PROTEINS	ANIMAL PROTEINS	TOTAL PROTEINS	FISH/ANIMAL PROTEINS	FISH/TOTAL PROTEINS
	(.......... metric tons in live weight ....)						(thousands)	(kilograms)	(grams per caput per day)			(%)	(%)
1961	...	...	...	...	...	...	9	...	...	...	...	...	...
1962	...	...	...	...	...	...	9	...	...	...	...	...	...
1963	...	...	...	...	...	...	9	...	...	...	...	...	...
1964	...	...	...	...	...	...	9	...	...	...	...	...	...
1965	...	...	...	...	...	...	9	...	...	...	...	...	...
1966	...	...	...	...	...	...	9	...	...	...	...	...	...
1967	...	...	...	...	...	...	9	...	...	...	...	...	...
1968	...	...	...	...	...	...	9	...	...	...	...	...	...
1969	...	...	...	...	...	...	10	...	...	...	...	...	...
1970	...	...	...	...	...	...	10	...	...	...	...	...	...
1971	...	...	...	...	...	...	11	...	...	...	...	...	...
1972	...	...	...	...	...	...	11	...	...	...	...	...	...
1973	...	...	...	...	...	...	12	...	...	...	...	...	...
1974	...	...	...	...	...	...	13	...	...	...	...	...	...
1975	...	...	...	...	...	...	14	...	...	...	...	...	...
1976	...	...	...	...	...	...	15	...	...	...	...	...	...
1977	...	...	...	...	...	...	15	...	...	...	...	...	...
1978	...	...	...	...	...	...	16	...	...	...	...	...	...
1979	...	...	...	...	...	...	16	...	...	...	...	...	...
1980	1,391	0	199	1,391	0	199	17	11.7	3.5	3.5	3.8	98.4	92.6
1981	3,992	0	306	3,992	0	306	18	17.3	5.2	5.3	5.5	99.4	95.6
1982	933	0	328	933	0	328	18	17.8	5.2	5.3	5.5	99.2	95.6
1983	655	0	360	655	0	360	19	18.7	5.6	5.6	5.8	99.1	96.4
1984	423	0	393	423	0	393	20	19.5	5.7	5.7	5.9	98.9	96.4
1985	396	0	392	396	0	392	21	18.7	5.6	5.6	5.8	99.0	96.1
1986	478	0	374	478	0	374	22	17.1	5.4	5.5	5.6	98.8	97.1
1987	1,097	0	721	1,102	0	716	23	31.2	10.1	10.1	10.2	99.9	99.1
1988	386	0	769	394	0	761	24	31.8	10.2	10.2	10.3	99.8	99.2
1989	615	0	793	565	0	843	25	33.7	10.7	10.7	10.7	99.9	99.7
1990	837	0	711	762	0	786	26	30.2	9.5	9.5	9.6	99.6	99.5
1991	825	0	791	773	0	843	27	31.2	8.9	8.9	8.9	99.8	99.8
1992	825	0	844	733	0	936	28	33.4	9.7	9.8	9.8	99.2	99.1
1993	625	0	687	564	0	748	29	25.8	7.5	7.5	7.6	99.3	99.1
1994	710	0	779	597	0	893	30	29.8	8.7	8.8	8.8	99.5	99.3
1995	612	0	811	233	0	1,191	31	38.4	10.7	10.8	10.8	99.6	99.4

37 CENTRAL AFRICAN REPUBLIC	- FOOD BALANCE SHEET OF FISH AND FISHERY PRODUCTS IN LIVE WEIGHT AND FISH CONTRIBUTION TO PROTEIN SUPPLY -												FAO-FIDI
YEAR	PRODUCTION	NON-FOOD USES	IMPORTS	EXPORTS	STOCKS VARIATIONS	TOTAL FOOD SUPPLY	POPULATION	PER CAPUT SUPPLY	FISH PROTEINS	ANIMAL PROTEINS	TOTAL PROTEINS	FISH/ANIMAL PROTEINS	FISH/TOTAL PROTEINS
	(.. metric tons in live weight ....)						(thousands)	(kilograms)	(grams per caput per day)			(%)	(%)
1961	2,000	0	400	...	0	2,400	1,561	1.5	0.4	5.9	32.8	7.5	1.3
1962	2,000	0	500	...	0	2,500	1,588	1.6	0.5	5.9	33.4	7.6	1.4
1963	2,500	0	600	...	0	3,100	1,617	1.9	0.6	6.1	32.7	9.0	1.7
1964	2,500	0	800	...	0	3,300	1,646	2.0	0.6	6.1	32.8	9.5	1.8
1965	3,000	0	600	...	0	3,600	1,677	2.2	0.6	6.1	32.9	10.2	1.9
1966	3,000	0	1,000	...	0	4,000	1,709	2.3	0.7	6.4	32.6	10.6	2.1
1967	3,500	0	600	...	0	4,100	1,742	2.4	0.7	6.7	33.6	10.2	2.0
1968	4,000	0	600	...	0	4,600	1,777	2.6	0.8	6.8	33.0	11.1	2.3
1969	5,000	0	1,000	...	0	6,000	1,812	3.3	1.0	7.1	34.5	13.4	2.8
1970	7,000	0	1,200	...	0	8,200	1,849	4.4	1.3	7.6	35.7	16.7	3.6
1971	8,000	0	1,100	...	0	9,100	1,888	4.8	1.4	7.9	34.0	17.4	4.1
1972	8,000	0	808	...	0	8,808	1,927	4.6	1.3	8.1	33.6	16.2	3.9
1973	10,000	0	800	...	0	10,800	1,968	5.5	1.6	8.6	33.9	18.3	4.6
1974	11,883	0	600	...	0	12,483	2,012	6.2	1.8	8.7	35.9	20.5	5.0
1975	8,044	0	900	...	0	8,944	2,057	4.4	1.3	8.6	35.4	14.6	3.6
1976	10,465	0	1,325	...	0	11,790	2,105	5.6	1.6	9.3	35.8	17.6	4.6
1977	13,000	0	1,146	...	0	14,146	2,155	6.6	1.9	9.2	34.9	20.6	5.5
1978	13,000	0	553	...	0	13,553	2,207	6.1	1.8	9.2	33.2	19.4	5.4
1979	13,000	0	315	...	0	13,315	2,260	5.9	1.7	9.0	34.0	18.9	5.0
1980	13,000	0	634	...	0	13,634	2,313	5.9	1.7	9.9	34.3	17.4	5.0
1981	13,000	0	36	...	0	13,036	2,367	5.5	1.6	10.3	36.0	15.4	4.4
1982	13,000	0	0	...	0	13,000	2,422	5.4	1.6	10.8	36.4	14.3	4.3
1983	13,000	0	521	...	0	13,521	2,478	5.5	1.6	12.1	36.9	13.1	4.3
1984	13,309	0	152	...	0	13,461	2,536	5.3	1.5	11.5	35.7	13.3	4.3
1985	13,288	0	1,107	...	0	14,395	2,597	5.5	1.6	11.9	36.5	13.7	4.5
1986	13,193	0	1,550	...	0	14,743	2,660	5.5	1.6	11.8	37.6	13.9	4.4
1987	13,088	0	1,585	...	0	14,673	2,725	5.4	1.6	12.6	38.4	12.7	4.1
1988	13,079	0	1,623	...	0	14,702	2,792	5.3	1.6	12.8	38.1	12.1	4.1
1989	13,082	0	1,756	...	0	14,838	2,860	5.2	1.5	12.9	39.0	11.9	3.9
1990	13,105	0	1,703	...	0	14,808	2,929	5.1	1.5	13.0	39.8	11.5	3.8
1991	13,605	0	593	...	0	14,198	2,997	4.7	1.4	12.9	39.7	10.7	3.5
1992	13,338	0	575	...	0	13,913	3,065	4.5	1.3	12.9	40.8	10.3	3.2
1993	13,501	0	943	...	0	14,444	3,134	4.6	1.3	13.0	41.3	10.2	3.2
1994	13,000	0	631	...	0	13,631	3,203	4.3	1.2	12.6	41.6	9.7	3.0
1995	13,300	0	500	...	0	13,800	3,273	4.2	1.2	12.9	43.1	9.5	2.8

39 CHAD	- FOOD BALANCE SHEET OF FISH AND FISHERY PRODUCTS IN LIVE WEIGHT AND FISH CONTRIBUTION TO PROTEIN SUPPLY -												FAO-FIDI
YEAR	PRODUCTION	NON-FOOD USES	IMPORTS	EXPORTS	STOCKS VARIATIONS	TOTAL FOOD SUPPLY	POPULATION	PER CAPUT SUPPLY	FISH PROTEINS	ANIMAL PROTEINS	TOTAL PROTEINS	FISH/ANIMAL PROTEINS	FISH/TOTAL PROTEINS
	(...... metric tons in live weight ......)						(thousands)	(kilograms)	(grams per caput per day)			(%)	(%)
1961	45,000	...	200	31,500	0	13,700	3,114	4.4	1.3	11.1	73.2	11.5	1.8
1962	45,000	...	200	39,000	0	6,200	3,166	2.0	0.6	11.2	75.4	5.0	0.7
1963	45,000	...	200	39,000	0	6,200	3,220	1.9	0.6	11.4	74.5	4.8	0.7
1964	50,000	...	200	42,000	0	8,200	3,276	2.5	0.7	11.5	81.1	6.3	0.9
1965	55,000	...	0	45,000	0	10,000	3,334	3.0	0.9	11.8	75.7	7.4	1.2
1966	60,000	...	200	45,000	0	15,200	3,393	4.5	1.3	11.7	72.1	11.2	1.8
1967	60,000	...	200	53,400	0	6,800	3,455	2.0	0.6	11.4	72.7	4.9	0.8
1968	65,000	...	200	53,400	0	11,800	3,518	3.4	1.0	11.6	72.5	8.4	1.4
1969	65,000	...	200	52,600	0	12,600	3,584	3.5	1.0	9.8	69.5	10.4	1.5
1970	70,000	53,500	200	1,700	0	15,000	3,652	4.1	1.2	10,1	67.3	11.8	1.8
1971	70,000	50,500	200	1,500	0	18,200	3,723	4.9	1.4	9.9	63.1	14.3	2.3
1972	70,000	48,500	200	3,000	0	18,700	3,796	4.9	1.4	10.7	59.5	13.4	2.4
1973	70,000	48,500	800	3,000	0	19,300	3,871	5.0	1.5	10.5	51.0	13.8	2.8
1974	75,000	55,500	1,100	3,000	0	17,600	3,949	4.5	1.3	8.8	50.4	14.7	2.6
1975	75,000	51,500	400	9,600	0	14,300	4,030	3.6	1.0	9.3	53.7	11.2	1.9
1976	75,000	60,000	...	...	0	15,000	4,113	3.7	1.1	9.1	52.6	11.6	2.0
1977	80,000	63,200	...	...	0	16,800	4,198	4.0	1.2	9.2	53.2	12.7	2.2
1978	75,000	56,500	...	...	0	18,500	4,286	4.3	1.3	9.8	54.3	12.8	2.3
1979	70,000	51,200	...	...	0	18,800	4,379	4.3	1.3	10.6	54.6	11.8	2.3
1980	60,000	43,000	...	...	0	17,000	4,477	3.8	1.1	10.3	50.2	10.8	2.2
1981	55,000	39,500	...	...	0	15,500	4,581	3.4	1.0	10.2	48.7	9.7	2.0
1982	30,000	18,300	...	...	0	11,700	4,690	2.5	0.7	10.3	42.4	7.1	1.7
1983	40,000	28,000	...	...	0	12,000	4,802	2.5	0.7	10.4	41.8	7.0	1.7
1984	50,000	35,300	...	...	0	14,700	4,913	3.0	0.9	8.1	41.3	10.8	2.1
1985	55,000	38,900	...	...	0	16,100	5,020	3.2	0.9	9.3	44.1	9.9	2.1
1986	60,000	42,100	...	...	0	17,900	5,122	3.5	1.0	8.8	50.3	11.6	2.0
1987	70,000	50,200	...	...	0	19,800	5,221	3.8	1.1	9.0	47.8	12.2	2.3
1988	58,000	36,800	...	...	0	21,200	5,322	4.0	1.2	9.2	47.0	12.5	2.5
1989	64,400	41,600	...	...	0	22,800	5,430	4.2	1.2	9.3	47.8	13.1	2.5
1990	70,000	45,000	...	...	0	25,000	5,552	4.5	1.3	9.3	46.0	14.1	2.8
1991	60,000	28,800	...	...	0	31,200	5,687	5.5	1.6	9.7	50.2	16.4	3.2
1992	80,000	46,200	...	...	0	33,800	5,835	5.8	1.7	9.8	54.8	17.1	3.1
1993	87,300	50,131	...	...	0	37,169	5,994	6.2	1.8	9.5	55.2	19.2	3.3
1994	80,000	42,000	1,197	...	0	39,181	6,161	6.4	1.9	9.3	53.9	19.9	3.5
1995	90,000	49,600	669	...	0	41,065	6,335	6.5	1.9	9.2	55.1	20.6	3.4

40 CHILE			- FOOD BALANCE SHEET OF FISH AND FISHERY PRODUCTS IN LIVE WEIGHT AND FISH CONTRIBUTION TO PROTEIN SUPPLY -										FAO-FIDI
YEAR	PRODUCTION	NON-FOOD USES	IMPORTS	EXPORTS	STOCKS VARIATIONS	TOTAL FOOD SUPPLY	POPULATION	PER CAPUT SUPPLY	FISH PROTEINS	ANIMAL PROTEINS	TOTAL PROTEINS	FISH/ANIMAL PROTEINS	FISH/TOTAL PROTEINS
	(.................................... metric tons in live weight					.....................................)	(thousands)	(kilograms)	(grams per caput per day)			(%)	(%)
1961	429,800	340,800	400	11,126	0	78,274	7,795	10.0	2.0	21.9	65.4	9.2	3.1
1962	638,600	533,200	490	14,175	0	91,715	7,987	11.5	2.2	23.4	66.1	9.4	3.3
1963	761,900	627,700	192	14,190	0	120,202	8,183	14.7	3.4	24.3	68.3	14.1	5.0
1964	1,160,900	1,016,400	0	15,523	0	128,977	8,378	15.4	2.9	22.5	67.3	13.0	4.3
1965	708,500	556,800	1,232	27,817	-327	124,787	8,572	14.6	2.9	23.3	68.5	12.3	4.2
1966	1,383,500	1,219,000	231	30,094	328	134,964	8,763	15.4	3.0	24.6	71.5	12.0	4.1
1967	1,052,900	876,500	0	33,677	0	142,723	8,951	15.9	3.0	23.9	69.8	12.7	4.4
1968	1,376,100	1,190,500	816	36,754	0	149,661	9,136	16.4	3.1	25.0	70.3	12.2	4.3
1969	1,076,900	913,100	445	41,302	0	122,943	9,318	13.2	2.6	24.5	68.7	10.6	3.8
1970	1,200,300	996,500	291	50,906	0	153,185	9,496	16.1	2.9	25.7	69.6	11.3	4.2
1971	1,487,000	1,308,100	505	47,622	-1,650	130,133	9,670	13.5	2.5	23.1	68.7	10.7	3.6
1972	791,800	582,100	1,183	33,032	-3,561	174,290	9,841	17.7	3.3	24.3	71.1	13.7	4.7
1973	663,900	467,000	1,058	19,627	-8,372	169,959	10,009	17.0	3.6	21.8	69.0	16.5	5.2
1974	1,123,963	950,400	0	24,533	-5,108	143,922	10,174	14.2	2.8	26.4	74.1	10.5	3.7
1975	896,305	711,383	75	32,029	-12,603	140,366	10,337	13.6	2.8	23.4	66.8	12.0	4.2
1976	1,375,754	1,136,700	1,041	74,748	-12,176	153,171	10,498	14.6	2.8	22.0	67.5	12.9	4.2
1977	1,317,028	1,055,900	2,619	116,212	23,405	170,940	10,658	16.0	3.8	22.4	66.0	16.9	5.7
1978	1,927,089	1,647,511	5,348	92,004	6,555	199,476	10,818	18.4	4.1	23.6	66.0	17.2	6.2
1979	2,630,345	2,282,924	3,237	108,853	-6,848	234,957	10,981	21.4	4.2	24.8	70.0	16.8	6.0
1980	2,816,768	2,445,998	10,073	152,697	-3,893	224,253	11,147	20.1	4.3	26.1	70.3	16.3	6.0
1981	3,393,456	3,007,375	12,267	160,645	2,376	240,079	11,316	21.2	4.7	28.1	72.7	16.8	6.5
1982	3,672,849	3,343,327	8,450	147,062	1,154	192,064	11,490	16.7	3.4	26.4	70.4	13.0	4.9
1983	3,977,044	3,584,057	4,721	175,991	-6,260	215,456	11,669	18.5	3.7	25.2	68.9	14.8	5.4
1984	4,499,272	4,097,350	2,966	201,772	6,873	209,989	11,855	17.7	3.2	23.8	67.5	13.4	4.7
1985	4,804,430	4,406,074	2,062	165,514	7,131	242,034	12,047	20.1	3.8	23.9	65.5	15.8	5.8
1986	5,571,638	5,077,127	323	253,247	-329	241,257	12,246	19.7	3.6	23.7	65.6	15.3	5.5
1987	4,814,360	4,270,489	2,304	289,476	-3,456	253,243	12,452	20.3	3.8	25.1	66.9	15.1	5.7
1988	5,209,844	4,598,076	1,318	333,540	-9,623	269,923	12,664	21.3	4.4	26.6	66.6	16.6	6.6
1989	6,454,160	5,770,341	6,401	380,297	2,023	311,946	12,880	24.2	5.3	29.0	68.0	18.5	7.9
1990	5,195,220	4,463,209	14,359	405,965	-6,041	334,369	13,100	25.5	6.0	30.9	70.4	19.4	8.5
1991	6,006,457	5,314,951	15,460	409,114	-7,861	289,989	13,322	21.8	5.0	30.2	70.4	16.7	7.1
1992	6,501,788	5,571,210	25,369	478,423	-4,804	472,716	13,547	34.9	9.5	36.7	77.3	25.8	12.3
1993	6,034,895	5,168,277	22,001	447,639	15,808	456,786	13,772	33.2	8.7	38.3	78.5	22.6	11.0
1994	7,838,538	6,916,998	23,801	517,908	16,117	443,552	13,994	31.7	7.5	38.1	78.2	19.6	9.6
1995	7,590,947	6,745,438	34,148	589,456	200	290,398	14,210	20.4	4.3	36.9	76.5	11.6	5.6

41 CHINA	- FOOD BALANCE SHEET OF FISH AND FISHERY PRODUCTS IN LIVE WEIGHT AND FISH CONTRIBUTION TO PROTEIN SUPPLY -	FAO-FIDI

YEAR	PRODUCTION	NON-FOOD USES	IMPORTS	EXPORTS	STOCKS VARIATIONS	TOTAL FOOD SUPPLY	POPULATION	PER CAPUT SUPPLY	FISH PROTEINS	ANIMAL PROTEINS	TOTAL PROTEINS	FISH/ANIMAL PROTEINS	FISH/TOTAL PROTEINS
	(............ metric tons in live weight ....)						(thousands)	(kilograms)	(grams per caput per day)			(%)	(%)
1961	2,835,600	0	...	...	0	2,835,600	658,421	4.3	1.2	3.3	42.6	36.6	2.8
1962	2,942,400	0	...	...	0	2,942,400	671,027	4.4	1.6	3.9	44.8	40.7	3.5
1963	2,942,400	0	...	...	0	2,942,400	684,854	4.3	1.6	4.6	45.9	34 6	3.5
1964	2,942,400	10,000	...	75,849	0	2,856,551	700,072	4.1	1.6	5.2	47.7	31.2	3.4
1965	3,149,200	10,000	...	82,010	0	3,057,190	716,749	4.3	1.7	5.5	48.7	30.8	3.5
1966	3,149,200	10,000	...	85,688	0	3,053,512	734,943	4.2	1.7	5.7	50.2	23.7	3.4
1967	3,149,200	10,000	...	72,854	0	3,066,346	754,527	4.1	1.6	5.7	48.5	28.4	3.3
1968	3,255,900	10,000	...	64,424	0	3,181,476	775,035	4.1	1.6	5.6	46.7	29.4	3.5
1969	3,255,900	10,000	...	90,974	0	3,154,926	795,676	4.0	1.5	5.3	45.3	28.2	3.3
1970	3,254,400	11,000	...	112,198	0	3,131,202	816,170	3.8	1.5	5.2	47.4	28.0	3.1
1971	3,554,000	10,500	...	129,219	0	3,414,281	836,584	4.1	1.6	5.6	48.1	28.0	3.3
1972	3,888,100	12,000	...	138,956	0	3,737,144	856,712	4.4	1.7	6.0	47.0	28.0	3.5
1973	3,977,800	14,500	...	133,734	0	3,829,566	876,205	4.4	1.7	5.9	48.3	27.9	3.4
1974	4,344,534	14,500	...	135,658	0	4,194,376	894,658	4.7	1.8	6.0	48.3	29.6	3.7
1975	4,528,632	3,000	...	123,319	0	4,402,313	911,807	4.8	1.8	6.1	48.6	29.8	3.8
1976	4,632,744	...	...	98,845	0	4,533,900	927,464	4.9	1.4	5.6	47.4	24.5	2.9
1977	4,817,097	...	...	94,096	0	4,723,001	941,777	5.0	1.4	5.7	48.0	24.8	2.9
1978	4,614,807	...	...	92,185	0	4,522,622	955,164	4.7	1.4	5.9	51.4	23.8	2.7
1979	4,247,606	...	...	109,999	0	4,137,607	968,159	4.3	1.3	6.5	52.7	20.0	2.5
1980	4,455,361	...	...	124,732	0	4,330,628	981,235	4.4	1.3	7.0	53.3	19.1	2.5
1981	4,608,045	...	...	138,322	0	4,469,723	994,436	4.5	1.3	7.1	54.4	18.2	2.4
1982	5,192,272	...	...	129,679	0	5,062,593	1,007,722	5.0	1.5	7.6	57.4	19.0	2.5
1983	5,628,123	250,000	...	130,467	0	5,247,656	1,021,399	5.1	1.4	7.8	58.3	18.3	2.5
1984	6,357,655	265,000	75,112	152,218	0	6,015,550	1,035,779	5.8	1.6	8.7	60.4	18.5	2.6
1985	7,329,015	272,500	104,014	153,449	0	7,007,080	1,051,040	6.7	1.8	9.8	60.9	18.5	3.0
1986	8,721,497	256,000	119,685	241,852	0	8,343,330	1,067,393	7.8	2.1	10.6	61.7	19.4	3.3
1987	10,243,780	284,000	135,094	326,342	0	9,768,526	1,084,652	9.0	2.3	11.2	61.7	20.4	3.7
1988	11,379,600	293,500	146,522	396,330	0	10,836,300	1,102,191	9.8	2.5	12.4	62.3	20.0	4.0
1989	12,210,240	450,000	217,655	451,235	0	11,526,660	1,119,204	10.3	2.7	12.9	63.0	20.6	4.2
1990	13,136,840	550,000	165,030	547,390	0	12,204,480	1,135,089	10.8	2.8	13.7	64.0	20.1	4.3
1991	14,253,600	500,000	174,307	558,139	0	13,369,760	1,149,595	11.6	3.0	15.0	63.7	20.1	4.7
1992	16,579,040	500,022	488,246	790,874	0	15,776,390	1,162,932	13.6	3.4	16.5	64.6	20.8	5.3
1993	19,708,200	710,000	616,484	866,410	0	18,748,270	1,175,344	16.0	4.0	18.5	68.1	21.5	5.8
1994	23,833,630	924,115	722,124	1,108,083	0	22,523,560	1,187,252	19.0	4.7	21.0	70.2	22.2	6.7
1995	28,418,360	1,305,017	764,598	1,212,717	0	26,665,230	1,199,019	22.2	5.4	23.1	72.5	23.1	7.4

Note: data exclude the Taiwan Province. Data for Hong Kong, which became part of China in 1997, are shown separately.

214 CHINA (TAIWAN PROVINCE)	- FOOD BALANCE SHEET OF FISH AND FISHERY PRODUCTS IN LIVE WEIGHT AND FISH CONTRIBUTION TO PROTEIN SUPPLY -												FAO-FIDI
YEAR	PRODUCTION	NON-FOOD USES	IMPORTS	EXPORTS	STOCKS VARIATIONS	TOTAL FOOD SUPPLY	POPULATION	PER CAPUT SUPPLY	FISH PROTEINS	ANIMAL PROTEINS	TOTAL PROTEINS	FISH/ANIMAL PROTEINS	FISH/TOTAL PROTEINS
	(............................ metric tons in live weight					............................)	(thousands)	(kilograms)	(grams per caput per day)			(%)	(%)
1961	311,300	10,700	400	500	-2,906	297,594	10,971	27.1	8.2	15.8	60.8	51.8	13.5
1962	329,900	11,100	1,000	500	-3,913	315,387	11,330	27.8	9.1	16.7	60.3	54.6	15.1
1963	349,900	12,500	600	1,200	-6,000	330,800	11,698	28.3	9.2	17.5	60.4	52.7	15.2
1964	375,400	7,700	1,800	2,398	-6,913	360,190	12,070	29.8	9.5	18.1	61.7	52.5	15.4
1965	383,900	5,900	1,400	2,709	-6,091	370,600	12,442	29.8	9.2	18.1	62.2	50.6	14.7
1966	423,600	5,600	1,700	4,200	-8,600	406,900	12,811	31.8	9.1	18.8	62.0	48.6	14.7
1967	456,200	9,800	1,700	6,500	-16,000	425,600	13,145	32.4	9.3	20.0	64.0	46.4	14.5
1968	527,500	34,300	500	8,708	-14,092	470,900	13,473	35.0	10.2	21.3	65.8	47.8	15.5
1969	557,800	33,500	1,697	15,316	-12,634	498,048	13,993	35.6	10.5	21.9	68.7	47.7	15.2
1970	610,370	34,900	1,914	112,293	-9,227	455,864	14,505	31.4	8.8	21.2	69.5	41.3	12.6
1971	646,764	27,900	2,796	127,144	-5,114	489,403	14,835	33.0	9.0	21.6	70.7	41.8	12.8
1972	689,749	34,300	1,575	153,011	-1,600	502,413	15,142	33.2	9.0	22.1	72.8	40.6	12.3
1973	751,437	32,500	1,647	173,950	5,000	551,635	15,427	35.8	9.8	24.2	73.1	40.3	13.3
1974	688,198	17,900	17,337	129,054	-8,465	550,116	15,709	35.0	9.3	23.3	73.0	39.8	12.7
1975	769,621	78,200	26,261	100,584	2,000	619,099	16,001	38.7	10.5	24.4	75.2	43.0	13.9
1976	800,392	44,207	13,486	202,374	8,999	576,295	16,329	35.3	9.5	25.6	75.5	37.2	12.6
1977	836,177	30,547	16,949	249,526	2,000	575,053	16,661	34.5	9.5	27.9	75.5	34.1	12.6
1978	870,591	30,462	25,462	274,330	-4,000	587,261	16,974	34.6	9.3	28.9	77.1	32.4	12.1
1979	916,225	51,603	31,017	285,388	17,001	627,252	17,308	36.2	10.0	31.9	79.1	31.3	12.6
1980	924,109	45,109	33,234	301,754	32,009	642,489	17,642	36.4	10.1	30.2	77.7	33.3	13.0
1981	898,821	11,553	15,998	290,826	15,500	627,940	17,970	34.9	9.9	30.1	76.5	32.8	12.9
1982	910,735	13,222	17,189	301,699	6,499	619,502	18,297	33.9	9.7	30.5	75.3	31.6	12.8
1983	918,722	4,260	18,777	350,812	6,502	588,930	18,595	31.7	8.8	31.9	75.5	27.6	11.7
1984	990,521	3,541	19,277	363,159	-5,821	637,277	18,873	33.8	9.5	33.8	77.1	27.9	12.3
1985	1,025,599	5,067	22,041	365,439	2,326	679,461	19,135	35.5	10.3	35.8	79.4	28.7	12.9
1986	1,086,033	24,965	26,630	464,613	0	623,085	19,356	32.2	9.3	35.4	79.5	26.2	11.7
1987	1,236,538	25,572	48,352	596,959	8,001	670,360	19,564	34.3	9.7	37.0	81.8	26.3	11.9
1988	1,353,520	74,395	68,894	621,980	2,001	728,040	19,789	36.8	9.9	37.4	82.7	26.4	11.9
1989	1,361,716	189,948	102,003	464,188	-2,000	807,584	20,006	40.4	10.9	38.7	85.1	28.1	12.8
1990	1,444,453	236,533	100,224	484,499	3,999	827,644	20,216	40.9	10.9	39.6	85.6	27.6	12.8
1991	1,307,034	119,680	106,620	467,237	0	826,737	20,443	40.4	11.2	40.8	85.0	27.4	13.2
1992	1,314,233	119,779	139,899	556,467	0	777,885	20,636	37.7	9.8	41.0	84.0	24.0	11.7
1993	1,410,887	46,988	167,644	688,930	0	842,614	20,823	40.5	10.5	43.2	85.2	24.4	12.3
1994	1,248,895	55,991	179,495	624,340	0	748,059	21,026	35.6	9.2	42.9	84.7	21.4	10.8
1995	1,288,406	45,463	182,097	603,742	-2,034	819,264	21,205	38.6	10.3	44.4	86.3	23.2	11.9

44 COLOMBIA	- FOOD BALANCE SHEET OF FISH AND FISHERY PRODUCTS IN LIVE WEIGHT AND FISH CONTRIBUTION TO PROTEIN SUPPLY -												FAO-FIDI
YEAR	PRODUCTION	NON-FOOD USES	IMPORTS	EXPORTS	STOCKS VARIATIONS	TOTAL FOOD SUPPLY	POPULATION	PER CAPUT SUPPLY	FISH PROTEINS	ANIMAL PROTEINS	TOTAL PROTEINS	FISH/ANIMAL PROTEINS	FISH/TOTAL PROTEINS
	(.......................... metric tons in live weight ....)						(thousands)	(kilograms)	(grams per caput per day)			(%)	(%)
1961	43,000	0	333	1,600	0	41,733	16,421	2.5	0.8	22.4	50.1	3.4	1.5
1962	51,700	0	327	1,800	0	50,227	16,918	3.0	0.9	23.1	50.5	3.8	1.8
1963	47,400	0	338	1,600	0	46,138	17,430	2.7	0.8	23.2	49.3	3.4	1.6
1964	50,300	0	167	938	0	49,529	17,959	2.8	0.8	22.7	50.6	3.6	1.6
1965	53,500	0	0	1,519	0	51,981	18,506	2.8	0.8	22.6	49.9	3.7	1.7
1966	59,200	0	0	1,454	0	57,746	19,074	3.0	0.9	21.1	47.3	4.3	1.9
1967	82,000	0	330	2,148	0	80,182	19,659	4.1	1.2	21.5	47.6	5.6	2.6
1968	75,000	0	1,000	2,214	0	73,786	20,246	3.6	1.1	21.2	47.7	5.3	2.4
1969	57,600	0	3,257	2,400	0	58,457	20,817	2.8	0.8	21.1	46.4	3.9	1.8
1970	54,500	0	2,167	2,500	0	54,167	21,360	2.5	0.8	21.4	47.4	3.6	1.6
1971	37,700	700	7,833	2,500	0	42,333	21,869	1.9	0.6	22.2	48.7	2.7	1.2
1972	110,700	0	5,341	3,500	0	112,541	22,348	5.0	1.5	21.6	48.0	6.9	3.1
1973	105,300	600	11,725	3,500	0	112,925	22,813	5.0	1.5	20.3	47.2	7.3	3.1
1974	62,418	200	15,200	3,400	0	74,018	23,283	3.2	0.9	19.4	47.0	4.8	2.0
1975	66,575	1,300	9,216	4,547	0	69,944	23,776	2.9	0.9	20.3	48.0	4.3	1.8
1976	75,107	0	16,178	4,606	-627	86,052	24,295	3.5	1.1	22.1	51.2	4.8	2.1
1977	63,965	0	22,612	4,428	0	82,149	24,836	3.3	1.0	22.4	50.5	4.6	2.0
1978	79,578	0	31,664	3,819	125	107,548	25,394	4.2	1.3	21.5	51.3	6.1	2.6
1979	63,393	0	30,379	3,197	250	90,824	25,959	3.5	1.1	22.1	51.9	5.0	2.1
1980	76,197	0	39,614	3,885	0	111,926	26,525	4.2	1.3	22.4	53.1	6.0	2.5
1981	94,690	0	38,705	4,743	0	128,651	27,090	4.8	1.5	23.5	54.0	6.4	2.8
1982	71,381	0	41,789	5,023	-28	108,120	27,657	3.9	1.2	22.8	54.0	5.4	2.3
1983	57,537	8	16,823	4,386	125	70,091	28,230	2.5	0.8	21.8	51.7	3.6	1.5
1984	79,264	0	31,363	4,602	0	106,025	28,815	3.7	1.1	23.0	54.3	5.0	2.1
1985	71,694	0	27,570	7,612	152	91,803	29,415	3.1	1.0	22.8	52.8	4.3	1.8
1986	83,641	0	38,344	8,358	0	113,627	30,032	3.8	1.2	23.3	53.2	5.1	2.2
1987	86,063	3	31,408	9,164	0	108,304	30,662	3.5	1.1	23.3	53.8	4.7	2.0
1988	89,925	0	34,053	21,560	-850	101,568	31,303	3.2	1.0	24.9	55.4	4.1	1.8
1989	98,419	4	27,821	37,422	-3,224	85,590	31,949	2.7	0.8	25.9	56.8	3.2	1.5
1990	127,918	1	29,827	51,431	0	106,313	32,596	3.3	1.0	27.2	59.4	3.8	1.8
1991	113,431	1	35,791	83,020	3,536	69,737	33,243	2.1	0.7	27.3	59.1	2.7	1.3
1992	164,653	14	38,371	80,214	0	122,796	33,891	3.6	1.2	26.6	60.3	4.3	1.9
1993	147,771	1	57,208	71,121	-15,792	118,065	34,536	3.4	1.1	27.4	63.1	3.9	1.7
1994	123,719	3	81,834	99,004	15,018	121,564	35,178	3.5	1.1	28.8	64.1	3.7	1.7
1995	167,080	0	109,583	108,090	0	168,574	35,814	4.7	1.5	30.3	66.8	4.9	2.2

45 COMOROS	- FOOD BALANCE SHEET OF FISH AND FISHERY PRODUCTS IN LIVE WEIGHT AND FISH CONTRIBUTION TO PROTEIN SUPPLY -												FAO-FIDI
YEAR	PRODUCTION	NON-FOOD USES	IMPORTS	EXPORTS	STOCKS VARIATIONS	TOTAL FOOD SUPPLY	POPULATION	PER CAPUT SUPPLY	FISH PROTEINS	ANIMAL PROTEINS	TOTAL PROTEINS	FISH/ANIMAL PROTEINS	FISH/TOTAL PROTEINS
	(..		metric tons in live weight	....)			(thousands)	(kilograms)	(grams per caput per day)			(%)	(%)
1961	1,200	0	...	0	0	1,200	219	5.5	1.8	5.1	30.6	34.9	5.8
1962	1,300	0	...	0	0	1,300	224	5.8	1.9	5.4	32.6	34.6	5.7
1963	1,300	0	...	0	0	1,300	229	5.7	1.8	5.5	33.0	33.1	5.5
1964	1,300	0	...	0	0	1,300	235	5.5	1.8	5.6	33.1	31.7	5.4
1965	1,400	0	...	0	0	1,400	240	5.8	1.9	5.9	34.4	32.5	5.6
1966	1,500	0	...	0	0	1,500	247	6.1	2.1	6.5	34.5	31.4	5.9
1967	1,500	0	...	0	0	1,500	253	5.9	2.0	6.5	34.8	30.6	5.7
1968	2,000	0	...	0	0	2,000	260	7.7	2.6	7.0	35.2	36.4	7.3
1969	2,000	0	...	0	0	2,000	267	7.5	2.5	7.0	34.6	35.3	7.2
1970	2,500	0	...	0	0	2,500	275	9.1	3.1	7.7	34.1	40.3	9.2
1971	3,000	0	...	0	0	3,000	282	10.6	3.6	8.3	36.4	43.7	10.0
1972	3,000	0	...	0	0	3,000	289	10.4	3.5	8.1	36.4	43.7	9.7
1973	3,000	0	...	0	0	3,000	297	10.1	3.5	8.0	36.2	43.1	9.5
1974	3,500	0	...	0	0	3,500	306	11.5	3.9	8.5	36.2	45.5	10.7
1975	3,850	0	...	0	0	3,850	316	12.2	4.1	8.8	36.0	47.0	11.5
1976	3,500	0	...	0	0	3,500	328	10.7	3.6	8.6	35.4	42.2	10.2
1977	4,000	0	...	0	0	4,000	341	11.7	4.0	9.3	39.1	42.4	10.2
1978	4,000	0	...	0	0	4,000	355	11.3	3.8	9.5	39.2	40.2	9.7
1979	4,100	0	...	0	0	4,100	369	11.1	3.8	9.0	39.3	41.6	9.6
1980	4,200	0	...	0	0	4,200	383	11.0	3.7	8.7	39.5	42.4	9.4
1981	4,400	0	376	0	0	4,776	396	12.1	4.1	8.3	39.6	49.1	10.3
1982	4,600	0	478	0	-105	4,973	409	12.2	4.1	8.8	40.7	47.0	10.1
1983	4,800	0	151	25	26	4,953	422	11.7	4.0	8.7	40.2	45.5	9.8
1984	5,000	0	404	28	35	5,411	435	12.4	4.2	9.1	40.6	45.9	10.3
1985	5,200	0	112	19	26	5,319	448	11.9	4.0	8.3	39.0	48.2	10.3
1986	5,250	0	246	18	18	5,496	462	11.9	4.0	8.1	40.1	49.5	10.0
1987	5,250	0	314	0	0	5,564	477	11.7	3.9	8.4	39.9	46.5	9.8
1988	5,700	0	413	0	0	6,113	491	12.4	4.2	8.6	39.9	48.9	10.6
1989	11,734	0	553	0	0	12,287	507	24.3	8.3	13.0	45.3	64.4	18.4
1990	12,234	0	602	0	0	12,836	523	24.6	8.4	12.6	44.0	66.5	19.1
1991	12,634	0	546	0	0	13,180	539	24.4	8.4	13.7	43.8	61.4	19.2
1992	11,767	0	518	0	0	12,285	557	22.1	7.5	12.2	42.8	61.3	17.5
1993	11,587	0	949	0	0	12,536	575	21.8	7.4	12.1	42.3	60.8	17.4
1994	12,976	0	920	0	0	13,896	593	23.4	7.9	12.8	43.6	61.5	18.1
1995	13,200	0	746	0	0	13,946	612	22.8	7.7	12.0	42.5	64.3	18.2

250 CONGO, DEMOCRATIC REP.	- FOOD BALANCE SHEET OF FISH AND FISHERY PRODUCTS IN LIVE WEIGHT AND FISH CONTRIBUTION TO PROTEIN SUPPLY -												FAO-FIDI
YEAR	PRODUCTION	NON-FOOD USES	IMPORTS	EXPORTS	STOCKS VARIATIONS	TOTAL FOOD SUPPLY	POPULATION	PER CAPUT SUPPLY	FISH PROTEINS	ANIMAL PROTEINS	TOTAL PROTEINS	FISH/ANIMAL PROTEINS	FISH/TOTAL PROTEINS
	(.......................... .metric tons in live weight ....)						(thousands)	(kilograms)	(grams per caput per day)			(%)	(%)
1961	75,000	0	85,006	0	0	160,006	15,736	10.2	3.0	10.5	37.5	28.8	8.1
1962	77,000	0	83,857	0	0	160,857	16,160	10.0	2.9	10.1	37.0	28.3	7.7
1963	69,500	0	56,175	0	0	125,675	16,605	7.6	2.2	9.3	36.6	23.2	5.9
1964	75,000	0	64,604	0	0	139,604	17,071	8.2	2.4	9.4	35.7	25.2	6.6
1965	81,000	0	69,415	0	0	150,415	17,558	8.6	2.5	9.3	36.3	27.3	7.0
1966	84,000	0	92,105	0	0	176,105	18,068	9.8	2.9	9.6	36.7	30.0	7.8
1967	93,700	0	77,050	0	0	170,750	18,600	9.2	2.7	9.1	36.2	29.8	7.5
1968	127,700	0	70,704	0	0	198,404	19,149	10.4	3.0	9.2	36.4	32.9	8.3
1969	149,300	0	76,520	0	0	225,820	19,707	11.5	3.5	9.4	36.3	36.8	9.6
1970	136,600	0	85,330	0	0	221,930	20,270	11.0	3.3	9.7	37.6	33.5	8.7
1971	123,900	0	83,542	0	0	207,443	20,835	10.0	2.9	9.0	37.1	32.6	7.9
1972	124,200	0	87,695	300	0	211,596	21,408	9.9	2.9	8.9	36.9	32.8	7.9
1973	156,900	0	87,861	0	0	244,761	21,995	11.1	3.3	9.0	37.8	36.4	8.7
1974	128,810	0	98,585	303	0	227,092	22,607	10.1	2.9	8.7	37.4	33.8	7.8
1975	113,380	0	85,721	0	0	199,101	23,251	8.6	2.5	8.0	36.2	31.1	6.8
1976	117,858	0	88,512	0	0	206,370	23,931	8.6	2.6	8.1	36.5	31.6	7.0
1977	107,000	0	58,999	0	0	165,999	24,647	6.7	2.0	7.1	35.1	28.7	5.8
1978	108,706	0	49,936	15	0	158,626	25,399	6.3	1.9	7.0	33.6	27.1	5.6
1979	115,182	0	53,798	0	0	168,980	26,186	6.5	2.0	6.9	33.9	28.2	5.7
1980	102,415	0	73,636	0	0	176,051	27,009	6.5	2.0	6.6	33.8	29.9	5.8
1981	102,640	0	100,464	0	0	203,104	27,871	7.3	2.2	6.7	34.3	32.1	6.3
1982	100,700	0	72,839	0	0	173,539	28,775	6.0	1.8	6.2	33.7	29.3	5.4
1983	102,000	0	83,157	0	0	185,157	29,717	6.2	1.8	6.2	33.5	29.2	5.4
1984	148,300	0	78,775	0	0	227,075	30,691	7.4	2.2	6.8	33.6	31.9	6.4
1985	148,456	0	176,869	0	0	325,325	31,693	10.3	3.0	7.9	35.1	38.3	8.6
1986	156,514	0	159,612	0	0	316,126	32,710	9.7	2.8	7.5	35.1	36.9	7.9
1987	162,000	0	119,646	0	0	281,646	33,750	8.3	2.4	7.0	34.4	34.4	7.0
1988	162,000	0	110,790	0	0	272,790	34,850	7.8	2.2	6.7	34.0	33.2	6.6
1989	166,000	0	119,249	0	0	285,249	36,058	7.9	2.3	6.7	33.9	34.0	6.7
1990	162,000	0	103,849	0	0	265,849	37,405	7.1	2.1	6.4	34.0	32.3	6.1
1991	166,550	0	108,261	0	0	274,811	38,917	7.1	1.9	6.0	33.0	30.9	5.6
1992	188,570	0	96,321	0	0	284,891	40,568	7.0	1.6	5.7	32.6	28.5	5.0
1993	197,489	0	168,228	0	0	365,717	42,276	8.7	1.5	5.5	32.4	26.8	4.5
1994	155,917	0	89,017	0	0	244,934	43,930	5.6	1.4	5.3	31.2	26.9	4.5
1995	158,627	0	108,340	102	0	266,865	45,453	5.9	1.4	5.2	30.9	26.5	4.4

46 CONGO, REPUBLIC	- FOOD BALANCE SHEET OF FISH AND FISHERY PRODUCTS IN LIVE WEIGHT AND FISH CONTRIBUTION TO PROTEIN SUPPLY -												FAO-FIDI
YEAR	PRODUCTION	NON-FOOD USES	IMPORTS	EXPORTS	STOCKS VARIATIONS	TOTAL FOOD SUPPLY	POPULATION	PER CAPUT SUPPLY	FISH PROTEINS	ANIMAL PROTEINS	TOTAL PROTEINS	FISH/ANIMAL PROTEINS	FISH/TOTAL PROTEINS
	(..........		metric tons in live weight		..........)		(thousands)	(kilograms)	(grams per caput per day)			(%)	(%)
1961	11,000	0	17,584	...	0	28,584	1,010	28.3	8.8	14.9	36.0	58.8	24.4
1962	12,600	0	12,387	...	0	24,987	1,034	24.2	7.3	13.5	36.2	54.1	20.2
1963	12,500	0	14,153	...	0	26,653	1,059	25.2	7.7	14.1	36.7	54.5	21.0
1964	13,600	0	17,972	...	0	31,572	1,084	29.1	8.9	15.1	37.9	58.9	23.4
1965	14,000	0	14,348	...	0	28,348	1,111	25.5	7.7	13.6	36.2	56.3	21.1
1966	14,200	0	16,189	...	0	30,389	1,139	26.7	8.0	13.7	35.8	58.6	22.4
1967	13,600	0	16,586	...	0	30,186	1,168	25.9	7.8	13.7	36.1	56.9	21.6
1968	14,100	0	18,078	...	0	32,178	1,198	26.9	8.1	14.4	37.9	56.3	21.4
1969	13,900	0	20,070	...	0	33,970	1,230	27.6	8.4	15.1	37.0	55.7	22.7
1970	14,200	0	14,405	...	0	28,605	1,263	22.7	6.8	13.3	35.3	51.4	19.3
1971	16,000	0	11,429	...	0	27,429	1,297	21.2	6.3	12.8	36.9	48.9	16.9
1972	22,800	0	14,410	...	0	37,210	1,332	27.9	8.2	14.6	40.2	55.9	20.3
1973	22,200	0	11,029	...	0	33,229	1,369	24.3	7.0	13.3	38.1	52.7	18.4
1974	23,717	0	5,470	...	0	29,187	1,407	20.7	6.0	11.5	36.3	51.6	16.4
1975	23,103	0	14,416	...	0	37,519	1,447	25.9	7.7	13.4	38.7	57.6	20.0
1976	26,869	0	11,355	...	0	38,224	1,489	25.7	7.8	14.0	38.2	56.0	20.5
1977	24,364	0	10,553	...	0	34,917	1,532	22.8	6.9	13.3	39.1	52.2	17.7
1978	26,297	0	19,621	30	0	45,888	1,576	29.1	8.9	15.6	41.3	57.5	21.7
1979	29,630	0	20,252	638	0	49,244	1,622	30.4	9.4	16.7	41.4	56.4	22.7
1980	31,965	0	20,573	2,350	0	50,188	1,669	30.1	9.2	16.2	41.7	56.8	22.1
1981	29,665	0	21,476	3,977	0	47,164	1,717	27.5	8.5	15.6	40.1	54.5	21.1
1982	30,835	0	25,169	3,910	0	52,094	1,766	29.5	9.2	17.3	44.3	53.5	20.9
1983	35,250	0	30,736	3,917	0	62,069	1,817	34.2	10.7	18.9	46.1	57.0	23.3
1984	32,844	0	51,479	9,008	0	75,315	1,869	40.3	12.8	21.3	49.6	60.2	25.8
1985	29,936	0	55,869	7,328	0	78,477	1,924	40.8	12.9	21.0	48.8	61.3	26.4
1986	32,085	0	48,900	5,439	0	75,546	1,981	38.1	12.2	19.8	49.1	61.7	24.8
1987	37,852	0	42,316	3,336	0	76,832	2,041	37.7	12.0	19.1	47.8	62.6	25.1
1988	41,977	0	37,600	2,751	0	76,826	2,102	36.5	11.5	19.8	47.4	58.2	24.3
1989	45,840	0	36,007	4,201	0	77,646	2,166	35.8	11.1	18.4	46.1	60.4	24.1
1990	48,210	0	36,029	3,903	0	80,336	2,232	36.0	11.1	17.5	45.1	63.7	24.7
1991	45,577	0	37,825	5,700	0	77,702	2,301	33.8	10.5	21.2	48.5	49.4	21.6
1992	40,182	0	44,655	7,210	-13	77,614	2,372	32.7	10.3	23.2	51.8	44.2	19.8
1993	36,898	0	45,487	8,729	0	73,656	2,444	30.1	9.5	21.3	48.9	44.4	19.4
1994	35,150	0	35,014	6,713	0	63,451	2,518	25.2	8.0	17.3	44.0	46.0	18.1
1995	36,824	0	21,942	31	13	58,747	2,593	22.7	7.1	15.7	42.9	44.9	16.4

47 COOK ISLANDS	- FOOD BALANCE SHEET OF FISH AND FISHERY PRODUCTS IN LIVE WEIGHT AND FISH CONTRIBUTION TO PROTEIN SUPPLY -	FAO-FIDI

YEAR	PRODUCTION	NON-FOOD USES	IMPORTS	EXPORTS	STOCKS VARIATIONS	TOTAL FOOD SUPPLY	POPULATION	PER CAPUT SUPPLY	FISH PROTEINS	ANIMAL PROTEINS	TOTAL PROTEINS	FISH/ANIMAL PROTEINS	FISH/TOTAL PROTEINS
	(.. metric tons in live weight					....)	(thousands)	(kilograms)	(grams per caput per day)			(%)	(%)
1961	850	100	...	0	0	750	18	41.1	11.6	26.6	52.3	43.6	22.2
1962	850	100	...	0	0	750	18	40.8	11.5	27.8	52.7	41.4	21.8
1963	800	100	...	0	0	700	19	37.8	10.7	28.9	55.5	36.9	19.2
1964	800	100	...	0	0	700	19	37.4	10.6	30.8	58.1	34.2	18.2
1965	850	100	...	0	0	750	19	39.5	11.1	30.9	58.6	36.1	19.0
1966	850	100	...	0	0	750	19	38.6	10.9	29.0	59.1	37.6	18.4
1967	900	100	120	0	0	920	20	46.1	13.0	31.7	61.5	41.1	21.2
1968	900	100	250	0	0	1,050	20	51.3	14.5	35.1	62.8	41.3	23.1
1969	900	100	417	0	0	1,217	21	58.4	16.5	43.4	69.3	38.0	23.8
1970	850	500	379	0	0	729	21	34.7	9.9	35.9	65.7	27.6	15.1
1971	850	400	370	0	0	820	21	39.3	11.2	34.3	63.5	32.7	17.6
1972	850	550	590	0	0	890	20	43.5	12.4	36.0	67.2	34.5	18.5
1973	900	300	310	0	0	910	20	45.6	12.8	36.7	68.0	35.0	18.9
1974	900	600	700	0	0	1,000	19	51.5	14.5	35.7	67.6	40.8	21.5
1975	900	100	100	0	0	900	19	47.4	13.3	38.2	68.7	34.8	19.4
1976	1,000	450	439	0	0	989	19	52.9	16.6	40.5	71.2	41.1	23.4
1977	1,000	600	614	0	0	1,014	18	54.9	17.9	45.4	76.7	39.5	23.4
1978	1,091	414	316	0	0	993	18	54.1	16.3	45.1	75.3	36.1	21.6
1979	830	176	348	0	0	1,002	18	55.1	16.7	48.1	76.3	34.8	21.9
1980	860	182	296	0	0	974	18	54.1	16.2	48.1	80.0	33.7	20.3
1981	900	190	391	0	0	1,101	18	62.0	18.9	44.6	76.6	42.4	24.7
1982	930	196	234	0	0	968	17	55.3	16.4	39.2	72.5	41.9	22.7
1983	960	202	404	0	0	1,162	17	67.4	20.3	44.4	76.4	45.7	26.5
1984	1,000	210	354	0	0	1,144	17	67.1	20.4	45.8	76.0	44.6	26.9
1985	1,031	216	338	0	0	1,153	17	67.8	20.5	47.1	76.3	43.5	26.8
1986	1,060	222	161	0	0	999	17	58.5	17.1	42.9	73.0	39.9	23.4
1987	1,100	230	288	0	0	1,158	17	67.2	20.1	44.9	74.6	44.8	27.0
1988	1,100	230	213	0	0	1,083	17	61.9	18.3	44.0	73.1	41.6	25.0
1989	1,150	240	316	0	0	1,226	18	69.1	20.8	43.4	71.6	47.9	29.0
1990	1,200	250	421	0	0	1,371	18	76.2	23.2	42.3	70.0	54.9	33.1
1991	1,100	229	368	0	0	1,239	18	68.0	20.2	44.1	70.5	45.8	28.7
1992	982	204	376	0	0	1,154	18	62.6	19.0	44.4	69.9	42.8	27.2
1993	1,000	208	367	0	0	1,159	19	62.3	18.9	49.2	73.4	38.4	25.7
1994	950	198	393	86	0	1,059	19	56.3	17.3	49.7	73.7	34.8	23.5
1995	1,124	196	380	83	0	1,225	19	64.5	18.6	44.9	68.2	41.3	27.2

48 COSTA RICA	- FOOD BALANCE SHEET OF FISH AND FISHERY PRODUCTS IN LIVE WEIGHT AND FISH CONTRIBUTION TO PROTEIN SUPPLY -												FAO-FIDI
YEAR	PRODUCTION	NON-FOOD USES	IMPORTS	EXPORTS	STOCKS VARIATIONS	TOTAL FOOD SUPPLY	POPULATION	PER CAPUT SUPPLY	FISH PROTEINS	ANIMAL PROTEINS	TOTAL PROTEINS	FISH/ANIMAL PROTEINS	FISH/TOTAL PROTEINS
	(............ metric tons in live weight ....)						(thousands)	(kilograms)	(grams per caput per day)			(%)	(%)
1961	2,400	0	1,690	1,336	0	2,754	1,283	2.2	0.7	21.7	52.3	3.4	1.4
1962	2,500	0	1,515	960	0	3,055	1,332	2.3	0.8	20.5	51.8	4.0	1.6
1963	2,400	0	2,216	1,020	0	3,596	1,381	2.6	0.9	22.5	53.2	4.1	1.7
1964	3,600	0	2,216	1,333	0	4,483	1,431	3.1	1.1	22.1	53.9	5.1	2.1
1965	3,400	0	2,216	1,333	0	4,283	1,482	2.9	0.9	22.4	57.5	4.2	1.6
1966	3,300	0	2,566	1,583	0	4,283	1,532	2.8	0.9	20.0	54.4	4.4	1.6
1967	3,500	0	2,391	1,491	0	4,400	1,583	2.8	0.9	20.5	54.8	4.3	1.6
1968	5,000	0	2,400	1,700	0	5,700	1,633	3.5	1.2	22.3	56.3	5.4	2.1
1969	4,100	0	1,303	1,446	0	3,956	1,683	2.4	0.8	22.0	55.8	3.6	1.4
1970	8,700	0	1,470	2,063	0	8,107	1,731	4.7	1.8	24.0	57.8	7.3	3.0
1971	9,200	0	1,289	2,619	0	7,870	1,777	4.4	1.7	25.3	58.6	6.8	2.9
1972	12,300	0	1,461	2,389	0	11,372	1,822	6.2	2.1	24.6	58.2	8.6	3.6
1973	12,500	0	788	2,907	0	10,381	1,868	5.6	2.0	25.8	58.0	7.7	3.4
1974	15,312	0	167	5,244	0	10,234	1,916	5.3	1.9	24.8	57.1	7.6	3.3
1975	15,907	0	740	4,725	0	11,922	1,968	6.1	1.7	27.4	59.8	6.3	2.9
1976	14,405	0	424	4,356	0	10,473	2,025	5.2	1.6	31.4	64.1	4.9	2.4
1977	14,228	0	1,341	5,238	0	10,331	2,086	5.0	1.4	31.7	63.0	4.5	2.3
1978	22,635	0	957	2,936	0	20,656	2,150	9.6	3.2	32.4	62.7	9.9	5.1
1979	22,303	0	2,954	7,103	0	18,154	2,217	8.2	2.6	33.7	65.1	7.8	4.1
1980	20,023	0	3,776	3,797	0	20,002	2,284	8.8	2.9	32.1	66.5	8.9	4.3
1981	17,424	0	1,800	4,964	0	14,260	2,353	6.1	2.0	28.9	62.5	7.0	3.3
1982	13,164	0	1,076	4,908	0	9,331	2,423	3.9	1.4	26.3	58.3	5.3	2.4
1983	10,567	0	2,997	4,818	0	8,746	2,494	3.5	1.0	28.9	63.0	3.6	1.6
1984	16,569	0	6,043	6,735	-3,769	12,108	2,567	4.7	1.4	29.2	64.3	4.6	2.1
1985	20,481	0	5,785	8,405	-3,075	14,786	2,642	5.6	1.8	32.8	66.1	5.3	2.7
1986	20,993	0	5,753	10,396	-1,800	14,550	2,719	5.4	1.7	29.9	62.9	5.8	2.8
1987	17,230	0	9,653	12,299	1,040	15,624	2,797	5.6	1.7	31.5	66.5	5.4	2.6
1988	17,059	3	10,091	14,039	-1,077	12,030	2,876	4.2	1.3	30.8	64.1	4.3	2.1
1989	17,380	1	13,208	17,767	653	13,473	2,955	4.6	1.5	31.5	68.6	4.8	2.2
1990	17,553	1	16,054	24,631	7,944	16,920	3,035	5.6	1.9	32.8	67.5	5.7	2.8
1991	17,905	2	15,622	21,186	0	12,339	3,113	4.0	1.1	32.2	66.0	3.5	1.7
1992	18,096	1	18,842	18,589	0	18,348	3,192	5.8	1.6	32.2	64.7	4.9	2.4
1993	18,895	2	16,819	19,617	0	16,095	3,270	4.9	1.5	32.3	64.5	4.6	2.3
1994	20,849	3	30,311	27,513	0	23,645	3,347	7.1	2.1	33.1	68.3	6.4	3.1
1995	27,928	1	24,901	32,001	0	20,851	3,424	6.1	1.9	34.2	72.8	5.6	2.6

107 CÔTE D'IVOIRE	- FOOD BALANCE SHEET OF FISH AND FISHERY PRODUCTS IN LIVE WEIGHT AND FISH CONTRIBUTION TO PROTEIN SUPPLY -	FAO-FIDI

YEAR	PRODUCTION	NON-FOOD USES	IMPORTS	EXPORTS	STOCKS VARIATIONS	TOTAL FOOD SUPPLY	POPULATION	PER CAPUT SUPPLY	FISH PROTEINS	ANIMAL PROTEINS	TOTAL PROTEINS	FISH/ANIMAL PROTEINS	FISH/TOTAL PROTEINS
	(...... metric tons in live weight ......)						(thousands)	(kilograms)	(grams per caput per day)			(%)	(%)
1961	44,100	0	12,732	1,695	0	55,137	3,930	14.0	4.5	11.4	41.8	39.4	10.7
1962	45,600	0	8,620	1,180	0	53,040	4,066	13.0	4.2	10.9	43.6	38.6	9.6
1963	47,600	0	10,429	1,540	0	56,489	4,210	13.4	4.3	11.0	47.9	38.8	8.9
1964	57,100	0	8,475	2,420	0	63,155	4,363	14.5	4.7	11.7	49.8	39.9	9.4
1965	61,500	0	13,101	2,745	0	71,856	4,527	15.9	5.1	12.1	49.7	41.9	10.2
1966	61,600	0	14,365	1,595	0	74,370	4,703	15.8	5.1	12.4	49.3	41.4	10.4
1967	66,900	0	12,415	1,275	0	78,040	4,890	16.0	5.1	12.7	50.1	40.1	10.2
1968	69,800	0	16,112	3,860	0	82,052	5,088	16.1	5.2	13.4	52.4	38.7	9.9
1969	71,000	0	20,921	3,195	0	88,726	5,297	16.8	5.5	13.7	52.5	40.0	10.4
1970	72,401	0	28,769	5,415	0	95,755	5,515	17.4	5.4	13.8	53.3	39.5	10.2
1971	76,500	0	30,466	5,400	0	101,566	5,744	17.7	5.5	13.9	51.6	39.9	10.7
1972	81,400	0	35,971	6,250	0	111,121	5,984	18.6	5.8	14.3	51.8	40.7	11.2
1973	65,500	0	54,871	9,600	0	110,771	6,233	17.8	5.5	13.3	48.2	41.0	11.3
1974	75,250	0	93,986	13,300	0	155,936	6,490	24.0	7.2	14.5	49.0	49.6	14.6
1975	68,470	0	101,629	17,650	0	152,449	6,755	22.6	6.7	13.8	51.3	48.5	13.0
1976	76,994	6,940	95,455	20,619	0	144,890	7,025	20.6	6.2	14.2	52.1	43.4	11.8
1977	83,407	8,900	86,501	33,398	0	127,611	7,303	17.5	5.4	14.6	53.0	36.7	10.1
1978	79,011	6,680	91,867	21,568	0	142,630	7,589	18.8	5.8	15.2	55.5	38.2	10.4
1979	86,613	11,110	117,275	36,021	0	156,758	7,885	19.9	6.1	15.4	57.9	39.6	10.5
1980	85,920	11,510	122,955	45,846	0	151,519	8,194	18.5	5.6	15.2	59.4	37.1	9.5
1981	91,995	13,100	145,712	48,331	0	176,276	8,515	20.7	6.3	15.8	60.9	40.3	10.4
1982	91,222	15,370	134,749	49,850	0	160,752	8,848	18.2	5.6	13.8	58.2	40.5	9.6
1983	94,627	18,580	118,014	59,070	0	134,990	9,190	14.7	4.5	12.7	55.7	35.1	8.0
1984	88,110	17,000	130,245	55,704	0	145,651	9,537	15.3	4.5	12.5	54.2	36.3	8.4
1985	110,662	13,270	144,157	63,962	0	177,587	9,886	18.0	5.4	13.0	56.2	41.6	9.6
1986	104,385	13,270	154,315	51,664	0	193,766	10,234	18.9	5.7	13.9	56.2	41.3	10.2
1987	101,682	18,600	171,011	62,603	0	191,490	10,581	18.1	5.5	13.9	55.7	39.7	9.9
1988	89,506	15,600	232,083	65,742	0	240,247	10,934	22.0	6.6	15.5	58.2	42.7	11.4
1989	99,171	14,700	230,245	78,815	0	235,901	11,299	20.9	6.4	14.2	54.2	45.1	11.8
1990	104,375	15,000	219,431	84,952	0	223,854	11,682	19.2	5.9	13.3	52.7	44.3	11.2
1991	85,008	23,200	207,439	95,716	0	173,532	12,086	14.4	4.4	11.6	51.7	38.3	8.6
1992	87,283	28,400	213,248	84,056	0	188,075	12,507	15.0	4.6	11.5	51.0	40.1	9.0
1993	70,174	29,700	225,890	100,653	0	165,711	12,929	12.8	4.0	10.6	50.4	37.2	7.9
1994	74,094	16,500	186,826	88,224	0	156,196	13,329	11.7	3.6	9.4	48.3	38.6	7.5
1995	70,526	19,100	249,567	114,869	0	166,124	13,694	13.6	4.2	9.9	50.3	42.4	7.9

98 CROATIA	- FOOD BALANCE SHEET OF FISH AND FISHERY PRODUCTS IN LIVE WEIGHT AND FISH CONTRIBUTION TO PROTEIN SUPPLY -	FAO-FIDI

YEAR	PRODUCTION	NON-FOOD USES	IMPORTS	EXPORTS	STOCKS VARIATIONS	TOTAL FOOD SUPPLY	POPULATION	PER CAPUT SUPPLY	FISH PROTEINS	ANIMAL PROTEINS	TOTAL PROTEINS	FISH/ANIMAL PROTEINS	FISH/TOTAL PROTEINS
	(.. metric tons in live weight					....)	(thousands)	(kilograms)	(grams per caput per day)			(%)	(%)
1961	0	0	0	0	0	0	4,061	0.0	0.0	0.0	0.0	0.0	0.0
1962	0	0	0	0	0	0	4,078	0.0	0.0	0.0	0.0	0.0	0.0
1963	0	0	0	0	0	0	4,094	0.0	0.0	0.0	0.0	0.0	0.0
1964	0	0	0	0	0	0	4,109	0.0	0.0	0.0	0.0	0.0	0.0
1965	0	0	0	0	0	0	4,122	0.0	0.0	0.0	0.0	0.0	0.0
1966	0	0	0	0	0	0	4,132	0.0	0.0	0.0	0.0	0.0	0.0
1967	0	0	0	0	0	0	4,140	0.0	0.0	0.0	0.0	0.0	0.0
1968	0	0	0	0	0	0	4,148	0.0	0.0	0.0	0.0	0.0	0.0
1969	0	0	0	0	0	0	4,157	0.0	0.0	0.0	0.0	0.0	0.0
1970	0	0	0	0	0	0	4,169	0.0	0.0	0.0	0.0	0.0	0.0
1971	0	0	0	0	0	0	4,184	0.0	0.0	0.0	0.0	0.0	0.0
1972	0	0	0	0	0	0	4,201	0.0	0.0	0.0	0.0	0.0	0.0
1973	0	0	0	0	0	0	4,221	0.0	0.0	0.0	0.0	0.0	0.0
1974	0	0	0	0	0	0	4,242	0.0	0.0	0.0	0.0	0.0	0.0
1975	0	0	0	0	0	0	4,263	0.0	0.0	0.0	0.0	0.0	0.0
1976	0	0	0	0	0	0	4,286	0.0	0.0	0.0	0.0	0.0	0.0
1977	0	0	0	0	0	0	4,308	0.0	0.0	0.0	0.0	0.0	0.0
1978	0	0	0	0	0	0	4,331	0.0	0.0	0.0	0.0	0.0	0.0
1979	0	0	0	0	0	0	4,354	0.0	0.0	0.0	0.0	0.0	0.0
1980	0	0	0	0	0	0	4,377	0.0	0.0	0.0	0.0	0.0	0.0
1981	0	0	0	0	0	0	4,398	0.0	0.0	0.0	0.0	0.0	0.0
1982	0	0	0	0	0	0	4,419	0.0	0.0	0.0	0.0	0.0	0.0
1983	0	0	0	0	0	0	4,438	0.0	0.0	0.0	0.0	0.0	0.0
1984	0	0	0	0	0	0	4,456	0.0	0.0	0.0	0.0	0.0	0.0
1985	0	0	0	0	0	0	4,471	0.0	0.0	0.0	0.0	0.0	0.0
1986	0	0	0	0	0	0	4,485	0.0	0.0	0.0	0.0	0.0	0.0
1987	0	0	0	0	0	0	4,497	0.0	0.0	0.0	0.0	0.0	0.0
1988	0	0	0	0	0	0	4,507	0.0	0.0	0.0	0.0	0.0	0.0
1989	0	0	0	0	0	0	4,513	0.0	0.0	0.0	0.0	0.0	0.0
1990	0	0	0	0	0	0	4,517	0.0	0.0	0.0	0.0	0.0	0.0
1991	0	0	0	0	0	0	4,518	0.0	0.0	0.0	0.0	0.0	0.0
1992	32,799	0	10,127	30,487	-2,306	10,134	4,517	2.2	0.7	32.6	64.7	2.2	1.1
1993	30,304	0	8,174	32,093	1,417	7,802	4,513	1.7	0.5	28.4	57.7	1.8	0.9
1994	21,313	0	12,247	22,201	889	12,248	4,509	2.7	0.8	28.5	55.8	2.7	1.4
1995	19,160	0	14,823	19,672	0	14,312	4,505	3.2	0.9	29.7	61.8	3.1	1.5

Note: for years before 1992 see Yugoslavia, S.F.R.

49 CUBA	- FOOD BALANCE SHEET OF FISH AND FISHERY PRODUCTS IN LIVE WEIGHT AND FISH CONTRIBUTION TO PROTEIN SUPPLY -												FAO-FIDI
YEAR	PRODUCTION	NON-FOOD USES	IMPORTS	EXPORTS	STOCKS VARIATIONS	TOTAL FOOD SUPPLY	POPULATION	PER CAPUT SUPPLY	FISH PROTEINS	ANIMAL PROTEINS	TOTAL PROTEINS	FISH/ANIMAL PROTEINS	FISH/TOTAL PROTEINS
	(........................ metric tons in live weight ....)						(thousands)	(kilograms)	(grams per caput per day)			(%)	(%)
1961	30,600	600	23,652	400	0	53,252	7,125	7.5	2.0	20.9	50.9	9.5	3.9
1962	35,000	300	14,773	1,200	-600	47,673	7,276	6.6	1.7	22.6	52.3	7.7	3.3
1963	35,600	300	43,262	1,200	-200	77,162	7,435	10.4	2.7	25.1	58.8	10.6	4.5
1964	37,100	100	64,580	640	-667	100,274	7,595	13.2	3.9	26.1	56.5	15.1	7.0
1965	40,900	200	37,661	4,038	1,013	75,336	7,754	9.7	2.5	24.5	55.2	10.3	4.6
1966	44,300	100	54,939	2,000	-200	96,938	7,908	12.3	3.6	26.1	54.7	13.9	6.7
1967	65,700	3,200	75,047	6,150	0	131,397	8,058	16.3	4.8	29.5	59.2	16.2	8.1
1968	66,000	7,500	70,141	14,764	340	114,216	8,208	13.9	3.7	33.6	63.1	11.1	5.9
1969	79,700	15,600	70,937	15,627	900	120,310	8,361	14.4	4.2	32.7	63.6	12.9	6.6
1970	104,300	15,300	63,550	22,898	0	129,652	8,520	15.2	3.8	35.1	67.9	11.0	5.7
1971	126,900	26,400	91,926	20,689	0	171,737	8,685	19.8	4.9	34.5	70.5	14.3	7.0
1972	139,400	28,200	68,160	20,860	-1,609	156,890	8,855	17.7	4.2	32.6	66.8	12.9	6.3
1973	150,200	24,000	81,782	25,612	-806	181,564	9,022	20.1	4.8	33.7	67.2	14.2	7.2
1974	164,379	24,500	84,668	33,451	-2,392	188,704	9,174	20.6	4.9	32.9	70.2	15.0	7.0
1975	143,053	29,600	110,685	26,041	-1,589	196,508	9,306	21.1	5.4	32.3	64.9	16.9	8.4
1976	192,190	48,000	72,465	24,749	0	191,906	9,413	20.4	5.1	33.1	68.6	15.3	7.4
1977	181,781	37,700	61,657	31,095	-3,874	170,769	9,500	18.0	4.9	32.4	67.1	15.2	7.3
1978	212,958	45,800	52,754	52,125	-6,250	161,537	9,573	16.9	4.6	32.0	67.0	14.3	6.8
1979	153,201	25,700	43,473	54,147	3,956	120,783	9,640	12.5	3.7	32.4	66.1	11.3	5.6
1980	186,405	31,211	68,621	61,888	12,500	174,427	9,710	18.0	5.4	34.8	72.4	15.6	7.5
1981	164,766	28,415	82,762	39,297	0	179,816	9,783	18.4	5.3	36.5	74.1	14.6	7.2
1982	194,615	39,982	60,498	34,508	0	180,623	9,859	18.3	5.2	36.2	73.6	14.5	7.1
1983	198,363	31,395	71,042	43,599	0	194,411	9,938	19.6	5.6	36.2	73.8	15.4	7.5
1984	199,807	31,262	45,038	34,654	0	178,929	10,024	17.9	4.9	35.9	74.4	13.5	6.5
1985	219,810	47,976	77,446	44,222	0	205,058	10,115	20.3	5.6	37.1	76.3	15.1	7.4
1986	244,422	53,207	60,466	45,929	0	205,752	10,214	20.1	5.6	36.2	73.1	15.5	7.7
1987	214,715	34,821	65,973	33,288	0	212,578	10,321	20.6	5.6	35.1	73.2	16.1	7.7
1988	231,250	48,439	66,503	33,871	0	215,443	10,429	20.7	5.9	35.7	72.9	16.4	8.0
1989	192,095	34,647	56,638	29,496	0	184,590	10,533	17.5	4.8	35.6	73.0	13.4	6.6
1990	188,220	29,081	52,880	23,154	0	188,865	10,628	17.8	4.9	34.1	68.9	14.2	7.0
1991	171,137	19,580	19,472	21,832	0	149,197	10,711	13.9	3.9	32.6	65.9	11.9	5.9
1992	121,288	1,100	36,484	14,856	0	141,816	10,784	13.2	3.2	24.8	60.2	13.0	5.4
1993	93,735	547	58,650	11,873	-322	139,643	10,848	12.9	3.0	23.3	54.4	13.1	5.6
1994	87,708	239	50,931	15,221	308	123,487	10,908	11.3	2.7	22.5	53.9	12.2	5.1
1995	94,195	161	59,969	15,949	0	138,054	10,964	12.6	3.0	23.0	53.2	13.1	5.7

50 CYPRUS	- FOOD BALANCE SHEET OF FISH AND FISHERY PRODUCTS IN LIVE WEIGHT AND FISH CONTRIBUTION TO PROTEIN SUPPLY -	FAO-FIDI

YEAR	PRODUCTION	NON-FOOD USES	IMPORTS	EXPORTS	STOCKS VARIATIONS	TOTAL FOOD SUPPLY	POPULATION	PER CAPUT SUPPLY	FISH PROTEINS	ANIMAL PROTEINS	TOTAL PROTEINS	FISH/ANIMAL PROTEINS	FISH/TOTAL PROTEINS
	(..		metric tons in live weight	....)			(thousands)	(kilograms)		(grams per caput per day)		(%)	(%)
1961	400	0	2,900	0	0	3,300	577	5.7	1.6	21.7	66.2	7.4	2.4
1962	600	0	3,000	0	0	3,600	578	6.2	1.8	21.6	67.2	8.1	2.6
1963	600	0	3,750	0	0	4,350	578	7.5	2.1	23.5	68.6	9.0	3.1
1964	600	0	3,550	0	0	4,150	579	7.2	2.0	24.5	72.4	8.2	2.8
1965	1,000	0	3,100	0	0	4,100	582	7.0	1.9	26.9	72.1	7.2	2.7
1966	1,000	0	3,200	0	0	4,200	588	7.1	2.0	29.7	74.1	6.6	2.7
1967	1,000	0	3,900	0	0	4,900	595	8.2	2.3	32.3	79.4	7.1	2.9
1968	1,400	0	3,450	0	0	4,850	604	8.0	2.2	35.1	82.7	6.2	2.7
1969	1,400	0	3,450	0	0	4,850	611	7.9	2.2	38.3	87.7	5.7	2.5
1970	1,314	0	2,930	0	0	4,244	615	6.9	1.9	39.0	85.4	4.9	2.2
1971	1,417	0	3,395	0	0	4,812	617	7.8	2.2	42.3	88.9	5.1	2.4
1972	1,406	0	3,240	0	0	4,646	616	7.5	2.1	45.1	85.3	4.6	2.4
1973	1,506	0	3,678	0	0	5,184	614	8.4	2.3	46.4	88.3	4.9	2.6
1974	1,216	0	2,029	0	0	3,245	611	5.3	1.4	37.5	77.6	3.7	1.8
1975	950	0	4,052	0	0	5,002	609	8.2	2.3	30.2	70.4	7.6	3.2
1976	1,083	0	2,360	8	0	3,436	608	5.7	1.6	31.0	73.5	5.2	2.2
1977	1,220	0	3,202	0	0	4,422	607	7.3	2.1	34.6	76.0	6.2	2.8
1978	1,276	0	5,136	195	0	6,217	607	10.3	3.1	38.1	76.8	8.0	4.0
1979	1,314	0	3,340	94	0	4,560	608	7.5	2.2	38.9	79.4	5.5	2.7
1980	1,336	0	3,901	78	0	5,159	611	8.4	2.4	40.1	79.8	5.9	3.0
1981	1,462	0	3,921	129	0	5,254	616	8.5	2.4	42.1	82.0	5.8	3.0
1982	1,602	0	5,903	0	0	7,505	623	12.0	2.9	43.8	83.2	6.7	3.5
1983	1,983	0	4,893	0	0	6,876	631	10.9	2.7	47.4	88.9	5.8	3.1
1984	2,264	0	5,375	0	-1	7,638	640	11.9	2.8	48.6	89.5	5.8	3.1
1985	2,436	0	6,402	0	0	8,838	647	13.7	3.3	50.1	91.6	6.5	3.6
1986	2,609	0	6,497	54	1	9,053	654	13.9	3.4	52.2	92.4	6.4	3.6
1987	2,605	0	9,095	50	0	11,650	659	17.7	4.0	54.7	96.4	7.3	4.2
1988	2,555	0	8,321	131	0	10,745	665	16.2	3.8	56.8	96.9	6.7	3.9
1989	2,648	0	10,218	352	-800	11,713	672	17.4	4.7	60.8	101.9	7.7	4.6
1990	2,744	0	9,777	507	0	12,014	681	17.7	5.0	62.4	103.2	8.1	4.9
1991	2,747	0	10,002	968	800	12,580	692	18.2	5.2	60.5	99.6	8.7	5.3
1992	2,833	0	12,694	452	0	15,075	705	21.4	5.9	64.8	108.1	9.2	5.5
1993	2,966	0	11,444	1,384	-5	13,021	719	18.1	5.0	64.7	106.2	7.7	4.7
1994	3,092	0	12,074	413	5	14,758	732	20.2	5.6	66.3	106.2	8.5	5.3
1995	3,045	0	14,177	550	0	16,671	745	22.4	6.1	68.2	110.2	8.9	5.5

51 CZECHOSLOVAKIA	- FOOD BALANCE SHEET OF FISH AND FISHERY PRODUCTS IN LIVE WEIGHT AND FISH CONTRIBUTION TO PROTEIN SUPPLY -	FAO-FIDI

YEAR	PRODUCTION	NON-FOOD USES	IMPORTS	EXPORTS	STOCKS VARIATIONS	TOTAL FOOD SUPPLY	POPULATION	PER CAPUT SUPPLY	FISH PROTEINS	ANIMAL PROTEINS	TOTAL PROTEINS	FISH/ANIMAL PROTEINS	FISH/TOTAL PROTEINS
	(..............................		metric tons in live weight			..............................)	(thousands)	(kilograms)		(grams per caput per day)		(%)	(%)
1961	9,900	0	78,898	0	0	88,798	13,791	6.4	1.7	41.4	89.2	4.0	1.9
1962	10,100	0	85,142	0	0	95,242	13,875	6.9	1.8	41.6	90.8	4.4	2.0
1963	9,700	0	102,288	0	0	111,988	13,950	8.0	2.1	42.0	92.2	4.9	2.2
1964	10,600	0	91,695	644	0	101,651	14,018	7.3	1.8	42.7	93.2	4.3	2.0
1965	11,000	0	91,590	667	0	101,923	14,079	7.2	1.8	45.1	94.1	4.1	2.0
1966	11,500	0	92,106	556	0	103,051	14,132	7.3	1.9	45.4	94.7	4.1	2.0
1967	11,800	0	95,171	1,000	0	105,971	14,181	7.5	1.9	47.2	95.9	4.1	2.0
1968	13,800	0	97,348	1,000	0	110,148	14,227	7.7	1.9	49.0	96.9	4.0	2.0
1969	14,000	0	107,663	1,111	0	120,552	14,277	8.4	2.0	49.8	95.4	3.9	2.1
1970	13,400	0	94,299	2,768	0	104,931	14,334	7.3	2.2	50.9	95.5	4.3	2.3
1971	13,900	0	95,482	1,500	0	107,882	14,398	7.5	2.6	52.1	96.3	5.0	2.7
1972	15,200	0	95,677	2,000	0	108,877	14,469	7.5	2.2	52.3	96.0	4.2	2.3
1973	15,400	0	97,674	2,447	0	110,628	14,548	7.6	2.3	53.2	96.6	4.3	2.4
1974	15,855	0	92,621	2,447	0	106,029	14,636	7.2	2.3	54.3	97.6	4.2	2.3
1975	16,940	0	92,794	2,444	0	107,290	14,733	7.3	2.3	55.7	98.7	4.1	2.3
1976	17,364	0	86,242	2,465	0	101,142	14,840	6.8	2.7	56.6	99.3	4.8	2.7
1977	17,958	0	75,126	2,459	0	90,625	14,955	6.1	2.4	56.7	98.2	4.3	2.5
1978	17,091	0	71,993	2,667	0	86,418	15,070	5.7	2.4	57.5	99.0	4.1	2.4
1979	16,817	0	56,885	3,111	0	70,591	15,174	4.7	2.0	57.3	98.4	3.5	2.1
1980	15,957	0	54,282	2,948	0	67,291	15,260	4.4	1.9	58.6	98.7	3.3	2.0
1981	16,503	0	60,553	3,114	0	73,942	15,323	4.8	2.1	58.8	99.9	3.5	2.1
1982	18,041	0	57,636	2,778	0	72,900	15,367	4.7	2.0	56.1	98.8	3.6	2.1
1983	19,525	0	47,991	3,000	0	64,516	15,397	4.2	1.8	57.6	99.4	3.1	1.8
1984	19,688	0	64,066	1,000	0	82,754	15,421	5.4	2.1	58.4	101.6	3.6	2.1
1985	20,034	0	63,168	970	0	82,231	15,445	5.3	2.1	58.7	101.9	3.5	2.0
1986	21,253	0	80,987	1,878	0	100,362	15,471	6.5	2.4	58.3	101.7	4.1	2.4
1987	20,736	0	84,909	1,456	0	104,189	15,497	6.7	2.5	59.7	103.9	4.1	2.4
1988	21,247	0	93,947	3,221	0	111,973	15,522	7.2	2.6	61.6	106.6	4.2	2.5
1989	21,574	0	86,857	2,895	0	105,536	15,544	6.8	2.5	61.6	107.3	4.0	2.3
1990	22,564	0	78,544	3,539	0	97,569	15,562	6.3	2.4	60.0	105.7	4.0	2.3
1991	22,046	0	62,319	5,276	0	79,089	15,576	5.1	1.7	49.4	93.0	3.5	1.9
1992	24,135	0	61,493	5,142	0	80,486	15,587	5.2	1.8	48.2	87.5	3.8	2.1

Note: data refer to the area that was formerly the Czechoslovak Socialist Republic. The new independent Republics are: Czech Republic and Slovakia, information for which is shown separately.

167 CZECH REPUBLIC	- FOOD BALANCE SHEET OF FISH AND FISHERY PRODUCTS IN LIVE WEIGHT AND FISH CONTRIBUTION TO PROTEIN SUPPLY -												FAO-FIDI
YEAR	PRODUCTION	NON-FOOD USES	IMPORTS	EXPORTS	STOCKS VARIATIONS	TOTAL FOOD SUPPLY	POPULATION	PER CAPUT SUPPLY	FISH PROTEINS	ANIMAL PROTEINS	TOTAL PROTEINS	FISH/ANIMAL PROTEINS	FISH/TOTAL PROTEINS
	(................................ metric tons in live weight ....)						(thousands)	(kilograms)	(grams per caput per day)			(%)	(%)
1961	0	0	0	0	0	0	9,592	0.0	0.0	0.0	0.0	0.0	0.0
1962	0	0	0	0	0	0	9,629	0.0	0.0	0.0	0.0	0.0	0.0
1963	0	0	0	0	0	0	9,663	0.0	0.0	0.0	0.0	0.0	0.0
1964	0	0	0	0	0	0	9,692	0.0	0.0	0.0	0.0	0.0	0.0
1965	0	0	0	0	0	0	9,717	0.0	0.0	0.0	0.0	0.0	0.0
1966	0	0	0	0	0	0	9,736	0.0	0.0	0.0	0.0	0.0	0.0
1967	0	0	0	0	0	0	9,752	0.0	0.0	0.0	0.0	0.0	0.0
1968	0	0	0	0	0	0	9,766	0.0	0.0	0.0	0.0	0.0	0.0
1969	0	0	0	0	0	0	9,783	0.0	0.0	0.0	0.0	0.0	0.0
1970	0	0	0	0	0	0	9,805	0.0	0.0	0.0	0.0	0.0	0.0
1971	0	0	0	0	0	0	9,832	0.0	0.0	0.0	0.0	0.0	0.0
1972	0	0	0	0	0	0	9,865	0.0	0.0	0.0	0.0	0.0	0.0
1973	0	0	0	0	0	0	9,902	0.0	0.0	0.0	0.0	0.0	0.0
1974	0	0	0	0	0	0	9,947	0.0	0.0	0.0	0.0	0.0	0.0
1975	0	0	0	0	0	0	9,997	0.0	0.0	0.0	0.0	0.0	0.0
1976	0	0	0	0	0	0	10,056	0.0	0.0	0.0	0.0	0.0	0.0
1977	0	0	0	0	0	0	10,121	0.0	0.0	0.0	0.0	0.0	0.0
1978	0	0	0	0	0	0	10,186	0.0	0.0	0.0	0.0	0.0	0.0
1979	0	0	0	0	0	0	10,242	0.0	0.0	0.0	0.0	0.0	0.0
1980	0	0	0	0	0	0	10,283	0.0	0.0	0.0	0.0	0.0	0.0
1981	0	0	0	0	0	0	10,307	0.0	0.0	0.0	0.0	0.0	0.0
1982	0	0	0	0	0	0	10,316	0.0	0.0	0.0	0.0	0.0	0.0
1983	0	0	0	0	0	0	10,314	0.0	0.0	0.0	0.0	0.0	0.0
1984	0	0	0	0	0	0	10,309	0.0	0.0	0.0	0.0	0.0	0.0
1985	0	0	0	0	0	0	10,305	0.0	0.0	0.0	0.0	0.0	0.0
1986	0	0	0	0	0	0	10,304	0.0	0.0	0.0	0.0	0.0	0.0
1987	0	0	0	0	0	0	10,305	0.0	0.0	0.0	0.0	0.0	0.0
1988	0	0	0	0	0	0	10,306	0.0	0.0	0.0	0.0	0.0	0.0
1989	0	0	0	0	0	0	10,308	0.0	0.0	0.0	0.0	0.0	0.0
1990	0	0	0	0	0	0	10,306	0.0	0.0	0.0	0.0	0.0	0.0
1991	0	0	0	0	0	0	10,301	0.0	0.0	0.0	0.0	0.0	0.0
1992	0	0	0	0	0	0	10,299	0.0	0.0	0.0	0.0	0.0	0.0
1993	23,427	0	52,792	13,418	0	62,801	10,285	6.1	2.4	59.8	93.7	4.0	2.6
1994	22,613	0	72,859	16,055	0	79,417	10,275	7.7	2.8	52.4	87.6	5.3	3.2
1995	22,608	0	67,659	13,563	0	76,704	10,263	7.5	2.7	53.7	90.5	4.9	2.9

Note: for years before 1993 see Czechoslovakia.

54 DENMARK	- FOOD BALANCE SHEET OF FISH AND FISHERY PRODUCTS IN LIVE WEIGHT AND FISH CONTRIBUTION TO PROTEIN SUPPLY -	FAO-FIDI

YEAR	PRODUCTION	NON-FOOD USES	IMPORTS	EXPORTS	STOCKS VARIATIONS	TOTAL FOOD SUPPLY	POPULATION	PER CAPUT SUPPLY	FISH PROTEINS	ANIMAL PROTEINS	TOTAL PROTEINS	FISH/ANIMAL PROTEINS	FISH/TOTAL PROTEINS
	(..		metric tons in live weight		....)		(thousands)	(kilograms)		(grams per caput per day)		(%)	(%)
1961	637,500	382,400	90,467	264,120	-5,044	76,404	4,614	16.6	5.0	47.3	82.2	10.5	6.0
1962	785,100	543,000	132,255	284,531	-8,404	81,420	4,649	17.5	5.3	44.4	78.5	11.8	6.7
1963	847,900	592,000	164,833	329,849	-6,453	84,431	4,685	18.0	5.3	46.3	79.6	11.5	6.7
1964	871,100	644,000	183,792	328,165	5,606	88,333	4,722	18.7	6.1	47.9	80.3	12.7	7.6
1965	840,800	624,300	207,039	333,485	-2,200	87,853	4,758	18.5	5.9	46.3	77.9	12.6	7.5
1966	850,800	639,900	199,710	314,177	-3,700	92,732	4,794	19.3	6.5	48.6	80.5	13.3	8.0
1967	1,070,400	853,100	212,802	332,812	-400	96,890	4,829	20.1	6.7	47.7	78.7	14.0	8.5
1968	1,466,800	1,210,200	197,343	351,282	-2,600	100,061	4,863	20.6	6.5	45.4	76.3	14.3	8.5
1969	1,275,400	1,015,800	156,205	306,434	0	109,371	4,897	22.3	7.5	48.0	78.3	15.7	9.6
1970	1,226,500	1,009,389	210,112	321,032	-1,900	104,291	4,929	21.2	7.0	46.2	77.8	15.2	9.0
1971	1,400,901	1,105,600	174,234	374,233	2,800	98,103	4,959	19.8	7.0	47.2	78.1	14.8	8.9
1972	1,442,900	1,166,853	231,071	403,357	1,884	105,645	4,988	21.2	7.3	47.5	78.4	15.3	9.3
1973	1,443,702	1,202,467	270,861	407,745	7,049	111,401	5,014	22.2	6.7	44.2	74.6	15.2	9.0
1974	1,835,370	1,657,787	302,401	369,371	-1,090	109,524	5,038	21.7	7.1	47.1	77.4	15.1	9.2
1975	1,767,039	1,539,140	253,787	391,636	10,648	100,699	5,060	19.9	6.9	49.8	79.1	13.9	8.8
1976	1,919,197	1,633,590	185,546	371,213	4,803	104,743	5,079	20.6	7.7	50.1	80.0	15.4	9.7
1977	1,806,612	1,532,492	206,951	377,252	2,098	105,916	5,095	20.8	7.5	51.5	81.8	14.5	9.2
1978	1,750,669	1,475,036	217,044	390,831	-5	101,841	5,108	19.9	9.4	55.8	86.4	16.8	10.9
1979	1,741,106	1,429,472	242,496	451,825	406	102,711	5,117	20.1	8.9	53.8	84.6	16.5	10.5
1980	2,031,673	1,721,559	284,463	487,932	-1,294	105,351	5,123	20.6	9.1	56.0	87.8	16.3	10.4
1981	1,858,943	1,514,627	300,788	536,963	-1,700	106,443	5,125	20.8	8.4	56.0	88.8	15.0	9.5
1982	1,930,773	1,575,159	320,085	571,150	1,000	105,549	5,123	20.6	10.0	55.0	87.6	18.1	11.4
1983	1,869,809	1,502,000	318,702	582,187	2,000	106,324	5,119	20.8	8.5	53.4	85.9	16.0	9.9
1984	1,851,484	1,464,807	350,257	630 369	-250	106,314	5,115	20.8	8.3	53.6	87.3	15.6	9.6
1985	1,796,988	1,372,180	373,986	682,286	-11,184	105,323	5,114	20.6	8.8	60.4	95.2	14.5	9.2
1986	1,849,767	1,465,471	419,110	700,800	4,464	107,071	5,114	20.9	10.8	64.6	99.6	16.7	10.8
1987	1,707,824	1,358,049	457,827	696,467	-888	110,247	5,117	21.6	9.9	66.3	101.8	15.0	9.8
1988	1,974,424	1,600,423	458,410	716,444	-3,076	112,894	5,122	22.0	10.6	68.2	103.0	15.5	10.3
1989	1,929,305	1,573,115	526,062	755,003	-11,894	115,358	5,129	22.5	9.9	68.6	103.3	14.5	9.6
1990	1,517,657	1,139,862	576,720	827,872	-14,417	112,220	5,140	21.8	8.1	64.8	99.1	12.5	8.2
1991	1,793,252	1,391,809	577,236	884,842	20,529	114,365	5,154	22.2	13.2	71.2	106.6	18.6	12.4
1992	1,997,098	1,594,992	573,827	845,647	-9,659	120,623	5,171	23.3	11.7	66.0	102.4	17.7	11.4
1993	1,657,137	1,259,511	563,813	866,527	21,351	116,262	5,189	22.4	9.9	63.6	100 8	15.5	9.8
1994	1,916,234	1,579,909	711,879	915,916	-6,000	126,286	5,207	24.3	9.9	63.5	100.9	15.6	9.8
1995	2,043,769	1,728,563	663,626	843,527	-7,500	127,805	5,223	24.5	9.5	66.2	104.2	14.4	9.2

72 DJIBOUTI	- FOOD BALANCE SHEET OF FISH AND FISHERY PRODUCTS IN LIVE WEIGHT AND FISH CONTRIBUTION TO PROTEIN SUPPLY -												FAO-FIDI
YEAR	PRODUCTION	NON-FOOD USES	IMPORTS	EXPORTS	STOCKS VARIATIONS	TOTAL FOOD SUPPLY	POPULATION	PER CAPUT SUPPLY	FISH PROTEINS	ANIMAL PROTEINS	TOTAL PROTEINS	FISH/ANIMAL PROTEINS	FISH/TOTAL PROTEINS
	(.............................. metric tons in live weight					..............................)	(thousands)	(kilograms)	(grams per caput per day)			(%)	(%)
1961	400	0	...	0	0	400	87	4.6	1.3	18.5	42.5	7.0	3.1
1962	400	0	...	0	0	400	91	4.4	1.2	18.1	45.0	6.8	2.8
1963	400	0	...	0	0	400	96	4.2	1.2	17.7	42.9	6.6	2.7
1964	300	0	...	0	0	300	102	2.9	0.8	17.0	44.7	4.9	1.9
1965	300	0	...	0	0	300	108	2.8	0.8	16.1	43.0	4.9	1.8
1966	300	0	...	0	0	300	115	2.6	0.7	16.5	43.8	4.5	1.7
1967	300	0	...	0	0	300	122	2.5	0.7	16.2	42.9	4.3	1.6
1968	300	0	...	0	0	300	130	2.3	0.7	16.0	42.4	4.1	1.5
1969	300	0	...	0	0	300	139	2.2	0.6	16.0	41.3	3.8	1.5
1970	300	0	...	0	0	300	148	2.0	0.6	15.6	42.6	3.7	1.3
1971	300	0	...	0	0	300	158	1.9	0.5	15.2	42.2	3.5	1.3
1972	100	0	...	0	0	100	169	0.6	0.2	14.7	40.3	1.1	0.4
1973	200	0	...	0	0	200	180	1.1	0.3	13.7	40.7	2.3	0.8
1974	380	0	...	0	0	380	192	2.0	0.6	15.4	42.0	3.6	1.3
1975	300	0	...	0	0	300	205	1.5	0.4	16.9	42.0	2.4	1.0
1976	230	0	...	0	0	230	218	1.1	0.3	19.1	43.0	1.6	0.7
1977	230	0	...	0	0	230	232	1.0	0.3	16.6	41.6	1.7	0.7
1978	230	0	...	0	0	230	247	0.9	0.3	17.5	40.8	1.5	0.6
1979	231	0	...	0	0	231	264	0.9	0.3	18.4	43.1	1.3	0.6
1980	251	0	...	0	0	251	281	0.9	0.3	19.5	47.0	1.3	0.5
1981	385	0	...	0	0	385	301	1.3	0.4	19.4	49.4	1.9	0.7
1982	426	0	536	0	0	962	321	3.0	0.9	18.7	50.6	4.6	1.7
1983	409	0	375	0	0	784	343	2.3	0.6	17.4	46.8	3.6	1.3
1984	409	0	696	0	0	1,105	367	3.0	0.8	16.5	44.4	5.0	1.9
1985	380	0	947	0	0	1,327	391	3.4	1.0	15.9	42.7	6.2	2.3
1986	409	0	426	0	0	835	417	2.0	0.6	15.3	43.4	3.7	1.3
1987	426	0	194	40	0	580	443	1.3	0.4	13.8	48.0	2.7	0.8
1988	454	0	450	24	0	880	469	1.9	0.5	13.8	49.3	3.5	1.0
1989	391	0	333	70	0	654	494	1.3	0.4	12.6	47.0	2.8	0.8
1990	359	0	253	83	0	529	517	1.0	0.3	12.2	45.2	2.4	0.6
1991	252	0	280	56	0	476	537	0.9	0.3	12.4	37.9	2.0	0.7
1992	275	0	780	23	0	1,032	554	1.9	0.6	12.8	38.8	4.5	1.5
1993	300	0	804	25	0	1,079	570	1.9	0.6	14.1	41.3	4.1	1.4
1994	320	0	784	30	0	1,074	585	1.8	0.6	13.4	40.6	4.4	1.5
1995	350	0	560	30	0	880	601	1.5	0.5	13.3	40.8	3.4	1.1

55 DOMINICA	- FOOD BALANCE SHEET OF FISH AND FISHERY PRODUCTS IN LIVE WEIGHT AND FISH CONTRIBUTION TO PROTEIN SUPPLY -	FAO-FIDI

YEAR	PRODUCTION	NON-FOOD USES	IMPORTS	EXPORTS	STOCKS VARIATIONS	TOTAL FOOD SUPPLY	POPULATION	PER CAPUT SUPPLY	FISH PROTEINS	ANIMAL PROTEINS	TOTAL PROTEINS	FISH/ANIMAL PROTEINS	FISH/TOTAL PROTEINS
	(..		metric tons in live weight	....)			(thousands)	(kilograms)	(grams per caput per day)			(%)	(%)
1961	500	0	750	0	0	1,250	61	20.5	6.6	18.3	41.5	36.3	16.0
1962	600	0	1,000	0	0	1,600	62	25.7	8.4	21.0	46.1	39.8	18.2
1963	600	0	1,000	0	0	1,600	64	25.2	8.2	21.0	46.5	39.0	17.7
1964	600	0	1,000	0	0	1,600	65	24.7	8.0	21.6	46.6	37.3	17.3
1965	500	0	1,500	0	0	2,000	66	30.3	9.6	23.1	48.3	41.7	19.9
1966	500	0	1,150	0	0	1,650	67	24.6	7.7	21.6	44.7	35.8	17.3
1967	500	0	1,150	0	0	1,650	68	24.3	7.6	21.7	44.8	35.1	17.0
1968	600	0	1,150	0	0	1,750	69	25.5	8.0	23.3	49.0	34.1	16.2
1969	600	0	1,300	0	0	1,900	69	27.4	8.7	25.8	50.9	33.9	17.2
1970	500	0	1,050	0	0	1,550	70	22.1	6.3	24.3	50.1	26.0	12.7
1971	600	0	1,050	0	0	1,650	71	23.4	6.7	24.7	50.0	27.1	13.4
1972	700	0	1,250	0	0	1,950	71	27.5	7.8	25.2	51.0	31.1	15.4
1973	800	0	1,250	0	0	2,050	71	28.8	8.2	22.9	48.2	35.8	17.0
1974	900	0	1,250	0	0	2,150	72	30.0	8.6	23.6	48.2	36.3	17.8
1975	1,001	0	800	0	0	1,801	72	25.0	7.1	22.4	47.1	31.8	15.1
1976	1,024	0	794	0	0	1,818	72	25.1	7.1	21.5	47.9	33.1	14.9
1977	1,047	0	804	0	0	1,851	73	25.4	7.2	25.8	53.1	27.8	13.5
1978	1,070	0	851	0	0	1,921	74	26.1	7.7	27.7	56.2	27.8	13.7
1979	642	0	728	0	0	1,370	74	18.5	5.5	23.8	55.5	23.1	9.9
1980	1,445	0	523	0	0	1,968	74	26.6	7.7	26.5	58.9	29.0	13.1
1981	1,514	0	387	0	0	1,901	74	25.7	7.3	27.4	59.6	26.6	12.2
1982	1,545	0	257	0	0	1,802	73	24.5	6.9	27.7	58.6	25.0	11.8
1983	800	0	531	0	0	1,331	73	18.2	5.2	27.3	58.4	18.9	8.8
1984	700	0	761	0	0	1,461	72	20.2	5.8	30.5	62.6	18.9	9.2
1985	640	0	776	0	0	1,416	72	19.7	5.6	30.7	64.4	18.3	8.7
1986	644	0	708	0	0	1,352	72	18.9	5.4	35.0	71.3	15.4	7.6
1987	500	0	601	0	0	1,101	71	15.4	4.4	36.6	73.3	12.0	6.0
1988	500	0	646	0	0	1,146	71	16.1	4.6	37.5	74.5	12.2	6.2
1989	500	0	852	0	0	1,352	71	19.0	5.4	40.2	76.7	13.5	7.1
1990	448	0	1,069	0	0	1,517	71	21.4	6.1	41.4	77.0	14.7	7.9
1991	552	0	1,073	0	0	1,625	71	22.9	6.5	42.3	76.4	15.4	8.5
1992	717	0	1,078	0	0	1,795	71	25.3	7.2	44.4	78.6	16.2	9.2
1993	797	0	1,121	0	0	1,918	71	27.0	7.7	43.9	76.9	17.5	10.0
1994	885	0	1,157	0	0	2,042	71	28.8	8.2	46.7	83.3	17.5	9.8
1995	842	0	1,214	0	0	2,056	71	29.0	8.3	42.6	76.9	19.4	10.7

56 DOMINICAN REPUBLIC	- FOOD BALANCE SHEET OF FISH AND FISHERY PRODUCTS IN LIVE WEIGHT AND FISH CONTRIBUTION TO PROTEIN SUPPLY -												FAO-FIDI
YEAR	PRODUCTION	NON-FOOD USES	IMPORTS	EXPORTS	STOCKS VARIATIONS	TOTAL FOOD SUPPLY	POPULATION	PER CAPUT SUPPLY	FISH PROTEINS	ANIMAL PROTEINS	TOTAL PROTEINS	FISH/ANIMAL PROTEINS	FISH/TOTAL PROTEINS
	(.. metric tons in live weight ....)						(thousands)	(kilograms)	(grams per caput per day)			(%)	(%)
1961	1,600	0	13,294	0	0	14,894	3,341	4.5	1.4	12.8	38.3	11.2	3.8
1962	1,700	0	27,903	0	0	29,603	3,454	8.6	2.8	15.0	40.6	18.6	6.9
1963	3,200	0	30,921	0	0	34,121	3,569	9.6	3.1	15.3	41.2	20.2	7.5
1964	4,200	0	39,512	0	0	43,712	3,686	11.9	3.7	17.0	43.0	21.9	8.7
1965	4,100	0	18,750	0	0	22,850	3,806	6.0	1.9	13.6	40.4	13.7	4.6
1966	3,500	0	35,259	0	0	38,759	3,927	9.9	3.1	16.1	42.7	19.6	7.4
1967	3,200	0	25,020	100	0	28,120	4,050	6.9	2.2	14.9	41.7	14.9	5.3
1968	4,500	0	31,459	200	0	35,759	4,174	8.6	2.7	16.4	42.7	16.8	6.4
1969	4,800	0	30,953	400	0	35,353	4,298	8.2	2.6	16.4	43.9	16.1	6.0
1970	5,200	0	30,485	400	0	35,285	4,423	8.0	2.5	16.5	43.7	15.3	5.8
1971	4,600	0	27,114	500	0	31,214	4,547	6.9	2.2	16.7	44.1	13.0	4.9
1972	5,600	0	23,624	700	0	28,524	4,672	6.1	1.9	15.8	43.5	11.9	4.3
1973	9,200	0	17,053	700	-120	25,433	4,797	5.3	1.6	15.3	44.7	10.7	3.6
1974	6,810	0	13,752	600	-280	19,682	4,922	4.0	1.2	14.6	45.6	8.2	2.6
1975	5,906	0	22,438	1,100	0	27,244	5,048	5.4	1.6	14.8	45.4	11.0	3.6
1976	7,053	0	24,527	958	280	30,903	5,176	6.0	1.9	15.5	44.6	11.9	4.1
1977	4,594	0	33,427	731	120	37,410	5,305	7.1	2.2	17.2	46.0	12.7	4.7
1978	5,086	0	30,019	1,041	0	34,065	5,434	6.3	1.9	16.6	45.5	11.6	4.2
1979	7,885	0	37,257	848	0	44,294	5,565	8.0	2.4	19.0	48.4	12.8	5.0
1980	10,658	0	42,109	1,600	0	51,167	5,697	9.0	2.7	20.1	51.4	13.5	5.3
1981	12,003	0	32,724	1,838	0	42,888	5,829	7.4	2.3	19.0	49.2	12.3	4.7
1982	13,169	0	32,192	1,813	0	43,548	5,963	7.3	2.2	18.4	48.4	12.1	4.6
1983	15,289	0	29,990	1,992	0	43,287	6,098	7.1	2.2	19.2	50.4	11.6	4.4
1984	14,664	0	19,384	1,742	0	32,307	6,235	5.2	1.6	18.1	49.9	8.6	3.1
1985	18,438	0	35,760	3,489	0	50,709	6,376	8.0	2.3	18.8	52.8	12.5	4.4
1986	17,282	0	37,029	2,030	0	52,281	6,520	8.0	2.3	18.0	50.1	13.0	4.7
1987	20,445	0	34,127	1,531	0	53,042	6,666	8.0	2.3	18.5	50.2	12.5	4.6
1988	12,912	0	38,967	916	0	50,963	6,815	7.5	2.2	19.4	51.4	11.1	4.2
1989	21,845	0	42,489	1,367	0	62,967	6,963	9.0	2.5	20.6	51.5	12.2	4.9
1990	20,050	0	43,072	540	0	62,582	7,110	8.8	2.4	20.0	48.0	12.2	5.1
1991	17,340	0	28,773	124	0	45,989	7,256	6.3	1.7	21.0	49.8	8.2	3.5
1992	13,595	0	30,938	218	0	44,315	7,400	6.0	1.8	21.9	49.1	8.0	3.6
1993	14,363	0	39,895	401	-17	53,841	7,543	7.1	2.1	22.7	50.0	9.2	4.2
1994	26,860	0	40,898	169	0	67,589	7,684	8.8	2.5	22.8	50.1	10.9	5.0
1995	20,183	0	43,378	1,790	17	61,787	7,823	7.9	2.3	23.0	50.5	10.2	4.6

58 ECUADOR	- FOOD BALANCE SHEET OF FISH AND FISHERY PRODUCTS IN LIVE WEIGHT AND FISH CONTRIBUTION TO PROTEIN SUPPLY -	FAO-FIDI

YEAR	PRODUCTION	NON-FOOD USES	IMPORTS	EXPORTS	STOCKS VARIATIONS	TOTAL FOOD SUPPLY	POPULATION	PER CAPUT SUPPLY	FISH PROTEINS	ANIMAL PROTEINS	TOTAL PROTEINS	FISH/ANIMAL PROTEINS	FISH/TOTAL PROTEINS
	(.....................		metric tons in live weight			.....................)	(thousands)	(kilograms)	(grams per caput per day)			(%)	(%)
1961	38,600	0	832	11,963	0	27,470	4,569	6.0	1.9	18.0	50.2	10.5	3.8
1962	42,500	0	1,490	14,471	0	29,520	4,705	6.3	1.8	17.8	50.3	10.4	3.7
1963	49,700	1,400	209	12,201	0	36,308	4,847	7.5	2.3	18.1	51.2	12.7	4.5
1964	46,300	4,500	452	12,898	0	29,355	4,993	5.9	1.7	17.5	50.4	9.8	3.4
1965	53,500	4,800	273	17,163	0	31,810	5,144	6.2	1.8	17.4	53.6	10.1	3.3
1966	48,200	5,700	0	13,926	0	28,574	5,300	5.4	1.5	17.2	52.3	9.0	3.0
1967	57,500	7,200	0	21,553	0	28,747	5,461	5.3	1.6	17.6	52.5	9.0	3.0
1968	68,900	18,300	0	17,347	0	33,253	5,626	5.9	1.7	17.8	51.6	9.3	3.2
1969	64,900	19,300	0	23,939	0	21,661	5,796	3.7	1.0	17.3	49.8	5.6	1.9
1970	91,020	29,000	198	17,689	-303	44,226	5,970	7.4	2.0	18.8	52.4	10.7	3.8
1971	102,580	28,000	0	30,404	304	44,480	6,148	7.2	1.9	18.1	49.4	10.7	3.9
1972	107,040	38,900	0	23,600	0	44,540	6,331	7.0	1.9	18.3	48.8	10.4	3.9
1973	152,070	68,800	0	28,261	0	55,009	6,518	8.4	2.3	18.7	47.1	12.3	4.9
1974	176,480	79,500	0	34,021	-1,017	61,942	6,710	9.2	2.6	19.2	47.3	13.3	5.4
1975	215,748	106,000	7	40,658	-2,547	66,550	6,907	9.6	2.6	19.3	48.4	13.6	5.4
1976	282,738	175,000	329	45,378	1,589	64,277	7,109	9.0	2.6	19.5	49.1	13.4	5.3
1977	425,360	292,000	698	58,426	980	76,612	7,316	10.5	3.0	20.7	47.5	14.3	6.2
1978	614,280	465,000	891	63,305	-1,530	85,335	7,527	11.3	3.3	21.8	47.9	14.9	6.8
1979	604,964	432,400	1,251	78,816	2,551	97,549	7,742	12.6	3.6	22.3	47.5	16.3	7.7
1980	639,157	426,000	1,312	112,057	-3,571	98,842	7,961	12.4	3.8	21.0	48.4	17.8	7.8
1981	538,840	348,000	352	102,398	510	89,304	8,184	10.9	3.0	19.7	45.7	15.4	6.6
1982	607,411	415,000	2	103,070	-1,531	87,812	8,410	10.4	2.9	19.8	46.0	14.6	6.3
1983	371,856	200,500	0	69,958	1,531	102,929	8,638	11.9	3.4	19.8	44.4	17.3	7.7
1984	882,780	678,200	13	80,081	-14,239	110,274	8,868	12.4	3.5	20.1	47.9	17.7	7.4
1985	1,087,005	900,000	114	96,001	-148	90,970	9,099	10.0	2.8	20.3	46.9	13.8	6.0
1986	1,003,560	811,300	0	118,139	1,102	75,224	9,330	8.1	2.2	19.6	45.8	11.4	4.9
1987	680,131	447,100	0	161,908	0	71,123	9,562	7.4	2.1	19.8	46.5	10.6	4.5
1988	882,000	654,500	0	142,150	2,044	87,394	9,794	8.9	2.7	20.3	50.8	13.4	5.4
1989	739,940	449,300	0	164,866	-23,051	102,724	10,028	10.2	3.4	20.6	48.2	16.5	7.0
1990	391,159	162,100	83	160,299	16,852	85,695	10,264	8.4	3.4	22.0	49.1	15.2	6.8
1991	373,987	130,600	7	179,514	1,019	64,899	10,502	6.2	1.8	21.1	46.3	8.4	3.8
1992	347,407	99,700	182	176,207	2,956	74,638	10,741	7.0	2.0	21.2	48.3	9.6	4.2
1993	331,372	72,800	627	171,963	3,438	90,674	10,980	8.3	2.3	21.9	48.6	10.3	4.6
1994	339,927	59,000	8,869	225,274	0	64,522	11,220	5.8	1.6	22.1	50.1	7.2	3.2
1995	591,560	236,000	14,716	281,622	0	88,654	11,460	7.7	2.1	24.1	50.0	8.6	4.1

59 EGYPT	- FOOD BALANCE SHEET OF FISH AND FISHERY PRODUCTS IN LIVE WEIGHT AND FISH CONTRIBUTION TO PROTEIN SUPPLY -												FAO-FIDI
YEAR	PRODUCTION	NON-FOOD USES	IMPORTS	EXPORTS	STOCKS VARIATIONS	TOTAL FOOD SUPPLY	POPULATION	PER CAPUT SUPPLY	FISH PROTEINS	ANIMAL PROTEINS	TOTAL PROTEINS	FISH/ANIMAL PROTEINS	FISH/TOTAL PROTEINS
	(.. metric tons in live weight ....)						(thousands)	(kilograms)	(grams per caput per day)			(%)	(%)
1961	92,000	0	17,280	1,649	0	107,631	28,548	3.8	1.1	8.3	58.8	13.6	1.9
1962	118,000	0	15,272	3,377	0	129,896	29,284	4.4	1.3	8.4	62.9	15.8	2.1
1963	115,000	0	8,333	3,733	0	119,600	30,040	4.0	1.2	8.4	61.9	13.9	1.9
1964	135,000	0	10,489	3,389	0	142,100	30,803	4.6	1.4	8.4	64.6	16.1	2.1
1965	102,400	0	10,835	2,557	0	110,678	31,563	3.5	1.0	8.2	63.2	12.4	1.6
1966	99,000	0	36,740	2,071	0	133,668	32,322	4.1	1.2	8.5	66.0	14.2	1.8
1967	80,300	0	25,832	1,600	0	104,532	33,081	3.2	0.9	7.9	62.0	11.5	1.5
1968	80,500	0	19,283	900	0	98,883	33,834	2.9	0.9	8.8	64.4	9.7	1.3
1969	80,000	0	14,511	400	0	94,111	34,569	2.7	0.8	8.5	63.8	9.3	1.2
1970	80,900	0	15,333	222	0	96,011	35,285	2.7	0.8	8.5	64.0	9.3	1.2
1971	87,600	0	15,933	111	0	103,422	35,976	2.9	0.8	8.6	63.3	9.8	1.3
1972	93,800	0	20,638	311	0	114,127	36,650	3.1	0.9	8.6	65.0	10.6	1.4
1973	93,500	0	15,178	131	0	108,547	37,332	2.9	0.9	8.4	63.8	10.1	1.3
1974	96,165	5	22,425	211	0	118,374	38,054	3.1	0.9	8.4	64.4	10.7	1.4
1975	106,570	8	43,294	311	0	149,545	38,841	3.9	1.1	9.0	65.2	12.3	1.7
1976	102,758	10	76,607	291	0	179,064	39,701	4.5	1.3	9.4	66.8	13.7	1.9
1977	104,536	9	49,400	414	0	153,512	40,630	3.8	1.1	9.6	67.5	11.2	1.6
1978	99,910	5	87,908	311	0	187,502	41,621	4.5	1.3	9.9	69.2	12.9	1.8
1979	137,480	1	47,634	601	0	184,512	42,664	4.3	1.3	9.8	70.9	13.0	1.8
1980	140,397	1	63,748	462	0	203,681	43,749	4.7	1.4	10.6	72.7	13.0	1.9
1981	141,710	2	97,756	441	0	239,022	44,873	5.3	1.6	11.5	75.3	13.8	2.1
1982	155,728	3	100,608	690	0	255,643	46,037	5.6	1.6	11.6	76.7	14.0	2.1
1983	157,302	2	132,063	312	0	289,050	47,238	6.1	1.8	12.9	78.7	13.8	2.3
1984	163,838	1	148,182	315	0	311,704	48,476	6.4	1.9	13.4	78.9	13.8	2.4
1985	215,920	4	103,851	288	0	319,479	49,749	6.4	1.9	13.2	79.0	14.3	2.4
1986	229,078	18	124,920	1,029	0	352,951	51,056	6.9	2.0	12.9	79.3	15.5	2.5
1987	231,030	8	116,226	930	0	346,317	52,391	6.6	1.9	13.1	81.9	14.4	2.3
1988	284,247	13	118,132	1,224	0	401,142	53,730	7.5	2.2	13.3	82.0	16.2	2.6
1989	293,604	100	132,726	2,509	0	423,720	55,044	7.7	2.2	13.1	81.9	17.0	2.7
1990	312,950	131	170,858	3,926	0	479,752	56,312	8.5	2.5	13.4	84.4	18.7	3.0
1991	299,915	130	120,983	2,847	0	417,922	57,524	7.3	2.1	12.8	84.4	16.6	2.5
1992	295,368	231	155,707	2,196	0	448,648	58,685	7.7	2.2	12.7	85.2	17.2	2.6
1993	302,829	0	112,871	1,607	-125	413,968	59,816	6.9	2.0	12.7	85.3	15.9	2.4
1994	305,727	0	169,825	2,190	0	473,363	60,946	7.8	2.3	14.0	87.6	16.6	2.7
1995	309,576	0	141,892	2,277	125	449,316	62,096	7.2	2.2	13.9	87.8	15.8	2.5

60 EL SALVADOR	- FOOD BALANCE SHEET OF FISH AND FISHERY PRODUCTS IN LIVE WEIGHT AND FISH CONTRIBUTION TO PROTEIN SUPPLY -	FAO-FIDI

YEAR	PRODUCTION	NON-FOOD USES	IMPORTS	EXPORTS	STOCKS VARIATIONS	TOTAL FOOD SUPPLY	POPULATION	PER CAPUT SUPPLY	FISH PROTEINS	ANIMAL PROTEINS	TOTAL PROTEINS	FISH/ANIMAL PROTEINS	FISH/TOTAL PROTEINS
	(.................................... metric tons in live weight					....)	(thousands)	(kilograms)	(grams per caput per day)			(%)	(%)
1961	8,000	0	1,200	6,200	0	4,000	2,656	1.5	0.4	13.3	45.6	3.3	1.0
1962	8,200	0	2,200	6,100	0	5,000	2,737	1.8	0.5	13.5	47.5	3.9	1.1
1963	8,400	0	2,400	5,800	0	5,200	2,821	1.8	0.5	13.7	48.0	3.9	1.1
1964	9,200	0	3,482	5,982	0	6,700	2,912	2.3	0.7	13.0	47.4	5.0	1.4
1965	9,600	0	2,600	4,343	0	7,857	3,012	2.6	0.8	13.0	47.6	5.9	1.6
1966	10,600	0	2,000	6,478	0	6,122	3,122	2.0	0.6	12.8	47.5	4.6	1.2
1967	9,900	1,000	2,000	5,600	0	5,300	3,239	1.6	0.5	12.3	46.5	3.8	1.0
1968	10,200	1,700	1,850	4,500	0	5,850	3,361	1.7	0.5	11.8	47.3	4.2	1.1
1969	10,300	900	4,250	5,100	0	8,550	3,482	2.5	0.6	11.7	46.6	5.0	1.3
1970	11,100	900	1,594	5,744	0	6,050	3,598	1.7	0.5	11.3	46.2	4.2	1.0
1971	10,700	1,000	2,189	5,000	0	6,889	3,707	1.9	0.5	11.4	47.0	4.7	1.1
1972	10,400	1,400	1,119	3,500	-1,200	5,419	3,809	1.4	0.4	11.0	46.2	3.8	0.9
1973	12,400	1,000	1,827	6,300	0	6,927	3,907	1.8	0.5	11.4	47.7	4.5	1.1
1974	9,654	1,000	1,894	3,148	-225	7,176	4,004	1.8	0.5	12.8	49.3	4.2	1.1
1975	8,730	315	2,918	3,698	-112	7,523	4,102	1.8	0.5	13.0	52.1	4.2	1.0
1976	7,201	352	3,005	3,760	300	6,394	4,203	1.5	0.4	14.6	52.7	3.0	0.8
1977	6,368	397	4,039	4,439	1,200	6,771	4,305	1.6	0.5	14.2	53.6	3.2	0.9
1978	9,540	212	4,174	5,918	0	7,584	4,401	1.7	0.5	15.1	54.5	3.3	0.9
1979	12,738	0	2,915	6,553	0	9,101	4,484	2.0	0.6	15.1	56.1	4.1	1.1
1980	13,958	380	3,037	6,889	0	9,726	4,547	2.1	0.6	15.7	57.2	4.1	1.1
1981	20,254	4,030	2,686	7,245	0	11,665	4,589	2.5	0.7	14.2	56.8	5.1	1.3
1982	13,517	830	1,790	5,702	0	8,774	4,613	1.9	0.5	12.8	55.2	4.1	1.0
1983	7,603	1,595	737	3,632	0	3,113	4,628	0.7	0.2	11.2	54.9	1.8	0.4
1984	12,189	0	3,618	6,009	-1,328	8,470	4,646	1.8	0.5	12.6	57.2	3.8	0.8
1985	16,098	0	1,447	5,507	831	12,869	4,676	2.8	0.8	13.0	57.5	6.0	1.3
1986	20,456	0	1,837	7,965	360	14,688	4,720	3.1	0.9	13.1	56.3	7.0	1.6
1987	21,544	0	3,231	6,489	-3,860	14,426	4,778	3.0	0.9	13.7	55.5	6.7	1.7
1988	11,733	40	2,887	5,702	1,002	9,880	4,850	2.0	0.6	13.9	57.1	4.3	1.1
1989	11,627	0	4,574	4,632	136	11,705	4,934	2.4	0.7	14.7	59.2	5.1	1.3
1990	9,159	0	4,336	4,396	971	10,070	5,031	2.0	0.6	14.2	60.0	4.4	1.0
1991	11,349	1	4,049	5,050	1,402	11,748	5,140	2.3	0.7	14.2	62.8	4.9	1.1
1992	12,778	0	6,145	5,194	559	14,288	5,262	2.7	0.8	14.9	62.4	5.6	1.3
1993	13,077	95	3,614	5,612	332	11,316	5,393	2.1	0.6	14.6	62.3	3.9	0.9
1994	13,066	2	5,346	6,097	12	12,325	5,528	2.2	0.6	15.1	63.1	4.0	1.0
1995	15,812	0	6,049	6,744	0	15,117	5,662	2.7	0.8	17.3	64.5	4.4	1.2

61 EQUATORIAL GUINEA	- FOOD BALANCE SHEET OF FISH AND FISHERY PRODUCTS IN LIVE WEIGHT AND FISH CONTRIBUTION TO PROTEIN SUPPLY -													FAO-FIDI
YEAR	PRODUCTION	NON-FOOD USES	IMPORTS	EXPORTS	STOCKS VARIATIONS	TOTAL FOOD SUPPLY	POPULATION	PER CAPUT SUPPLY	FISH PROTEINS	ANIMAL PROTEINS	TOTAL PROTEINS	FISH/ANIMAL PROTEINS	FISH/TOTAL PROTEINS
	(.. metric tons in live weight					....)	(thousands)	(kilograms)	(grams per caput per day)			(%)	(%)
1961	500	0	...	0	0	500	255	2.0	0.6	1.0	8.1	58.8	7.0
1962	800	0	...	0	0	800	258	3.1	0.9	1.3	9.0	67.0	9.7
1963	1,000	0	...	0	0	1,000	261	3.8	1.2	1.6	9.0	73.1	12.8
1964	1,000	0	...	0	0	1,000	265	3.8	1.1	1.6	9.2	72.2	12.4
1965	600	0	...	0	0	600	270	2.2	0.7	1.1	8.6	59.2	7.6
1966	1,200	0	...	0	0	1,200	276	4.4	1.2	1.6	8.9	72.6	12.9
1967	1,100	0	...	0	0	1,100	284	3.9	1.0	1.4	9.3	69.5	10.5
1968	3,700	0	...	0	0	3,700	291	12.7	3.2	3.6	11.7	88.5	27.5
1969	3,500	0	...	0	0	3,500	294	11.9	3.0	3.5	11.4	86.9	26.3
1970	4,000	0	...	0	0	4,000	291	13.7	3.5	4.0	12.6	88.3	27.6
1971	4,000	0	...	0	0	4,000	283	14.1	3.6	4.1	13.3	88.1	27.0
1972	4,000	0	...	0	0	4,000	270	14.8	3.8	4.3	13.1	87.5	28.8
1973	4,000	0	...	0	0	4,000	253	15.8	4.0	4.6	13.7	87.4	29.2
1974	4,000	0	...	0	0	4,000	238	16.8	4.3	4.9	15.3	87.4	28.0
1975	4,000	0	...	0	0	4,000	225	17.8	4.5	5.2	17.5	86.7	25.8
1976	4,000	0	...	0	0	4,000	216	18.6	4.7	5.4	21.6	86.6	21.9
1977	4,000	0	...	0	0	4,000	209	19.1	4.9	5.6	18.3	86.5	26.5
1978	4,000	0	...	0	0	4,000	207	19.3	4.9	5.7	21.4	86.1	23.0
1979	4,000	0	...	0	0	4,000	209	19.1	4.9	6.1	20.8	80.2	23.4
1980	2,500	0	1,970	0	0	4,470	217	20.6	5.7	9.5	23.9	60.3	23.9
1981	2,500	0	3,388	2	0	5,886	231	25.5	7.0	12.0	27.3	58.1	25.5
1982	2,019	0	4,087	3	0	6,103	251	24.4	6.7	12.0	25.5	56.1	26.3
1983	2,341	0	2,168	207	0	4,302	273	15.8	4.3	5.8	18.5	74.5	23.1
1984	4,000	0	2,112	194	0	5,918	294	20.1	5.4	7.2	18.1	75.7	30.1
1985	3,600	0	5,616	149	0	9,067	312	29.1	8.1	10.4	24.1	77.6	33.6
1986	4,400	0	2,563	0	0	6,963	324	21.5	5.9	8.0	20.8	73.8	28.5
1987	4,000	0	5,044	0	0	9,044	333	27.2	7.5	9.3	22.0	80.9	34.2
1988	4,000	0	4,248	0	0	8,248	339	24.3	6.8	9.8	24.1	69.8	28.3
1989	4,000	0	4,514	257	0	8,257	345	24.0	6.7	9.0	23.5	74.7	28.7
1990	3,700	0	5,688	40	0	9,348	352	26.6	7.5	10.9	25.4	68.5	29.5
1991	3,500	0	4,784	0	0	8,284	360	23.0	6.5	14.1	26.5	45.8	24.4
1992	3,600	0	4,750	0	0	8,350	369	22.6	6.4	13.0	24.8	49.1	25.7
1993	3,800	0	4,995	0	0	8,795	379	23.2	6.6	12.8	23.3	51.0	28.1
1994	3,700	0	5,040	0	0	8,740	389	22.5	6.4	8.9	18.3	71.6	34.8
1995	3,380	0	5,886	120	0	9,146	400	22.9	6.6	9.9	19.1	66.9	34.7

178 ERITREA	- FOOD BALANCE SHEET OF FISH AND FISHERY PRODUCTS IN LIVE WEIGHT AND FISH CONTRIBUTION TO PROTEIN SUPPLY -	FAO-FIDI

YEAR	PRODUCTION	NON-FOOD USES	IMPORTS	EXPORTS	STOCKS VARIATIONS	TOTAL FOOD SUPPLY	POPULATION	PER CAPUT SUPPLY	FISH PROTEINS	ANIMAL PROTEINS	TOTAL PROTEINS	FISH/ANIMAL PROTEINS	FISH/TOTAL PROTEINS
	(...................................... metric tons in live weight					....)	(thousands)	(kilograms)	(grams per caput per day)			(%)	(%)
1961	0	0	0	0	0	0	1,455	0.0	0.0	0.0	0.0	0.0	0.0
1962	0	0	0	0	0	0	1,491	0.0	0.0	0.0	0.0	0.0	0.0
1963	0	0	0	0	0	0	1,529	0.0	0.0	0.0	0.0	0.0	0.0
1964	0	0	0	0	0	0	1,568	0.0	0.0	0.0	0.0	0.0	0.0
1965	0	0	0	0	0	0	1,608	0.0	0.0	0.0	0.0	0.0	0.0
1966	0	0	0	0	0	0	1,650	0.0	0.0	0.0	0.0	0.0	0.0
1967	0	0	0	0	0	0	1,693	0.0	0.0	0.0	0.0	0.0	0.0
1968	0	0	0	0	0	0	1,738	0.0	0.0	0.0	0.0	0.0	0.0
1969	0	0	0	0	0	0	1,784	0.0	0.0	0.0	0.0	0.0	0.0
1970	0	0	0	0	0	0	1,831	0.0	0.0	0.0	0.0	0.0	0.0
1971	0	0	0	0	0	0	1,880	0.0	0.0	0.0	0.0	0.0	0.0
1972	0	0	0	0	0	0	1,930	0.0	0.0	0.0	0.0	0.0	0.0
1973	0	0	0	0	0	0	1,981	0.0	0.0	0.0	0.0	0.0	0.0
1974	0	0	0	0	0	0	2,034	0.0	0.0	0.0	0.0	0.0	0.0
1975	0	0	0	0	0	0	2,089	0.0	0.0	0.0	0.0	0.0	0.0
1976	0	0	0	0	0	0	2,145	0.0	0.0	0.0	0.0	0.0	0.0
1977	0	0	0	0	0	0	2,203	0.0	0.0	0.0	0.0	0.0	0.0
1978	0	0	0	0	0	0	2,262	0.0	0.0	0.0	0.0	0.0	0.0
1979	0	0	0	0	0	0	2,321	0.0	0.0	0.0	0.0	0.0	0.0
1980	0	0	0	0	0	0	2,382	0.0	0.0	0.0	0.0	0.0	0.0
1981	0	0	0	0	0	0	2,442	0.0	0.0	0.0	0.0	0.0	0.0
1982	0	0	0	0	0	0	2,503	0.0	0.0	0.0	0.0	0.0	0.0
1983	0	0	0	0	0	0	2,566	0.0	0.0	0.0	0.0	0.0	0.0
1984	0	0	0	0	0	0	2,631	0.0	0.0	0.0	0.0	0.0	0.0
1985	0	0	0	0	0	0	2,698	0.0	0.0	0.0	0.0	0.0	0.0
1986	0	0	0	0	0	0	2,769	0.0	0.0	0.0	0.0	0.0	0.0
1987	0	0	0	0	0	0	2,843	0.0	0.0	0.0	0.0	0.0	0.0
1988	0	0	0	0	0	0	2,920	0.0	0.0	0.0	0.0	0.0	0.0
1989	0	0	0	0	0	0	2,844	0.0	0.0	0.0	0.0	0.0	0.0
1990	0	0	0	0	0	0	2,881	0.0	0.0	0.0	0.0	0.0	0.0
1991	0	0	0	0	0	0	2,921	0.0	0.0	0.0	0.0	0.0	0.0
1992	0	0	0	0	0	0	3,255	0.0	0.0	0.0	0.0	0.0	0.0
1993	417	0	0	0	0	417	3,017	0.1	0.0	5.7	47.6	0.7	0.1
1994	3,018	0	0	0	0	3,018	3,084	1.0	0.2	6.4	55.2	3.8	0.4
1995	3,826	0	0	0	0	3,826	3,171	1.2	0.3	6.3	50.0	5.1	0.6

Note: for years before 1993 see Ethiopia, P.D.R.

63 ESTONIA	- FOOD BALANCE SHEET OF FISH AND FISHERY PRODUCTS IN LIVE WEIGHT AND FISH CONTRIBUTION TO PROTEIN SUPPLY -												FAO-FIDI
YEAR	PRODUCTION	NON-FOOD USES	IMPORTS	EXPORTS	STOCKS VARIATIONS	TOTAL FOOD SUPPLY	POPULATION	PER CAPUT SUPPLY	FISH PROTEINS	ANIMAL PROTEINS	TOTAL PROTEINS	FISH/ANIMAL PROTEINS	FISH/TOTAL PROTEINS
	(.................. metric tons in live weight ....)						(thousands)	(kilograms)	(grams per caput per day)			(%)	(%)
1961	0	0	0	0	0	0	1,229	0.0	0.0	0.0	0.0	0.0	0.0
1962	0	0	0	0	0	0	1,243	0.0	0.0	0.0	0.0	0.0	0.0
1963	0	0	0	0	0	0	1,258	0.0	0.0	0.0	0.0	0.0	0.0
1964	0	0	0	0	0	0	1,276	0.0	0.0	0.0	0.0	0.0	0.0
1965	0	0	0	0	0	0	1,291	0.0	0.0	0.0	0.0	0.0	0.0
1966	0	0	0	0	0	0	1,303	0.0	0.0	0.0	0.0	0.0	0.0
1967	0	0	0	0	0	0	1,314	0.0	0.0	0.0	0.0	0.0	0.0
1968	0	0	0	0	0	0	1,327	0.0	0.0	0.0	0.0	0.0	0.0
1969	0	0	0	0	0	0	1,345	0.0	0.0	0.0	0.0	0.0	0.0
1970	0	0	0	0	0	0	1,365	0.0	0.0	0.0	0.0	0.0	0.0
1971	0	0	0	0	0	0	1,383	0.0	0.0	0.0	0.0	0.0	0.0
1972	0	0	0	0	0	0	1,398	0.0	0.0	0.0	0.0	0.0	0.0
1973	0	0	0	0	0	0	1,411	0.0	0.0	0.0	0.0	0.0	0.0
1974	0	0	0	0	0	0	1,423	0.0	0.0	0.0	0.0	0.0	0.0
1975	0	0	0	0	0	0	1,432	0.0	0.0	0.0	0.0	0.0	0.0
1976	0	0	0	0	0	0	1,441	0.0	0.0	0.0	0.0	0.0	0.0
1977	0	0	0	0	0	0	1,451	0.0	0.0	0.0	0.0	0.0	0.0
1978	0	0	0	0	0	0	1,461	0.0	0.0	0.0	0.0	0.0	0.0
1979	0	0	0	0	0	0	1,469	0.0	0.0	0.0	0.0	0.0	0.0
1980	0	0	0	0	0	0	1,480	0.0	0.0	0.0	0.0	0.0	0.0
1981	0	0	0	0	0	0	1,490	0.0	0.0	0.0	0.0	0.0	0.0
1982	0	0	0	0	0	0	1,501	0.0	0.0	0.0	0.0	0.0	0.0
1983	0	0	0	0	0	0	1,513	0.0	0.0	0.0	0.0	0.0	0.0
1984	0	0	0	0	0	0	1,525	0.0	0.0	0.0	0.0	0.0	0.0
1985	0	0	0	0	0	0	1,536	0.0	0.0	0.0	0.0	0.0	0.0
1986	0	0	0	0	0	0	1,547	0.0	0.0	0.0	0.0	0.0	0.0
1987	0	0	0	0	0	0	1,558	0.0	0.0	0.0	0.0	0.0	0.0
1988	0	0	0	0	0	0	1,568	0.0	0.0	0.0	0.0	0.0	0.0
1989	0	0	0	0	0	0	1,570	0.0	0.0	0.0	0.0	0.0	0.0
1990	0	0	0	0	0	0	1,571	0.0	0.0	0.0	0.0	0.0	0.0
1991	358,147	...	...	...	-60,000	...	1,564	...	...	...	...	...	...
1992	132,040	...	2,800	76,409	0	58,431	1,548	37.7	12.3	72.9	103.2	16.9	11.9
1993	146,884	...	8,840	103,936	-20	51,769	1,528	33.9	10.7	63.3	96.8	16.9	11.1
1994	124,097	...	40,005	144,090	30,000	50,012	1,507	33.2	10.0	62.2	96.6	16.1	10.4
1995	132,343	...	33,354	144,126	30,020	51,591	1,488	34.7	8.8	56.1	93.1	15.6	9.4

Note: 1991 data of apparent consumption not shown as trade statistics not available. For years before 1991 see USSR.

62 ETHIOPIA, P.D.R.	- FOOD BALANCE SHEET OF FISH AND FISHERY PRODUCTS IN LIVE WEIGHT AND FISH CONTRIBUTION TO PROTEIN SUPPLY -											FAO-FIDI
YEAR	PRODUCTION	NON-FOOD USES	IMPORTS	EXPORTS	STOCKS VARIATIONS	TOTAL FOOD SUPPLY	POPULATION	PER CAPUT SUPPLY	FISH PROTEINS	ANIMAL PROTEINS	TOTAL PROTEINS	FISH/ANIMAL PROTEINS	FISH/TOTAL PROTEINS
	(.. metric tons in live weight					....)	(thousands)	(kilograms)	(grams per caput per day)			(%)	(%)
1961	8,500	4,000	...	3,000	0	1,500	24,743	0.1	0.0	12.0	61.5	0.2	0.0
1962	9,000	4,000	...	2,100	0	2,900	25,314	0.1	0.0	12.0	61.4	0.3	0.1
1963	8,900	4,000	...	3,000	0	1,900	25,903	0.1	0.0	11.6	61.1	0.2	0.0
1964	11,700	7,500	...	1,800	0	2,400	26,515	0.1	0.0	11.5	61.2	0.2	0.0
1965	15,700	5,000	400	2,900	0	8,206	27,150	0.3	0.1	11.5	61.5	0.8	0.2
1966	16,000	4,400	600	4,173	0	8,027	27,811	0.3	0.1	11.5	58.9	0.8	0.2
1967	16,000	4,400	600	3,300	0	8,900	28,496	0.3	0.1	11.4	60.4	1.0	0.2
1968	14,400	800	600	3,000	0	11,200	29,199	0.4	0.1	11.3	57.5	1.0	0.2
1969	14,200	600	600	3,900	0	10,300	29,910	0.3	0.1	11.2	59.1	0.9	0.2
1970	17,200	4,000	877	5,962	0	8,115	30,622	0.3	0.1	11.2	59.0	0.6	0.1
1971	19,100	2,500	400	4,979	-4,736	7,286	31,333	0.2	0.1	10.0	57.3	0.6	0.1
1972	14,800	1,500	600	7,400	0	6,500	32,047	0.2	0.1	9.7	53.5	0.5	0.1
1973	4,700	2,000	1,400	1,250	3,000	5,850	32,773	0.2	0.1	9.5	52.4	0.5	0.1
1974	5,900	500	1,100	500	-1,250	4,750	33,524	0.1	0.0	9.5	50.1	0.4	0.1
1975	3,930	0	1,150	2,950	1,250	3,380	34,309	0.1	0.0	9.1	49.8	0.3	0.1
1976	2,970	0	681	340	500	3,811	35,137	0.1	0.0	9.0	51.6	0.4	0.1
1977	2,472	0	1,009	210	500	3,771	36,006	0.1	0.0	9.1	50.4	0.3	0.1
1978	2,515	0	680	0	250	3,445	36,906	0.1	0.0	9.1	54.0	0.3	0.1
1979	2,800	0	53	0	0	2,853	37,823	0.1	0.0	9.1	61.3	0.3	0.0
1980	3,507	0	8	0	0	3,515	38,749	0.1	0.0	9.0	60.2	0.3	0.1
1981	3,828	0	136	11	0	3,953	39,680	0.1	0.0	9.0	58.6	0.3	0.1
1982	3,750	0	4	12	0	3,742	40,626	0.1	0.0	8.9	58.9	0.3	0.1
1983	3,900	0	6	0	0	3,906	41,612	0.1	0.0	8.8	55.7	0.3	0.1
1984	4,300	0	28	30	0	4,298	42,673	0.1	0.0	8.4	48.1	0.4	0.1
1985	4,000	0	163	32	0	4,131	43,833	0.1	0.0	9.1	48.5	0.3	0.1
1986	4,100	0	572	30	0	4,642	45,104	0.1	0.0	10.2	54.0	0.3	0.1
1987	4,000	0	4	0	0	4,004	46,478	0.1	0.0	7.8	49.8	0.3	0.1
1988	4,067	0	56	6	0	4,117	47,938	0.1	0.0	7.9	52.0	0.3	0.1
1989	4,269	0	471	36	0	4,704	49,458	0.1	0.0	7.9	52.0	0.4	0.1
1990	4,981	0	40	33	0	4,988	51,020	0.1	0.0	8.0	49.6	0.4	0.1
1991	4,262	0	7	2	0	4,267	52,617	0.1	0.0	7.6	51.4	0.3	0.1
1992	4,607	0	47	0	0	4,654	54,256	0.1	0.0	7.4	50.6	0.4	0.1

Note: data refer to the area that was formerly the People's Democratic Republic of Ethiopia. The new independent Republics are Ethiopia and Eritrea, information for which is shown separately.

238 ETHIOPIA	- FOOD BALANCE SHEET OF FISH AND FISHERY PRODUCTS IN LIVE WEIGHT AND FISH CONTRIBUTION TO PROTEIN SUPPLY -	FAO-FIDI

YEAR	PRODUCTION	NON-FOOD USES	IMPORTS	EXPORTS	STOCKS VARIATIONS	TOTAL FOOD SUPPLY	POPULATION	PER CAPUT SUPPLY	FISH PROTEINS	ANIMAL PROTEINS	TOTAL PROTEINS	FISH/ANIMAL PROTEINS	FISH/TOTAL PROTEINS
	(.....................................	metric tons in live weight				.....................................)	(thousands)	(kilograms)		(grams per caput per day)		(%)	(%)
1961	0	0	0	0	0	0	23,289	0.0	0.0	0.0	0.0	0.0	0.0
1962	0	0	0	0	0	0	23,823	0.0	0.0	0.0	0.0	0.0	0.0
1963	0	0	0	0	0	0	24,375	0.0	0.0	0.0	0.0	0.0	0.0
1964	0	0	0	0	0	0	24,947	0.0	0.0	0.0	0.0	0.0	0.0
1965	0	0	0	0	0	0	25,542	0.0	0.0	0.0	0.0	0.0	0.0
1966	0	0	0	0	0	0	26,161	0.0	0.0	0.0	0.0	0.0	0.0
1967	0	0	0	0	0	0	26,803	0.0	0.0	0.0	0.0	0.0	0.0
1968	0	0	0	0	0	0	27,462	0.0	0.0	0.0	0.0	0.0	0.0
1969	0	0	0	0	0	0	28,126	0.0	0.0	0.0	0.0	0.0	0.0
1970	0	0	0	0	0	0	28,791	0.0	0.0	0.0	0.0	0.0	0.0
1971	0	0	0	0	0	0	29,454	0.0	0.0	0.0	0.0	0.0	0.0
1972	0	0	0	0	0	0	30,117	0.0	0.0	0.0	0.0	0.0	0.0
1973	0	0	0	0	0	0	30,791	0.0	0.0	0.0	0.0	0.0	0.0
1974	0	0	0	0	0	0	31,489	0.0	0.0	0.0	0.0	0.0	0.0
1975	0	0	0	0	0	0	32,221	0.0	0.0	0.0	0.0	0.0	0.0
1976	0	0	0	0	0	0	32,991	0.0	0.0	0.0	0.0	0.0	0.0
1977	0	0	0	0	0	0	33,799	0.0	0.0	0.0	0.0	0.0	0.0
1978	0	0	0	0	0	0	34,637	0.0	0.0	0.0	0.0	0.0	0.0
1979	0	0	0	0	0	0	35,495	0.0	0.0	0.0	0.0	0.0	0.0
1980	0	0	0	0	0	0	36,368	0.0	0.0	0.0	0.0	0.0	0.0
1981	0	0	0	0	0	0	37,252	0.0	0.0	0.0	0.0	0.0	0.0
1982	0	0	0	0	0	0	38,155	0.0	0.0	0.0	0.0	0.0	0.0
1983	0	0	0	0	0	0	39,091	0.0	0.0	0.0	0.0	0.0	0.0
1984	0	0	0	0	0	0	40,080	0.0	0.0	0.0	0.0	0.0	0.0
1985	0	0	0	0	0	0	41,137	0.0	0.0	0.0	0.0	0.0	0.0
1986	0	0	0	0	0	0	42,265	0.0	0.0	0.0	0.0	0.0	0.0
1987	0	0	0	0	0	0	43,463	0.0	0.0	0.0	0.0	0.0	0.0
1988	0	0	0	0	0	0	44,727	0.0	0.0	0.0	0.0	0.0	0.0
1989	0	0	0	0	0	0	46,614	0.0	0.0	0.0	0.0	0.0	0.0
1990	0	0	0	0	0	0	48,140	0.0	0.0	0.0	0.0	0.0	0.0
1991	0	0	0	0	0	0	49,696	0.0	0.0	0.0	0.0	0.0	0.0
1992	0	0	0	0	0	0	50,329	0.0	0.0	0.0	0.0	0.0	0.0
1993	4,203	0	55	0	0	4,258	52,934	0.1	0.0	7.3	52.1	0.3	0.1
1994	5,318	0	96	0	0	5,414	54,635	0.1	0.0	7.1	54.9	0.4	0.1
1995	6,380	0	237	18	0	6,599	56,404	0.1	0.0	6.9	56.9	0.5	0.1

Note: for years before 1993 see Ethiopia, P.D.R.

64 FAEROE ISLANDS	- FOOD BALANCE SHEET OF FISH AND FISHERY PRODUCTS IN LIVE WEIGHT AND FISH CONTRIBUTION TO PROTEIN SUPPLY -	FAO-FIDI

YEAR	PRODUCTION	NON-FOOD USES	IMPORTS	EXPORTS	STOCKS VARIATIONS	TOTAL FOOD SUPPLY	POPULATION	PER CAPUT SUPPLY	FISH PROTEINS	ANIMAL PROTEINS	TOTAL PROTEINS	FISH/ANIMAL PROTEINS	FISH/TOTAL PROTEINS
	(.......................... metric tons in live weight ....)						(thousands)	(kilograms)	(gram's per caput per day)			(%)	(%)
1961	120,100	400	400	103,782	-13,462	2,856	35	82.6	19.9	35.2	66.8	56.4	29.7
1962	143,500	1,400	1,900	144,647	3,487	2,840	35	80.7	18.7	37.4	74.2	50.0	25.2
1963	137,000	400	700	132,742	-1,669	2,889	36	80.6	21.4	41.4	76.5	51.6	27.9
1964	139,100	300	900	127,737	-8,955	3,008	36	82.5	17.2	38.0	65.7	45.4	26.2
1965	144,700	12,100	1,900	139,806	8,310	3,004	37	81.2	19.2	42.5	69.2	45.1	27.7
1966	165,600	34,501	1,904	109,861	-20,053	3,090	37	82.4	13.3	42.9	76.2	31.0	17.5
1967	173,300	58,089	200	116,413	4,200	3,198	38	84.4	12.3	43.9	73.5	27.9	16.7
1968	166,200	58,700	1,900	104,685	-1,434	3,280	38	85.7	12.7	46.1	78.1	27.6	16.3
1969	176,300	54,950	1,829	129,828	9,987	3,338	39	86.4	12.2	45.0	75.0	27.2	16.3
1970	207,800	73,942	900	138,742	7,306	3,323	39	85.2	13.1	46.8	76.5	27.9	17.1
1971	207,100	79,500	7,518	123,458	-8,296	3,364	39	85.4	16.4	52.9	83.6	31.1	19.7
1972	208,100	71,550	7,848	139,321	-1,669	3,408	40	85.7	19.4	55.2	84.1	35.2	23.1
1973	246,300	116,450	10,200	145,727	9,088	3,410	40	84.8	14.3	49.8	87.7	28.7	16.3
1974	246,635	119,729	9,500	130,437	-2,522	3,447	41	84.9	19.5	58.3	94.5	33.5	20.7
1975	285,618	140,986	2,600	146,684	2,909	3,457	41	84.3	25.8	61.0	100.6	42.2	25.6
1976	341,962	191,574	4,039	150,767	-178	3,483	41	84.2	20.7	60.3	96.9	34.3	21.3
1977	310,490	127,095	244	182,560	2,443	3,523	42	84.4	18.7	61.2	99.9	30.6	18.8
1978	316,675	133,208	4,028	161,857	-22,091	3,546	42	84.3	19.9	61.4	95.6	32.5	20.8
1979	266,745	110,328	1,618	153,389	-1,028	3,618	43	85.1	21.1	64.7	104.4	32.6	20.2
1980	274,652	114,311	2,209	165,759	6,890	3,681	43	85.6	24.7	67.8	103.4	36.4	23.9
1981	243,343	102,328	6,390	165,648	21,954	3,712	44	85.2	19.7	61.3	98.3	32.1	20.0
1982	250,878	91,476	5,433	162,811	1,700	3,723	44	84.2	19.7	59.6	97.9	33.0	20.1
1983	329,955	129,702	5,886	203,219	806	3,726	45	83.0	16.8	55.0	93.6	30.5	18.0
1984	347,352	135,730	4,652	212,608	138	3,802	46	83.6	20.2	59.6	94.7	33.8	21.3
1985	373,958	159,431	2,727	212,676	-700	3,878	46	84.3	24.7	63.8	98.8	38.7	25.0
1986	352,195	163,716	7,878	192,422	0	3,935	46	84.9	21.6	63.0	107.9	34.3	20.0
1987	391,031	200,631	7,190	193,607	0	3,983	47	85.4	22.2	63.5	105.2	34.9	21.1
1988	359,901	160,179	7,427	200,519	-2,575	4,055	47	86.6	22.9	61.4	97.3	37.3	23.5
1989	308,602	94,396	5,577	219,577	3,837	4,044	47	86.2	25.6	66.1	100.3	38.7	25.5
1990	285,870	59,065	9,958	232,759	0	4,005	47	85.2	22.0	61.7	93.8	35.6	23.4
1991	246,599	34,572	12,115	219,181	-964	3,997	47	85.0	22.5	62.0	94.3	36.3	23.9
1992	267,528	71,514	16,288	206,294	-2,000	4,008	47	85.3	29.6	68.5	100.4	43.3	29.5
1993	261,274	80,560	12,319	188,314	-697	4,021	47	85.7	28.1	61.3	92.3	45.8	30.4
1994	252,177	70,951	14,263	190,447	-1,023	4,019	47	85.6	28.0	59.5	91.5	47.1	30.6
1995	297,375	60,021	14,635	246,645	-1,277	4,067	47	86.5	23.2	55.1	84.3	42.1	27.5

65 FALKLAND ISLANDS (MALVINAS)	- FOOD BALANCE SHEET OF FISH AND FISHERY PRODUCTS IN LIVE WEIGHT AND FISH CONTRIBUTION TO PROTEIN SUPPLY -											FAO-FIDI
YEAR	PRODUCTION	NON-FOOD USES	IMPORTS	EXPORTS	STOCKS VARIATIONS	TOTAL FOOD SUPPLY	POPULATION	PER CAPUT SUPPLY	FISH PROTEINS	ANIMAL PROTEINS	TOTAL PROTEINS	FISH/ANIMAL PROTEINS	FISH/TOTAL PROTEINS
	(....................................... metric tons in live weight ....)						(thousands)	(kilograms)	(grams per caput per day)			(%)	(%)
1961	0	0	0	0	0	0	2	0.0	0.0	0.0	0.0	0.0	0.0
1962	0	0	0	0	0	0	2	0.0	0.0	0.0	0.0	0.0	0.0
1963	0	0	0	0	0	0	2	0.0	0.0	0.0	0.0	0.0	0.0
1964	0	0	0	0	0	0	2	0.0	0.0	0.0	0.0	0.0	0.0
1965	0	0	0	0	0	0	2	0.0	0.0	0.0	0.0	0.0	0.0
1966	0	0	0	0	0	0	2	0.0	0.0	0.0	0.0	0.0	0.0
1967	0	0	0	0	0	0	2	0.0	0.0	0.0	0.0	0.0	0.0
1968	0	0	0	0	0	0	2	0.0	0.0	0.0	0.0	0.0	0.0
1969	0	0	0	0	0	0	2	0.0	0.0	0.0	0.0	0.0	0.0
1970	0	0	0	0	0	0	2	0.0	0.0	0.0	0.0	0.0	0.0
1971	0	0	0	0	0	0	2	0.0	0.0	0.0	0.0	0.0	0.0
1972	0	0	0	0	0	0	2	0.0	0.0	0.0	0.0	0.0	0.0
1973	0	0	0	0	0	0	2	0.0	0.0	0.0	0.0	0.0	0.0
1974	0	0	0	0	0	0	2	0.0	0.0	0.0	0.0	0.0	0.0
1975	0	0	0	0	0	0	2	0.0	0.0	0.0	0.0	0.0	0.0
1976	0	0	0	0	0	0	2	0.0	0.0	0.0	0.0	0.0	0.0
1977	0	0	0	0	0	0	2	0.0	0.0	0.0	0.0	0.0	0.0
1978	0	0	0	0	0	0	2	0.0	0.0	0.0	0.0	0.0	0.0
1979	0	0	0	0	0	0	2	0.0	0.0	0.0	0.0	0.0	0.0
1980	0	0	0	0	0	0	2	0.0	0.0	0.0	0.0	0.0	0.0
1981	0	0	0	0	0	0	2	0.0	0.0	0.0	0.0	0.0	0.0
1982	0	0	0	0	0	0	2	0.0	0.0	0.0	0.0	0.0	0.0
1983	0	0	0	0	0	0	2	0.0	0.0	0.0	0.0	0.0	0.0
1984	0	0	0	0	0	0	2	0.0	0.0	0.0	0.0	0.0	0.0
1985	8	0	0	0	0	8	2	4.0	1.0		...	...	...
1986	8	0	0	0	0	8	2	4.0	1.0	...	...	...	...
1987	23	0	0	0	0	23	2	11.5	2.8	...	...	...	...
1988	2,628	0	0	2,598	0	30	2	15.0	3.9	...	...	...	...
1989	4,624	0	25	4,622	0	27	2	13.6	2.5	...	...	...	...
1990	5,994	0	19	5,982	0	31	2	15.3	3.6	...	...	...	...
1991	1,476	0	23	1,472	0	27	2	13.7	2.9	...	...	...	
1992	1,855	0	14	1,824	0	45	2	22.5	5.7	...	...	...	
1993	1,985	55	23	1,904	0	49	2	24.5	6.2	...	...	...	...
1994	8,769	75	0	8,630	0	64	2	32.0	8.9	...	...	...	...
1995	29,065	490	118	28,567	0	126	2	63.2	16.8	...	...	...	...

66 FIJI	- FOOD BALANCE SHEET OF FISH AND FISHERY PRODUCTS IN LIVE WEIGHT AND FISH CONTRIBUTION TO PROTEIN SUPPLY -											FAO-FIDI
YEAR	PRODUCTION	NON-FOOD USES	IMPORTS	EXPORTS	STOCKS VARIATIONS	TOTAL FOOD SUPPLY	POPULATION	PER CAPUT SUPPLY	FISH PROTEINS	ANIMAL PROTEINS	TOTAL PROTEINS	FISH/ANIMAL PROTEINS	FISH/TOTAL PROTEINS
	(............ metric tons in live weight ....)						(thousands)	(kilograms)	(grams per caput per day)			(%)	(%)
1961	3,000	0	3,400	...	0	6,400	408	15.7	4.5	13.8	51.2	32.5	8.8
1962	3,000	0	2,900	...	0	5,900	422	14.0	4.0	13.6	52.1	29.4	7.7
1963	3,200	0	4,300	...	0	7,500	436	17.2	5.0	15.9	55.2	31.2	9.0
1964	3,200	0	7,200	2,000	0	8,400	451	18.6	5.3	18.8	58.8	28.2	9.0
1965	3,300	0	8,200	3,700	0	7,800	464	16.8	4.8	17.4	57.7	27.8	8.4
1966	3,300	0	10,600	6,200	0	7,700	476	16.2	4.6	16.9	55.7	27.6	8.3
1967	3,300	0	12,500	6,000	0	9,800	488	20.1	5.9	17.7	55.2	33.0	10.6
1968	3,500	0	11,700	5,500	0	9,700	499	19.5	5.6	16.9	54.2	33.4	10.4
1969	3,500	0	15,100	8,800	0	9,800	509	19.2	5.6	18.1	54.5	30.8	10.3
1970	3,600	0	14,633	8,578	0	9,656	520	18.6	5.4	18.9	52.7	28.4	10.2
1971	3,600	0	17,556	9,756	0	11,400	531	21.5	6.1	19.4	55.0	31.4	11.1
1972	4,200	0	19,333	12,000	0	11,533	542	21.3	6.2	19.6	54.8	31.5	11.3
1973	4,100	0	18,144	8,156	0	14,089	554	25.5	7.3	22.3	57.7	32.8	12.7
1974	4,529	0	20,044	5,200	-6,000	13,373	565	23.7	6.7	21.4	56.6	31.4	11.9
1975	4,818	0	8,469	2,938	4,000	14,349	576	24.9	7.1	23.9	58.0	29.5	12.2
1976	5,186	650	12,570	4,538	1,959	14,527	587	24.7	7.1	22.1	55.8	32.1	12.7
1977	7,583	1,520	16,102	8,594	-1,955	11,615	598	19.4	5.2	22.8	55.1	22.9	9.5
1978	8,974	2,200	20,190	12,577	0	14,386	609	23.6	6.2	24.4	56.3	25.4	11.0
1979	20,227	5,380	16,389	12,908	2,000	20,328	621	32.7	8.7	26.3	59.6	32.9	14.5
1980	24,195	7,280	18,821	10,781	0	24,955	634	39.4	11.8	28.6	64.1	41.2	18.4
1981	24,225	3,200	17,612	12,283	0	26,353	647	40.7	11.1	29.0	63.4	38.4	17.5
1982	28,017	9,580	14,319	6,286	0	26,469	662	40.0	10.9	28.7	64.4	38.0	16.9
1983	27,838	3,710	11,773	10,797	0	25,105	676	37.1	9.4	29.1	64.4	32.4	14.6
1984	27,914	3,100	13,234	9,963	0	28,086	689	40.8	10.0	29.0	64.4	34.6	15.6
1985	27,630	8,960	17,035	7,663	0	28,042	699	40.1	10.4	27.6	62.2	37.6	16.7
1986	27,009	7,800	20,272	10,108	0	29,373	707	41.6	11.0	30.2	66.9	36.4	16.4
1987	35,275	10,100	15,903	12,508	0	28,569	712	40.2	10.4	29.3	65.1	35.4	15.9
1988	32,455	12,600	28,149	18,474	0	29,530	715	41.3	11.4	29.9	66.7	38.2	17.1
1989	32,791	10,600	25,191	18,576	0	28,806	720	40.0	10.3	29.1	67.4	35.5	15.3
1990	35,048	6,400	19,518	19,239	0	28,927	726	39.8	9.8	32.6	70.0	29.9	13.9
1991	31,095	8,500	26,103	19,694	0	29,004	735	39.5	10.6	33.2	70.7	32.0	15.0
1992	31,351	6,500	14,646	13,226	0	26,271	746	35.2	9.1	32.2	68.6	28.3	13.3
1993	31,399	1,450	24,975	24,567	0	30,356	759	40.0	11.1	35.4	75.1	31.4	14.8
1994	32,000	50	16,123	24,087	0	23,986	771	31.1	7.4	29.7	72.3	24.9	10.2
1995	34,577	3,180	19,324	22,530	0	28,191	784	36.0	9.1	32.3	72.9	28.1	12.5

67 FINLAND	- FOOD BALANCE SHEET OF FISH AND FISHERY PRODUCTS IN LIVE WEIGHT AND FISH CONTRIBUTION TO PROTEIN SUPPLY -	FAO-FIDI

YEAR	PRODUCTION	NON-FOOD USES	IMPORTS	EXPORTS	STOCKS VARIATIONS	TOTAL FOOD SUPPLY	POPULATION	PER CAPUT SUPPLY	FISH PROTEINS	ANIMAL PROTEINS	TOTAL PROTEINS	FISH/ANIMAL PROTEINS	FISH/TOTAL PROTEINS
	(..		metric tons in live weight		....)		(thousands)	(kilograms)		(grams per caput per day)		(%)	(%)
1961	67,000	11,876	22,821	0	0	77,945	4,463	17.5	5.4	53.4	95.5	10.1	5.6
1962	64,900	10,030	31,580	0	0	86,449	4,494	19.2	5.7	52.7	92.3	10.8	6.2
1963	80,500	8,997	25,618	0	0	97,121	4,523	21.5	6.5	54.7	94.3	11.9	6.9
1964	62,900	8,194	29,395	0	0	84,101	4,546	18.5	5.5	54.5	91.3	10.1	6.0
1965	73,200	25,703	35,261	0	0	82,758	4,564	18.1	5.2	54.6	91.3	9.6	5.7
1966	70,800	26,015	34,975	90	0	79,670	4,576	17.4	5.1	55.0	91.3	9.3	5.6
1967	73,700	25,904	29,372	180	0	76,988	4,583	16.8	4.9	55.0	89.5	8.9	5.5
1968	92,200	26,006	34,749	180	0	100,764	4,588	22.0	6.6	55.4	89.8	11.9	7.3
1969	86,500	26,604	37,086	288	0	96,695	4,595	21.0	6.1	56.1	91.2	10.9	6.7
1970	80,700	19,100	40,355	330	0	101,626	4,606	22.1	6.4	54.1	88.7	11.7	7.2
1971	90,200	21,800	36,874	339	0	104,936	4,623	22.7	6.6	56.8	89.7	11.7	7.4
1972	87,600	26,500	42,237	477	0	102,860	4,644	22.2	6.3	57.0	90.2	11.1	7.0
1973	100,000	31,400	46,631	692	0	114,539	4,669	24.5	7.0	59.3	92.2	11.8	7.6
1974	110,127	29,900	41,911	475	-250	121,413	4,692	25.9	7.5	58.7	91.7	12.8	8.2
1975	109,808	31,400	42,954	662	0	120,700	4,711	25.6	7.3	61.3	94.5	11.9	7.7
1976	117,235	43,517	55,737	729	-250	128,476	4,727	27.2	7.4	60.4	92.7	12.3	8.0
1977	115,595	30,006	47,249	905	-125	131,809	4,739	27.8	7.9	60.3	91.2	13.1	8.7
1978	139,166	54,008	47,504	2,704	0	129,958	4,750	27.4	7.6	60.7	93.6	12.6	8.1
1979	133,928	46,014	45,898	2,710	-250	130,852	4,763	27.5	7.7	61.4	94.6	12.6	8.2
1980	172,581	84,024	51,581	5,571	0	134,567	4,780	28.2	7.9	61.1	94.5	12.9	8.3
1981	164,439	70,040	50,375	8,017	-95	136,662	4,801	28.5	7.9	60.0	93.1	13.2	8.5
1982	173,125	80,275	51,026	6,331	-250	137,295	4,827	28.4	7.9	62.3	95.9	12.7	8.3
1983	184,598	81,313	52,612	11,251	325	144,970	4,853	29.9	8.5	59.0	90.7	14.3	9.3
1984	189,984	87,657	55,921	9,552	845	149,540	4,879	30.7	8.5	60.4	94.0	14.1	9.0
1985	190,591	89,255	59,212	7,979	30	152,599	4,902	31.1	8.7	59.3	92.6	14.7	9.4
1986	171,828	79,264	66,333	2,669	-91	156,137	4,921	31.7	8.7	60.6	93.7	14.4	9.3
1987	170,667	79,605	70,481	2,928	-259	158,357	4,937	32.1	8.8	61.0	94.0	14.5	9.4
1988	179,609	81,850	69,402	5,686	-210	161,266	4,952	32.6	9.1	63.6	96.6	14.3	9.4
1989	169,086	77,363	73,665	3,988	-62	161,338	4,968	32.5	8.9	63.9	98.7	13.9	9.0
1990	160,444	61,776	69,590	4,569	77	163,766	4,986	32.8	9.0	62.2	96.3	14.5	9.4
1991	147,440	52,188	70,176	2,144	112	163,395	5,009	32.6	8.9	61.8	95.4	14.5	9.4
1992	169,662	69,337	68,110	3,151	-438	164,846	5,034	32.8	9.2	62.2	96.1	14.9	9.6
1993	173,820	76,432	69,281	3,179	-96	163,393	5,060	32.3	9.0	59.8	93.3	15.0	9.6
1994	180,951	127,570	128,705	10,138	13	171,961	5,084	33.8	8.5	60.8	94.1	14.0	9.0
1995	184,288	60,051	52,320	8,567	725	168,716	5,107	33.0	9.6	64.2	94.8	15.0	10.1

Note: starting 1980 production includes recreational fishing in inland waters.

68 FRANCE	- FOOD BALANCE SHEET OF FISH AND FISHERY PRODUCTS IN LIVE WEIGHT AND FISH CONTRIBUTION TO PROTEIN SUPPLY -	FAO-FIDI

YEAR	PRODUCTION	NON-FOOD USES	IMPORTS	EXPORTS	STOCKS VARIATIONS	TOTAL FOOD SUPPLY	POPULATION	PER CAPUT SUPPLY	FISH PROTEINS	ANIMAL PROTEINS	TOTAL PROTEINS	FISH/ANIMAL PROTEINS	FISH/TOTAL PROTEINS
	(..		metric tons in live weight		....)		(thousands)	(kilograms)	(grams per caput per day)			(%)	(%)
1961	707,958	900	218,393	96,680	0	828,771	46,277	17.9	4.5	57.0	102.6	7.8	4.3
1962	688,372	700	258,430	89,593	0	856,509	46,917	18.3	4.4	57.9	103.6	7.6	4.3
1963	696,270	600	297,260	75,119	0	917,811	47,572	19.3	4.6	58.5	103.3	7.9	4.5
1964	728,149	800	280,620	68,501	0	939,468	48,196	19.5	4.8	59.0	101.9	8.2	4.7
1965	766,069	500	299,054	72,370	0	992,254	48,758	20.4	4.8	59.7	102.5	8.1	4.7
1966	797,547	500	307,264	95,289	0	1,009,021	49,243	20.5	5.1	59.9	101.5	8.5	5.0
1967	818,063	800	299,651	91,439	0	1,025,475	49,661	20.7	5.2	62.0	101.9	8.4	5.1
1968	826,454	700	348,245	84,451	0	1,089,548	50,034	21.8	5.4	64.1	104.2	8.4	5.1
1969	761,293	700	334,499	88,504	0	1,006,588	50,396	20.0	4.9	64.1	103.3	7.6	4.7
1970	762,691	6,200	355,980	77,425	0	1,035,045	50,772	20.4	5.1	65.6	104.4	7.7	4.9
1971	752,920	5,800	404,635	95,335	0	1,056,420	51,171	20.7	5.2	65.7	103.0	7.9	5.0
1972	778,364	9,100	434,057	105,505	0	1,097,817	51,581	21.3	5.1	66.0	103.1	7.7	4.9
1973	803,480	9,000	440,846	129,611	0	1,105,715	51,988	21.3	4.9	66.5	103.0	7.3	4.7
1974	805,559	9,157	451,547	115,060	-1,000	1,131,889	52,366	21.6	4.9	66.6	103.2	7.4	4.8
1975	800,041	9,058	462,865	103,329	-200	1,150,319	52,699	21.8	4.8	66.8	103.4	7.2	4.6
1976	782,585	95,464	506,234	133,451	-2,200	1,057,704	52,983	20.0	4.4	68.3	104.2	6.5	4.2
1977	753,252	115,202	592,199	121,827	3,400	1,111,822	53,228	20.9	4.7	70.2	108.0	6.6	4.3
1978	774,698	125,373	620,059	132,747	0	1,136,636	53,446	21.3	4.7	71.8	109.2	6.6	4.3
1979	746,368	110,369	650,791	133,309	0	1,153,482	53,658	21.5	4.8	74.7	111.9	6.4	4.3
1980	790,034	69,686	686,039	151,302	0	1,255,086	53,880	23.3	5.1	75.1	112.3	6.8	4.5
1981	799,287	80,317	725,412	159,218	0	1,285,163	54,115	23.8	5.2	75.6	112.8	6.9	4.6
1982	779,604	88,083	753,880	177,578	0	1,267,822	54,360	23.3	5.0	75.0	113.2	6.7	4.4
1983	803,173	92,259	807,738	195,465	0	1,323,187	54,618	24.2	5.2	75.1	113.6	6.9	4.6
1984	788,066	107,890	794,233	188,420	0	1,285,989	54,888	23.4	5.1	76.1	114.0	6.7	4.5
1985	834,049	85,817	838,368	231,121	0	1,355,480	55,170	24.6	5.3	78.7	117.2	6.7	4.5
1986	875,700	99,227	932,423	269,067	0	1,439,829	55,467	26.0	5.5	77.4	115.9	7.1	4.7
1987	850,550	95,332	1,055,587	290,664	-250	1,519,891	55,776	27.3	5.8	77.3	115.7	7.6	5.1
1988	891,672	105,416	1,125,986	301,430	250	1,611,062	56,092	28.7	6.2	76.5	115.9	8.2	5.4
1989	855,549	110,619	1,196,890	351,734	0	1,590,086	56,408	28.2	6.1	76.1	115.6	8.0	5.3
1990	869,375	107,466	1,367,005	391,838	0	1,737,083	56,718	30.6	6.5	77.2	116.3	8.4	5.6
1991	819,134	101,238	1,364,109	385,740	0	1,696,265	57,019	29.8	6.3	79.8	118.8	8.0	5.3
1992	820,965	108,063	1,348,166	406,619	0	1,654,449	57,311	28.9	6.2	77.1	115.3	8.1	5.4
1993	860,415	111,681	1,284,461	440,825	0	1,592,369	57,591	27.7	6.2	74.8	114.7	8.3	5.4
1994	852,786	114,665	1,308,597	464,386	0	1,582,332	57,856	27.4	6.3	74.6	114.1	8.4	5.5
1995	793,413	111,503	1,315,500	460,234	0	1,537,176	58,104	26.5	6.1	75.8	115.8	8.1	5.3

69 FRENCH GUIANA	- FOOD BALANCE SHEET OF FISH AND FISHERY PRODUCTS IN LIVE WEIGHT AND FISH CONTRIBUTION TO PROTEIN SUPPLY -												FAO-FIDI
YEAR	PRODUCTION	NON-FOOD USES	IMPORTS	EXPORTS	STOCKS VARIATIONS	TOTAL FOOD SUPPLY	POPULATION	PER CAPUT SUPPLY	FISH PROTEINS	ANIMAL PROTEINS	TOTAL PROTEINS	FISH/ANIMAL PROTEINS	FISH/TOTAL PROTEINS
	(................................ metric tons in live weight					................................)	(thousands)	(kilograms)	(grams per caput per day)			(%)	(%)
1961	600	0	500	0	0	1,100	34	32.2	10.0	29.4	64.4	33.8	15.5
1962	600	0	425	0	0	1,025	35	29.0	8.9	30.5	64.1	29.1	13.9
1963	600	0	425	0	0	1,025	37	27.8	8.6	29.3	61.8	29.4	13.9
1964	700	0	425	100	0	1,025	38	26.7	8.3	32.0	63.1	25.9	13.1
1965	600	0	1,725	1,200	0	1,125	40	28.1	8.6	32.2	65.8	26.6	13.0
1966	700	0	1,825	1,700	0	825	42	19.8	6.3	30.8	63.6	20.5	9.9
1967	800	0	2,025	1,800	0	1,025	44	23.5	7.4	32.5	67.7	22.6	10.9
1968	800	0	2,025	1,900	0	925	45	20.4	6.4	35.3	70.1	18.2	9.2
1969	900	0	3,625	3,300	0	1,225	47	25.9	7.9	36.3	70.1	21.8	11.3
1970	1,200	0	3,525	2,800	-400	1,525	49	31.1	9.3	37.9	73.4	24.5	12.7
1971	1,100	0	2,176	1,700	0	1,576	51	31.1	9.4	36.9	69.6	25.4	13.5
1972	1,500	0	2,500	2,000	-300	1,700	52	32.6	9.6	37.8	72.0	25.4	13.3
1973	1,200	0	2,400	1,900	0	1,700	54	31.0	9.2	35.8	67.3	25.6	13.6
1974	1,142	0	1,126	400	-150	1,718	55	31.0	9.3	32.9	62.1	28.2	14.9
1975	1,241	0	939	414	0	1,766	57	31.0	9.0	37.3	67.2	24.1	13.4
1976	1,207	0	1,697	1,003	-150	1,751	59	29.8	8.8	36.8	65.1	23.9	13.5
1977	1,266	0	1,964	1,876	260	1,614	60	26.7	8.6	45.6	79.2	18.9	10.9
1978	1,240	0	1,832	962	-280	1,830	62	29.3	9.6	46.6	78.1	20.5	12.2
1979	1,186	0	3,137	2,237	-230	1,856	65	28.6	9.2	48.1	82.2	19.2	11.2
1980	1,150	0	3,604	3,080	300	1,975	68	29.0	9.6	48.7	83.2	19.7	11.6
1981	1,436	0	3,764	3,669	650	2,180	72	30.4	10.5	50.5	83.8	20.9	12.6
1982	1,992	0	3,867	3,637	100	2,322	76	30.5	9.9	50.3	83.4	19.7	11.8
1983	2,118	0	3,653	2,818	-300	2,653	81	32.8	9.8	49.8	87.7	19.8	11.2
1984	2,194	0	3,325	2,739	-11	2,769	86	32.2	9.7	48.9	85.2	19.9	11.4
1985	2,491	0	3,168	2,707	0	2,952	91	32.4	10.1	51.3	88.3	19.6	11.4
1986	3,326	0	2,902	3,003	-89	3,136	96	32.7	10.4	47.5	83.7	21.8	12.4
1987	5,250	0	2,493	3,835	-600	3,308	101	32.7	9.8	49.5	86.0	19.8	11.4
1988	5,540	0	2,319	3,584	392	4,667	106	43.9	11.4	51.5	87.5	22.2	13.0
1989	6,499	0	2,760	4,512	-198	4,550	112	40.8	11.3	53.5	87.7	21.1	12.9
1990	6,548	0	1,936	4,505	94	4,073	117	34.8	10.3	53.3	90.5	19.4	11.4
1991	7,119	0	1,823	5,114	588	4,415	123	36.0	10.3	52.0	89.3	19.8	11.5
1992	7,617	0	2,191	4,753	295	5,350	129	41.6	11.9	53.0	89.9	22.5	13.3
1993	6,771	0	1,972	4,235	-172	4,336	135	32.2	9.0	46.0	82.3	19.6	10.9
1994	7,813	0	2,290	5,131	96	5,068	141	36.0	10.2	49.3	88.2	20.6	11.5
1995	7,737	0	2,464	4,950	106	5,357	147	36.4	10.2	50.8	87.0	20.1	11.7

70 FRENCH POLYNESIA	- FOOD BALANCE SHEET OF FISH AND FISHERY PRODUCTS IN LIVE WEIGHT AND FISH CONTRIBUTION TO PROTEIN SUPPLY -	FAO-FIDI

YEAR	PRODUCTION	NON-FOOD USES	IMPORTS	EXPORTS	STOCKS VARIATIONS	TOTAL FOOD SUPPLY	POPULATION	PER CAPUT SUPPLY	FISH PROTEINS	ANIMAL PROTEINS	TOTAL PROTEINS	FISH/ANIMAL PROTEINS	FISH/TOTAL PROTEINS
	(............ metric tons in live weight ....)						(thousands)	(kilograms)	(grams per caput per day)			(%)	(%)
1961	2,700	0	1,000	0	0	3,700	82	45.3	13.6	30.4	68.4	44.8	19.9
1962	3,000	0	1,000	0	0	4,000	84	47.5	14.6	31.9	67.9	45.6	21.4
1963	2,500	0	800	0	0	3,300	87	38.0	11.5	29.4	65.6	39.0	17.5
1964	2,500	0	1,400	0	0	3,900	90	43.5	13.0	31.1	67.1	41.9	19.4
1965	1,900	0	2,300	0	0	4,200	93	45.2	12.6	33.0	71.8	38.2	17.6
1966	2,100	0	2,600	0	0	4,700	96	48.9	13.2	38.9	78.4	33.9	16.8
1967	2,000	0	2,200	0	0	4,200	100	42.1	11.8	40.4	78.2	29.1	15.0
1968	2,100	0	3,000	0	0	5,100	103	49.3	13.9	40.8	79.4	33.9	17.4
1969	2,300	0	1,340	0	0	3,640	107	34.0	9.9	38.0	75.8	26.0	13.1
1970	2,300	0	2,000	0	0	4,300	111	38.8	10.9	38.8	76.2	28.2	14.4
1971	2,200	0	1,900	0	0	4,100	115	35.8	10.1	37.0	74.2	27.4	13.7
1972	2,400	0	2,100	0	0	4,500	118	38.0	10.6	39.4	76.5	26.9	13.8
1973	2,600	0	2,500	0	0	5,100	122	41.8	11.7	40.9	77.9	28.5	15.0
1974	2,386	0	2,700	0	0	5,086	126	40.4	11.3	36.3	74.0	31.1	15.3
1975	2,169	0	1,500	0	0	3,669	130	28.3	8.1	39.3	76.6	20.6	10.6
1976	2,828	0	2,495	68	0	5,255	134	39.3	10.6	39.7	76.8	26.8	13.9
1977	2,704	0	2,054	171	0	4,587	138	33.2	8.6	37.4	74.8	23.0	11.5
1978	2,994	0	2,833	48	-13	5,767	142	40.5	11.2	39.9	77.1	28.0	14.5
1979	2,484	0	3,167	332	-13	5,307	147	36.2	9.7	41.8	79.6	23.3	12.2
1980	2,545	1	2,957	2	0	5,499	151	36.4	9.9	37.8	75.2	26.3	13.2
1981	2,279	0	3,145	34	0	5,390	156	34.6	9.4	38.6	74.6	24.2	12.6
1982	2,371	0	3,038	7	13	5,415	160	33.8	9.2	38.4	75.1	23.9	12.2
1983	2,291	0	3,106	6	13	5,404	165	32.8	9.0	39.7	77.0	22.6	11.7
1984	2,889	1	3,015	5	0	5,898	169	34.8	9.6	40.6	77.1	23.6	12.4
1985	2,398	1	3,364	4	0	5,757	174	33.1	9.1	41.3	77.6	22.0	11.7
1986	2,269	0	3,572	26	-56	5,759	179	32.2	9.2	44.0	81.7	20.9	11.3
1987	2,310	0	3,299	34	-22	5,553	183	30.3	8.8	43.6	81.3	20.2	10.8
1988	2,859	0	3,693	58	-38	6,457	188	34.3	9.9	46.3	83.7	21.3	11.8
1989	3,130	0	4,301	5	0	7,427	193	38.5	10.7	46.0	82.0	23.1	13.0
1990	2,937	1	4,261	141	31	7,086	197	35.9	9.5	45.3	82.8	21.0	11.5
1991	2,407	0	4,356	145	-38	6,581	202	32.6	8.6	53.7	90.8	16.0	9.4
1992	2,455	0	4,149	142	35	6,497	206	31.5	8.8	53.0	88.3	16.5	9.9
1993	8,082	1	4,193	255	-33	11,985	210	57.0	16.6	60.4	98.5	27.5	16.8
1994	8,832	2	4,533	220	96	13,239	215	61.7	17.9	61.4	99.0	29.1	18.1
1995	9,081	2	4,135	258	25	12,981	219	59.3	17.3	63.5	99.2	27.2	17.4

Note: up to 1992 data refer only to quantities sold at the principal fish markets. Increased catch for 1993-95 due to complete data collection.

74 GABON	- FOOD BALANCE SHEET OF FISH AND FISHERY PRODUCTS IN LIVE WEIGHT AND FISH CONTRIBUTION TO PROTEIN SUPPLY -	FAO-FIDI

YEAR	PRODUCTION	NON-FOOD USES	IMPORTS	EXPORTS	STOCKS VARIATIONS	TOTAL FOOD SUPPLY	POPULATION	PER CAPUT SUPPLY	FISH PROTEINS	ANIMAL PROTEINS	TOTAL PROTEINS	FISH/ANIMAL PROTEINS	FISH/TOTAL PROTEINS
	(.. metric tons in live weight ....)						(thousands)	(kilograms)	(grams per caput per day)			(%)	(%)
1961	6,500	0	6,655	0	0	13,155	488	27.0	8.8	28.4	51.6	31.1	17.1
1962	7,000	0	5,745	0	0	12,745	490	26.0	8.5	29.1	52.8	29.1	16.0
1963	7,000	0	5,842	0	0	12,842	492	26.1	8.4	30.5	53.5	27.7	15.8
1964	7,000	0	5,233	0	0	12,233	494	24.8	8.0	31.1	54.2	25.6	14.7
1965	8,000	0	4,624	303	0	12,321	495	24.9	8.0	31.5	55.2	25.3	14.5
1966	8,000	0	4,424	303	0	12,121	495	24.5	7.9	32.4	57.1	24.3	13.8
1967	8,000	0	3,618	303	0	11,315	494	22.9	7.4	32.2	58.2	22.9	12.6
1968	9,000	0	4,218	0	0	13,218	494	26.8	8.6	33.9	60.1	25.4	14.3
1969	9,000	0	4,818	0	0	13,818	497	27.8	8.9	34.9	61.0	25.5	14.6
1970	10,000	0	4,521	0	0	14,521	504	28.8	9.3	34.8	61.5	26.8	15.2
1971	10,000	0	4,715	0	0	14,715	516	28.5	9.2	33.7	60.0	27.2	15.3
1972	10,000	0	4,915	0	0	14,915	532	28.0	9.0	33.1	58.3	27.1	15.4
1973	12,000	0	4,309	0	0	16,309	552	29.6	9.5	32.8	56.3	28.9	16.8
1974	12,000	0	3,149	0	0	15,149	572	26.5	8.6	32.5	60.7	26.4	14.1
1975	14,900	0	8,318	0	0	23,218	593	39.2	12.1	36.6	65.8	33.0	18.4
1976	14,820	0	6,098	27	0	20,891	612	34.1	10.6	33.7	66.7	31.4	15.8
1977	14,010	0	10,017	224	0	23,803	632	37.7	11.6	36.6	68.1	31.7	17.0
1978	17,800	0	7,475	430	0	24,845	651	38.2	11.5	37.9	69.9	30.4	16.5
1979	21,000	0	10,718	1,025	0	30,693	671	45.8	13.7	39.5	70.8	34.5	19.3
1980	19,800	0	13,081	166	0	32,715	691	47.3	14.1	40.8	71.2	34.5	19.7
1981	20,030	0	14,203	2,079	0	32,154	712	45.1	13.6	42.7	72.1	31.7	18.8
1982	20,605	0	13,855	1,422	0	33,038	734	45.0	13.4	43.9	74.8	30.5	17.9
1983	19,449	0	12,509	1,701	0	30,257	756	40.0	12.1	41.6	72.8	29.0	16.5
1984	21,002	0	17,613	1,357	0	37,258	779	47.9	14.4	45.6	77.3	31.6	18.6
1985	20,955	0	16,464	1,489	0	35,930	803	44.8	13.8	46.8	80.2	29.5	17.2
1986	20,226	0	16,683	1,755	0	35,154	828	42.5	12.2	42.1	75.6	29.0	16.1
1987	22,188	0	17,328	2,082	0	37,434	854	43.9	12.6	41.6	75.0	30.3	16.8
1988	22,093	0	15,779	1,390	0	36,482	881	41.4	12.0	40.2	75.4	29.9	15.9
1989	20,502	0	14,192	1,386	0	33,308	908	36.7	10.6	36.3	69.5	29.3	15.3
1990	20,002	0	11,882	924	0	30,961	935	33.1	9.8	32.2	65.0	30.4	15.0
1991	22,003	0	12,819	604	0	34,218	963	35.6	10.3	37.1	73.8	27.7	13.9
1992	24,005	0	11,653	865	0	34,793	990	35.2	9.8	38.0	73.7	25.7	13.3
1993	31,794	0	16,941	739	0	47,996	1,018	47.2	13.2	42.1	76.9	31.4	17.2
1994	31,020	0	11,873	912	-311	41,670	1,046	39.8	11.3	38.3	73.0	29.4	15.4
1995	40,397	0	6,630	1,515	189	45,701	1,076	42.5	12.1	39.2	75.4	30.8	16.0

75 GAMBIA	- FOOD BALANCE SHEET OF FISH AND FISHERY PRODUCTS IN LIVE WEIGHT AND FISH CONTRIBUTION TO PROTEIN SUPPLY -	FAO-FIDI

YEAR	PRODUCTION	NON-FOOD USES	IMPORTS	EXPORTS	STOCKS VARIATIONS	TOTAL FOOD SUPPLY	POPULATION	PER CAPUT SUPPLY	FISH PROTEINS	ANIMAL PROTEINS	TOTAL PROTEINS	FISH/ANIMAL PROTEINS	FISH/TOTAL PROTEINS
	(............		metric tons in live weight			............)	(thousands)	(kilograms)		(grams per caput per day)		(%)	(%)
1961	4,800	0	0	0	0	4,800	362	13.3	4.5	10.9	58.5	41.4	7.7
1962	5,300	0	200	0	0	5,500	372	14.8	5.0	11.6	58.8	43.4	8.5
1963	5,300	0	200	0	0	5,500	382	14.4	5.0	11.6	58.6	43.6	8.6
1964	5,300	0	200	0	0	5,500	393	14.0	4.9	11.4	57.9	43.2	8.5
1965	6,200	0	0	0	0	6,200	404	15.4	5.3	11.8	58.0	44.9	9.1
1966	6,200	0	0	1,000	0	5,200	415	12.5	4.0	10.8	56.4	37.0	7.1
1967	6,200	0	200	1,050	0	5,350	426	12.6	4.0	10.9	54.7	36.4	7.2
1968	6,700	0	200	623	0	6,277	438	14.3	4.6	11.7	55.5	39.2	8.2
1969	6,700	0	200	1,454	0	5,446	450	12.1	3.8	10.8	55.0	35.1	6.9
1970	7,500	0	82	1,600	0	5,982	464	12.9	4.2	11.6	54.8	36.0	7.6
1971	7,700	0	130	1,650	0	6,180	479	12.9	4.1	11.4	56.2	36.3	7.4
1972	7,900	0	276	2,253	0	5,923	495	12.0	3.7	10.9	54.9	34.1	6.8
1973	12,100	2,000	413	2,984	0	7,529	512	14.7	4.7	11.9	55.8	39.7	8.5
1974	12,695	1,400	3,015	6,192	0	8,118	530	15.3	4.8	11.9	51.5	40.1	9.2
1975	12,695	500	3,050	7,585	0	7,660	548	14.0	4.4	11.4	43.2	38.4	10.1
1976	12,695	0	9,752	14,047	0	8,400	566	14.9	4.5	11.5	45.0	39.2	10.0
1977	18,092	0	11,138	19,971	0	9,259	584	15.8	4.7	11.2	41.9	42.4	11.3
1978	14,972	0	7,652	12,177	0	10,447	603	17.3	4.7	11.2	47.5	42.1	10.0
1979	11,143	0	200	3,400	0	7,943	622	12.8	3.4	9.8	44.7	34.7	7.7
1980	13,265	0	468	2,860	0	10,873	641	17.0	4.5	10.5	42.0	43.2	10.8
1981	14,068	0	678	2,150	0	12,596	659	19.1	5.3	12.9	47.2	41.1	11.2
1982	9,212	0	313	2,519	0	7,006	678	10.3	3.0	9.7	46.7	30.7	6.4
1983	11,653	0	1,226	1,352	0	11,527	697	16.5	5.0	12.4	49.9	40.4	10.0
1984	11,882	0	1,835	1,110	0	12,607	719	17.5	5.3	11.8	48.5	44.5	10.8
1985	10,712	0	700	1,629	0	9,783	745	13.1	4.0	12.1	55.3	32.9	7.2
1986	13,262	0	600	1,125	0	12,737	775	16.4	5.2	11.8	56.4	44.0	9.2
1987	14,646	0	400	1,647	0	13,399	808	16.6	4.7	12.0	52.9	38.9	8.8
1988	13,852	0	0	1,841	0	12,011	844	14.2	4.6	12.0	53.1	38.3	8.6
1989	19,842	0	7	2,005	0	17,844	882	20.2	6.8	12.5	53.3	54.0	12.7
1990	17,922	0	18	2,147	0	15,793	921	17.2	5.4	10.8	50.4	50.1	10.7
1991	23,743	0	123	2,201	0	21,666	960	22.6	7.5	13.7	52.0	54.6	14.4
1992	22,718	0	176	2,367	0	20,527	1,000	20.5	6.5	11.6	50.3	56.2	12.9
1993	20,479	0	669	2,685	0	18,463	1,039	17.8	5.7	10.9	45.6	52.4	12.5
1994	22,318	0	248	3,373	0	19,193	1,077	17.8	5.9	11.0	46.2	53.3	12.7
1995	23,098	0	317	3,342	0	20,073	1,111	18.1	5.6	10.2	47.8	55.1	11.8

76 GAZA STRIP	- FOOD BALANCE SHEET OF FISH AND FISHERY PRODUCTS IN LIVE WEIGHT AND FISH CONTRIBUTION TO PROTEIN SUPPLY -	FAO-FIDI

YEAR	PRODUCTION	NON-FOOD USES	IMPORTS	EXPORTS	STOCKS VARIATIONS	TOTAL FOOD SUPPLY	POPULATION	PER CAPUT SUPPLY	FISH PROTEINS	ANIMAL PROTEINS	TOTAL PROTEINS	FISH/ANIMAL PROTEINS	FISH/TOTAL PROTEINS
	(.. metric tons in live weight					....)	(thousands)	(kilograms)	(grams per caput per day)			(%)	(%)
1961	0	0	0	0	0	0	312	0.0	0.0	0.0	0.0	0.0	0.0
1962	0	0	0	0	0	0	322	0.0	0.0	0.0	0.0	0.0	0.0
1963	0	0	0	0	0	0	331	0.0	0.0	0.0	0.0	0.0	0.0
1964	0	0	0	0	0	0	338	0.0	0.0	0.0	0.0	0.0	0.0
1965	0	0	0	0	0	0	343	0.0	0.0	0.0	0.0	0.0	0.0
1966	0	0	0	0	0	0	345	0.0	0.0	0.0	0.0	0.0	0.0
1967	0	0	0	0	0	0	344	0.0	0.0	0.0	0.0	0.0	0.0
1968	0	0	0	0	0	0	342	0.0	0.0	0.0	0.0	0.0	0.0
1969	0	0	0	0	0	0	341	0.0	0.0	0.0	0.0	0.0	0.0
1970	0	0	0	0	0	0	343	0.0	0.0	0.0	0.0	0.0	0.0
1971	0	0	0	0	0	0	349	0.0	0.0	0.0	0.0	0.0	0.0
1972	0	0	0	0	0	0	358	0.0	0.0	0.0	0.0	0.0	0.0
1973	0	0	0	0	0	0	369	0.0	0.0	0.0	0.0	0.0	0.0
1974	0	0	0	0	0	0	381	0.0	0.0	0.0	0.0	0.0	0.0
1975	0	0	0	0	0	0	393	0.0	0.0	0.0	0.0	0.0	0.0
1976	0	0	0	0	0	0	404	0.0	0.0	0.0	0.0	0.0	0.0
1977	0	0	0	0	0	0	415	0.0	0.0	0.0	0.0	0.0	0.0
1978	0	0	0	0	0	0	427	0.0	0.0	0.0	0.0	0.0	0.0
1979	0	0	0	0	0	0	438	0.0	0.0	0.0	0.0	0.0	0.0
1980	0	0	0	0	0	0	451	0.0	0.0	0.0	0.0	0.0	0.0
1981	0	0	0	0	0	0	464	0.0	0.0	0.0	0.0	0.0	0.0
1982	0	0	0	0	0	0	478	0.0	0.0	0.0	0.0	0.0	0.0
1983	0	0	0	0	0	0	492	0.0	0.0	0.0	0.0	0.0	0.0
1984	0	0	0	0	0	0	508	0.0	0.0	0.0	0.0	0.0	0.0
1985	0	0	0	0	0	0	524	0.0	0.0	0.0	0.0	0.0	0.0
1986	0	0	0	0	0	0	541	0.0	0.0	0.0	0.0	0.0	0.0
1987	0	0	0	0	0	0	560	0.0	0.0	0.0	0.0	0.0	0.0
1988	0	0	0	0	0	0	580	0.0	0.0	0.0	0.0	0.0	0.0
1989	0	0	0	0	0	0	602	0.0	0.0	0.0	0.0	0.0	0.0
1990	0	0	0	0	0	0	627	0.0	0.0	0.0	0.0	0.0	0.0
1991	0	0	0	0	0	0	656	0.0	0.0	0.0	0.0	0.0	0.0
1992	0	0	0	0	0	0	688	0.0	0.0	0.0	0.0	0.0	0.0
1993	0	0	0	0	0	0	721	0.0	0.0	0.0	0.0	0.0	0.0
1994	0	0	0	0	0	0	757	0.0	0.0	0.0	0.0	0.0	0.0
1995	1,229	0	...	...	0	1,229	792	1.6	0.4	11.1	84.1	4.0	0.5

Note: up to 1994 production and trade statistics included with those of Israel.

73 GEORGIA	- FOOD BALANCE SHEET OF FISH AND FISHERY PRODUCTS IN LIVE WEIGHT AND FISH CONTRIBUTION TO PROTEIN SUPPLY -	FAO-FIDI

YEAR	PRODUCTION	NON-FOOD USES	IMPORTS	EXPORTS	STOCKS VARIATIONS	TOTAL FOOD SUPPLY	POPULATION	PER CAPUT SUPPLY	FISH PROTEINS	ANIMAL PROTEINS	TOTAL PROTEINS	FISH/ANIMAL PROTEINS	FISH/TOTAL PROTEINS
	(.. metric tons in live weight ....)						(thousands)	(kilograms)	(grams per caput per day)			(%)	(%)
1961	0	0	0	0	0	0	4,224	0.0	0.0	0.0	0.0	0.0	0.0
1962	0	0	0	0	0	0	4,292	0.0	0.0	0.0	0.0	0.0	0.0
1963	0	0	0	0	0	0	4,357	0.0	0.0	0.0	0.0	0.0	0.0
1964	0	0	0	0	0	0	4,420	0.0	0.0	0.0	0.0	0.0	0.0
1965	0	0	0	0	0	0	4,476	0.0	0.0	0.0	0.0	0.0	0.0
1966	0	0	0	0	0	0	4,529	0.0	0.0	0.0	0.0	0.0	0.0
1967	0	0	0	0	0	0	4,576	0.0	0.0	0.0	0.0	0.0	0.0
1968	0	0	0	0	0	0	4,618	0.0	0.0	0.0	0.0	0.0	0.0
1969	0	0	0	0	0	0	4,662	0.0	0.0	0.0	0.0	0.0	0.0
1970	0	0	0	0	0	0	4,708	0.0	0.0	0.0	0.0	0.0	0.0
1971	0	0	0	0	0	0	4,754	0.0	0.0	0.0	0.0	0.0	0.0
1972	0	0	0	0	0	0	4,798	0.0	0.0	0.0	0.0	0.0	0.0
1973	0	0	0	0	0	0	4,837	0.0	0.0	0.0	0.0	0.0	0.0
1974	0	0	0	0	0	0	4,876	0.0	0.0	0.0	0.0	0.0	0.0
1975	0	0	0	0	0	0	4,908	0.0	0.0	0.0	0.0	0.0	0.0
1976	0	0	0	0	0	0	4,940	0.0	0.0	0.0	0.0	0.0	0.0
1977	0	0	0	0	0	0	4,973	0.0	0.0	0.0	0.0	0.0	0.0
1978	0	0	0	0	0	0	5,001	0.0	0.0	0.0	0.0	0.0	0.0
1979	0	0	0	0	0	0	5,028	0.0	0.0	0.0	0.0	0.0	0.0
1980	0	0	0	0	0	0	5,048	0.0	0.0	0.0	0.0	0.0	0.0
1981	0	0	0	0	0	0	5,085	0.0	0.0	0.0	0.0	0.0	0.0
1982	0	0	0	0	0	0	5,125	0.0	0.0	0.0	0.0	0.0	0.0
1983	0	0	0	0	0	0	5,167	0.0	0.0	0.0	0.0	0.0	0.0
1984	0	0	0	0	0	0	5,210	0.0	0.0	0.0	0.0	0.0	0.0
1985	0	0	0	0	0	0	5,252	0.0	0.0	0.0	0.0	0.0	0.0
1986	0	0	0	0	0	0	5,292	0.0	0.0	0.0	0.0	0.0	0.0
1987	0	0	0	0	0	0	5,330	0.0	0.0	0.0	0.0	0.0	0.0
1988	0	0	0	0	0	0	5,365	0.0	0.0	0.0	0.0	0.0	0.0
1989	0	0	0	0	0	0	5,439	0.0	0.0	0.0	0.0	0.0	0.0
1990	0	0	0	0	0	0	5,460	0.0	0.0	0.0	0.0	0.0	0.0
1991	60,813	0	...	...	0	60,813	5,471	11.1	3.8	29.7	88.6	12.7	4.3
1992	40,493	0	...	...	0	40,493	5,472	7.4	2.5	20.8	71.4	12.1	3.5
1993	19,721	0	1,198	9	0	20,910	5,467	3.8	1.3	18.0	59.1	7.2	2.2
1994	7,953	0	539	255	0	8,237	5,458	1.5	0.5	14.2	63.4	3.6	0.8
1995	3,156	0	2,613	136	0	5,633	5,450	1.0	0.3	19.2	67.3	1.8	0.5

Note: for years before 1991 see USSR.

79 GERMANY	- FOOD BALANCE SHEET OF FISH AND FISHERY PRODUCTS IN LIVE WEIGHT AND FISH CONTRIBUTION TO PROTEIN SUPPLY -												FAO-FIDI
YEAR	PRODUCTION	NON-FOOD USES	IMPORTS	EXPORTS	STOCKS VARIATIONS	TOTAL FOOD SUPPLY	POPULATION	PER CAPUT SUPPLY	FISH PROTEINS	ANIMAL PROTEINS	TOTAL PROTEINS	FISH/ANIMAL PROTEINS	FISH/TOTAL PROTEINS
	(.......... metric tons in live weight					..........)	(thousands)	(kilograms)	(grams per caput per day)			(%)	(%)
1961	748,000	168,800	288,626	164,097	0	703,729	73,321	9.6	3.3	45.1	79.6	7.3	4.1
1962	781,700	184,100	293,177	166,588	0	724,189	74,032	9.8	3.4	45.9	80.6	7.4	4.2
1963	836,800	183,700	302,099	168,718	0	786,482	74,760	10.5	3.5	46.1	80.2	7.7	4.4
1964	849,200	196,500	315,219	170,326	0	797,593	75,442	10.6	3.5	46.3	82.3	7.6	4.3
1965	863,800	184,500	334,136	165,117	0	848,318	76,031	11.2	3.6	47.0	83.5	7.6	4.3
1966	880,400	205,100	291,407	175,893	0	790,812	76,508	10.3	3.4	46.8	83.2	7.2	4.0
1967	949,700	204,000	272,027	182,122	0	835,604	76,885	10.9	3.5	48.3	84.1	7.3	4.2
1968	984,300	196,800	294,326	195,350	0	886,471	77,187	11.5	3.6	50.0	85.4	7.1	4.2
1969	967,400	194,000	298,843	198,192	0	874,055	77,452	11.3	3.7	50.8	86.2	7.3	4.3
1970	934,614	194,400	366,779	192,272	0	914,723	77,709	11.8	3.7	52.5	87.9	7.1	4.2
1971	845,801	156,500	374,183	163,206	0	900,276	77,966	11.6	3.7	53.4	88.6	6.8	4.1
1972	754,306	95,200	364,923	188,152	0	835,875	78,211	10.7	3.7	54.0	88.8	6.9	4.2
1973	843,002	153,600	441,055	156,550	0	973,915	78,427	12.4	4.0	54.5	89.4	7.3	4.5
1974	888,866	151,385	396,734	145,575	0	988,634	78,588	12.6	3.9	55.1	89.5	7.2	4.4
1975	817,871	134,430	418,477	161,125	0	940,797	78,679	12.0	4.0	55.8	89.8	7.1	4.4
1976	733,738	55,997	425,421	185,744	0	917,416	78,701	11.7	3.9	56.1	90.6	6.9	4.3
1977	644,281	69,175	434,881	192,412	-1,111	816,457	78,667	10.4	3.6	55.6	90.1	6.4	3.9
1978	613,576	71,098	448,454	174,493	-2,222	814,218	78,582	10.4	3.5	57.8	93.5	6.1	3.8
1979	578,574	59,493	485,405	190,214	-2,222	812,051	78,457	10.4	3.5	58.3	94.1	6.0	3.7
1980	542,194	40,140	564,469	199,920	-2,222	864,389	78,304	11.0	3.8	60.1	96.0	6.4	4.0
1981	575,339	42,376	524,559	191,254	-2,222	864,048	78,120	11.1	3.8	59.4	95.3	6.3	3.9
1982	549,792	27,778	536,706	205,875	0	852,837	77,918	11.0	3.7	59.6	95.9	6.2	3.8
1983	542,850	18,785	561,526	190,112	1,111	896,598	77,741	11.5	3.8	61.4	98.0	6.3	3.9
1984	553,182	19,847	574,479	216,544	8,889	900,167	77,645	11.6	3.8	61.8	99.6	6.1	3.8
1985	426,941	12,040	623,100	190,331	0	847,670	77,668	10.9	3.6	62.0	100.1	5.8	3.6
1986	415,388	8,830	691,846	186,513	0	911,893	77,828	11.7	3.8	62.2	100.4	6.1	3.8
1987	399,288	5,176	633,290	199,810	0	827,590	78,109	10.6	3.5	62.9	101.3	5.6	3.5
1988	393,530	5,216	711,057	228,287	0	871,074	78,485	11.1	3.7	63.6	102.1	5.7	3.6
1989	412,260	5,866	837,943	292,664	0	951,679	78,915	12.1	4.0	62.0	100.7	6.4	3.9
1990	390,737	3,479	1,006,485	347,330	0	1,046,408	79,365	13.2	4.4	61.5	98.5	7.1	4.4
1991	303,396	5,889	1,086,755	402,514	0	981,748	79,827	12.3	3.9	59.7	98.1	6.6	4.0
1992	307,239	4,846	1,112,878	388,802	0	1,026,461	80,303	12.8	4.0	58.9	97.7	6.8	4.1
1993	316,374	4,757	1,046,670	371,099	0	987,188	80,771	12.2	3.7	55.9	95.7	6.6	3.9
1994	272,783	5,484	1,207,007	413,845	0	1,060,449	81,207	13.1	4.0	57.7	94.9	6.9	4.2
1995	298,017	5,442	1,179,309	419,618	0	1,052,267	81,594	12.9	3.9	57.6	93.3	6.8	4.2

81 GHANA	- FOOD BALANCE SHEET OF FISH AND FISHERY PRODUCTS IN LIVE WEIGHT AND FISH CONTRIBUTION TO PROTEIN SUPPLY -												FAO-FIDI
YEAR	PRODUCTION	NON-FOOD USES	IMPORTS	EXPORTS	STOCKS VARIATIONS	TOTAL FOOD SUPPLY	POPULATION	PER CAPUT SUPPLY	FISH PROTEINS	ANIMAL PROTEINS	TOTAL PROTEINS	FISH/ANIMAL PROTEINS	FISH/TOTAL PROTEINS
	(.. metric tons in live weight ....)						(thousands)	(kilograms)	(grams per caput per day)			(%)	(%)
1961	36,500	0	82,875	1,400	0	117,975	6,994	16.9	6.2	12.1	43.9	51.0	14.1
1962	44,700	0	65,435	4,700	0	105,435	7,216	14.6	5.6	11.0	42.6	50.4	13.1
1963	58,800	0	73,060	6,700	0	125,160	7,436	16.8	6.6	11.9	40.4	55.4	16.3
1964	75,200	0	48,480	5,800	0	117,880	7,643	15.4	5.8	11.0	38.9	52.8	14.9
1965	78,500	0	45,100	6,800	0	116,800	7,831	14.9	5.7	11.1	37.9	50.8	14.9
1966	83,200	0	60,525	7,800	0	135,925	7,995	17.0	6.5	11.9	40.9	54.8	16.0
1967	95,700	0	44,900	...	0	140,600	8,140	17.3	6.5	12.0	45.9	54.1	14.2
1968	70,800	0	34,125	...	0	104,925	8,279	12.7	4.8	10.3	43.0	46.8	11.2
1969	128,300	0	30,717	...	0	159,017	8,433	18.9	6.9	12.7	43.2	54.0	15.9
1970	171,500	0	75,148	600	0	246,048	8,616	28.6	10.2	16.1	48.8	63.3	21.0
1971	232,700	0	76,733	3,300	0	306,133	8,835	34.7	12.2	18.2	54.4	67.1	22.4
1972	281,189	0	78,697	3,500	0	356,386	9,087	39.2	13.5	18.8	52.9	71.6	25.4
1973	222,947	0	91,553	11,152	0	303,348	9,354	32.4	11.0	16.2	50.2	68.2	22.0
1974	219,513	0	89,254	3,300	0	305,467	9,608	31.8	10.9	16.4	52.6	66.5	20.7
1975	254,641	0	70,626	2,200	0	323,067	9,835	32.9	11.4	16.3	50.1	69.9	22.8
1976	236,490	0	153,744	3,112	0	387,122	10,025	38.6	13.3	18.3	48.8	72.5	27.1
1977	272,739	0	76,484	8,826	0	340,397	10,188	33.4	11.6	16.8	46.6	68.9	24.8
1978	273,471	0	76,936	39,880	0	310,527	10,352	30.0	10.5	15.8	44.5	66.3	23.6
1979	236,534	0	70,338	40,471	0	266,402	10,550	25.3	8.8	14.0	42.3	63.0	20.8
1980	231,891	0	36,594	29,892	0	238,593	10,808	22.1	7.5	12.9	43.9	58.1	17.1
1981	239,894	0	29,755	34,818	0	234,831	11,135	21.1	7.2	12.6	41.2	57.4	17.6
1982	242,680	0	21,806	33,278	0	231,209	11,521	20.1	6.9	12.5	39.2	54.9	17.5
1983	250,898	0	25,590	23,590	0	252,898	11,950	21.2	7.2	13.5	40.0	53.8	18.1
1984	269,551	0	15,320	17,675	0	267,197	12,395	21.6	7.4	13.3	41.5	55.5	17.8
1985	276,406	0	32,903	30,045	0	279,264	12,838	21.8	7.5	13.1	45.2	57.4	16.6
1986	319,882	0	43,487	35,024	0	328,344	13,273	24.7	8.5	13.8	44.8	61.4	18.9
1987	382,195	0	4,672	21,788	0	365,079	13,706	26.6	9.1	14.3	45.5	63.5	19.9
1988	362,503	0	14,126	26,006	0	350,623	14,138	24.8	8.5	13.6	45.8	62.5	18.5
1989	362,084	0	23,448	13,906	0	371,626	14,575	25.5	8.8	14.0	47.7	63.1	18.5
1990	394,820	0	22,698	26,586	0	390,932	15,018	26.0	8.9	14.0	43.5	63.2	20.4
1991	366,584	0	26,576	31,682	0	361,478	15,468	23.4	8.1	13.8	44.9	58.4	18.0
1992	426,717	0	34,274	23,844	0	437,147	15,923	27.5	9.5	15.2	50.0	62.6	19.0
1993	375,684	0	36,983	24,948	0	387,719	16,385	23.7	8.1	13.3	47.7	61.1	17.0
1994	336,986	0	18,827	33,153	0	322,660	16,856	19.1	6.4	11.4	51.3	56.4	12.5
1995	344,460	0	25,662	36,566	0	333,556	17,338	19.2	6.7	11.6	50.0	57.8	13.4

84 GREECE	- FOOD BALANCE SHEET OF FISH AND FISHERY PRODUCTS IN LIVE WEIGHT AND FISH CONTRIBUTION TO PROTEIN SUPPLY -												FAO-FIDI
YEAR	PRODUCTION	NON-FOOD USES	IMPORTS	EXPORTS	STOCKS VARIATIONS	TOTAL FOOD SUPPLY	POPULATION	PER CAPUT SUPPLY	FISH PROTEINS	ANIMAL PROTEINS	TOTAL PROTEINS	FISH/ANIMAL PROTEINS	FISH/TOTAL PROTEINS
	(...................... metric tons in live weight ....)						(thousands)	(kilograms)	(grams per caput per day)			(%)	(%)
1961	85,000	19	51,230	428	0	135,783	8,380	16.2	5.3	27.6	83.8	19.3	6.4
1962	80,000	35	54,032	610	0	133,387	8,426	15.8	5.2	30.0	83.7	17.5	6.3
1963	80,000	20	63,893	1,389	0	142,484	8,468	16.8	5.6	32.4	86.9	17.2	6.4
1964	83,500	20	56,717	3,328	0	136,870	8,509	16.1	5.2	32.6	88.6	15.9	5.9
1965	89,600	9	68,874	3,939	0	154,526	8,551	18.1	5.8	36.6	92.8	15.8	6.2
1966	90,900	17	65,963	3,368	0	153,478	8,597	17.9	5.5	38.2	93.2	14.5	5.9
1967	93,100	27	58,132	2,835	0	148,370	8,647	17.2	5.3	39.7	96.1	13.4	5.5
1968	99,800	23	67,067	2,849	0	163,995	8,698	18.9	5.8	39.9	95.2	14.6	6.1
1969	94,100	15	63,129	5,005	0	152,209	8,747	17.4	5.4	41.7	97.5	12.8	5.5
1970	98,500	0	48,273	5,157	0	141,616	8,793	16.1	5.0	42.5	99.6	11.7	5.0
1971	107,100	0	47,806	8,600	0	146,306	8,834	16.6	5.1	45.6	100.1	11.2	5.1
1972	95,670	0	37,646	12,811	0	120,505	8,871	13.6	4.2	45.2	100.0	9.2	4.2
1973	96,148	0	48,497	6,682	0	137,963	8,912	15.5	4.8	49.3	106.9	9.7	4.5
1974	92,055	0	38,475	5,479	0	125,051	8,969	13.9	4.3	48.1	105.3	9.0	4.1
1975	94,752	0	44,313	8,700	0	130,365	9,047	14.4	4.7	49.2	106.5	9.6	4.4
1976	105,627	0	45,616	10,095	0	141.148	9,150	15.4	4.9	52.6	107.3	9.3	4.5
1977	106,433	0	41,091	11,272	0	136,252	9,275	14.7	4.7	51.6	105.6	9.0	4.4
1978	102,074	0	56,499	9,566	0	149,007	9,409	15.8	4.9	54.0	108.2	9.1	4.5
1979	104,924	0	60,851	11,525	0	154,250	9,535	16.2	5.0	54.5	108.1	9.2	4.6
1980	105,214	0	55,206	9,842	0	150,579	9,643	15.6	4.7	53.0	104.1	8.9	4.6
1981	101,695	0	73,823	8,431	-1,413	165,673	9,726	17.0	4.9	52.1	106.5	9.4	4.6
1982	104,660	0	83,791	8,624	-1,667	178,161	9,790	18.2	5.0	55.2	109.2	9.1	4.6
1983	99,935	0	67,471	11,936	-27	155,442	9,839	15.8	4.7	56.1	109.2	8.4	4.3
1984	108,166	0	71,885	15,368	986	165,668	9,884	16.8	5.1	56.5	111.8	9.0	4.5
1985	114,758	0	72,967	18,877	-585	168,264	9,934	16.9	5.2	56.9	109.9	9.1	4.7
1986	124,655	0	77,095	23,724	946	178,971	9,989	17.9	5.3	57.2	106.3	9.3	5.0
1987	134,599	0	79,761	29,469	1,348	186,238	10,047	18.5	5.5	58.7	110.6	9.3	5.0
1988	126,939	15	71,397	23,171	333	175,483	10,106	17.4	5.1	55.4	108.4	9.2	4.7
1989	140,127	10	91,074	35,182	333	196,342	10,165	19.3	5.6	59.8	113.1	9.4	5.0
1990	146,367	6	91,199	24,311	0	213,249	10,220	20.9	6.0	59.3	112.7	10.1	5.3
1991	156,766	15	85,633	23,758	-20	218,606	10,272	21.3	6.0	57.8	112.7	10.4	5.3
1992	179,196	11	84,334	31,108	20	232,431	10,322	22.5	6.1	60.3	113.8	10.2	5.4
1993	199,108	7	85,872	38,559	-520	245,894	10,369	23.7	6.5	59.2	111.5	10.9	5.8
1994	223,351	15	93,472	37,882	429	279,355	10,413	26.8	7.3	59.9	114.5	12.2	6.4
1995	198,217	56	100,328	44,154	91	254,427	10,454	24.3	6.6	60.3	113.2	11.0	5.9

85 GREENLAND	- FOOD BALANCE SHEET OF FISH AND FISHERY PRODUCTS IN LIVE WEIGHT AND FISH CONTRIBUTION TO PROTEIN SUPPLY -											FAO-FIDI
YEAR	PRODUCTION	NON-FOOD USES	IMPORTS	EXPORTS	STOCKS VARIATIONS	TOTAL FOOD SUPPLY	POPULATION	PER CAPUT SUPPLY	FISH PROTEINS	ANIMAL - PROTEINS	TOTAL PROTEINS	FISH/ANIMAL PROTEINS	FISH/TOTAL PROTEINS
	(.. metric tons in live weight ....)						(thousands)	(kilograms)	(grams per caput per day)			(%)	(%)
1961	41,800	10,000	200	29,371	0	2,629	34	76.5	25.6	48.2	72.7	53.0	35.2
1962	43,300	9,895	200	30,810	0	2,795	36	78.1	27.5	47.9	76.8	57.4	35.8
1963	33,300	0	300	30,710	0	2,890	37	77.7	26.0	47.1	74.4	55.3	35.0
1964	38,200	9,940	300	25,547	0	3,013	39	78.0	17.9	44.7	73.1	40.1	24.5
1965	40,600	9,883	250	27,842	0	3,125	40	78.1	21.3	50.7	80.3	42.0	26.5
1966	44,600	9,721	250	31,898	0	3,231	41	78.2	23.0	49.9	80.9	46.0	28.4
1967	45,000	9,121	8,143	40,695	0	3,326	43	78.1	24.9	56.8	88.4	43.9	28.2
1968	34,300	4,950	5,686	31,505	0	3,531	44	80.7	26.2	54.2	84.0	48.4	31.2
1969	39,500	11,350	2,600	26,890	-200	3,660	45	81.5	25.3	53.5	80.6	47.2	31.3
1970	38,500	5,300	1,937	28,440	-2,974	3,723	46	80.9	27.6	59.3	86.3	46.6	32.0
1971	38,300	7,330	1,043	26,736	-1,462	3,816	47	81.1	27.2	58.1	85.5	46.8	31.8
1972	41,800	10,000	800	28,640	-100	3,860	48	80.4	23.7	57.7	87.2	41.0	27.2
1973	44,100	8,900	200	31,383	-55	3,962	49	81.1	25.0	52.8	83.3	47.4	30.1
1974	51,260	14,100	387	33,505	0	4,042	50	81.6	18.7	46.4	70.9	40.3	26.4
1975	47,593	8,550	355	35,348	0	4,049	50	81.0	21.4	50.6	74.4	42.3	28.8
1976	44,644	6,530	436	37,316	2,800	4,034	50	80.4	19.9	51.4	81.7	38.7	24.4
1977	59,865	10,090	333	43,077	-3,000	4,031	50	80.4	21.9	56.7	87.2	38.7	25.2
1978	68,233	17,890	903	45,578	-1,590	4,079	50	81.6	22.1	61.1	91.5	36.2	24.2
1979	83,135	12,585	395	67,263	435	4,117	50	82.5	21.2	66.8	98.0	31.8	21.6
1980	103,738	15,485	462	86,314	1,770	4,170	50	83.4	23.6	71.6	106.6	32.9	22.1
1981	107,486	15,400	511	91,187	2,862	4,272	50	84.9	24.1	73.6	109.6	32.8	22.0
1982	105,977	22,070	507	79,698	416	4,300	51	84.6	21.5	68.6	104.9	31.3	20.5
1983	99,310	17,985	430	77,655	216	4,316	52	83.7	21.6	69.4	110.3	31.1	19.6
1984	86,274	20,510	1,552	61,920	-1,054	4,342	52	83.0	23.5	70.3	109.1	33.5	21.6
1985	94,949	7,750	569	82,742	-622	4,404	53	83.1	22.8	71.2	103.1	32.1	22.1
1986	102,906	11,055	586	87,899	-109	4,429	54	82.5	23.6	70.6	104.2	33.5	22.7
1987	100,413	3,770	1,078	93,125	-146	4,450	54	81.9	21.2	72.7	109.4	29.1	19.4
1988	120,337	105	724	118,384	1,952	4,524	55	82.4	22.9	73.1	103.3	31.4	22.2
1989	168,786	8,450	1,112	157,443	620	4,625	55	83.4	20.9	71.7	103.7	29.2	20.2
1990	143,331	8,370	687	131,815	817	4,649	56	83.0	21.5	64.7	91.6	33.2	23.5
1991	114,164	2,130	2,645	109,117	-900	4,662	56	82.6	18.2	62.2	94.1	29.3	19.4
1992	113,267	0	4,563	114,515	1,301	4,616	57	81.2	14.8	56.1	82.9	26.3	17.8
1993	115,094	10,300	3,637	103,710	0	4,721	57	82.4	18.3	53.0	70.1	34.5	26.1
1994	117,445	10,841	2,574	104,359	0	4,819	58	83.6	19.0	52.2	67.6	36.4	28.1
1995	129,018	15,550	2,603	111,172	0	4,900	58	84.5	18.9	49.4	62.2	38.3	30.4

86 GRENADA	- FOOD BALANCE SHEET OF FISH AND FISHERY PRODUCTS IN LIVE WEIGHT AND FISH CONTRIBUTION TO PROTEIN SUPPLY -	FAO-FIDI

YEAR	PRODUCTION	NON-FOOD USES	IMPORTS	EXPORTS	STOCKS VARIATIONS	TOTAL FOOD SUPPLY	POPULATION	PER CAPUT SUPPLY	FISH PROTEINS	ANIMAL PROTEINS	TOTAL PROTEINS	FISH/ANIMAL PROTEINS	FISH/TOTAL PROTEINS
	(.. metric tons in live weight ....)						(thousands)	(kilograms)	(grams per caput per day)			(%)	(%)
1961	500	0	1,450	0	0	1,950	91	21.3	7.1	16.2	42.4	43.5	16.6
1962	900	0	1,450	0	0	2,350	93	25.2	8.2	18.4	44.4	44.6	18.5
1963	600	0	1,700	0	0	2,300	95	24.2	8.1	19.2	45.7	42.3	17.7
1964	1,100	0	1,450	0	0	2,550	96	26.5	8.8	20.8	47.4	42.4	18.6
1965	1,300	0	1,450	0	0	2,750	97	28.4	9.1	21.5	48.8	42.2	18.6
1966	1,200	0	1,700	0	0	2,900	97	29.9	9.8	24.5	52.9	40.0	18.5
1967	1,300	0	1,150	0	0	2,450	97	25.4	8.1	22.7	51.4	35.6	15.7
1968	1,300	0	700	0	0	2,000	96	20.9	6.6	23.8	52.3	27.9	12.7
1969	1,400	0	1,650	0	0	3,050	95	32.2	10.5	27.6	57.4	37.8	18.2
1970	1,300	0	1,650	0	0	2,950	94	31.4	10.5	28.9	58.5	36.3	17.9
1971	1,500	0	1,900	0	0	3,400	93	36.4	12.2	31.4	58.8	38.9	20.8
1972	1,700	0	1,700	0	0	3,400	93	36.5	12.3	30.8	58.1	39.8	21.1
1973	1,700	0	1,650	300	0	3,050	93	32.9	10.9	30.5	57.6	35.6	18.9
1974	1,634	0	1,650	300	0	2,984	92	32.3	10.7	29.4	56.6	36.2	18.8
1975	1,700	0	1,650	300	0	3,050	92	33.2	11.0	30.3	57.5	36.1	19.0
1976	1,700	0	1,650	0	0	3,350	91	36.7	12.2	32.3	60.4	37.6	20.1
1977	2,958	0	1,650	0	0	4,608	91	50.8	16.9	38.5	68.0	44.0	24.9
1978	2,238	3	1,121	10	0	3,346	90	37.2	12.6	35.4	65.8	35.7	19.2
1979	1,491	0	1,113	11	0	2,593	89	29.0	9.9	33.9	63.9	29.1	15.5
1980	1,419	4	1,162	16	0	2,561	89	28.8	9.9	34.2	63.8	29.0	15.5
1981	788	3	1,105	11	0	1,880	89	21.1	7.2	33.5	62.9	21.4	11.4
1982	982	2	1,076	0	0	2,056	89	23.1	7.7	33.1	60.2	23.4	12.9
1983	1,484	1	823	0	0	2,306	89	25.8	8.6	32.4	59.9	26.5	14.3
1984	1,559	3	1,178	0	0	2,734	90	30.5	10.1	35.3	61.7	28.7	16.4
1985	1,731	5	1,303	0	0	3,029	90	33.7	11.0	39.7	66.3	27.7	16.6
1986	2,703	4	1,050	0	0	3,749	90	41.5	13.4	44.3	70.3	30.2	19.0
1987	2,225	8	1,385	24	0	3,578	90	39.6	12.1	45.3	69.9	26.8	17.4
1988	2,001	7	1,387	117	0	3,265	91	36.0	11.2	42.8	69.8	26.2	16.1
1989	1,710	5	1,495	71	0	3,129	91	34.5	10.8	38.7	67.6	27.9	16.0
1990	1,784	4	1,397	26	0	3,151	91	34.6	11.1	40.4	69.8	27.5	16.0
1991	1,990	13	1,567	72	0	3,473	91	38.1	12.2	41.8	72.2	29.2	16.9
1992	2,052	12	1,172	72	0	3,140	91	34.4	11.3	39.1	70.4	28.9	16.1
1993	2,105	13	1,047	195	0	2,944	92	32.2	10.6	37.3	67.2	28.4	15.8
1994	1,635	40	1,098	286	0	2,407	92	26.2	8.8	37.3	67.3	23.5	13.0
1995	1,486	7	869	423	0	1,926	92	20.9	6.8	36.7	66.7	18.6	10.3

87 GUADELOUPE	- FOOD BALANCE SHEET OF FISH AND FISHERY PRODUCTS IN LIVE WEIGHT AND FISH CONTRIBUTION TO PROTEIN SUPPLY -	FAO-FIDI

YEAR	PRODUCTION	NON-FOOD USES	IMPORTS	EXPORTS	STOCKS VARIATIONS	TOTAL FOOD SUPPLY	POPULATION	PER CAPUT SUPPLY	FISH PROTEINS	ANIMAL PROTEINS	TOTAL PROTEINS	FISH/ANIMAL PROTEINS	FISH/TOTAL PROTEINS
	(...................................... metric tons in live weight ....)						(thousands)	(kilograms)	(grams per caput per day)			(%)	(%)
1961	3,300	0	6,158	0	0	9,458	281	33.6	11.1	20.8	54.0	53.3	20.5
1962	3,000	0	5,855	0	0	8,855	287	30.9	10.2	20.2	53.8	50.4	19.0
1963	3,200	0	6,158	0	0	9,358	291	32.1	10.6	19.9	54.7	53.6	19.4
1964	3,300	0	6,683	0	0	9,983	296	33.7	11.1	21.4	57.6	52.1	19.3
1965	3,100	0	5,863	0	0	8,963	300	29.9	9.6	21.6	57.5	44.4	16.7
1966	2,600	0	5,663	0	0	8,263	305	27.1	8.8	23.8	60.4	37.1	14.6
1967	3,500	0	5,368	0	0	8,868	309	28.7	9.2	25.1	61.9	36.6	14.8
1968	3,600	0	4,768	0	0	8,368	313	26.7	8.6	25.5	62.6	33.8	13.8
1969	4,500	0	5,479	0	0	9,979	317	31.5	10.0	28.5	65.6	35.1	15.2
1970	4,700	0	5,376	100	0	9,976	320	31.2	9.8	29.8	67.7	33.1	14.5
1971	4,900	0	4,873	100	0	9,673	323	30.0	9.4	30.7	70.2	30.7	13.4
1972	5,000	0	5,073	0	0	10,073	325	31.0	9.7	31.3	69.0	31.0	14.1
1973	4,620	0	4,489	0	0	9,109	327	27.9	8.6	32.1	69.3	26.8	12.4
1974	6,000	0	4,608	0	0	10,608	328	32.3	10.1	32.8	70.3	30.7	14.3
1975	4,740	0	5,197	0	0	9,937	329	30.3	9.5	36.4	71.9	26.0	13.2
1976	5,060	0	5,014	0	0	10,074	328	30.7	9.8	37.0	72.6	26.6	13.5
1977	9,525	0	5,076	13	0	14,588	327	44.6	13.7	41.9	76.4	32.7	17.9
1978	9,000	0	5,196	8	0	14,187	326	43.6	13.4	41.4	74.7	32.5	18.0
1979	8,500	0	5,592	22	0	14,070	326	43.2	13.4	45.4	79.7	29.5	16.8
1980	8,000	0	5,676	14	0	13,662	327	41.8	12.9	47.0	82.6	27.5	15.6
1981	8,300	0	6,139	43	0	14,396	330	43.6	13.5	46.4	80.6	29.1	16.8
1982	8,730	0	5,457	33	0	14,154	335	42.2	13.1	47.1	81.8	27.7	16.0
1983	8,836	0	6,992	19	0	15,809	341	46.3	14.3	49.3	85.1	28.9	16.7
1984	8,980	0	6,265	13	0	15,232	348	43.8	13.6	47.7	82.8	28.4	16.4
1985	8,421	0	6,289	69	0	14,641	355	41.2	12.8	48.8	84.5	26.2	15.1
1986	8,546	0	6,684	63	0	15,167	362	41.9	13.0	47.3	83.1	27.4	15.6
1987	8,606	0	6,797	14	0	15,389	369	41.7	12.8	46.9	85.5	27.3	15.0
1988	8,233	0	8,255	16	0	16,472	376	43.8	13.0	47.1	86.0	27.5	15.1
1989	8,544	0	8,462	56	0	16,950	384	44.2	13.3	46.6	86.2	28.5	15.4
1990	8,642	0	7,824	31	0	16,435	391	42.1	12.4	46.7	83.5	26.6	14.9
1991	8,544	2	8,928	29	0	17,441	398	43.9	12.7	47.0	84.9	27.1	15.0
1992	8,535	1	9,167	18	0	17,683	404	43.7	12.7	45.9	84.2	27.7	15.1
1993	8,640	1	9,306	43	0	17,902	411	43.6	12.5	48.3	85.1	25.9	14.7
1994	8,826	1	10,416	20	-4	19,217	418	46.0	13.1	50.9	86.5	25.8	15.2
1995	9,530	1	11,700	15	-4	21,210	424	50.0	14.0	51.1	88.8	27.5	15.8

88 GUAM	- FOOD BALANCE SHEET OF FISH AND FISHERY PRODUCTS IN LIVE WEIGHT AND FISH CONTRIBUTION TO PROTEIN SUPPLY -												FAO-FIDI
YEAR	PRODUCTION	NON-FOOD USES	IMPORTS	EXPORTS	STOCKS VARIATIONS	TOTAL FOOD SUPPLY	POPULATION	PER CAPUT SUPPLY	FISH PROTEINS	ANIMAL PROTEINS	TOTAL PROTEINS	FISH/ANIMAL PROTEINS	FISH/TOTAL PROTEINS
	(........................ metric tons in live weight ....)						(thousands)	(kilograms)	(grams per caput per day)			(%)	(%)
1961	100	0	0	0	0	100	69	1.5	0.4	15.1	28.1	2.7	1.5
1962	100	0	0	0	0	100	70	1.4	0.4	15.9	29.3	2.5	1.4
1963	100	0	0	0	0	100	72	1.4	0.4	17.1	30.6	2.3	1.3
1964	100	0	0	0	0	100	74	1.4	0.4	18.9	32.6	2.0	1.2
1965	100	0	0	0	0	100	76	1.3	0.4	17.7	30.0	2.1	1.2
1966	100	0	0	0	0	100	78	1.3	0.4	20.9	40.4	1.7	0.9
1967	100	0	0	0	0	100	80	1.3	0.4	15.6	35.2	2.3	1.0
1968	100	0	0	0	0	100	82	1.2	0.4	16.0	34.7	2.2	1.0
1969	100	0	0	0	0	100	84	1.2	0.3	21.2	39.5	1.6	0.9
1970	100	0	0	0	0	100	85	1.2	0.3	35.9	57.8	0.9	0.6
1971	100	0	0	0	0	100	87	1.1	0.3	34.8	58.8	0.9	0.6
1972	100	0	0	0	0	100	89	1.1	0.3	36.6	62.1	0.9	0.5
1973	100	0	0	0	0	100	91	1.1	0.3	35.4	61.2	0.9	0.5
1974	91	0	0	0	0	91	93	1.0	0.3	35.8	58.6	0.8	0.5
1975	131	0	0	0	0	131	95	1.4	0.4	33.5	53.5	1.2	0.7
1976	95	0	0	0	0	95	97	1.0	0.3	31.4	51.0	0.9	0.5
1977	184	0	1,374	1,374	0	184	100	1.9	0.5	33.6	55.3	1.6	0.9
1978	200	0	2,611	2,611	0	200	102	2.0	0.6	37.7	61.6	1.5	0.9
1979	164	0	5,071	5,071	0	164	104	1.6	0.5	52.8	77.8	0.9	0.6
1980	236	0	3,551	3,551	0	236	107	2.2	0.6	51.1	81.4	1.2	0.8
1981	293	0	0	0	0	293	109	2.7	0.8	49.9	76.0	1.5	1.0
1982	327	0	0	0	0	327	112	2.9	0.8	50.0	74.3	1.7	1.1
1983	324	0	4,146	4,146	0	324	114	2.8	0.9	50.6	74.9	1.7	1.2
1984	432	0	0	0	0	432	117	3.7	1.2	50.6	74.0	2.3	1.6
1985	539	0	0	0	0	539	119	4.5	1.4	52.0	76.4	2.7	1.8
1986	556	0	0	0	0	556	122	4.6	1.4	39.4	61.8	3.5	2.2
1987	512	0	0	0	0	512	125	4.1	1.3	42.4	64.3	2.9	1.9
1988	572	0	0	0	0	572	128	4.5	1.4	39.1	60.6	3.6	2.3
1989	597	0	0	0	0	597	131	4.6	1.4	41.2	62.3	3.5	2.3
1990	706	0	0	0	0	706	134	5.3	1.7	38.4	57.9	4.3	2.9
1991	752	0	0	0	0	752	137	5.5	1.8	30.1	48.9	5.9	3.6
1992	532	0	0	0	0	532	140	3.8	1.2	29.2	47.6	4.2	2.6
1993	642	0	0	0	0	642	144	4.5	1.4	27.7	47.0	5.2	3.1
1994	650	0	0	0	0	650	147	4.4	1.4	29.0	49.1	4.8	2.8
1995	401	0	0	0	0	401	150	2.7	0.8	25.7	47.2	3.3	1.8

89 GUATEMALA	- FOOD BALANCE SHEET OF FISH AND FISHERY PRODUCTS IN LIVE WEIGHT AND FISH CONTRIBUTION TO PROTEIN SUPPLY -											FAO-FIDI
YEAR	PRODUCTION	NON-FOOD USES	IMPORTS	EXPORTS	STOCKS VARIATIONS	TOTAL FOOD SUPPLY	POPULATION	PER CAPUT SUPPLY	FISH PROTEINS	ANIMAL PROTEINS	TOTAL PROTEINS	FISH/ANIMAL PROTEINS	FISH/TOTAL PROTEINS
	(.......... metric tons in live weight					..........)	(thousands)	(kilograms)	(grams per caput per day)			(%)	(%)
1961	1,300	0	1,200	300	0	2,200	4,079	0.5	0.2	11.6	53.5	1.4	0.3
1962	2,100	0	960	1,280	0	1,780	4,197	0.4	0.1	11.0	52.9	1.1	0.2
1963	2,500	0	800	1,333	0	1,967	4,318	0.5	0.1	12.2	55.2	1.1	0.2
1964	2,700	0	1,573	1,938	0	2,335	4,442	0.5	0.2	12.6	56.8	1.2	0.3
1965	2,100	0	2,333	1,533	0	2,900	4,568	0.6	0.2	11.3	55.5	1.6	0.3
1966	3,000	0	1,529	1,186	-1,150	2,193	4,698	0.5	0.1	10.7	54.9	1.3	0.3
1967	2,500	0	1,927	3,309	1,145	2,264	4,830	0.5	0.1	11.1	55.7	1.2	0.2
1968	1,900	0	1,471	1,300	0	2,071	4,965	0.4	0.1	11.4	56.1	1.1	0.2
1969	3,000	0	1,000	1,700	0	2,300	5,104	0.5	0.1	11.5	56.5	1.2	0.3
1970	3,400	0	1,620	2,580	0	2,440	5,246	0.5	0.1	11.9	56.5	1.2	0.3
1971	3,300	0	1,515	2,277	0	2,538	5,393	0.5	0.1	11.8	55.9	1.2	0.3
1972	3,100	0	2,325	2,042	0	3,383	5,544	0.6	0.2	11.8	55.5	1.5	0.3
1973	3,700	0	2,018	2,141	0	3,576	5,699	0.6	0.2	10.9	56.0	1.9	0.4
1974	3,881	0	2,485	2,290	0	4,076	5,859	0.7	0.2	10.5	55.1	2.1	0.4
1975	4,530	0	2,424	2,527	0	4,427	6,023	0.7	0.2	10.9	56.3	2.2	0.4
1976	3,653	5	2,781	2,217	0	4,212	6,191	0.7	0.2	10.6	55.2	2.0	0.4
1977	3,074	6	2,631	2,106	0	3,593	6,364	0.6	0.2	10.4	55.4	1.7	0.3
1978	5,504	7	2,441	3,062	0	4,877	6,543	0.8	0.2	10.9	55.9	2.2	0.4
1979	4,898	10	3,553	2,029	0	6,412	6,727	1.0	0.3	10.5	55.0	3.0	0.6
1980	3,507	7	3,331	2,241	0	4,591	6,917	0.7	0.2	10.3	54.8	2.3	0.4
1981	4,265	6	3,070	2,894	0	4,435	7,113	0.6	0.2	9.5	60.9	2.4	0.4
1982	4,284	6	2,401	2,432	0	4,247	7,315	0.6	0.2	9.8	57.5	2.4	0.4
1983	2,376	0	2,114	1,703	0	2,787	7,524	0.4	0.1	9.9	55.3	1.2	0.2
1984	3,024	0	2,163	1,838	-720	2,629	7,740	0.3	0.1	9.5	51.3	1.1	0.2
1985	3,166	1	2,000	2,224	-417	2,523	7,963	0.3	0.1	9.5	56.1	1.0	0.2
1986	2,647	1	1,058	1,651	-151	1,902	8,195	0.2	0.1	10.2	57.7	0.7	0.1
1987	3,077	5	2,341	1,943	-120	3,349	8,434	0.4	0.1	11.0	60.4	1.1	0.2
1988	3,725	3	5,279	3,773	555	5,782	8,681	0.7	0.2	11.2	60.3	1.7	0.3
1989	4,327	0	3,934	3,571	89	4,780	8,935	0.5	0.2	11.2	58.8	1.4	0.3
1990	7,798	7	5,063	3,839	-796	8,219	9,197	0.9	0.3	10.8	59.6	2.4	0.4
1991	7,344	6	3,357	5,297	916	6,313	9,467	0.7	0.2	10.0	57.4	2.0	0.3
1992	7,813	0	3,638	4,012	207	7,646	9,744	0.8	0.2	10.6	57.0	2.2	0.4
1993	10,868	0	3,464	6,312	183	8,203	10,029	0.8	0.2	11.0	57.0	2.2	0.4
1994	11,604	0	4,059	6,982	150	8,831	10,322	0.9	0.3	12.1	57.4	2.0	0.4
1995	11,928	0	4,370	4,441	-1,014	10,843	10,621	1.0	0.3	12.4	58.7	2.6	0.5

90 GUINEA	- FOOD BALANCE SHEET OF FISH AND FISHERY PRODUCTS IN LIVE WEIGHT AND FISH CONTRIBUTION TO PROTEIN SUPPLY -	FAO-FIDI

YEAR	PRODUCTION	NON-FOOD USES	IMPORTS	EXPORTS	STOCKS VARIATIONS	TOTAL FOOD SUPPLY	POPULATION	PER CAPUT SUPPLY	FISH PROTEINS	ANIMAL PROTEINS	TOTAL PROTEINS	FISH/ANIMAL PROTEINS	FISH/TOTAL PROTEINS
	(..		metric tons in live weight			....)	(thousands)	(kilograms)		(grams per caput per day)		(%)	(%)
1961	4,000	0	...	0	0	4,000	3,202	1.3	0.4	3.9	49.1	10.1	0.8
1962	4,000	0	...	0	0	4,000	3,269	1.2	0.4	3.9	49.0	9.9	0.8
1963	4,000	0	...	0	0	4,000	3,338	1.2	0.4	4.0	47.6	9.6	0.8
1964	5,000	0	...	0	0	5,000	3,411	1.5	0.5	4.1	47.2	11.3	1.0
1965	5,000	0	...	0	0	5,000	3,488	1.4	0.5	4.1	50.0	11.1	0.9
1966	5,000	0	...	0	0	5,000	3,572	1.4	0.4	4.1	46.0	10.8	1.0
1967	6,000	0	...	0	0	6,000	3,659	1.6	0.5	4.0	46.5	13.0	1.1
1968	6,000	0	...	0	0	6,000	3,747	1.6	0.5	4.0	48.9	12.2	1.0
1969	6,000	0	3,000	0	0	9,000	3,829	2.4	0.7	4.0	47.1	18.4	1.6
1970	6,000	0	3,000	0	0	9,000	3,900	2.3	0.7	4.0	47.8	18.3	1.5
1971	8,300	0	3,106	0	0	11,406	3,961	2.9	0.9	4.1	47.3	22.4	1.9
1972	8,300	0	7,099	0	0	15,399	4,012	3.8	1.2	4.6	47.3	25.4	2.5
1973	9,500	0	8,162	0	0	17,662	4,057	4.4	1.4	4.7	48.3	28.8	2.8
1974	12,100	0	8,343	0	0	20,443	4,101	5.0	1.6	5.1	48.6	30.2	3.2
1975	13,370	0	8,931	0	0	22,301	4,149	5.4	1.7	5.3	48.9	31.6	3.4
1976	9,920	0	11,094	0	0	21,014	4,200	5.0	1.5	5.2	49.2	28.6	3.0
1977	9,120	0	12,850	0	0	21,970	4,255	5.2	1.5	5.2	48.9	28.8	3.1
1978	10,000	0	11,760	0	0	21,760	4,316	5.0	1.5	5.2	49.2	28.4	3.0
1979	18,453	0	9,661	0	0	28,114	4,385	6.4	1.9	5.6	49.3	34.5	3.9
1980	20,000	0	10,757	0	0	30,757	4,461	6.9	2.1	5.7	49.4	36.5	4.2
1981	22,000	0	9,985	0	0	31,985	4,548	7.0	2.1	5.6	52.3	37.6	4.1
1982	24,000	0	11,671	0	0	35,671	4,647	7.7	2.3	5.8	51.0	39.6	4.5
1983	26,000	0	12,453	0	0	38,453	4,756	8.1	2.4	6.0	50.2	40.3	4.8
1984	28,003	0	13,732	0	0	41,735	4,870	8.6	2.6	6.2	50.9	41.3	5.1
1985	30,003	0	8,645	0	0	38,648	4,987	7.8	2.4	5.7	50.3	41.6	4.7
1986	33,002	0	14,505	0	0	47,507	5,103	9.3	2.8	5.5	48.8	51.0	5.7
1987	35,002	0	8,878	0	0	43,880	5,219	8.4	2.6	5.8	47.6	44.2	5.4
1988	38,001	0	9,386	0	0	47,387	5,352	8.9	2.7	5.6	48.2	47.4	5.6
1989	41,001	0	7,979	0	0	48,980	5,527	8.9	2.7	5.6	46.3	47.4	5.8
1990	44,001	0	8,862	0	0	52,863	5,755	9.2	2.8	5.6	46.0	48.7	6.0
1991	49,501	0	8,736	0	0	58,237	6,050	9.6	2.9	5.8	46.3	49.7	6.2
1992	55,005	0	7,840	0	0	62,845	6,398	9.8	2.9	6.0	47.3	48.4	6.2
1993	60,005	0	8,574	1,048	0	67,531	6,763	10.0	3.0	6.0	46.3	49.1	6.4
1994	64,405	0	8,458	1,050	0	71,813	7,092	10.1	3.0	5.9	46.2	50.8	6.5
1995	68,764	0	7,454	5,000	0	71,218	7,349	9.7	2.9	5.9	45.8	48.9	6.3

175 GUINEA-BISSAU	- FOOD BALANCE SHEET OF FISH AND FISHERY PRODUCTS IN LIVE WEIGHT AND FISH CONTRIBUTION TO PROTEIN SUPPLY -	FAO-FIDI

YEAR	PRODUCTION	NON-FOOD USES	IMPORTS	EXPORTS	STOCKS VARIATIONS	TOTAL FOOD SUPPLY	POPULATION	PER CAPUT SUPPLY	FISH PROTEINS	ANIMAL PROTEINS	TOTAL PROTEINS	FISH/ANIMAL PROTEINS	FISH/TOTAL PROTEINS
	(............ metric tons in live weight					............)	(thousands)	(kilograms)	(grams per caput per day)			(%)	(%)
1961	700	0	503	...	0	1,203	541	2.2	0.7	7.9	37.4	8.4	1.8
1962	600	0	503	...	0	1,103	538	2.1	0.6	8.0	37.2	7.7	1.7
1963	600	0	606	...	0	1,206	533	2.3	0.7	8.2	38.4	8.6	1.8
1964	800	0	303	...	0	1,103	529	2.1	0.6	8.2	39.4	7.7	1.6
1965	900	0	303	...	0	1,203	524	2.3	0.7	8.3	40.2	8.2	1.7
1966	700	0	303	...	0	1,003	520	1.9	0.6	8.4	41.9	6.9	1.4
1967	700	0	352	...	0	1,052	517	2.0	0.6	8.7	41.8	6.8	1.4
1968	1,300	0	303	...	0	1,603	516	3.1	0.9	9.2	39.6	9.9	2.3
1969	1,400	0	303	...	0	1,703	518	3.3	1.0	9.5	42.8	10.2	2.3
1970	1,500	0	803	...	0	2,303	525	4.4	1.5	10.3	40.9	14.2	3.6
1971	1,400	0	903	...	0	2,303	537	4.3	1.4	10.3	44.1	13.8	3.2
1972	1,700	0	903	...	0	2,603	553	4.7	1.6	10.3	39.6	15.1	3.9
1973	1,700	0	1,103	...	0	2,803	574	4.9	1.6	10.1	42.5	15.9	3.8
1974	1,700	0	300	...	0	2,000	599	3.3	1.1	9.1	40.0	12.3	2.8
1975	1,669	0	200	...	0	1,869	627	3.0	1.0	8.9	42.7	11.4	2.4
1976	3,575	0	1,034	1,249	0	3,360	660	5.1	1.7	9.9	47.5	16.8	3.5
1977	4,275	0	485	2,353	0	2,407	696	3.5	1.2	9.4	37.4	12.3	3.1
1978	4,283	0	176	1,670	0	2,789	732	3.8	1.3	9.3	47.7	14.0	2.7
1979	2,936	0	225	1,870	0	1,291	766	1.7	0.6	8.4	41.0	6.7	1.4
1980	4,166	0	340	891	0	3,615	795	4.6	1.5	8.8	38.6	17.5	4.0
1981	3,643	0	507	2,663	-130	1,357	817	1.7	0.6	7.8	47.1	7.2	1.2
1982	3,977	0	140	1,290	-70	2,757	834	3.3	1.0	8.3	47.4	11.8	2.1
1983	2,671	0	200	530	-100	2,241	847	2.7	0.8	8.4	43.8	9.9	1.9
1984	2,751	0	257	2,235	300	1,073	859	1.3	0.3	7.8	46.1	3.9	0.7
1985	3,743	0	626	2,232	0	2,137	873	2.5	0.7	8.2	47.3	8.2	1.4
1986	3,700	0	540	543	0	3,697	889	4.2	1.2	9.2	47.0	13.3	2.6
1987	4,100	0	500	120	0	4,480	906	4.9	1.5	9.5	49.7	15.3	2.9
1988	4,690	0	330	480	0	4,540	925	4.9	1.5	9.2	48.8	15.9	3.0
1989	5,400	0	360	720	0	5,040	944	5.3	1.6	9.6	48.8	16.6	3.3
1990	5,400	0	768	1,580	0	4,588	964	4.8	1.4	9.5	49.1	14.6	2.8
1991	5,000	0	600	2,460	0	3,140	984	3.2	0.9	8.9	50.8	9.7	1.7
1992	5,200	0	600	220	0	5,580	1,005	5.6	1.7	10.4	51.4	15.9	3.2
1993	5,350	0	620	1,020	0	4,950	1,026	4.8	1.4	10.2	50.4	13.6	2.8
1994	6,000	0	600	1,480	0	5,120	1,047	4.9	1.4	9.4	48.6	14.6	2.8
1995	6,329	0	685	1,241	0	5,773	1,069	5.4	1.6	9.4	48.3	16.7	3.2

91 GUYANA	- FOOD BALANCE SHEET OF FISH AND FISHERY PRODUCTS IN LIVE WEIGHT AND FISH CONTRIBUTION TO PROTEIN SUPPLY -	FAO-FIDI

YEAR	PRODUCTION	NON-FOOD USES	IMPORTS	EXPORTS	STOCKS VARIATIONS	TOTAL FOOD SUPPLY	POPULATION	PER CAPUT SUPPLY	FISH PROTEINS	ANIMAL PROTEINS	TOTAL PROTEINS	FISH/ANIMAL PROTEINS	FISH/TOTAL PROTEINS
	(.. metric tons in live weight					....)	(thousands)	(kilograms)	(grams per caput per day)			(%)	(%)
1961	7,400	0	9,079	1,900	0	14,579	585	24.9	7.9	21.2	55.5	37.2	14.2
1962	7,900	0	7,561	2,300	0	13,161	600	21.9	6.9	20.3	54.4	33.7	12.6
1963	9,200	0	8,061	2,800	0	14,461	615	23.5	7.3	21.2	54.1	34.3	13.5
1964	10,400	0	8,358	3,100	0	15,658	630	24.8	7.7	22.2	56.8	34.6	13.5
1965	12,000	0	7,358	3,600	0	15,758	645	24.4	7.5	22.4	55.2	33.7	13.6
1966	12,300	0	8,267	4,300	0	16,267	660	24.7	7.6	22.9	57.0	33.3	13.4
1967	13,900	0	7,558	4,000	0	17,458	674	25.9	7.9	23.4	58.1	33.9	13.6
1968	16,000	0	6,655	4,200	0	18,455	688	26.8	8.1	24.0	57.6	33.8	14.1
1969	16,600	0	7,661	4,400	0	19,861	700	28.4	8.7	25.6	59.3	34.1	14.7
1970	17,400	0	8,852	5,000	-300	20,952	709	29.5	8.8	25.8	57.4	34.0	15.3
1971	15,500	0	5,945	4,900	-200	16,345	717	22.8	6.8	24.1	55.9	28.3	12.2
1972	17,600	0	503	5,800	500	12,803	722	17.7	5.0	22.3	53.9	22.5	9.3
1973	19,000	0	0	5,400	0	13,600	726	18.7	5.3	22.2	55.4	23.8	9.5
1974	23,617	0	0	5,100	-500	18,017	730	24.7	7.1	22.8	59.7	31.2	11.9
1975	20,123	0	0	4,500	500	16,123	734	22.0	6.3	23.1	61.7	27.1	10.2
1976	20,956	0	239	3,810	0	17,385	738	23.6	7.1	25.5	62.6	27.6	11.3
1977	31,907	0	16	3,406	0	28,517	743	38.4	11.6	28.6	66.5	40.5	17.4
1978	35,978	0	0	4,074	-2,000	29,904	748	40.0	12.1	27.7	63.5	43.8	19.1
1979	35,288	0	0	3,663	-1,000	30,625	753	40.7	11.9	26.8	61.6	44.2	19.2
1980	31,530	0	0	3,405	1,500	29,626	759	39.0	11.1	28.0	62.9	39.6	17.6
1981	32,798	0	2	1,926	0	30,874	766	40.3	11.5	27.8	63.4	41.3	18.1
1982	32,794	0	2	3,074	1,000	30,722	774	39.7	11.3	22.3	54.6	50.5	20.6
1983	35,366	0	0	2,894	500	32,972	782	42.2	12.0	21.2	53.8	56.3	22.2
1984	37,256	0	0	2,048	0	35,208	789	44.6	12.8	20.3	52.7	62.8	24.2
1985	37,606	0	0	2,029	0	35,577	793	44.9	12.8	19.5	52.7	65.6	24.2
1986	37,395	0	0	2,983	-50	34,362	795	43.2	12.3	18.5	55.8	66.4	22.1
1987	36,764	0	0	2,987	-55	33,722	795	42.4	12.0	19.7	56.2	61.0	21.4
1988	36,523	0	0	3,235	15	33,303	794	42.0	11.9	19.2	56.0	61.8	21.2
1989	35,339	0	0	3,545	60	31,854	793	40.2	11.3	18.0	55.3	62.8	20.5
1990	36,922	0	0	3,799	30	33,153	795	41.7	11.8	19.1	56.8	61.6	20.7
1991	40,806	0	142	5,698	0	35,250	799	44.1	12.5	20.5	58.5	61.3	21.4
1992	41,337	0	86	5,130	0	36,293	805	45.1	13.1	25.5	63.1	51.6	20.8
1993	44,263	0	468	5,423	0	39,308	813	48.4	13.8	28.3	66.5	48.8	20.8
1994	46,577	0	596	4,850	-100	42,223	821	51.4	14.7	28.5	66.2	51.5	22.2
1995	44,620	0	256	6,885	100	38,091	830	45.9	13.0	27.9	66.4	46.5	19.5

93 HAITI	- FOOD BALANCE SHEET OF FISH AND FISHERY PRODUCTS IN LIVE WEIGHT AND FISH CONTRIBUTION TO PROTEIN SUPPLY -												FAO-FIDI
YEAR	PRODUCTION	NON-FOOD USES	IMPORTS	EXPORTS	STOCKS VARIATIONS	TOTAL FOOD SUPPLY	POPULATION	PER CAPUT SUPPLY	FISH PROTEINS	ANIMAL PROTEINS	TOTAL PROTEINS	FISH/ANIMAL PROTEINS	FISH/TOTAL PROTEINS
	(.. metric tons in live weight ....)						(thousands)	(kilograms)	(grams per caput per day)			(%)	(%)
1961	3,000	0	1,750	100	0	4,650	3,868	1.2	0.3	5.9	44.1	5.8	0.8
1962	3,000	0	1,250	100	0	4,150	3,935	1.1	0.3	5.7	44.7	5.2	0.7
1963	3,100	0	2,000	100	0	5,000	4,003	1.3	0.4	5.8	45.0	6.1	0.8
1964	3,300	0	2,250	100	0	5,450	4,072	1.3	0.4	6.1	45.9	6.3	0.8
1965	3,400	0	2,250	100	0	5,550	4,143	1.3	0.4	5.9	45.8	6.5	0.8
1966	3,400	0	4,500	100	0	7,800	4,216	1.9	0.5	5.9	45.4	9.0	1.2
1967	3,500	0	3,179	100	0	6,579	4,291	1.5	0.4	6.2	45.5	7.1	1.0
1968	3,500	0	3,143	100	0	6,543	4,366	1.5	0.4	6.4	44.1	6.8	1.0
1969	3,500	0	3,643	100	0	7,043	4,443	1.6	0.5	6.5	44.8	7.0	1.0
1970	4,000	0	1,500	100	0	5,400	4,520	1.2	0.3	6.6	44.7	5.2	0.8
1971	4,000	0	750	99	0	4,651	4,599	1.0	0.3	6.7	44.7	4.3	0.6
1972	4,000	0	2,917	100	0	6,817	4,678	1.5	0.4	6.9	45.5	6.2	0.9
1973	4,000	0	3,306	70	0	7,236	4,758	1.5	0.5	6.9	45.9	6.5	1.0
1974	4,000	0	3,756	76	0	7,680	4,839	1.6	0.5	7.1	46.1	6.6	1.0
1975	4,000	0	5,005	100	0	8,905	4,920	1.8	0.5	7.1	46.2	7.4	1.2
1976	4,000	0	6,631	107	0	10,524	5,003	2.1	0.7	7.6	46.3	8.8	1.4
1977	4,150	0	7,194	139	0	11,205	5,087	2.2	0.7	7.4	44.6	9.5	1.6
1978	4,300	0	11,519	819	0	14,999	5,173	2.9	0.9	8.2	47.6	11.6	2.0
1979	4,500	0	10,990	162	0	15,328	5,261	2.9	0.9	7.6	48.6	12.3	1.9
1980	5,000	0	15,300	227	0	20,073	5,353	3.8	1.2	8.0	48.6	15.4	2.5
1981	5,500	0	12,382	239	0	17,643	5,447	3.2	1.0	7.5	48.1	13.9	2.2
1982	6,000	0	14,892	470	0	20,422	5,544	3.7	1.2	8.0	47.2	15.0	2.5
1983	6,500	0	14,296	416	0	20,380	5,645	3.6	1.2	8.1	49.0	14.5	2.4
1984	6,900	0	21,441	380	0	27,961	5,751	4.9	1.6	8.6	49.7	18.3	3.2
1985	6,400	0	20,924	227	0	27,097	5,861	4.6	1.5	9.0	50.2	16.6	3.0
1986	6,000	0	21,947	128	0	27,819	5,976	4.7	1.5	9.0	49.6	16.9	3.1
1987	5,750	0	21,161	197	0	26,714	6,096	4.4	1.4	9.6	48.6	14.8	2.9
1988	5,500	0	17,675	183	0	22,992	6,220	3.7	1.2	7.9	46.7	14.9	2.5
1989	5,500	0	20,004	172	0	25,332	6,346	4.0	1.2	7.2	44.7	17.3	2.8
1990	5,150	0	19,764	175	0	24,739	6,473	3.8	1.2	6.9	44.0	17.1	2.7
1991	5,200	0	11,663	154	0	16,709	6,600	2.5	0.8	6.6	43.4	11.7	1.8
1992	5,000	0	12,060	120	0	16,940	6,729	2.5	0.8	6.5	44.3	11.9	1.7
1993	5,600	0	13,612	138	0	19,074	6,859	2.8	0.9	6.7	42.6	12.6	2.0
1994	5,200	0	12,674	140	0	17,734	6,991	2.5	0.8	6.4	42.2	11.9	1.8
1995	5,500	0	15,416	119	0	20,797	7,124	2.9	0.9	6.7	42.2	12.8	2.0

95 HONDURAS	- FOOD BALANCE SHEET OF FISH AND FISHERY PRODUCTS IN LIVE WEIGHT AND FISH CONTRIBUTION TO PROTEIN SUPPLY -	FAO-FIDI

YEAR	PRODUCTION	NON-FOOD USES	IMPORTS	EXPORTS	STOCKS VARIATIONS	TOTAL FOOD SUPPLY	POPULATION	PER CAPUT SUPPLY	FISH PROTEINS	ANIMAL PROTEINS	TOTAL PROTEINS	FISH/ANIMAL PROTEINS	FISH/TOTAL PROTEINS
	(....................................		metric tons in live weight		....	)	(thousands)	(kilograms)		(grams per caput per day)		(%)	(%)
1961	1,500	0	600	200	0	1,900	1,961	1.0	0.3	13.8	54.0	1.8	0.5
1962	1,500	0	0	267	0	1,233	2,031	0.6	0.2	13.2	52.2	1.2	0.3
1963	1,500	0	400	600	0	1,300	2,102	0.6	0.2	13.4	52.9	1.2	0.3
1964	1,700	0	850	450	0	2,100	2,174	1.0	0.3	13.3	52.7	1.8	0.5
1965	1,700	0	700	1,000	0	1,400	2,245	0.6	0.1	12.8	52.8	1.1	0.3
1966	2,200	0	742	1,842	0	1,100	2,313	0.5	0.1	12.4	53.5	1.1	0.3
1967	2,500	0	800	554	0	2,746	2,381	1.2	0.4	13.9	55.8	2.5	0.6
1968	3,600	0	1,584	711	0	4,474	2,449	1.8	0.6	15.2	55.7	3.7	1.0
1969	4,000	0	852	1,524	0	3,329	2,519	1.3	0.4	14.5	55.5	2.7	0.7
1970	5,000	0	1,350	2,267	0	4,083	2,592	1.6	0.4	14.6	56.0	3.0	0.8
1971	5,700	0	1,050	3,638	0	3,112	2,669	1.2	0.3	15.0	54.7	2.2	0.6
1972	4,800	0	1,700	3,846	0	2,654	2,749	1.0	0.3	14.5	53.0	1.8	0.5
1973	3,600	0	900	3,000	0	1,500	2,833	0.5	0.1	14.0	52.7	1.0	0.3
1974	3,221	0	1,800	2,084	0	2,937	2,922	1.0	0.3	14.3	51.8	2.1	0.6
1975	3,958	0	1,410	3,084	0	2,284	3,017	0.8	0.2	13.8	52.9	1.5	0.4
1976	3,807	1	2,065	3,089	0	2,782	3,118	0.9	0.3	13.7	52.2	2.0	0.5
1977	5,189	2	2,872	3,605	-25	4,429	3,225	1.4	0.4	13.9	51.5	2.7	0.7
1978	6,375	2	1,985	3,614	25	4,770	3,337	1.4	0.4	15.2	52.1	2.5	0.7
1979	7,302	2	2,336	4,108	0	5,528	3,452	1.6	0.4	14.6	51.7	2.9	0.8
1980	6,409	0	3,699	4,539	-263	5,306	3,569	1.5	0.4	15.4	52.7	2.8	0.8
1981	6,300	0	3,166	4,985	0	4,481	3,688	1.2	0.4	15.5	53.0	2.3	0.7
1982	8,023	2	1,583	4,485	25	5,145	3,808	1.4	0.4	14.2	51.0	3.0	0.9
1983	11,432	2	944	5,844	150	6,680	3,932	1.7	0.5	13.2	51.3	3.9	1.0
1984	8,381	2	1,440	4,957	0	4,862	4,057	1.2	0.3	13.6	51.2	2.2	0.6
1985	9,693	2	1,429	3,723	25	7,422	4,186	1.8	0.5	14.3	52.4	3.5	0.9
1986	20,559	2	2,071	12,082	-908	9,638	4,319	2.2	0.6	13.1	50.6	4.3	1.1
1987	24,601	2	2,512	21,220	957	6,848	4,454	1.5	0.4	14.7	51.9	2.8	0.8
1988	19,951	0	2,435	14,647	-266	7,472	4,592	1.6	0.4	14.9	53.4	2.9	0.8
1989	17,434	2	2,221	14,220	0	5,433	4,734	1.2	0.3	15.0	54.5	2.1	0.6
1990	15,592	0	1,441	10,474	-1,333	5,225	4,879	1.1	0.3	15.2	55.2	1.9	0.5
1991	24,699	1	1,582	19,602	13	6,690	5,028	1.3	0.3	15.4	55.3	2.2	0.6
1992	22,741	0	1,431	11,583	-5,239	7,350	5,180	1.4	0.4	16.4	55.0	2.1	0.6
1993	31,684	1	2,413	22,615	0	11,481	5,335	2.2	0.6	15.9	55.2	3.8	1.1
1994	27,516	0	1,847	23,509	5,695	11,549	5,493	2.1	0.6	16.4	55.2	3.6	1.1
1995	26,994	1	3,857	15,663	-33	15,153	5,654	2.7	0.7	15.9	54.6	4.6	1.3

96 HONG KONG	- FOOD BALANCE SHEET OF FISH AND FISHERY PRODUCTS IN LIVE WEIGHT AND FISH CONTRIBUTION TO PROTEIN SUPPLY -												FAO-FIDI
YEAR	PRODUCTION	NON-FOOD USES	IMPORTS	EXPORTS	STOCKS VARIATIONS	TOTAL FOOD SUPPLY	POPULATION	PER CAPUT SUPPLY	FISH PROTEINS	ANIMAL PROTEINS	TOTAL PROTEINS	FISH/ANIMAL PROTEINS	FISH/TOTAL PROTEINS
	(... metric tons in live weight ....)						(thousands)	(kilograms)	(grams per caput per day)			(%)	(%)
1961	54,700	0	48,891	7,200	0	96,391	3,207	30.1	9.6	28.5	67.9	33.6	14.1
1962	59,300	0	53,991	7,983	0	105,308	3,344	31.5	9.5	30.3	69.7	31.5	13.7
1963	60,800	0	57,616	9,642	0	108,774	3,477	31.3	9.2	31.0	67.9	29.8	13.6
1964	63,600	0	74,186	9,642	0	128,145	3,596	35.6	10.1	33.0	70.1	30.8	14.5
1965	66,500	0	76,174	10,617	0	132,057	3,692	35.8	10.0	33.6	70.2	29.6	14.2
1966	64,600	2,000	85,183	14,592	0	133,191	3,763	35.4	11.1	36.5	71.7	30.3	15.4
1967	67,800	4,000	83,381	12,908	0	134,273	3,812	35.2	10.7	35.6	73.9	30.0	14.5
1968	111,900	4,000	77,877	11,975	0	173,802	3,850	45.2	13.4	40.2	78.4	33.3	17.1
1969	117,100	4,000	74,921	13,975	0	174,046	3,889	44.8	13.6	41.1	79.9	33.0	17.0
1970	135,400	4,000	78,997	9,517	0	200,881	3,942	51.0	15.0	44.4	83.8	33.9	17.9
1971	127,100	3,713	90,845	11,342	0	202,890	4,009	50.6	14.9	47.1	85.9	31.7	17.3
1972	134,700	3,368	91,797	15,708	0	207,421	4,088	50.7	15.3	48.5	88.0	31.6	17.4
1973	127,900	5,854	93,156	18,367	0	196,835	4,180	47.1	14.7	48.1	87.3	30.6	16.9
1974	142,016	5,700	92,190	20,133	0	208,372	4,283	48.7	14.9	46.2	85.5	32.3	17.5
1975	150,882	8,403	96,985	28,000	0	211,464	4,396	48.1	14.5	46.0	84.5	31.5	17.1
1976	156,896	9,434	112,308	58,324	-125	201,321	4,519	44.6	14.0	47.4	85.7	29.6	16.4
1977	158,362	9,153	107,203	62,260	-525	193,627	4,652	41.6	12.6	48.0	84.2	26.3	15.0
1978	162,594	8,548	114,497	72,183	456	196,816	4,788	41.1	12.9	49.4	86.6	26.0	14.8
1979	189,802	9,391	127,845	86,737	88	221,606	4,919	45.1	14.1	51.2	87.7	27.6	16.1
1980	194,565	11,382	137,585	103,743	75	217,100	5,039	43.1	13.2	51.6	88.7	25.6	14.9
1981	182,335	11,579	132,747	97,845	38	205,696	5,145	40.0	12.3	51.1	87.9	24.0	14.0
1982	180,990	12,327	156,879	101,736	-600	223,206	5,238	42.6	13.6	52.9	89.6	25.7	15.2
1983	189,278	10,008	173,314	117,329	-1,120	234,135	5,321	44.0	13.9	53.9	89.9	25.7	15.4
1984	200,388	12,086	171,698	127,319	1,833	234,514	5,393	43.5	13.6	52.6	88.0	25.8	15.4
1985	198,196	13,378	200,479	152,364	-56	232,878	5,456	42.7	13.0	52.4	88.1	24.8	14.7
1986	213,557	15,737	241,149	177,186	0	261,782	5,510	47.5	14.3	55.9	94.2	25.6	15.2
1987	228,094	27,737	281,068	209,757	-1,144	270,524	5,555	48.7	14.6	57.0	93.7	25.7	15.6
1988	238,164	35,026	338,784	260,123	0	281,799	5,597	50.4	14.4	58.6	94.0	24.6	15.4
1989	242,510	34,105	337,675	283,725	1,200	263,555	5,645	46.7	13.9	56.7	91.2	24.4	15.2
1990	234,493	33,582	369,892	265,335	0	305,468	5,705	53.6	15.6	59.8	93.5	26.1	16.7
1991	230,910	34,820	415,623	271,528	0	340,184	5,778	58.9	17.0	60.4	94.0	28.2	18.1
1992	229,514	32,142	457,327	302,600	-5,375	346,724	5,864	59.1	16.9	63.2	95.6	26.7	17.6
1993	226,843	32,828	424,841	277,628	5,250	346,478	5,955	58.2	16.8	61.5	94.6	27.3	17.8
1994	219,976	27,102	454,532	293,974	125	353,556	6,044	58.5	16.9	71.4	101.6	23.7	16.6
1995	203,570	18,002	432,780	269,329	0	349,019	6,123	57.0	17.4	78.7	109.2	22.1	15.9

97 HUNGARY	- FOOD BALANCE SHEET OF FISH AND FISHERY PRODUCTS IN LIVE WEIGHT AND FISH CONTRIBUTION TO PROTEIN SUPPLY -											FAO-FIDI
YEAR	PRODUCTION	NON-FOOD USES	IMPORTS	EXPORTS	STOCKS VARIATIONS	TOTAL FOOD SUPPLY	POPULATION	PER CAPUT SUPPLY	FISH PROTEINS	ANIMAL PROTEINS	TOTAL PROTEINS	FISH/ANIMAL PROTEINS	FISH/TOTAL PROTEINS
	(.. metric tons in live weight ....)						(thousands)	(kilograms)	(grams per caput per day)			(%)	(%)
1961	19,300	0	400	0	0	19,700	10,012	2.0	0.6	34.5	86.6	1.7	0.7
1962	21,000	0	400	0	0	21,400	10,043	2.1	0.6	35.1	88.0	1.8	0.7
1963	22,000	0	4,113	0	0	26,113	10,077	2.6	0.8	34.9	87.4	2.3	0.9
1964	22,700	0	6,238	3,300	0	25,638	10,112	2.5	0.7	36.0	88.2	2.0	0.8
1965	24,800	0	8,850	2,800	0	30,850	10,148	3.0	0.9	35.3	88.8	2.4	1.0
1966	26,600	0	15,088	2,900	0	38,788	10,184	3.8	1.0	35.4	88.0	2.8	1.1
1967	28,200	0	17,363	3,200	0	42,363	10,221	4.1	1.1	38.0	90.3	2.9	1.2
1968	29,900	0	36,088	2,600	0	63,388	10,260	6.2	1.7	39.7	90.9	4.2	1.9
1969	27,800	0	17,113	2,500	0	42,413	10,299	4.1	1.1	41.1	92.1	2.7	1.2
1970	26,000	0	12,828	3,000	0	35,828	10,338	3.5	1.0	42.2	92.5	2.4	1.1
1971	25,800	0	14,294	3,500	0	36,594	10,376	3.5	1.0	42.2	92.3	2.5	1.1
1972	28,900	0	12,850	2,900	0	38,850	10,414	3.7	1.1	44.0	93.6	2.5	1.2
1973	29,100	0	14,022	3,400	0	39,722	10,452	3.8	1.1	44.3	93.3	2.6	1.2
1974	30,159	0	12,311	3,800	0	38,670	10,491	3.7	1.1	45.6	94.8	2.4	1.1
1975	30,788	0	15,022	2,900	0	42,910	10,532	4.1	1.2	47.9	96.5	2.5	1.2
1976	31,855	0	16,340	3,588	0	44,607	10,576	4.2	1.3	48.6	96.4	2.6	1.3
1977	34,661	0	13,838	4,200	0	44,299	10,621	4.2	1.3	47.8	95.2	2.6	1.3
1978	32,584	0	12,830	4,206	0	41,208	10,662	3.9	1.2	49.2	96.8	2.3	1.2
1979	32,828	0	14,807	4,674	0	42,961	10,693	4.0	1.2	49.1	95.7	2.4	1.2
1980	33,713	0	15,192	6,740	0	42,165	10,707	3.9	1.2	49.5	96.4	2.3	1.2
1981	39,269	0	11,387	8,338	0	42,318	10,705	4.0	1.2	50.8	96.8	2.3	1.2
1982	42,042	0	17,143	7,136	0	52,049	10,686	4.9	1.4	51.7	97.4	2.8	1.5
1983	43,857	0	19,913	4,830	0	58,940	10,656	5.5	1.6	53.2	98.5	3.0	1.6
1984	39,167	0	19,384	5,430	0	53,121	10,618	5.0	1.5	52.6	98.5	2.8	1.5
1985	36,927	0	14,605	5,222	0	46,310	10,579	4.4	1.3	57.9	104.2	2.2	1.2
1986	36,062	0	19,754	7,852	0	47,964	10,540	4.6	1.3	58.8	104.7	2.3	1.3
1987	36,759	0	27,684	6,031	0	58,412	10,499	5.6	1.6	60.2	107.0	2.7	1.5
1988	38,294	0	24,676	3,987	0	58,983	10,456	5.6	1.6	58.5	105.0	2.8	1.6
1989	35,517	0	24,141	6,074	0	53,584	10,412	5.2	1.5	59.9	106.0	2.5	1.4
1990	33,888	0	18,804	6,237	0	46,455	10,365	4.5	1.3	55.9	101.1	2.3	1.3
1991	29,378	47	17,309	4,091	-122	42,427	10,316	4.1	1.2	53.9	96.3	2.2	1.2
1992	29,436	9	16,257	3,038	-60	42,586	10,266	4.2	1.2	54.2	97.2	2.2	1.2
1993	23,404	53	18,339	2,025	28	39,694	10,215	3.9	1.1	51.3	92.0	2.1	1.2
1994	23,957	14	22,094	3,472	153	42,719	10,162	4.2	1.2	49.8	90.1	2.4	1.3
1995	22,866	14	20,781	3,123	153	40,663	10,106	4.0	1.1	47.9	87.7	2.4	1.3

99 ICELAND	- FOOD BALANCE SHEET OF FISH AND FISHERY PRODUCTS IN LIVE WEIGHT AND FISH CONTRIBUTION TO PROTEIN SUPPLY -	FAO-FIDI

YEAR	PRODUCTION	NON-FOOD USES	IMPORTS	EXPORTS	STOCKS VARIATIONS	TOTAL FOOD SUPPLY	POPULATION	PER CAPUT SUPPLY	FISH PROTEINS	ANIMAL PROTEINS	TOTAL PROTEINS	FISH/ANIMAL PROTEINS	FISH/TOTAL PROTEINS
	(.. metric tons in live weight					....)	(thousands)	(kilograms)	(grams per caput per day)			(%)	(%)
1961	716,000	245,400	200	452,842	-7,230	10,728	179	59.8	18.7	92.7	120.3	20.2	15.5
1962	832,400	370,400	400	475,676	24,531	11,255	183	61.6	17.4	93.3	123.1	18.7	14.2
1963	782,500	293,900	600	476,681	-900	11,619	186	62.4	19.0	92.0	117.3	20.6	16.2
1964	974,100	484,800	200	499,785	22,452	12,167	189	64.3	23.2	94.9	123.4	24.5	18.8
1965	1,197,800	724,800	0	461,061	756	12,696	192	66.1	19.5	91.0	119.6	21.4	16.3
1966	1,240,200	804,800	0	430,884	9,324	13,840	195	71.1	25.6	97.8	123.4	26.2	20.7
1967	897,100	479,800	0	375,022	-29,321	12,957	197	65.8	14.9	84.9	110.7	17.5	13.4
1968	599,300	138,350	0	441,514	-5,283	14,153	199	71.0	15.3	82.0	109.5	18.7	14.0
1969	689,200	181,100	300	509,177	14,281	13,503	202	67.0	18.2	84.9	111.5	21.5	16.4
1970	733,800	209,785	200	543,933	34,391	14,672	204	71.9	32.0	98.1	124.2	32.6	25.8
1971	684,900	196,908	200	461,257	-11,735	15,200	207	73.5	24.6	91.6	118.3	26.9	20.8
1972	726,500	282,338	1,300	451,276	21,528	15,714	210	74.9	26.9	96.8	124.3	27.8	21.6
1973	902,000	442,500	1,600	453,242	8,334	16,192	213	76.1	15.1	86.4	112.6	17.4	13.4
1974	944,849	467,763	6,089	464,016	-2,338	16,821	215	78.1	12.9	81.7	108.2	15.8	11.9
1975	994,791	474,618	3,217	501,080	-4,683	17,627	218	80.8	27.1	101.5	125.6	26.7	21.6
1976	986,137	475,213	4,114	490,115	-6,734	18,189	220	82.6	21.8	90.4	117.7	24.2	18.6
1977	1,374,432	833,102	216	486,560	-36,125	18,861	222	84.9	24.8	91.5	116.8	27.1	21.2
1978	1,566,749	971,775	887	600,172	22,961	18,649	224	83.2	28.5	97.6	123.5	29.2	23.1
1979	1,645,253	984,802	725	614,789	-26,937	19,450	226	86.1	28.4	97.3	123.2	29.2	23.1
1980	1,514,874	787,576	528	691,004	-17,380	19,442	228	85.2	38.3	104.6	131.1	36.6	29.2
1981	1,441,747	681,863	491	736,893	-3,629	19,853	231	86.1	43.2	112.9	140.0	38.3	30.9
1982	788,659	22,941	4,379	701,449	-48,340	20,308	233	87.1	32.0	99.9	127.0	32.0	25.2
1983	839,244	164,151	4,293	659,219	557	20,725	236	87.9	30.2	101.0	127.7	29.9	23.6
1984	1,534,951	879,700	10,407	652,052	7,893	21,500	239	90.1	25.7	94.3	122.8	27.2	20.9
1985	1,680,405	935,107	10,641	790,672	56,641	21,907	241	90.8	29.1	99.7	126.8	29.2	22.9
1986	1,658,551	858,114	2,083	805,105	24,531	21,946	244	89.9	35.0	104.8	133.5	33.4	26.2
1987	1,632,707	824,749	3,340	808,301	19,158	22,158	247	89.8	25.4	90.5	122.9	28.1	20.7
1988	1,757,676	964,646	8,090	723,659	-54,658	22,802	249	91.4	20.8	81.2	115.8	25.6	18.0
1989	1,502,394	692,582	16,640	806,674	2,894	22,673	252	89.9	23.2	83.0	119.6	27.9	19.4
1990	1,508,141	723,824	23,952	814,947	29,682	23,004	255	90.3	21.6	80.3	117.4	26.9	18.4
1991	1,050,268	304,016	18,160	748,493	7,235	23,155	258	89.9	23.3	81.8	118.9	28.5	19.6
1992	1,577,206	881,727	16,144	717,972	29,821	23,472	260	90.2	22.8	81.6	118.3	28.0	19.3
1993	1,718,498	1,027,424	27,133	689,401	-4,701	24,101	263	91.6	26.8	85.3	120.0	31.4	22.4
1994	1,560,184	880,518	25,237	687,067	6,199	24,032	266	90.4	26.1	82.0	122.8	31.8	21.3
1995	1,616,033	967,447	33,503	665,133	7,634	24,586	269	91.6	21.6	76.1	118.2	28.4	18.3

100 INDIA	- FOOD BALANCE SHEET OF FISH AND FISHERY PRODUCTS IN LIVE WEIGHT AND FISH CONTRIBUTION TO PROTEIN SUPPLY -	FAO-FIDI

YEAR	PRODUCTION	NON-FOOD USES	IMPORTS	EXPORTS	STOCKS VARIATIONS	TOTAL FOOD SUPPLY	POPULATION	PER CAPUT SUPPLY	FISH PROTEINS	ANIMAL PROTEINS	TOTAL PROTEINS	FISH/ANIMAL PROTEINS	FISH/TOTAL PROTEINS
	(..................... metric tons in live weight					.....................)	(thousands)	(kilograms)	(grams per caput per day)			(%)	(%)
1961	961,000	80,700	18,643	47,911	0	851,032	452,476	1.9	0.5	6.2	53.4	8.3	1.0
1962	973,900	81,900	33,271	19,929	0	905,343	462,780	2.0	0.5	6.1	53.4	8.7	1.0
1963	1,045,800	52,400	30,471	36,971	0	986,900	473,293	2.1	0.6	6.1	50.6	9.5	1.1
1964	1,321,600	88,900	19,281	47,896	0	1,204,085	484,071	2.5	0.7	6.0	51.5	11.4	1.3
1965	1,331,300	59,000	6,000	32,678	0	1,245,622	495,157	2.5	0.7	5.9	51.7	12.0	1.4
1966	1,367,200	63,900	0	41,267	0	1,262,033	506,547	2.5	0.7	5.9	48.1	11.8	1.5
1967	1,400,400	103,000	0	43,323	0	1,254,078	518,221	2.4	0.7	5.7	47.3	12.3	1.5
1968	1,525,600	103,100	0	48,191	0	1,374,310	530,176	2.6	0.8	6.0	50.7	12.6	1.5
1969	1,605,000	103,800	0	54,194	0	1,447,006	542,410	2.7	0.8	6.1	50.7	12.7	1.5
1970	1,758,500	128,600	0	72,354	0	1,557,546	554,911	2.8	0.8	5.8	52.4	13.9	1.6
1971	1,853,000	158,900	195	65,167	0	1,629,128	567,704	2.9	0.8	6.2	49.8	13.7	1.7
1972	1,639,100	134,900	200	67,334	0	1,437,066	580,777	2.5	0.7	6.0	50.4	11.9	1.4
1973	1,960,700	178,300	3,300	79,679	0	1,706,021	594,039	2.9	0.8	6.1	50.5	13.2	1.6
1974	2,256,991	183,300	4,292	69,301	0	2,008,682	607,372	3.3	0.9	6.4	48.4	14.4	1.9
1975	2,267,192	184,261	1,966	76,773	0	2,008,124	620,701	3.2	0.9	6.3	48.5	14.5	1.9
1976	2,177,096	127,850	73	89,043	0	1,960,276	633,969	3.1	0.9	6.4	48.7	13.4	1.8
1977	2,314,703	123,063	54	90,595	0	2,101,098	647,230	3.3	0.9	6.5	50.2	13.6	1.8
1978	2,309,709	132,846	79	109,938	0	2,067,004	660,653	3.1	0.9	6.5	52.6	13.5	1.7
1979	2,343,292	153,251	19	118,752	0	2,071,308	674,474	3.1	0.9	6.5	51.9	13.1	1.6
1980	2,445,337	171,883	927	99,644	-4,000	2,170,736	688,856	3.2	0.9	6.7	48.0	13.1	1.8
1981	2,448,342	189,226	1,138	96,311	0	2,163,943	703,840	3.1	0.9	7.0	52.5	12.4	1.7
1982	2,369,320	153,682	0	102,313	4,000	2,117,325	719,353	2.9	0.9	7.4	51.4	11.5	1.7
1983	2,508,598	133,031	0	118,742	0	2,256,825	735,299	3.1	0.9	7.7	54.0	11.4	1.6
1984	2,874,519	229,175	0	134,264	0	2,511,080	751,536	3.3	1.0	8.1	55.0	11.7	1.7
1985	2,839,307	198,606	0	122,558	0	2,518,144	767,940	3.3	0.9	8.3	52.1	11.0	1.8
1986	2,937,470	264,125	0	125,385	-559	2,547,402	784,498	3.3	0.9	8.4	55.5	11.1	1.7
1987	2,921,885	228,009	0	124,306	240	2,569,810	801,193	3.2	0.9	8.4	54.9	10.6	1.6
1988	3,125,962	199,820	0	134,346	140	2,791,936	817,900	3.4	1.0	8.4	54.6	11.5	1.8
1989	3,640,689	377,158	0	154,924	0	3,108,608	834,469	3.7	1.0	8.7	58.4	11.6	1.7
1990	3,794,707	385,740	0	165,904	160	3,243,223	850,793	3.8	1.1	8.9	55.0	12.2	2.0
1991	4,045,494	380,499	5	239,272	0	3,425,727	866,804	4.0	1.1	8.9	57.3	12.5	1.9
1992	4,232,746	403,092	57	251,401	0	3,578,310	882,520	4.1	1.2	9.2	58.4	12.9	2.0
1993	4,545,717	460,434	185	284,165	0	3,801,304	898,035	4.2	1.3	9.4	54.5	13.3	2.3
1994	4,737,765	495,507	447	333,222	0	3,909,483	913,495	4.3	1.3	9.5	57.2	13.7	2.3
1995	4,905,921	583,166	795	318,754	0	4,004,796	929,005	4.3	1.3	9.7	58.3	13.7	2.3

101 INDONESIA	- FOOD BALANCE SHEET OF FISH AND FISHERY PRODUCTS IN LIVE WEIGHT AND FISH CONTRIBUTION TO PROTEIN SUPPLY -	FAO-FIDI

YEAR	PRODUCTION	NON-FOOD USES	IMPORTS	EXPORTS	STOCKS VARIATIONS	TOTAL FOOD SUPPLY	POPULATION	PER CAPUT SUPPLY	FISH PROTEINS	ANIMAL PROTEINS	TOTAL PROTEINS	FISH/ANIMAL PROTEINS	FISH/TOTAL PROTEINS
	(.. metric tons in live weight					....)	(thousands)	(kilograms)	(grams per caput per day)			(%)	(%)
1961	906,800	0	28,401	855	0	934,346	98,256	9.5	2.9	4.8	35.1	60.0	8.2
1962	943,200	0	4,132	955	0	946,376	100,345	9.4	2.8	4.8	34.8	59.6	8.2
1963	936,200	9,000	800	0	0	928,000	102,485	9.1	2.7	4.7	34.2	58.5	8.0
1964	1,000,000	9,000	1,600	0	0	992,600	104,708	9.5	2.9	4.8	34.9	59.5	8.2
1965	1,066,800	10,000	200	0	0	1,057,000	107,041	9.9	3.0	4.8	35.1	61.6	8.5
1966	1,201,600	12,000	600	500	0	1,189,700	109,483	10.9	3.3	5.1	35.9	64.5	9.1
1967	1,180,400	12,000	5,034	3,478	0	1,169,956	112,026	10.4	3.3	5.1	36.0	65.1	9.2
1968	1,159,000	12,000	6,879	16,177	0	1,137,702	114,673	9.9	3.1	5.0	35.9	63.0	8.8
1969	1,214,800	14,100	5,289	19,583	0	1,186,407	117,425	10.1	3.2	5.1	36.6	63.1	8.7
1970	1,225,200	20,900	5,632	19,729	0	1,190,202	120,280	9.9	3.4	5.4	38.5	62.9	8.9
1971	1,241,300	20,800	10,989	35,988	0	1,195,501	123,243	9.7	3.4	5.3	38.8	63.3	8.6
1972	1,266,800	18,200	7,240	52,767	0	1,203,074	126,305	9.5	3.3	5.4	39.7	60.8	8.3
1973	1,261,900	0	12,946	61,586	0	1,213,260	129,427	9.4	3.2	5.4	41.1	59.9	7.9
1974	1,331,160	0	11,853	65,725	0	1,277,288	132,560	9.6	3.4	5.6	41.6	60.0	8.1
1975	1,381,648	848	12,388	59,232	0	1,333,956	135,666	9.8	3.3	5.5	41.7	60.4	8.0
1976	1,478,926	11,279	39,546	79,167	0	1,428,026	138,726	10.3	3.5	5.7	41.4	60.9	8.4
1977	1,567,754	12,138	30,783	63,621	0	1,522,778	141,748	10.7	3.7	5.8	41.6	63.6	8.9
1978	1,642,042	4,135	16,382	74,414	0	1,579,875	144,764	10.9	3.8	5.9	43.8	63.1	8.6
1979	1,742,452	4,494	4,478	82,275	0	1,660,160	147,823	11.2	4.1	6.6	45.5	62.3	9.1
1980	1,841,808	3,503	6,683	82,352	0	1,762,636	150,958	11.7	4.2	7.0	45.9	60.4	9.2
1981	1,907,254	5,518	10,677	70,510	0	1,841,904	154,178	12.0	4.3	7.1	47.9	60.3	8.9
1982	1,990,062	6,871	12,154	90,969	0	1,904,376	157,463	12.1	4.4	7.4	46.8	60.1	9.4
1983	2,204,076	14,286	6,397	91,149	-14,805	2,091,034	160,779	13.0	4.9	8.2	48.7	59.5	10.0
1984	2,251,897	9,324	5,840	76,781	5,417	2,177,049	164,081	13.3	4.7	8.2	50.8	57.6	9.3
1985	2,332,688	6,021	4,288	82,131	4,392	2,253,217	167,332	13.5	4.7	8.2	51.0	57.4	9.3
1986	2,456,972	3,421	7,225	104,108	-400	2,356,268	170,525	13.8	4.8	8.5	53.3	56.3	9.0
1987	2,583,874	4,394	2,682	146,403	50	2,435,809	173,666	14.0	5.0	8.8	53.1	57.2	9.4
1988	2,795,208	17,206	6,453	197,196	790	2,588,050	176,758	14.6	5.1	9.0	55.8	56.8	9.1
1989	2,948,406	20,617	7,256	244,012	100	2,691,135	179,804	15.0	5.3	9.3	56.5	57.0	9.3
1990	3,044,180	20,395	7,732	341,325	210	2,690,402	182,812	14.7	5.1	9.3	58.2	55.0	8.8
1991	3,352,473	33,783	4,067	438,707	-13,562	2,870,488	185,774	15.5	5.4	9.8	60.4	54.6	8.9
1992	3,439,412	27,660	8,164	446,668	3,729	2,976,976	188,694	15.8	5.5	10.4	63.4	53.4	8.7
1993	3,685,449	26,893	13,537	559,231	8,151	3,121,014	191,595	16.3	5.7	10.7	63.4	52.8	8.9
1994	3,917,151	33,947	16,511	579,839	-3,593	3,316,284	194,509	17.1	6.2	11.7	66.1	52.9	9.4
1995	4,144,715	140,311	15,910	588,679	8,718	3,440,353	197,460	17.4	5.9	11.5	67.5	51.4	8.7

102 IRAN	- FOOD BALANCE SHEET OF FISH AND FISHERY PRODUCTS IN LIVE WEIGHT AND FISH CONTRIBUTION TO PROTEIN SUPPLY -												FAO-FIDI
YEAR	PRODUCTION	NON-FOOD USES	IMPORTS	EXPORTS	STOCKS VARIATIONS	TOTAL FOOD SUPPLY	POPULATION	PER CAPUT SUPPLY	FISH PROTEINS	ANIMAL PROTEINS	TOTAL PROTEINS	FISH/ANIMAL PROTEINS	FISH/TOTAL PROTEINS
	(.. metric tons in live weight ....)						(thousands)	(kilograms)	(grams per caput per day)			(%)	(%)
1961	20,000	0	569	9,700	0	10,869	22,126	0.5	0.1	13.4	50.7	1.1	0.3
1962	20,000	0	0	4,900	0	15,100	22,725	0.7	0.2	13.6	49.5	1.4	0.4
1963	18,000	0	0	5,100	0	12,900	23,349	0.6	0.2	12.6	46.1	1.3	0.3
1964	18,000	0	0	9,200	0	8,800	23,996	0.4	0.1	12.5	44.6	0.8	0.2
1965	20,000	0	0	8,000	0	12,000	24,663	0.5	0.1	12.4	51.7	1.1	0.3
1966	21,100	0	100	9,200	0	12,000	25,351	0.5	0.1	12.1	55.3	1.1	0.2
1967	22,500	0	0	12,700	0	9,800	26,063	0.4	0.1	12.5	53.6	0.9	0.2
1968	24,000	0	0	6,286	0	17,714	26,807	0.7	0.2	12.8	56.0	1.5	0.3
1969	24,000	0	0	5,600	0	18,400	27,593	0.7	0.2	13.0	53.5	1.5	0.4
1970	26,000	0	560	4,200	0	22,360	28,429	0.8	0.2	13.3	52.8	1.8	0.4
1971	24,300	0	560	6,440	0	18,420	29,325	0.6	0.2	13.0	57.6	1.4	0.3
1972	19,500	0	280	6,780	0	13,000	30,282	0.4	0.1	13.3	56.2	1.0	0.2
1973	20,000	0	100	3,060	0	17,040	31,284	0.5	0.2	13.6	58.6	1.2	0.3
1974	68,994	0	1,300	4,120	0	66,174	32,309	2.1	0.6	14.6	65.4	4.1	0.9
1975	66,739	0	532	6,380	0	60,891	33,344	1.8	0.5	15.1	69.0	3.6	0.8
1976	69,816	0	4,600	3,707	0	70,709	34,376	2.1	0.6	16.2	66.9	3.7	0.9
1977	69,575	0	10,598	8,347	0	71,826	35,423	2.0	0.6	17.8	69.7	3.4	0.9
1978	67,101	0	5,253	4,549	0	67,805	36,539	1.9	0.5	17.3	65.7	3.1	0.8
1979	65,269	750	7,220	3,817	0	67,921	37,799	1.8	0.5	16.9	66.6	3.1	0.8
1980	43,529	850	7,144	1,708	0	48,115	39,254	1.2	0.4	17.1	68.5	2.0	0.5
1981	44,582	1,000	12,938	1,134	0	55,385	40,920	1.4	0.4	18.2	71.6	2.1	0.5
1982	95,724	1,200	6,110	1,443	0	99,192	42,773	2.3	0.7	18.1	72.9	3.6	0.9
1983	111,889	8,400	14,498	1,491	0	116,496	44,767	2.6	0.7	18.4	75.3	4.0	1.0
1984	115,771	5,650	5,315	1,678	0	113,758	46,832	2.4	0.7	17.6	72.1	3.9	1.0
1985	118,541	6,950	132	1,429	0	110,293	48,916	2.3	0.6	16.4	68.4	3.9	0.9
1986	151,695	11,300	202	909	0	139,688	51,004	2.7	0.8	15.8	68.4	4.9	1.1
1987	210,955	10,550	251	1,222	0	199,434	53,096	3.8	1.1	15.6	69.0	6.8	1.5
1988	235,005	6,950	40	1,221	0	226,874	55,173	4.1	1.2	15.3	69.3	7.6	1.7
1989	260,187	6,700	247	1,488	0	252,246	57,219	4.4	1.3	15.5	69.4	8.1	1.8
1990	269,506	12,800	0	1,312	0	255,394	59,219	4.3	1.2	15.9	71.3	7.8	1.7
1991	275,729	19,925	28,404	2,088	0	282,120	61,170	4.6	1.3	16.4	74.5	8.0	1.8
1992	334,213	22,650	30,000	2,225	0	339,338	63,069	5.4	1.5	16.8	76.9	9.1	2.0
1993	318,055	13,800	27,778	6,391	0	325,642	64,905	5.0	1.4	17.2	77.4	8.3	1.9
1994	332,417	20,400	17,793	11,596	0	318,214	66,671	4.8	1.4	16.7	76.4	8.2	1.8
1995	368,300	36,600	17,320	7,092	-333	341,595	68,365	5.0	1.4	16.5	77.5	8.7	1.9

103 IRAQ	- FOOD BALANCE SHEET OF FISH AND FISHERY PRODUCTS IN LIVE WEIGHT AND FISH CONTRIBUTION TO PROTEIN SUPPLY -	FAO-FIDI

YEAR	PRODUCTION	NON-FOOD USES	IMPORTS	EXPORTS	STOCKS VARIATIONS	TOTAL FOOD SUPPLY	POPULATION	PER CAPUT SUPPLY	FISH PROTEINS	ANIMAL PROTEINS	TOTAL PROTEINS	FISH/ANIMAL PROTEINS	FISH/TOTAL PROTEINS
	(..........		metric tons in live weight			..........)	(thousands)	(kilograms)		(grams per caput per day)		(%)	(%)
1961	9,000	0	400	0	0	9,400	7,056	1.3	0.4	14.6	50.9	2.7	0.8
1962	9,000	0	600	0	0	9,600	7,273	1.3	0.4	14.2	54.4	2.7	0.7
1963	11,300	0	400	0	0	11,700	7,498	1.6	0.5	13.8	56.4	3.4	0.8
1964	19,200	0	800	0	0	20,000	7,732	2.6	0.8	13.9	57.7	5.5	1.3
1965	16,800	0	400	100	0	17,100	7,976	2.1	0.6	12.9	57.1	4.9	1.1
1966	14,100	0	400	0	0	14,500	8,231	1.8	0.5	13.1	57.7	3.9	0.9
1967	10,200	0	400	500	0	10,100	8,496	1.2	0.4	12.7	58.0	2.7	0.6
1968	19,100	0	0	3,100	0	16,000	8,772	1.8	0.5	13.4	59.0	4.0	0.9
1969	22,000	0	400	1,900	0	20,500	9,058	2.3	0.7	13.2	59.3	5.1	1.1
1970	32,100	0	200	2,100	0	30,200	9,356	3.2	1.0	13.8	61.0	6.9	1.6
1971	27,200	0	0	2,100	0	25,100	9,665	2.6	0.8	13.4	61.1	5.8	1.3
1972	32,000	0	0	2,100	0	29,900	9,985	3.0	0.9	13.5	58.7	6.6	1.5
1973	27,700	0	200	3,600	0	24,300	10,317	2.4	0.7	12.8	58.6	5.4	1.2
1974	24,246	0	400	1,000	0	23,646	10,662	2.2	0.7	12.9	59.2	5.0	1.1
1975	21,832	0	5,252	...	0	27,084	11,020	2.5	0.7	13.3	60.9	5.3	1.2
1976	28,283	0	...	...	0	28,283	11,391	2.5	0.7	13.3	61.0	5.5	1.2
1977	26,101	0	...	...	0	26,101	11,777	2.2	0.7	13.8	61.7	4.7	1.1
1978	26,100	0	...	...	0	26,100	12,175	2.1	0.6	14.0	65.8	4.5	1.0
1979	55,228	0	...	...	0	55,228	12,585	4.4	1.3	16.4	70.4	8.0	1.9
1980	53,245	0	...	...	0	53,245	13,007	4.1	1.3	17.6	74.5	7.4	1.8
1981	26,219	0	...	...	0	26,219	13,437	2.0	0.6	18.9	77.6	3.1	0.7
1982	24,000	0	...	...	0	24,000	13,875	1.7	0.5	19.0	81.1	2.7	0.6
1983	22,500	0	...	...	0	22,500	14,329	1.6	0.5	19.6	83.0	2.4	0.6
1984	21,000	0	...	...	0	21,000	14,808	1.4	0.4	21.3	86.9	2.0	0.5
1985	21,500	0	...	...	0	21,500	15,317	1.4	0.4	20.3	88.0	2.0	0.5
1986	20,564	0	...	...	0	20,564	15,863	1.3	0.4	18.9	87.4	2.0	0.4
1987	20,500	0	...	...	0	20,500	16,441	1.3	0.4	18.2	86.4	2.0	0.4
1988	18,045	0	...	...	0	18,045	17,025	1.1	0.3	18.6	86.4	1.7	0.4
1989	18,550	0	...	...	0	18,550	17,578	1.1	0.3	16.9	83.9	1.8	0.4
1990	16,000	0	...	...	0	16,000	18,078	0.9	0.3	14.4	81.2	1.8	0.3
1991	14,400	0	...	...	0	14,400	18,514	0.8	0.2	5.7	59.7	4.0	0.4
1992	23,542	0	...	...	0	23,542	18,898	1.3	0.4	6.3	54.7	5.9	0.7
1993	23,500	0	...	...	0	23,500	19,261	1.2	0.4	6.0	52.2	6.0	0.7
1994	22,000	0	...	...	0	22,000	19,650	1.1	0.3	5.4	49.1	6.1	0.7
1995	22,550	0	...	...	0	22,550	20,095	1.1	0.3	5.4	46.6	6.1	0.7

104 IRELAND	- FOOD BALANCE SHEET OF FISH AND FISHERY PRODUCTS IN LIVE WEIGHT AND FISH CONTRIBUTION TO PROTEIN SUPPLY -	FAO-FIDI

YEAR	PRODUCTION	NON-FOOD USES	IMPORTS	EXPORTS	STOCKS VARIATIONS	TOTAL FOOD SUPPLY	POPULATION	PER CAPUT SUPPLY	FISH PROTEINS	ANIMAL PROTEINS	TOTAL PROTEINS	FISH/ANIMAL PROTEINS	FISH/TOTAL PROTEINS
	(.......................... metric tons in live weight ....)						(thousands)	(kilograms)	(grams per caput per day)			(%)	(%)
1961	32,100	4,900	7,312	13,131	-1,312	20,069	2,833	7.1	1.8	55.6	100.9	3.3	1.8
1962	29,000	3,600	8,130	13,945	1,064	20,648	2,839	7.3	1.9	57.0	102.1	3.3	1.9
1963	27,600	2,500	9,240	11,452	235	23,123	2,850	8.1	2.1	58.3	102.6	3.7	2.1
1964	31,900	4,700	7,681	9,500	-800	24,581	2,863	8.6	2.3	58.3	101.2	3.9	2.2
1965	35,700	5,600	9,053	11,659	0	27,494	2,876	9.6	2.7	59.7	102.7	4.6	2.6
1966	39,700	4,600	8,835	16,952	822	27,804	2,887	9.6	2.6	60.2	102.6	4.4	2.6
1967	50,200	6,700	8,385	21,164	-600	30,122	2,897	10.4	2.8	63.1	104.3	4.4	2.7
1968	53,100	6,100	9,592	27,769	-231	28,592	2,910	9.8	2.5	61.7	103.4	4.0	2.4
1969	66,500	12,800	11,801	36,253	235	29,483	2,928	10.1	2.6	63.7	104.4	4.1	2.5
1970	78,900	13,900	11,376	41,246	-2,065	33,065	2,954	11.2	2.9	65.4	105.8	4.5	2.8
1971	74,000	14,300	11,713	41,517	835	30,731	2,989	10.3	2.5	65.2	105.3	3.9	2.4
1972	92,000	13,500	10,219	55,054	472	34,137	3,031	11.3	2.9	64.8	105.1	4.4	2.7
1973	90,800	13,400	11,154	57,507	-1,882	29,165	3,078	9.5	2.4	64.0	104.5	3.8	2.3
1974	89,537	11,842	12,025	54,272	2,490	37,938	3,128	12.1	3.2	68.4	109.7	4.7	2.9
1975	88,317	11,513	12,873	47,171	-200	42,306	3,177	13.3	3.3	69.8	110.0	4.7	3.0
1976	94,880	15,634	17,482	41,261	0	55,467	3,225	17.2	5.1	69.0	108.6	7.3	4.7
1977	93,641	11,471	14,674	44,190	-2,500	50,154	3,271	15.3	4.6	68.5	110.9	6.7	4.1
1978	104,021	22,390	16,306	52,003	-3,500	42,434	3,316	12.8	3.5	72.6	113.2	4.8	3.1
1979	92,836	9,471	17,349	61,042	1,000	40,672	3,359	12.1	3.2	69.0	111.0	4.6	2.9
1980	149,392	9,644	18,884	102,638	0	55,993	3,401	16.5	4.6	75.4	115.3	6.1	4.0
1981	190,522	7,902	25,266	155,661	4,000	56,225	3,442	16.3	4.3	73.1	113.7	5.9	3.8
1982	212,197	9,097	38,295	186,909	-3,234	51,253	3,480	14.7	4.1	68.8	111.0	6.0	3.7
1983	203,400	6,848	38,717	188,777	4,000	50,492	3,514	14.4	4.1	67.8	110.2	6.1	3.7
1984	210,149	6,313	33,814	185,209	1,235	53,676	3,538	15.2	4.4	68.0	109.3	6.5	4.0
1985	230,692	6,477	32,488	202,009	0	54,694	3,552	15.4	4.3	70.0	111.7	6.2	3.9
1986	231,648	9,849	31,637	187,524	0	65,913	3,553	18.6	5.0	73.7	116.3	6.8	4.3
1987	250,955	18,352	49,613	221,115	0	61,101	3,542	17.3	4.6	76.3	118.5	6.1	3.9
1988	260,652	28,064	50,945	220,561	-4,469	58,504	3,526	16.6	4.6	74.8	117.6	6.1	3.9
1989	208,069	2,562	38,765	187,761	-800	55,711	3,511	15.9	4.5	71.1	114.5	6.3	3.9
1990	239,302	25,595	57,623	205,204	-8,800	57,326	3,503	16.4	4.6	71.7	115.6	6.4	4.0
1991	264,907	18,262	51,706	250,992	6,429	53,789	3,503	15.4	4.4	74.7	118.7	5.9	3.7
1992	275,748	13,167	53,284	259,431	4,178	60,612	3,511	17.3	4.8	69.7	114.9	6.8	4.2
1993	308,080	6,020	39,706	277,775	1,607	65,598	3,523	18.6	4.4	70.7	114.5	6.2	3.8
1994	319,198	16,512	36,181	272,682	1,336	67,521	3,536	19.1	4.9	69.5	112.9	7.1	4.4
1995	412,722	40,009	39,279	342,175	517	70,334	3,546	19.8	5.3	70.8	113.3	7.5	4.7

105 ISRAEL	- FOOD BALANCE SHEET OF FISH AND FISHERY PRODUCTS IN LIVE WEIGHT AND FISH CONTRIBUTION TO PROTEIN SUPPLY -												FAO-FIDI
YEAR	PRODUCTION	NON-FOOD USES	IMPORTS	EXPORTS	STOCKS VARIATIONS	TOTAL FOOD SUPPLY	POPULATION	PER CAPUT SUPPLY	FISH PROTEINS	ANIMAL PROTEINS	TOTAL PROTEINS	FISH/ANIMAL PROTEINS	FISH/TOTAL PROTEINS
	(..		metric tons in live weight		....	)	(thousands)	(kilograms)	(grams per caput per day)			(%)	(%)
1961	14,900	0	24,986	0	0	39,886	2,198	18.2	4.7	38.3	89.1	12.4	5.3
1962	16,400	0	27,556	0	0	43,956	2,288	19.2	4.9	42.1	91.4	11.6	5.3
1963	17,500	0	24,785	0	0	42,285	2,381	17.8	4.5	40.5	88.5	11.2	5.1
1964	18,400	0	19,618	0	0	38,019	2,474	15.4	4.0	42.9	92.2	9.4	4.4
1965	19,500	0	21,009	0	0	40,509	2,563	15.8	4.1	43.2	90.2	9.5	4.6
1966	24,500	0	20,632	0	0	45,132	2,647	17.1	4.5	46.1	93.3	9.8	4.8
1967	23,700	0	19,435	150	0	42,985	2,727	15.8	4.2	43.8	92.3	9.7	4.6
1968	28,000	0	19,662	610	0	47,052	2,806	16.8	4.4	45.9	94.7	9.6	4.7
1969	22,200	0	19,888	757	0	41,331	2,887	14.3	3.7	45.8	94.7	8.1	3.9
1970	27,100	0	18,504	986	0	44,618	2,974	15.0	4.1	47.5	97.0	8.6	4.2
1971	31,100	0	23,333	657	0	53,776	3,066	17.5	4.6	49.9	101.9	9.2	4.5
1972	33,600	0	17,585	695	0	50,491	3,163	16.0	4.2	50.4	100.7	8.4	4.2
1973	31,500	0	21,924	1,910	0	51,514	3,262	15.8	4.1	51.9	101.7	7.9	4.0
1974	27,454	0	26,427	979	0	52,902	3,360	15.7	4.0	50.5	100.5	7.9	4.0
1975	28,851	0	16,055	931	0	43,975	3,455	12.7	3.7	52.0	103.1	7.0	3.5
1976	29,909	0	15,678	1,402	0	44,185	3,546	12.5	3.7	53.5	103.7	6.9	3.5
1977	31,680	0	17,760	1,435	0	48,005	3,635	13.2	3.9	55.9	105.2	7.0	3.8
1978	32,617	0	21,943	685	0	53,875	3,720	14.5	3.9	51.1	98.4	7.6	4.0
1979	30,065	0	44,157	806	0	73,416	3,801	19.3	4.7	57.6	104.4	8.2	4.5
1980	27,783	0	46,143	603	0	73,322	3,879	18.9	4.5	49.9	97.8	8.9	4.6
1981	27,474	0	37,956	592	0	64,837	3,953	16.4	4.2	51.9	100.9	8.1	4.2
1982	26,099	0	29,737	629	0	55,207	4,025	13.7	3.7	55.1	104.2	6.6	3.5
1983	22,994	0	43,281	3,236	0	63,039	4,095	15.4	3.8	55.9	105.3	6.7	3.6
1984	24,442	0	37,755	769	0	61,428	4,164	14.8	3.8	57.4	103.6	6.6	3.6
1985	27,339	0	46,652	1,019	0	72,972	4,233	17.2	4.2	56.0	102.7	7.5	4.1
1986	25,223	0	61,490	932	-2,222	83,559	4,300	19.4	4.8	52.5	100.1	9.1	4.8
1987	28,410	0	59,154	916	0	86,648	4,368	19.8	4.9	53.4	99.9	9.2	4.9
1988	28,198	0	57,749	531	2,137	87,553	4,444	19.7	4.8	57.3	106.5	8.4	4.5
1989	26,491	0	68,512	580	0	94,423	4,539	20.8	4.9	54.2	103.4	9.0	4.7
1990	23,618	0	71,791	511	0	94,898	4,660	20.4	4.8	52.7	103.3	9.1	4.6
1991	20,723	0	83,133	627	0	103,229	4,812	21.5	4.9	54.3	107.7	9.1	4.6
1992	19,070	0	94,469	727	86	112,897	4,989	22.6	5.0	54.1	103.9	9.2	4.8
1993	19,043	0	99,086	457	0	117,672	5,179	22.7	5.0	53.9	104.3	9.2	4.8
1994	20,565	0	92,571	679	0	112,458	5,362	21.0	4.6	56.5	109.1	8.2	4.2
1995	20,941	0	100,373	423	0	120,890	5,525	21.9	4.9	55.3	109.5	8.9	4.5

Note: until 1994 includes Gaza Strip.

106 ITALY	- FOOD BALANCE SHEET OF FISH AND FISHERY PRODUCTS IN LIVE WEIGHT AND FISH CONTRIBUTION TO PROTEIN SUPPLY -	FAO-FIDI

YEAR	PRODUCTION	NON-FOOD USES	IMPORTS	EXPORTS	STOCKS VARIATIONS	TOTAL FOOD SUPPLY	POPULATION	PER CAPUT SUPPLY	FISH PROTEINS	ANIMAL PROTEINS	TOTAL PROTEINS	FISH/ANIMAL PROTEINS	FISH/TOTAL PROTEINS
	(.......... metric tons in live weight ....)						(thousands)	(kilograms)	(grams per caput per day)			(%)	(%)
1961	239,600	0	358,741	3,995	0	594,346	50,564	11.8	3.4	28.6	81.3	11.8	4.1
1962	220,700	0	333,333	2,454	0	551,579	50,948	10.8	3.3	29.8	82.0	11.0	4.0
1963	232,200	0	345,737	3,396	0	574,540	51,343	11.2	3.3	31.8	84.7	10.3	3.9
1964	324,600	0	314,996	3,314	0	636,282	51,734	12.3	3.6	32.7	85.9	11.0	4.2
1965	353,800	0	377,102	8,339	0	722,563	52,112	13.9	3.9	34.2	87.7	11.3	4.4
1966	369,000	0	345,022	10,415	0	703,607	52,473	13.4	3.9	35.5	89.0	10.9	4.4
1967	373,100	0	318,300	11,016	0	680,384	52,818	12.9	3.8	37.1	91.1	10.1	4.1
1968	363,800	0	324,016	15,067	0	672,749	53,152	12.7	3.7	38.0	92.6	9.8	4.0
1969	370,900	0	308,851	18,210	0	661,541	53,485	12.4	3.8	38.7	93.6	9.9	4.1
1970	395,894	0	338,488	33,910	0	700,473	53,822	13.0	3.9	41.8	97.2	9.2	4.0
1971	397,692	0	356,291	49,118	0	704,864	54,164	13.0	3.8	42.7	97.2	9.0	3.9
1972	421,227	0	349,571	70,352	0	700,446	54,506	12.9	3.7	44.0	97.9	8.5	3.8
1973	401,418	0	351,207	75,218	0	677,407	54,840	12.4	3.8	46.4	101.8	8.1	3.7
1974	450,930	0	321,681	101,554	0	671,058	55,154	12.2	3.6	45.5	98.3	8.0	3.7
1975	438,719	0	330,726	89,430	0	680,015	55,441	12.3	3.6	45.3	98.3	7.9	3.6
1976	467,413	6,100	386,075	85,007	0	762,382	55,698	13.7	3.9	46.8	98.6	8.4	4.0
1977	426,759	5,150	386,067	68,936	0	738,740	55,927	13.2	3.8	47.9	99.6	7.9	3.8
1978	452,019	24,150	416,715	90,644	0	753,940	56,126	13.4	3.9	49.7	102.5	7.8	3.8
1979	479,766	33,250	480,090	127,772	0	798,834	56,295	14.2	4.1	52.0	106.6	8.0	3.9
1980	507,565	31,500	461,379	100,088	0	837,355	56,434	14.8	4.4	53.7	107.6	8.1	4.1
1981	515,116	37,820	417,267	111,541	0	783,022	56,541	13.9	4.0	52.9	104.6	7.5	3.8
1982	547,679	45,460	485,820	105,077	0	882,963	56,619	15.6	4.4	54.1	103.0	8.2	4.3
1983	552,701	16,455	497,830	92,129	0	941,946	56,676	16.6	4.7	55.7	102.8	8.4	4.5
1984	580,485	20,260	531,703	111,148	0	980,780	56,724	17.3	4.9	55.5	104.6	8.8	4.7
1985	591,395	15,900	693,879	137,979	0	1,131,396	56,771	19.9	5.6	57.2	105.6	9.8	5.3
1986	570,702	23,200	640,209	97,708	0	1,090,003	56,821	19.2	5.3	56.7	106.7	9.4	5.0
1987	564,998	22,500	766,499	64,329	0	1,244,668	56,873	21.9	5.9	57.8	107.4	10.2	5.5
1988	580,445	20,000	724,486	133,888	0	1,151,044	56,925	20.2	5.8	57.9	107.8	10.1	5.4
1989	551,494	21,064	809,151	135,716	0	1,203,865	56,976	21.1	6.1	58.2	108.3	10.4	5.6
1990	525,522	22,190	840,544	120,627	0	1,223,250	57,023	21.5	6.1	60.0	109.9	10.1	5.5
1991	552,191	22,694	890,734	122,115	0	1,298,117	57,067	22.8	6.5	60.7	112.0	10.7	5.8
1992	558,264	10,042	843,455	136,555	0	1,255,122	57,108	22.0	6.3	60.2	110.4	10.4	5.7
1993	564,693	14,076	818,228	143,286	0	1,225,559	57,145	21.5	6.0	59.2	107.5	10.2	5.6
1994	575,998	13,422	832,560	164,628	-400	1,230,108	57,177	21.5	6.2	57.2	106.1	10.8	5.8
1995	609,768	13,674	794,815	146,280	400	1,245,029	57,204	21.8	6.1	58.0	108.4	10.5	5.6

109 JAMAICA	- FOOD BALANCE SHEET OF FISH AND FISHERY PRODUCTS IN LIVE WEIGHT AND FISH CONTRIBUTION TO PROTEIN SUPPLY -												FAO-FIDI
YEAR	PRODUCTION	NON-FOOD USES	IMPORTS	EXPORTS	STOCKS VARIATIONS	TOTAL FOOD SUPPLY	POPULATION	PER CAPUT SUPPLY	FISH PROTEINS	ANIMAL PROTEINS	TOTAL PROTEINS	FISH/ANIMAL PROTEINS	FISH/TOTAL PROTEINS
	(.. metric tons in live weight ....)						(thousands)	(kilograms)	(grams per caput per day)			(%)	(%)
1961	11,500	0	42,345	0	0	53,845	1,652	32.6	10.5	23.1	51.8	45.3	20.2
1962	11,900	0	44,411	0	0	56,311	1,679	33.5	10.8	24.0	54.4	45.0	19.8
1963	10,000	0	55,742	0	0	65,742	1,707	38.5	12.5	26.1	55.9	47.8	22.3
1964	10,000	0	48,105	0	0	58,105	1,735	33.5	10.7	25.6	57.7	41.9	18.6
1965	9,000	0	46,289	0	0	55,289	1,760	31.4	10.1	24.7	56.8	40.8	17.7
1966	9,000	0	52,597	0	0	61,597	1,783	34.5	11.1	27.1	58.7	41.0	19.0
1967	9,000	0	50,626	0	0	59,626	1,804	33.1	10.6	27.8	61.9	38.2	17.2
1968	8,500	0	54,217	0	0	62,717	1,824	34.4	11.0	29.0	62.9	37.8	17.4
1969	8,500	0	50,174	0	0	58,674	1,845	31.8	10.0	29.2	61.7	34.3	16.2
1970	8,500	0	41,683	0	0	50,183	1,869	26.9	8.1	29.7	66.6	27.4	12.2
1971	9,300	0	40,729	0	0	50,029	1,896	26.4	8.1	30.3	68.3	26.6	11.8
1972	9,500	0	65,556	0	0	75,056	1,925	39.0	11.9	33.7	72.1	35.2	16.5
1973	9,600	0	43,423	0	0	53,023	1,956	27.1	8.3	29.1	66.8	28.5	12.4
1974	10,100	0	36,901	200	0	46,801	1,985	23.6	7.2	29.3	67.7	24.6	10.7
1975	10,100	0	42,729	200	0	52,629	2,013	26.2	8.1	32.9	71.5	24.7	11.4
1976	10,130	0	38,710	22	0	48,818	2,037	24.0	7.1	32.9	70.1	21.7	10.2
1977	10,107	0	23,948	4	0	34,052	2,059	16.5	4.8	28.4	63.8	17.1	7.6
1978	9,647	0	32,481	6	0	42,122	2,080	20.3	6.0	31.2	68.3	19.1	8.7
1979	9,651	0	36,195	29	0	45,817	2,104	21.8	6.2	27.3	63.8	22.7	9.7
1980	9,050	0	28,579	20	-60	37,549	2,133	17.6	5.1	26.2	62.8	19.4	8.1
1981	7,606	0	32,952	90	26	40,695	2,168	18.8	5.5	26.9	63.9	20.5	8.6
1982	7,884	0	35,411	40	0	43,255	2,207	19.6	5.9	27.7	63.4	21.2	9.3
1983	8,661	0	30,631	53	33	39,272	2,247	17.5	5.4	26.7	62.6	20.2	8.6
1984	9,669	0	34,608	62	-61	44,153	2,283	19.3	6.1	26.9	64.4	22.7	9.5
1985	10,499	0	26,950	551	0	36,898	2,311	16.0	4.8	23.5	60.6	20.6	8.0
1986	10,837	0	33,920	541	0	44,216	2,330	19.0	5.7	27.0	63.6	21.2	9.0
1987	10,677	0	31,494	540	44	41,675	2,341	17.8	5.2	28.2	64.8	18.6	8.1
1988	10,305	0	36,041	544	0	45,802	2,348	19.5	5.8	28.5	63.5	20.3	9.1
1989	11,415	0	36,702	365	-9	47,744	2,355	20.3	6.1	29.8	65.0	20.5	9.4
1990	11,382	0	31,214	387	18	42,227	2,366	17.9	5.3	30.5	65.2	17.5	8.2
1991	11,590	13	27,257	390	0	38,443	2,381	16.2	4.8	29.0	63.0	16.4	7.6
1992	12,217	14	24,127	551	0	35,779	2,400	14.9	4.3	29.2	65.2	14.8	6.6
1993	13,010	15	28,786	601	0	41,180	2,422	17.0	4.9	31.4	68.5	15.7	7.2
1994	13,460	0	27,935	482	-83	40,830	2,445	16.7	4.7	28.3	67.9	16.7	7.0
1995	13,617	0	31,872	2,814	83	42,758	2,468	17.3	5.1	28.5	68.4	17.8	7.4

110 JAPAN	- FOOD BALANCE SHEET OF FISH AND FISHERY PRODUCTS IN LIVE WEIGHT AND FISH CONTRIBUTION TO PROTEIN SUPPLY -	FAO-FIDI

YEAR	PRODUCTION	NON-FOOD USES	IMPORTS	EXPORTS	STOCKS VARIATIONS	TOTAL FOOD SUPPLY	POPULATION	PER CAPUT SUPPLY	FISH PROTEINS	ANIMAL PROTEINS	TOTAL PROTEINS	FISH/ANIMAL PROTEINS	FISH/TOTAL PROTEINS
	(........ metric tons in live weight ....)						(thousands)	(kilograms)	(grams per caput per day)			(%)	(%)
1961	6,260,300	1,043,800	30,705	438,494	-153,397	4,655,316	94,984	49.0	16.0	23.5	72.3	67.9	22.1
1962	6,344,500	1,024,700	43,356	537,100	36,708	4,862,766	95,918	50.7	16.5	25.3	73.5	65.2	22.4
1963	6,248,000	797,100	68,899	547,888	-39,234	4,932,679	96,889	50.9	16.8	26.0	74.6	64.5	22.5
1964	5,989,700	924,800	139,687	596,661	158,039	4,765,966	97,881	48.7	14.2	24.5	72.6	57.9	19.5
1965	6,501,400	925,100	193,041	586,553	3,399	5,186,188	98,881	52.5	17.2	28.4	75.6	60.8	22.8
1966	6,695,200	939,000	299,822	652,263	-12,401	5,391,358	99,881	54.0	17.8	29.9	77.3	59.6	23.1
1967	7,394,900	1,269,000	280,711	606,561	-12,574	5,787,478	100,892	57.4	18.8	31.8	79.3	59.0	23.7
1968	8,166,900	1,680,800	287,325	631,973	36,865	6,178,318	101,945	60.6	20.0	33.6	80.6	59.6	24.9
1969	8,168,600	1,947,400	327,348	667,527	0	5,881,022	103,083	57.1	19.1	34.0	80.2	56.0	23.8
1970	8,824,800	1,858,400	394,942	719,955	-17,178	6,624,212	104,331	63.5	20.1	36.1	81.6	55.7	24.7
1971	9,359,600	2,062,000	489,248	712,115	-13,440	7,061,298	105,703	66.8	21.2	38.0	83.4	55.9	25.5
1972	9,709,300	2,198,900	617,048	820,155	30,583	7,337,878	107,176	68.5	21.2	38.8	84.4	54.5	25.1
1973	10,091,860	2,363,000	830,889	847,549	0	7,712,201	108,689	71.0	22.3	40.4	86.1	55.1	25.9
1974	10,100,630	2,334,790	809,478	793,497	0	7,781,830	110,159	70.6	21.8	39.9	84.3	54.8	25.9
1975	9,896,170	2,166,662	906,253	757,399	0	7,878,366	111,524	70.6	22.3	40.3	84.7	55.2	26.3
1976	9,996,833	2,463,888	1,016,841	813,470	0	7,736,314	112,761	68.6	23.8	42.7	86.3	55.9	27.6
1977	10,127,640	3,036,907	1,148,802	689,543	0	7,549,999	113,882	66.3	23.1	42.9	85.8	53.9	27.0
1978	10,187,280	2,962,588	1,244,996	747,653	0	7,722,036	114,909	67.2	23.8	44.6	86.9	53.3	27.4
1979	9,950,330	3,129,119	1,452,479	741,399	0	7,532,292	115,875	65.0	23.0	44.5	86.9	51.6	26.4
1980	10,434,000	3,137,041	1,172,646	820,140	0	7,649,465	116,807	65.5	23.2	44.8	86.7	51.8	26.8
1981	10,740,120	3,676,132	1,299,710	704,740	0	7,658,956	117,708	65.1	23.6	45.3	86.8	52.1	27.2
1982	10,826,540	3,939,095	1,409,182	658,276	0	7,638,354	118,566	64.4	23.8	45.9	87.6	51.9	27.2
1983	11,254,720	4,123,196	1,483,667	602,738	0	8,012,457	119,377	67.1	25.2	47.7	89.1	52.9	28.3
1984	12,021,540	4,803,558	1,652,644	662,200	0	8,208,426	120,136	68.3	25.3	48.2	89.5	52.4	28.2
1985	11,409,360	4,269,829	1,857,394	603,523	0	8,393,398	120,837	69.5	25.8	49.2	90.6	52.5	28.5
1986	11,976,490	5,070,595	2,157,141	557,747	0	8,505,292	121,483	70.0	25.5	49.5	91.2	51.4	27.9
1987	11,857,650	5,048,578	2,448,580	503,797	0	8,753,858	122,078	71.7	26.4	51.6	93.3	51.2	28.3
1988	11,966,120	5,403,728	2,861,974	538,200	0	8,886,168	122,620	72.5	26.4	52.5	94.4	50.3	27.9
1989	11,173,490	4,550,785	2,795,343	532,099	0	8,885,953	123,107	72.2	26.9	53.2	95.2	50.5	28.2
1990	10,354,270	4,130,210	3,010,330	440,421	0	8,793,976	123,537	71.2	26.6	52.9	94.4	50.2	28.2
1991	9,301,160	3,936,268	3,359,901	459,519	0	8,265,289	123,911	66.7	25.0	52.0	94.1	48.1	26.6
1992	8,502,229	3,313,302	3,492,204	432,469	0	8,248,666	124,233	66.4	24.5	52.1	94.5	47.0	25.9
1993	8,080,857	2,935,928	3,713,459	411,915	0	8,446,487	124,520	67.8	25.2	53.1	95.7	47.5	26.3
1994	7,396,316	2,086,466	3,891,910	337,292	0	8,864,474	124,793	71.0	25.8	53.8	96.1	47.9	26.9
1995	6,757,570	1,582,991	3,967,511	272,294	0	8,869,796	125,068	70.9	26.0	55.2	97.5	47.1	26.6

112 JORDAN	- FOOD BALANCE SHEET OF FISH AND FISHERY PRODUCTS IN LIVE WEIGHT AND FISH CONTRIBUTION TO PROTEIN SUPPLY -	FAO-FIDI

YEAR	PRODUCTION	NON-FOOD USES	IMPORTS	EXPORTS	STOCKS VARIATIONS	TOTAL FOOD SUPPLY	POPULATION	PER CAPUT SUPPLY	FISH PROTEINS	ANIMAL PROTEINS	TOTAL PROTEINS	FISH/ANIMAL PROTEINS	FISH/TOTAL PROTEINS
	(..		metric tons in live weight		....)		(thousands)	(kilograms)		(grams per caput per day)		(%)	(%)
1961	100	0	2,750	0	0	2,850	841	3.4	0.5	7.8	52.8	6.4	0.9
1962	200	0	2,700	0	0	2,900	916	3.2	0.5	8.5	53.1	5.7	0.9
1963	200	0	3,250	0	0	3,450	947	3.6	0.6	9.0	54.7	6.3	1.0
1964	200	0	2,850	0	0	3,050	978	3.1	0.5	9.4	54.5	5.1	0.9
1965	200	0	3,100	0	0	3,300	1,011	3.3	0.5	10.7	54.3	4.6	0.9
1966	200	0	2,550	0	0	2,750	1,045	2.6	0.4	10.7	48.9	3.7	0.8
1967	100	0	2,050	0	0	2,150	1,212	1.8	0.3	8.0	48.6	3.7	0.6
1968	100	0	1,550	0	0	1,650	1,386	1.2	0.4	14.0	58.9	2.5	0.6
1969	100	0	2,050	0	0	2,150	1,433	1.5	0.4	16.9	67.7	2.6	0.7
1970	100	0	2,000	0	0	2,100	1,483	1.4	0.4	14.0	64.8	2.9	0.6
1971	200	0	3,000	0	0	3,200	1,535	2.1	0.6	16.2	67.9	3.8	0.9
1972	100	0	2,450	0	0	2,550	1,590	1.6	0.5	15.5	67.3	3.0	0.7
1973	100	0	4,150	0	0	4,250	1,646	2.6	0.7	18.2	68.3	4.1	1.1
1974	92	0	5,000	0	0	5,092	1,705	3.0	0.9	14.4	64.9	6.0	1.3
1975	65	0	5,000	0	0	5,065	1,773	2.9	0.8	15.8	66.7	5.1	1.2
1976	49	0	5,089	534	0	4,604	1,850	2.5	0.8	15.7	64.5	5.0	1.2
1977	31	0	6,077	209	0	5,898	1,930	3.1	1.0	17.2	65.7	5.6	1.5
1978	31	0	6,664	295	0	6,399	2,015	3.2	1.1	20.1	68.2	5.2	1.5
1979	36	0	5,789	144	0	5,681	2,095	2.7	0.9	19.6	66.3	4.5	1.3
1980	56	0	6,763	56	0	6,763	2,176	3.1	1.0	19.9	68.1	5.2	1.5
1981	35	0	6,896	99	0	6,832	2,276	3.0	1.0	21.1	69.3	4.7	1.4
1982	34	0	8,228	151	0	8,111	2,364	3.4	1.2	23.2	71.9	5.1	1.6
1983	32	0	9,525	56	0	9,501	2,456	3.9	1.3	23.5	74.0	5.5	1.8
1984	132	0	10,030	0	0	10,162	2,551	4.0	1.3	24.9	75.3	5.3	1.8
1985	197	0	13,923	21	0	14,099	2,650	5.3	1.8	28.4	78.8	6.2	2.2
1986	267	0	8,094	0	0	8,361	2,753	3.0	1.0	25.7	75.2	3.9	1.3
1987	322	0	14,205	98	0	14,429	2,860	5.1	1.6	25.0	77.0	6.5	2.1
1988	372	0	8,091	157	0	8,306	2,971	2.8	0.9	25.9	76.8	3.5	1.2
1989	407	0	8,923	3	0	9,328	3,086	3.0	1.0	19.5	66.2	4.9	1.5
1990	412	0	8,975	55	0	9,332	3,306	2.8	0.9	20.5	67.7	4.4	1.3
1991	372	0	11,625	1,116	0	10,881	3,585	3.0	1.0	24.3	74.5	4.0	1.3
1992	374	0	13,048	2,687	0	10,736	3,773	2.9	0.9	22.7	73.7	4.0	1.2
1993	412	0	15,045	559	0	14,898	3,919	3.8	1.2	26.4	70.1	4.5	1.7
1994	438	0	15,788	855	0	15,371	4,066	3.8	1.0	24.1	73.1	4.1	1.4
1995	522	0	19,894	475	0	19,940	4,215	4.7	1.2	22.4	75.1	5.2	1.5

108 KAZAKHSTAN	- FOOD BALANCE SHEET OF FISH AND FISHERY PRODUCTS IN LIVE WEIGHT AND FISH CONTRIBUTION TO PROTEIN SUPPLY -	FAO-FIDI

YEAR	PRODUCTION	NON-FOOD USES	IMPORTS	EXPORTS	STOCKS VARIATIONS	TOTAL FOOD SUPPLY	POPULATION	PER CAPUT SUPPLY	FISH PROTEINS	ANIMAL PROTEINS	TOTAL PROTEINS	FISH/ANIMAL PROTEINS	FISH/TOTAL PROTEINS
	(................................. metric tons in live weight					.................................)	(thousands)	(kilograms)	(grams per caput per day)			(%)	(%)
1961	0	0	0	0	0	0	10,458	0.0	0.0	0.0	0.0	0.0	0.0
1962	0	0	0	0	0	0	10,935	0.0	0.0	0.0	0.0	0.0	0.0
1963	0	0	0	0	0	0	11,297	0.0	0.0	0.0	0.0	0.0	0.0
1964	0	0	0	0	0	0	11,586	0.0	0.0	0.0	0.0	0.0	0.0
1965	0	0	0	0	0	0	11,911	0.0	0.0	0.0	0.0	0.0	0.0
1966	0	0	0	0	0	0	12,187	0.0	0.0	0.0	0.0	0.0	0.0
1967	0	0	0	0	0	0	12,457	0.0	0.0	0.0	0.0	0.0	0.0
1968	0	0	0	0	0	0	12,696	0.0	0.0	0.0	0.0	0.0	0.0
1969	0	0	0	0	0	0	12,906	0.0	0.0	0.0	0.0	0.0	0.0
1970	0	0	0	0	0	0	13,110	0.0	0.0	0.0	0.0	0.0	0.0
1971	0	0	0	0	0	0	13,321	0.0	0.0	0.0	0.0	0.0	0.0
1972	0	0	0	0	0	0	13,534	0.0	0.0	0.0	0.0	0.0	0.0
1973	0	0	0	0	0	0	13,742	0.0	0.0	0.0	0.0	0.0	0.0
1974	0	0	0	0	0	0	13,955	0.0	0.0	0.0	0.0	0.0	0.0
1975	0	0	0	0	0	0	14,136	0.0	0.0	0.0	0.0	0.0	0.0
1976	0	0	0	0	0	0	14,279	0.0	0.0	0.0	0.0	0.0	0.0
1977	0	0	0	0	0	0	14,425	0.0	0.0	0.0	0.0	0.0	0.0
1978	0	0	0	0	0	0	14,593	0.0	0.0	0.0	0.0	0.0	0.0
1979	0	0	0	0	0	0	14,771	0.0	0.0	0.0	0.0	0.0	0.0
1980	0	0	0	0	0	0	14,907	0.0	0.0	0.0	0.0	0.0	0.0
1981	0	0	0	0	0	0	15,070	0.0	0.0	0.0	0.0	0.0	0.0
1982	0	0	0	0	0	0	15,240	0.0	0.0	0.0	0.0	0.0	0.0
1983	0	0	0	0	0	0	15,415	0.0	0.0	0.0	0.0	0.0	0.0
1984	0	0	0	0	0	0	15,596	0.0	0.0	0.0	0.0	0.0	0.0
1985	0	0	0	0	0	0	15,780	0.0	0.0	0.0	0.0	0.0	0.0
1986	0	0	0	0	0	0	15,970	0.0	0.0	0.0	0.0	0.0	0.0
1987	0	0	0	0	0	0	16,164	0.0	0.0	0.0	0.0	0.0	0.0
1988	0	0	0	0	0	0	16,353	0.0	0.0	0.0	0.0	0.0	0.0
1989	0	0	0	0	0	0	16,617	0.0	0.0	0.0	0.0	0.0	0.0
1990	0	0	0	0	0	0	16,742	0.0	0.0	0.0	0.0	0.0	0.0
1991	82,690	0	...	...	0	82,690	16,815	4.9	1.5	45.0	91.7	3.3	1.6
1992	68,503	0	...	...	0	68,502	16,842	4.1	1.3	44.5	93.2	2.8	1.4
1993	59,628	0	...	...	0	59,628	16,838	3.5	1.1	41.4	97.8	2.6	1.1
1994	47,594	0	3,841	11,254	0	40,181	16,824	2.4	0.8	42.2	102.7	1.8	0.7
1995	49,602	0	10,025	19,940	-625	39,062	16,817	2.3	0.8	39.4	97.7	1.9	0.8

Note: for years before 1991 see USSR.

114 KENYA	- FOOD BALANCE SHEET OF FISH AND FISHERY PRODUCTS IN LIVE WEIGHT AND FISH CONTRIBUTION TO PROTEIN SUPPLY -	FAO-FIDI

YEAR	PRODUCTION	NON-FOOD USES	IMPORTS	EXPORTS	STOCKS VARIATIONS	TOTAL FOOD SUPPLY	POPULATION	PER CAPUT SUPPLY	FISH PROTEINS	ANIMAL PROTEINS	TOTAL PROTEINS	FISH/ANIMAL PROTEINS	FISH/TOTAL PROTEINS
	(..		metric tons in live weight		....	)	(thousands)	(kilograms)		(grams per caput per day)		(%)	(%)
1961	13,500	0	1,722	0	0	15,222	8,592	1.8	0.5	15.8	64.4	3.2	0.8
1962	18,400	0	1,830	0	0	20,230	8,864	2.3	0.7	15.5	64.2	4.4	1.1
1963	20,100	0	1,683	0	0	21,783	9,147	2.4	0.8	14.0	62.5	5.4	1.2
1964	20,700	0	2,425	246	0	22,879	9,442	2.4	0.8	14.6	63.9	5.2	1.2
1965	23,400	0	2,300	300	0	25,400	9,749	2.6	0.8	15.0	65.3	5.0	1.2
1966	27,900	0	1,400	0	0	29,300	10,069	2.9	0.8	14.5	64.7	5.8	1.3
1967	27,300	0	600	0	0	27,900	10,401	2.7	0.8	14.2	64.7	5.5	1.2
1968	28,100	0	400	708	0	27,792	10,749	2.6	0.8	14.3	64.0	5.3	1.2
1969	31,900	0	400	695	0	31,605	11,114	2.8	0.8	14.2	64.9	5.8	1.3
1970	33,600	0	9,536	4,946	0	38,190	11,498	3.3	1.0	14.6	64.1	6.9	1.6
1971	28,200	0	4,322	7,305	0	25,217	11,901	2.1	0.7	14.8	62.9	4.6	1.1
1972	29,900	0	4,453	5,391	0	28,963	12,323	2.4	0.7	14.4	64.1	4.8	1.1
1973	28,800	0	5,985	6,520	0	28,266	12,769	2.2	0.7	13.7	62.3	4.8	1.0
1974	28,481	0	8,965	4,827	0	32,619	13,240	2.5	0.7	13.5	62.1	5.4	1.2
1975	27,269	0	4,945	4,443	0	27,771	13,741	2.0	0.6	13.2	62.2	4.4	0.9
1976	40,972	0	3,452	1,020	-1,625	41,779	14,273	2.9	0.9	13.3	63.9	6.5	1.4
1977	42,762	0	1,868	824	0	43,806	14,833	3.0	0.9	15.0	64.1	5.8	1.4
1978	46,372	0	1,530	1,039	250	47,113	15,417	3.1	0.9	15.3	63.4	5.9	1.4
1979	51,743	0	1,843	2,335	-30	51,221	16,019	3.2	1.0	15.4	58.9	6.2	1.6
1980	47,723	0	1,414	1,138	625	48,625	16,632	2.9	0.9	14.8	56.5	5.9	1.5
1981	57,701	0	788	2,921	625	56,193	17,254	3.3	1.0	13.5	54.9	7.1	1.8
1982	81,529	0	739	1,815	30	80,483	17,887	4.5	1.3	14.8	55.4	9.0	2.4
1983	98,138	0	1,055	1,679	125	97,638	18,532	5.3	1.5	15.7	56.2	9.8	2.8
1984	91,046	0	510	820	0	90,736	19,193	4.7	1.4	15.6	53.6	8.8	2.6
1985	106,056	0	786	1,644	0	105,198	19,872	5.3	1.6	15.6	56.9	10.0	2.7
1986	119,901	0	630	2,387	0	118,145	20,569	5.7	1.7	16.8	56.9	10.0	3.0
1987	131,265	0	309	7,151	0	124,423	21,280	5.9	1.7	17.7	54.9	9.6	3.1
1988	138,350	0	296	12,024	0	126,622	22,003	5.8	1.7	18.3	54.0	9.1	3.1
1989	146,571	0	779	19,489	0	127,862	22,736	5.6	1.7	18.1	53.2	9.1	3.1
1990	202,134	0	188	26,800	0	175,522	23,475	7.5	2.2	18.7	52.0	11.7	4.2
1991	198,902	0	1,584	32,378	-2,500	165,607	24,225	6.8	2.0	17.9	50.9	11.3	4.0
1992	163,346	0	1,400	33,692	-2,625	128,429	24,981	5.1	1.5	16.4	49.2	9.1	3.0
1993	183,197	0	3,541	36,879	-7,500	142,358	25,732	5.5	1.6	15.5	46.6	10.4	3.5
1994	203,962	0	10,103	52,041	7,600	169,624	26,459	6.4	1.9	15.4	53.5	12.1	3.5
1995	193,790	0	6,463	46,248	5,025	159,030	27,150	5.9	1.7	15.5	52.9	11.0	3.2

83 KIRIBATI	- FOOD BALANCE SHEET OF FISH AND FISHERY PRODUCTS IN LIVE WEIGHT AND FISH CONTRIBUTION TO PROTEIN SUPPLY -	FAO-FIDI

YEAR	PRODUCTION	NON-FOOD USES	IMPORTS	EXPORTS	STOCKS VARIATIONS	TOTAL FOOD SUPPLY	POPULATION	PER CAPUT SUPPLY	FISH PROTEINS	ANIMAL PROTEINS	TOTAL PROTEINS	FISH/ANIMAL PROTEINS	FISH/TOTAL PROTEINS
	(..		metric tons in live weight			....)	(thousands)	(kilograms)	(grams per caput per day)			(%)	(%)
1961	5,000	3,000	0	0	0	2,000	41	48.9	13.3	19.6	45.2	68.1	29.5
1962	6,000	4,000	0	0	0	2,000	42	47.7	13.7	22.6	47.8	60.5	28.6
1963	6,000	4,000	0	0	0	2,000	43	46.6	13.4	22.4	49.7	59.5	26.9
1964	6,000	3,700	0	0	0	2,300	44	52.3	14.6	24.5	51.3	59.5	28.5
1965	6,000	3,500	0	0	0	2,500	45	55.6	15.3	24.5	53.1	62.2	28.7
1966	7,500	4,850	0	0	0	2,650	46	57.6	16.2	23.7	52.8	68.2	30.6
1967	7,500	4,800	0	0	0	2,700	47	57.4	16.1	24.9	54.4	64.6	29.5
1968	7,500	4,700	0	0	0	2,800	48	58.3	16.2	25.1	54.5	64.6	29.7
1969	7,500	4,650	0	0	0	2,850	49	58.2	16.1	24.1	52.9	66.8	30.4
1970	8,800	5,860	0	0	0	2,940	50	58.8	15.7	24.3	56.7	64.6	27.8
1971	8,900	5,880	0	0	0	3,020	51	59.3	15.8	24.5	56.2	64.3	28.1
1972	9,100	6,010	0	0	0	3,090	52	59.5	15.8	25.4	57.7	62.3	27.4
1973	9,200	6,000	0	0	0	3,200	53	60.5	16.0	26.3	58.9	60.8	27.1
1974	9,475	6,358	175	0	0	3,292	54	61.1	16.6	26.2	58.5	63.2	28.3
1975	9,650	6,477	175	0	0	3,348	55	60.9	16.5	25.7	59.9	64.2	27.6
1976	9,824	6,507	207	0	0	3,524	56	62.8	17.0	27.1	60.3	62.7	28.2
1977	10,053	6,657	198	0	0	3,594	57	62.7	16.9	28.2	60.8	60.1	27.8
1978	10,606	7,207	368	0	0	3,767	59	64.4	17.7	27.8	61.2	63.8	29.0
1979	10,838	7,397	525	133	0	3,833	60	64.1	17.8	28.4	62.0	62.4	28.6
1980	11,217	7,357	443	223	0	4,080	61	66.9	18.0	28.0	60.4	64.3	29.8
1981	11,690	7,214	551	826	0	4,201	62	67.5	16.7	25.9	58.9	64.7	28.4
1982	11,197	6,705	472	642	0	4,322	63	68.1	16.9	25.8	59.9	65.3	28.2
1983	20,673	14,400	240	2,061	0	4,452	65	68.8	18.1	26.4	58.4	68.5	31.0
1984	16,430	9,767	363	2,543	0	4,483	66	68.1	20.1	28.7	60.5	70.1	33.3
1985	21,047	15,481	229	1,152	0	4,643	67	69.3	20.4	28.5	60.2	71.6	33.9
1986	27,489	21,219	218	1,681	0	4,807	68	70.6	20.9	28.2	60.9	74.0	34.3
1987	36,090	30,816	396	731	0	4,939	69	71.6	21.2	30.4	63.0	69.7	33.6
1988	23,222	17,032	396	1,611	0	4,975	70	71.1	20.9	29.4	64.0	71.3	32.7
1989	29,208	21,598	423	2,852	0	5,181	71	73.0	21.4	31.6	64.4	67.8	33.3
1990	25,047	19,286	465	951	0	5,275	72	73.3	21.4	29.4	63.4	72.6	33.7
1991	25,178	20,123	490	161	0	5,384	73	73.7	21.6	30.7	64.8	70.3	33.4
1992	25,985	20,679	651	426	0	5,531	74	74.6	22.3	31.7	66.4	70.3	33.6
1993	24,920	19,588	614	322	0	5,624	75	74.6	22.5	31.3	66.1	71.8	34.0
1994	24,430	19,157	465	73	0	5,665	77	73.9	21.9	31.1	67.0	70.5	32.7
1995	24,685	19,528	746	107	0	5,796	78	74.3	21.8	33.7	69.9	64.6	31.1

116 KOREA, DEM. PEOP. DEM.	- FOOD BALANCE SHEET OF FISH AND FISHERY PRODUCTS IN LIVE WEIGHT AND FISH CONTRIBUTION TO PROTEIN SUPPLY -	FAO-FIDI

YEAR	PRODUCTION	NON-FOOD USES	IMPORTS	EXPORTS	STOCKS VARIATIONS	TOTAL FOOD SUPPLY	POPULATION	PER CAPUT SUPPLY	FISH PROTEINS	ANIMAL PROTEINS	TOTAL PROTEINS	FISH/ANIMAL PROTEINS	FISH/TOTAL PROTEINS
	(..........		metric tons in live weight			..........)	(thousands)	(kilograms)	(grams per caput per day)			(%)	(%)
1961	400,000	100,000	...	...	0	300,000	10,851	27.7	7.8	11.8	67.3	66.2	11.6
1962	400,000	100,000	...	...	0	300,000	11,142	26.9	7.6	11.6	66.3	65.8	11.5
1963	400,000	100,000	...	...	0	300,000	11,418	26.3	7.5	11.4	66.9	65.3	11.1
1964	500,000	170,000	...	...	0	330,000	11,716	28.2	8.0	12.1	69.5	65.8	11.5
1965	500,000	170,000	...	...	0	330,000	12,061	27.4	7.8	11.9	69.1	65.5	11.2
1966	500,000	165,000	...	...	0	335,000	12,457	26.9	7.6	11.7	68.4	65.3	11.2
1967	500,000	155,000	...	...	0	345,000	12,889	26.8	7.6	11.7	67.4	65.0	11.3
1968	500,000	145,000	...	...	0	355,000	13,345	26.6	7.5	11.6	67.6	65.3	11.2
1969	500,000	137,000	...	...	0	363,000	13,808	26.3	7.5	11.5	67.0	64.7	11.1
1970	470,000	90,000	...	2,100	0	377,900	14,263	26.5	7.5	11.6	66.5	65.0	11.3
1971	700,000	295,000	...	4,400	0	400,600	14,706	27.2	7.7	11.8	67.1	65.2	11.5
1972	840,000	420,000	...	8,560	0	411,440	15,140	27.2	7.7	11.9	68.2	64.8	11.3
1973	910,000	470,000	...	12,550	0	427,450	15,556	27.5	7.8	12.1	67.6	64.6	11.5
1974	980,000	517,200	...	13,450	0	449,350	15,945	28.2	8.0	12.4	67.9	64.8	11.8
1975	1,050,000	570,000	...	11,200	0	468,800	16,301	28.8	8.2	12.7	70.5	64.4	11.6
1976	1,120,000	603,494	...	22,482	0	494,024	16,621	29.7	8.4	13.2	69.7	64.0	12.1
1977	1,190,000	650,520	...	12,926	0	526,554	16,907	31.1	8.9	13.9	69.7	63.8	12.7
1978	1,260,000	704,854	...	11,806	0	543,340	17,168	31.7	9.0	14.3	71.0	62.8	12.7
1979	1,330,000	725,174	...	24,123	0	580,703	17,418	33.3	9.5	15.1	71.9	62.8	13.2
1980	1,400,000	771,786	...	24,644	0	603,570	17,666	34.2	9.7	15.5	73.5	62.5	13.2
1981	1,500,000	826,540	...	25,331	0	648,129	17,916	36.2	10.3	16.2	72.6	63.6	14.2
1982	1,550,000	831,626	...	28,993	0	689,381	18,168	37.9	10.7	16.6	73.9	64.5	14.5
1983	1,600,000	829,720	...	31,704	0	738,576	18,423	40.1	11.3	17.5	74.0	64.6	15.3
1984	1,650,300	846,948	...	36,667	0	766,685	18,681	41.0	11.6	18.0	73.9	64.3	15.7
1985	1,700,400	870,162	...	34,183	0	796,055	18,942	42.0	11.9	18.5	75.8	64.2	15.7
1986	1,700,839	835,164	...	32,753	0	832,921	19,207	43.4	12.2	19.0	77.7	64.1	15.7
1987	1,712,952	769,260	...	100,688	0	843,004	19,479	43.3	12.1	19.1	77.2	63.3	15.7
1988	1,718,000	744,006	...	99,721	0	874,273	19,760	44.2	12.3	19.4	77.6	63.5	15.9
1989	1,738,600	774,446	...	72,796	0	891,358	20,054	44.5	12.4	19.5	77.6	63.5	16.0
1990	1,753,800	781,318	...	51,774	0	920,708	20,363	45.2	12.5	19.6	78.1	63.5	15.9
1991	1,744,600	751,938	...	46,634	0	946,028	20,687	45.7	12.7	20.1	79.3	63.2	16.0
1992	1,780,100	751,882	...	65,335	0	962,883	21,025	45.8	12.7	19.7	80.7	64.2	15.7
1993	1,782,000	747,000	1,172	70,418	0	965,754	21,375	45.2	12.4	19.1	74.7	65.0	16.7
1994	1,802,300	731,964	1,098	66,008	0	1,005,426	21,734	46.3	12.7	19.1	73.0	66.7	17.4
1995	1,850,000	746,976	1,206	72,631	0	1,031,599	22,097	46.7	12.9	18.9	67.9	68.3	19.0

117 KOREA, REPUBLIC	- FOOD BALANCE SHEET OF FISH AND FISHERY PRODUCTS IN LIVE WEIGHT AND FISH CONTRIBUTION TO PROTEIN SUPPLY -	FAO-FIDI

YEAR	PRODUCTION	NON-FOOD USES	IMPORTS	EXPORTS	STOCKS VARIATIONS	TOTAL FOOD SUPPLY	POPULATION	PER CAPUT SUPPLY	FISH PROTEINS	ANIMAL PROTEINS	TOTAL PROTEINS	FISH/ANIMAL PROTEINS	FISH/TOTAL PROTEINS
	(......... metric tons in live weight					)	(thousands)	(kilograms)	(grams per caput per day)			(%)	(%)
1961	384,500	9,300	0	34,228	0	340,972	25,745	13.2	4.0	6.0	57.3	65.9	6.9
1962	417,500	13,300	220	41,515	0	362,905	26,460	13.7	3.9	5.8	56.4	67.0	6.9
1963	422,800	15,900	0	42,574	0	364,327	27,155	13.4	4.2	6.7	56.2	62.3	7.4
1964	538,100	13,600	0	74,903	0	449,597	27,841	16.2	3.7	6.4	56.1	57.1	6.5
1965	578,700	9,300	0	68,997	0	500,403	28,530	17.5	4.5	7.3	60.5	61.5	7.4
1966	633,500	12,000	0	94,529	0	526,972	29,216	18.0	4.7	7.4	61.9	63.3	7.6
1967	672,000	9,700	700	87,916	0	575,084	29,895	19.2	4.8	7.3	65.2	65.7	7.3
1968	721,200	6,500	724	89,942	0	625,482	30,569	20.5	5.4	8.2	69.1	66.4	7.9
1969	765,100	6,400	1,438	122,036	0	638,102	31,244	20.4	5.4	8.8	70.9	61.0	7.6
1970	753,985	17,200	2,276	163,189	0	575,872	31,923	18.0	4.4	7.8	70.3	56.3	6.3
1971	954,161	17,900	831	172,877	0	764,215	32,608	23.4	5.8	9.4	73.5	61.7	7.9
1972	1,213,220	26,300	7,002	194,575	0	999,347	33,296	30.0	7.8	11.6	76.2	67.3	10.2
1973	1,460,887	31,200	18,551	292,419	0	1,155,820	33,979	34.0	8.2	12.0	78.7	68.7	10.5
1974	1,688,387	50,800	14,695	316,550	0	1,335,732	34,643	38.6	10.1	14.1	80.8	71.3	12.5
1975	1,887,071	58,600	18,066	485,725	0	1,360,813	35,281	38.6	9.3	13.9	84.3	67.0	11.0
1976	2,117,755	184,665	31,690	374,184	0	1,590,596	35,887	44.3	11.7	16.4	86.1	71.2	13.5
1977	2,085,346	120,763	35,888	681,624	0	1,318,848	36,465	36.2	8.7	14.1	79.0	61.6	11.0
1978	2,091,921	128,300	66,240	640,999	0	1,388,862	37,022	37.5	9.2	15.6	82.6	58.7	11.1
1979	2,162,989	133,325	83,780	577,643	0	1,535,802	37,572	40.9	9.8	18.0	83.9	54.4	11.7
1980	2,091,134	87,444	58,709	501,846	-8,325	1,552,228	38,124	40.7	9.9	17.8	81.1	55.7	12.3
1981	2,365,990	194,333	55,642	467,411	-8,324	1,751,564	38,681	45.3	12.6	20.1	82.9	62.5	15.2
1982	2,280,810	148,801	93,032	455,791	16,743	1,785,994	39,240	45.5	12.3	20.7	79.9	59.4	15.4
1983	2,400,342	140,688	81,148	551,867	0	1,788,935	39,789	45.0	12.4	21.9	79.4	56.5	15.6
1984	2,476,876	211,122	102,754	478,302	-10,002	1,880,204	40,314	46.6	13.7	23.8	82.9	57.8	16.6
1985	2,649,976	277,521	108,982	516,148	-13,348	1,951,942	40,806	47.8	12.6	23.0	79.5	54.5	15.8
1986	3,103,486	476,104	132,033	687,955	14,986	2,086,447	41,259	50.6	14.6	25.0	81.2	58.6	18.0
1987	2,876,703	389,808	291,678	709,278	-2,128	2,067,170	41,681	49.6	13.4	24.8	84.4	54.1	15.9
1988	2,732,365	395,285	403,272	728,617	6,002	2,017,742	42,080	48.0	13.5	25.4	84.1	53.1	16.0
1989	2,840,946	480,606	305,366	576,193	-28,169	2,061,356	42,473	48.5	14.3	26.9	80.0	52.9	17.8
1990	2,843,760	486,993	269,969	537,693	-56,010	2,033,047	42,869	47.4	14.2	27.8	85.1	51.1	16.7
1991	2,513,086	523,864	362,658	557,500	76,655	1,871,035	43,273	43.2	13.1	27.9	81.6	47.1	16.1
1992	2,696,067	435,942	338,506	498,245	1,333	2,101,714	43,680	48.1	14.4	31.2	81.9	46.4	17.6
1993	2,649,368	342,378	380,876	422,410	-12,730	2,252,724	44,091	51.1	15.4	32.9	85.9	46.8	17.9
1994	2,700,530	385,966	419,313	394,770	-32,228	2,306,878	44,501	51.8	16.0	34.4	86.4	46.7	18.6
1995	2,688,083	395,389	447,543	498,729	53,735	2,295,243	44,909	51.1	15.8	35.2	87.1	44.9	18.1

118 KUWAIT	- FOOD BALANCE SHEET OF FISH AND FISHERY PRODUCTS IN LIVE WEIGHT AND FISH CONTRIBUTION TO PROTEIN SUPPLY -	FAO-FIDI

YEAR	PRODUCTION	NON-FOOD USES	IMPORTS	EXPORTS	STOCKS VARIATIONS	TOTAL FOOD SUPPLY	POPULATION	PER CAPUT SUPPLY	FISH PROTEINS	ANIMAL PROTEINS	TOTAL PROTEINS	FISH/ANIMAL PROTEINS	FISH/TOTAL PROTEINS
	(............ metric tons in live weight ....)						(thousands)	(kilograms)	(grams per caput per day)			(%)	(%)
1961	3,500	0	500	1,500	0	2,500	307	8.1	2.5	33.2	72.8	7.6	3.5
1962	3,500	0	400	1,000	0	2,900	341	8.5	2.6	31.4	74.8	8.1	3.4
1963	3,500	0	2,403	3,338	0	2,565	380	6.8	2.0	32.0	73.6	6.3	2.8
1964	4,000	0	2,506	2,400	0	4,106	424	9.7	2.8	34.3	79.4	8.3	3.6
1965	4,000	0	2,606	2,400	0	4,206	471	8.9	2.7	36.1	80.4	7.3	3.3
1966	4,000	0	3,206	2,400	0	4,806	522	9.2	2.7	34.6	79.3	7.9	3.4
1967	4,500	0	3,606	2,800	0	5,306	577	9.2	2.8	34.0	78.2	8.1	3.5
1968	4,500	0	3,706	2,800	0	5,406	633	8.5	2.6	34.6	77.5	7.4	3.3
1969	4,500	0	4,003	2,500	0	6,003	690	8.7	2.7	33.0	74.7	8.1	3.6
1970	4,700	0	4,306	3,000	0	6,006	744	8.1	2.3	33.0	75.4	6.9	3.0
1971	5,700	0	4,903	4,000	0	6,603	796	8.3	2.4	32.6	73.5	7.4	3.3
1972	5,000	0	3,706	3,000	0	5,706	845	6.8	1.9	34.7	75.4	5.5	2.5
1973	6,100	0	3,503	4,400	0	5,203	894	5.8	1.6	33.7	74.4	4.9	2.2
1974	5,502	0	4,306	2,100	0	7,708	947	8.1	2.4	35.0	75.7	6.7	3.1
1975	5,934	0	3,634	1,700	0	7,868	1,007	7.8	2.2	36.6	77.1	6.1	2.9
1976	4,648	0	9,371	4,239	-1	9,779	1,074	9.1	2.1	37.8	79.2	5.4	2.6
1977	5,913	0	18,378	4,149	-153	19,990	1,147	17.4	3.4	43.2	84.8	7.9	4.0
1978	6,489	0	9,664	3,535	-2	12,616	1,224	10.3	2.5	45.0	87.0	5.5	2.9
1979	3,065	0	15,705	4,940	156	13,985	1,301	10.8	2.2	46.8	89.2	4.7	2.5
1980	3,689	0	20,262	4,405	0	19,546	1,375	14.2	3.3	49.7	91.7	6.7	3.6
1981	3,714	0	21,446	7,089	0	18,070	1,442	12.5	2.3	49.0	91.3	4.8	2.6
1982	6,628	0	18,536	6,176	0	18,988	1,504	12.6	3.1	50.6	93.1	6.1	3.3
1983	8,722	0	16,206	4,849	0	20,079	1,566	12.8	3.0	48.7	91.6	6.1	3.2
1984	9,639	0	17,621	2,591	0	24,669	1,636	15.1	3.3	50.6	93.5	6.6	3.6
1985	10,118	0	13,598	1,186	0	22,530	1,720	13.1	3.2	48.6	91.3	6.5	3.5
1986	7,633	0	16,084	978	0	22,739	1,824	12.5	2.9	49.5	91.8	5.9	3.2
1987	7,704	0	17,301	1,517	0	23,488	1,941	12.1	2.8	47.4	90.6	6.0	3.1
1988	10,796	0	17,552	2,000	0	26,348	2,051	12.8	3.0	46.3	90.4	6.5	3.3
1989	7,653	0	18,666	1,437	0	24,882	2,125	11.7	2.6	45.8	90.6	5.6	2.8
1990	4,454	0	6,632	1,167	0	9,919	2,143	4.6	1.3	32.7	67.3	4.1	2.0
1991	2,034	0	6,536	210	0	8,360	2,095	4.0	2.7	33.8	61.0	8.1	4.5
1992	7,871	0	10,408	526	0	17,753	1,992	8.9	2.8	37.9	76.9	7.5	3.7
1993	8,466	0	10,437	887	0	18,015	1,864	9.7	2.9	49.3	91.6	5.8	3.1
1994	7,752	0	12,767	267	0	20,252	1,754	11.6	2.7	50.0	93.7	5.3	2.9
1995	8,706	0	13,670	331	0	22,045	1,691	13.0	3.0	50.7	95.8	5.9	3.1

113 KYRGYZSTAN	- FOOD BALANCE SHEET OF FISH AND FISHERY PRODUCTS IN LIVE WEIGHT AND FISH CONTRIBUTION TO PROTEIN SUPPLY -	FAO-FIDI

YEAR	PRODUCTION	NON-FOOD USES	IMPORTS	EXPORTS	STOCKS VARIATIONS	TOTAL FOOD SUPPLY	POPULATION	PER CAPUT SUPPLY	FISH PROTEINS	ANIMAL PROTEINS	TOTAL PROTEINS	FISH/ANIMAL PROTEINS	FISH/TOTAL PROTEINS
	(..		metric tons in live weight		....)		(thousands)	(kilograms)		(grams per caput per day)		(%)	(%)
1961	0	0	0	0	0	0	2,256	0.0	0.0	0.0	0.0	0.0	0.0
1962	0	0	0	0	0	0	2,334	0.0	0.0	0.0	0.0	0.0	0.0
1963	0	0	0	0	0	0	2,414	0.0	0.0	0.0	0.0	0.0	0.0
1964	0	0	0	0	0	0	2,495	0.0	0.0	0.0	0.0	0.0	0.0
1965	0	0	0	0	0	0	2,581	0.0	0.0	0.0	0.0	0.0	0.0
1966	0	0	0	0	0	0	2,663	0.0	0.0	0.0	0.0	0.0	0.0
1967	0	0	0	0	0	0	2,745	0.0	0.0	0.0	0.0	0.0	0.0
1968	0	0	0	0	0	0	2,826	0.0	0.0	0.0	0.0	0.0	0.0
1969	0	0	0	0	0	0	2,905	0.0	0.0	0.0	0.0	0.0	0.0
1970	0	0	0	0	0	0	2,965	0.0	0.0	0.0	0.0	0.0	0.0
1971	0	0	0	0	0	0	3,029	0.0	0.0	0.0	0.0	0.0	0.0
1972	0	0	0	0	0	0	3,094	0.0	0.0	0.0	0.0	0.0	0.0
1973	0	0	0	0	0	0	3,160	0.0	0.0	0.0	0.0	0.0	0.0
1974	0	0	0	0	0	0	3,231	0.0	0.0	0.0	0.0	0.0	0.0
1975	0	0	0	0	0	0	3,299	0.0	0.0	0.0	0.0	0.0	0.0
1976	0	0	0	0	0	0	3,365	0.0	0.0	0.0	0.0	0.0	0.0
1977	0	0	0	0	0	0	3,430	0.0	0.0	0.0	0.0	0.0	0.0
1978	0	0	0	0	0	0	3,495	0.0	0.0	0.0	0.0	0.0	0.0
1979	0	0	0	0	0	0	3,559	0.0	0.0	0.0	0.0	0.0	0.0
1980	0	0	0	0	0	0	3,617	0.0	0.0	0.0	0.0	0.0	0.0
1981	0	0	0	0	0	0	3,685	0.0	0.0	0.0	0.0	0.0	0.0
1982	0	0	0	0	0	0	3,759	0.0	0.0	0.0	0.0	0.0	0.0
1983	0	0	0	0	0	0	3,840	0.0	0.0	0.0	0.0	0.0	0.0
1984	0	0	0	0	0	0	3,923	0.0	0.0	0.0	0.0	0.0	0.0
1985	0	0	0	0	0	0	3,990	0.0	0.0	0.0	0.0	0.0	0.0
1986	0	0	0	0	0	0	4,065	0.0	0.0	0.0	0.0	0.0	0.0
1987	0	0	0	0	0	0	4,139	0.0	0.0	0.0	0.0	0.0	0.0
1988	0	0	0	0	0	0	4,213	0.0	0.0	0.0	0.0	0.0	0.0
1989	0	0	0	0	0	0	4,340	0.0	0.0	0.0	0.0	0.0	0.0
1990	0	0	0	0	0	0	4,395	0.0	0.0	0.0	0.0	0.0	0.0
1991	1,199	0	...	0	0	1,198	4,429	0.3	0.1	39.9	81.9	0.2	0.1
1992	843	0	...	0	0	844	4,447	0.2	0.1	36.2	81.5	0.2	0.1
1993	350	0	...	0	0	349	4,453	0.1	0.0	34.0	78.2	0.1	0.0
1994	294	0	737	0	0	1,031	4,455	0.2	0.1	33.4	77.3	0.2	0.1
1995	364	0	460	0	0	824	4,460	0.2	0.1	31.6	78.8	0.2	0.1

Note: for years before 1991 see USSR.

120 LAOS	- FOOD BALANCE SHEET OF FISH AND FISHERY PRODUCTS IN LIVE WEIGHT AND FISH CONTRIBUTION TO PROTEIN SUPPLY -	FAO-FIDI

YEAR	PRODUCTION	NON-FOOD USES	IMPORTS	EXPORTS	STOCKS VARIATIONS	TOTAL FOOD SUPPLY	POPULATION	PER CAPUT SUPPLY	FISH PROTEINS	ANIMAL PROTEINS	TOTAL PROTEINS	FISH/ANIMAL PROTEINS	FISH/TOTAL PROTEINS
	(............		metric tons in live weight			............)	(thousands)	(kilograms)		(grams per caput per day)		(%)	(%)
1961	12,000	0	...	...	0	12,000	2,227	5.4	1.6	5.0	47.4	32.0	3.4
1962	15,000	0	2,200	...	0	17,200	2,277	7.6	2.3	5.8	48.4	39.5	4.7
1963	15,000	0	1,150	...	0	16,150	2,328	6.9	2.1	5.5	49.0	38.0	4.3
1964	15,000	0	200	...	0	15,200	2,379	6.4	1.9	5.8	51.4	32.9	3.7
1965	15,000	0	1,620	...	0	16,620	2,432	6.8	2.0	6.3	52.0	31.6	3.8
1966	18,000	0	1,317	...	0	19,317	2,486	7.8	2.3	6.8	52.4	33.8	4.4
1967	18,000	0	1,161	...	0	19,161	2,540	7.5	2.3	7.0	53.5	32.1	4.2
1968	18,000	0	1,664	...	0	19,664	2,595	7.6	2.3	7.0	52.3	32.1	4.3
1969	18,000	0	864	...	0	18,864	2,652	7.1	2.1	6.5	52.7	32.8	4.0
1970	20,000	0	561	...	0	20,561	2,713	7.6	2.3	6.9	53.4	33.0	4.2
1971	20,000	0	311	...	0	20,311	2,777	7.3	2.2	6.4	52.1	34.3	4.2
1972	20,000	0	1,175	...	0	21,175	2,845	7.4	2.2	6.1	51.5	36.4	4.3
1973	20,000	0	511	...	0	20,511	2,912	7.0	2.1	5.6	51.2	37.9	4.1
1974	20,000	0	711	...	0	20,711	2,973	7.0	2.1	5.1	50.1	41.0	4.2
1975	22,000	0	783	...	0	22,783	3,024	7.5	2.3	5.2	48.0	43.3	4.7
1976	22,000	0	...	...	0	22,000	3,065	7.2	2.1	5.2	47.3	41.1	4.5
1977	22,000	0	...	...	0	22,000	3,097	7.1	2.1	4.7	48.4	44.8	4.4
1978	22,000	0	...	...	0	22,000	3,127	7.0	2.1	5.7	50.6	36.8	4.2
1979	24,000	0	...	...	0	24,000	3,161	7.6	2.3	6.3	51.5	36.2	4.4
1980	24,000	0	...	...	0	24,000	3,205	7.5	2.2	6.4	51.2	34.8	4.4
1981	24,000	0	...	...	0	24,000	3,262	7.4	2.2	6.2	49.7	35.2	4.4
1982	24,000	0	...	...	0	24,000	3,330	7.2	2.2	6.2	51.7	34.7	4.2
1983	26,000	0	...	...	0	26,000	3,408	7.6	2.3	6.5	50.8	35.2	4.5
1984	26,000	0	...	...	0	26,000	3,497	7.4	2.2	6.6	49.3	33.8	4.5
1985	26,000	0	...	...	0	26,000	3,594	7.2	2.2	6.2	48.6	34.7	4.5
1986	26,000	0	...	...	0	26,000	3,702	7.0	2.1	6.5	48.2	32.2	4.4
1987	28,000	0	...	...	0	28,000	3,819	7.3	2.2	6.6	46.7	33.2	4.7
1988	28,000	0	...	...	0	28,000	3,943	7.1	2.1	6.2	47.5	34.0	4.5
1989	28,000	0	...	...	0	28,000	4,072	6.9	2.1	6.2	46.5	33.1	4.4
1990	28,000	0	...	...	0	28,000	4,202	6.7	2.0	6.1	49.3	32.7	4.0
1991	29,000	0	...	...	0	29,000	4,332	6.7	2.0	6.3	49.7	31.9	4.0
1992	30,000	0	...	...	0	30,000	4,463	6.7	2.0	6.3	50.0	31.7	4.0
1993	30,500	0	689	...	0	31,189	4,597	6.8	2.0	6.4	48.8	31.3	4.1
1994	35,000	0	4,623	...	0	39,623	4,736	8.4	2.5	7.0	50.4	35.3	4.9
1995	40,250	0	4,545	...	0	44,795	4,882	9.2	2.7	7.2	49.8	37.7	5.4

119 LATVIA	- FOOD BALANCE SHEET OF FISH AND FISHERY PRODUCTS IN LIVE WEIGHT AND FISH CONTRIBUTION TO PROTEIN SUPPLY -	FAO-FIDI

YEAR	PRODUCTION	NON-FOOD USES	IMPORTS	EXPORTS	STOCKS VARIATIONS	TOTAL FOOD SUPPLY	POPULATION	PER CAPUT SUPPLY	FISH PROTEINS	ANIMAL PROTEINS	TOTAL PROTEINS	FISH/ANIMAL PROTEINS	FISH/TOTAL PROTEINS
	(.................... metric tons in live weight ....)						(thousands)	(kilograms)	(grams per caput per day)			(%)	(%)
1961	0	0	0	0	0	0	2,159	0.0	0.0	0.0	0.0	0.0	0.0
1962	0	0	0	0	0	0	2,186	0.0	0.0	0.0	0.0	0.0	0.0
1963	0	0	0	0	0	0	2,213	0.0	0.0	0.0	0.0	0.0	0.0
1964	0	0	0	0	0	0	2,241	0.0	0.0	0.0	0.0	0.0	0.0
1965	0	0	0	0	0	0	2,267	0.0	0.0	0.0	0.0	0.0	0.0
1966	0	0	0	0	0	0	2,292	0.0	0.0	0.0	0.0	0.0	0.0
1967	0	0	0	0	0	0	2,314	0.0	0.0	0.0	0.0	0.0	0.0
1968	0	0	0	0	0	0	2,335	0.0	0.0	0.0	0.0	0.0	0.0
1969	0	0	0	0	0	0	2,355	0.0	0.0	0.0	0.0	0.0	0.0
1970	0	0	0	0	0	0	2,374	0.0	0.0	0.0	0.0	0.0	0.0
1971	0	0	0	0	0	0	2,394	0.0	0.0	0.0	0.0	0.0	0.0
1972	0	0	0	0	0	0	2,414	0.0	0.0	0.0	0.0	0.0	0.0
1973	0	0	0	0	0	0	2,434	0.0	0.0	0.0	0.0	0.0	0.0
1974	0	0	0	0	0	0	2,455	0.0	0.0	0.0	0.0	0.0	0.0
1975	0	0	0	0	0	0	2,474	0.0	0.0	0.0	0.0	0.0	0.0
1976	0	0	0	0	0	0	2,488	0.0	0.0	0.0	0.0	0.0	0.0
1977	0	0	0	0	0	0	2,500	0.0	0.0	0.0	0.0	0.0	0.0
1978	0	0	0	0	0	0	2,510	0.0	0.0	0.0	0.0	0.0	0.0
1979	0	0	0	0	0	0	2,521	0.0	0.0	0.0	0.0	0.0	0.0
1980	0	0	0	0	0	0	2,534	0.0	0.0	0.0	0.0	0.0	0.0
1981	0	0	0	0	0	0	2,548	0.0	0.0	0.0	0.0	0.0	0.0
1982	0	0	0	0	0	0	2,563	0.0	0.0	0.0	0.0	0.0	0.0
1983	0	0	0	0	0	0	2,579	0.0	0.0	0.0	0.0	0.0	0.0
1984	0	0	0	0	0	0	2,596	0.0	0.0	0.0	0.0	0.0	0.0
1985	0	0	0	0	0	0	2,613	0.0	0.0	0.0	0.0	0.0	0.0
1986	0	0	0	0	0	0	2,631	0.0	0.0	0.0	0.0	0.0	0.0
1987	0	0	0	0	0	0	2,649	0.0	0.0	0.0	0.0	0.0	0.0
1988	0	0	0	0	0	0	2,664	0.0	0.0	0.0	0.0	0.0	0.0
1989	0	0	0	0	0	0	2,682	0.0	0.0	0.0	0.0	0.0	0.0
1990	0	0	0	0	0	0	2,684	0.0	0.0	0.0	0.0	0.0	0.0
1991	416,588	325,000	...	...	0	91,588	2,671	34.3	10.1	57.5	95.1	17.5	10.6
1992	157,949	78,800	7,293	8,937	0	77,505	2,645	29.3	10.0	53.4	92.9	18.7	10.8
1993	142,229	34,300	12,950	46,410	-111	74,357	2,610	28.5	8.2	50.8	91.1	16.2	9.1
1994	138,665	20,600	35,982	71,283	-2,389	80,376	2,572	31.3	9.7	52.5	93.3	18.5	10.4
1995	149,719	22,900	37,417	99,569	1,250	65,917	2,536	26.0	8.1	50.6	90.1	15.9	9.0

Note: for years before 1991 see USSR.

121 LEBANON	- FOOD BALANCE SHEET OF FISH AND FISHERY PRODUCTS IN LIVE WEIGHT AND FISH CONTRIBUTION TO PROTEIN SUPPLY -												FAO-FIDI
YEAR	PRODUCTION	NON-FOOD USES	IMPORTS	EXPORTS	STOCKS VARIATIONS	TOTAL FOOD SUPPLY	POPULATION	PER CAPUT SUPPLY	FISH PROTEINS	ANIMAL PROTEINS	TOTAL PROTEINS	FISH/ANIMAL PROTEINS	FISH/TOTAL PROTEINS
	(.. metric tons in live weight ....)						(thousands)	(kilograms)	(grams per caput per day)			(%)	(%)
1961	1,900	0	5,478	0	0	7,378	1,913	3.9	1.1	17.2	61.3	6.2	1.7
1962	2,000	0	5,833	0	0	7,833	1,970	4.0	1.1	18.9	63.2	5.8	1.7
1963	2,200	0	6,089	0	0	8,289	2,029	4.1	1.1	20.2	62.4	5.6	1.8
1964	2,000	0	5,389	0	0	7,389	2,090	3.5	1.0	20.8	63.5	4.7	1.5
1965	2,300	0	5,378	0	0	7,678	2,151	3.6	1.0	21.1	64.4	4.6	1.5
1966	2,500	0	5,267	278	0	7,489	2,213	3.4	0.9	22.2	64.9	4.2	1.4
1967	1,800	0	7,481	222	0	9,058	2,275	4.0	1.1	21.5	64.6	4.9	1.6
1968	2,500	0	7,422	222	0	9,700	2,337	4.2	1.1	21.2	64.2	5.1	1.7
1969	3,000	0	6,208	222	0	8,986	2,402	3.7	1.0	20.0	61.2	4.8	1.6
1970	2,200	0	6,231	222	0	8,208	2,469	3.3	0.9	19.9	59.3	4.6	1.5
1971	2,100	0	7,894	778	0	9,217	2,541	3.6	1.0	19.9	59.3	5.0	1.7
1972	1,800	0	7,894	778	0	8,917	2,615	3.4	0.9	18.7	57.8	5.0	1.6
1973	2,500	0	7,894	778	0	9,617	2,683	3.6	1.0	19.4	58.1	5.1	1.7
1974	3,200	0	7,894	778	0	10,317	2,737	3.8	1.1	19.3	58.8	5.4	1.8
1975	2,500	0	7,894	778	0	9,617	2,767	3.5	1.0	17.7	56.4	5.4	1.7
1976	1,800	0	7,761	...	0	9,561	2,772	3.5	1.0	17.8	57.7	5.5	1.7
1977	1,700	0	12,914	...	0	14,614	2,754	5.3	1.5	20.8	62.4	7.2	2.4
1978	1,800	0	...	...	0	1,800	2,722	0.7	0.2	20.5	64.2	1.0	0.3
1979	1,750	0	...	...	0	1,750	2,691	0.7	0.2	24.7	70.4	0.8	0.3
1980	1,800	0	...	...	0	1,800	2,669	0.7	0.2	29.1	76.8	0.7	0.3
1981	1,600	0	...	...	0	1,600	2,662	0.6	0.2	29.1	77.3	0.7	0.2
1982	1,500	0	...	...	0	1,500	2,667	0.6	0.2	29.8	79.9	0.6	0.2
1983	1,400	0	...	...	0	1,400	2,676	0.5	0.2	30.3	80.4	0.5	0.2
1984	1,500	0	...	...	0	1,500	2,679	0.6	0.2	27.7	77.9	0.6	0.2
1985	1,700	0	...	...	0	1,700	2,668	0.6	0.2	26.1	77.8	0.8	0.3
1986	1,875	0	...	...	0	1,875	2,639	0.7	0.2	26.0	78.6	0.9	0.3
1987	1,900	0	...	...	0	1,900	2,598	0.7	0.2	24.3	78.3	0.9	0.3
1988	1,800	0	...	...	0	1,800	2,559	0.7	0.2	25.1	80.5	0.9	0.3
1989	1,800	0	...	...	0	1,800	2,540	0.7	0.2	25.3	81.8	0.9	0.3
1990	1,500	0	...	...	0	1,500	2,555	0.6	0.2	25.3	81.9	0.7	0.2
1991	1,800	0	8,039	...	0	9,839	2,610	3.8	1.1	26.5	85.2	4.0	1.2
1992	1,850	0	6,259	...	0	8,109	2,698	3.0	0.9	26.6	85.3	3.2	1.0
1993	2,200	0	7,837	...	0	10,037	2,806	3.6	1.0	26.4	84.4	3.8	1.2
1994	2,425	0	10,303	...	0	12,728	2,915	4.4	1.2	26.1	83.7	4.8	1.5
1995	4,385	0	9,895	...	0	14,280	3,009	4.8	1.4	26.2	84.0	5.2	1.6

122 LESOTHO	- FOOD BALANCE SHEET OF FISH AND FISHERY PRODUCTS IN LIVE WEIGHT AND FISH CONTRIBUTION TO PROTEIN SUPPLY -	FAO-FIDI

YEAR	PRODUCTION	NON-FOOD USES	IMPORTS	EXPORTS	STOCKS VARIATIONS	TOTAL FOOD SUPPLY	POPULATION	PER CAPUT SUPPLY	FISH PROTEINS	ANIMAL PROTEINS	TOTAL PROTEINS	FISH/ANIMAL PROTEINS	FISH/TOTAL PROTEINS
	(.. metric tons in live weight					....)	(thousands)	(kilograms)	(grams per caput per day)			(%)	(%)
1961	0	0	...	0	0	...	888	...	...	10.7	63.3	...	...
1962	0	0	...	0	0	...	906	...	...	11.0	61.9	...	...
1963	0	0	...	0	0	...	924	...	...	10.2	60.9	...	...
1964	0	0	...	0	0	...	943	...	...	10.1	61.2	...	...
1965	0	0	...	0	0	...	963	...	...	10.7	63.1	...	...
1966	0	0	...	0	0	...	982	...	...	8.5	59.2	...	...
1967	0	0	...	0	0	...	1,002	...	...	9.0	59.8	...	...
1968	0	0	...	0	0	...	1,022	...	...	8.3	59.3	...	...
1969	0	0	...	0	0	...	1,043	...	...	8.6	58.4	...	...
1970	0	0	...	0	0	...	1,064	...	...	9.4	60.1	...	...
1971	0	0	...	0	0	...	1,086	...	...	10.6	62.8	...	...
1972	0	0	...	0	0	...	1,109	...	...	11.2	59.7	...	...
1973	0	0	...	0	0	...	1,132	...	...	10.8	58.2	...	...
1974	10	0	...	0	0	...	1,158	...	...	10.7	63.6	...	...
1975	23	0	...	0	0	...	1,187	...	...	10.2	60.2	...	...
1976	27	0	...	0	0	...	1,219	...	...	10.1	62.2	...	...
1977	22	0	...	0	0	...	1,254	...	...	10.4	71.6	...	...
1978	24	0	...	0	0	...	1,291	...	...	11.2	69.5	...	...
1979	27	0	1,318	0	0	1,345	1,329	1.0	0.3	11.3	67.2	3.0	0.5
1980	23	0	2,411	0	0	2,434	1,367	1.8	0.6	12.1	64.4	5.2	1.0
1981	15	0	2,456	0	0	2,471	1,405	1.8	0.6	11.6	64.8	5.4	1.0
1982	19	0	3,072	0	0	3,091	1,444	2.1	0.8	14.1	65.5	5.4	1.2
1983	20	0	4,238	0	0	4,258	1,483	2.9	1.0	16.8	70.6	6.1	1.5
1984	13	0	3,365	0	0	3,378	1,523	2.2	0.8	15.0	65.3	5.2	1.2
1985	23	0	3,158	0	0	3,181	1,563	2.0	0.7	13.5	67.1	5.4	1.1
1986	25	0	2,982	0	0	3,007	1,605	1.9	0.7	12.4	64.7	5.4	1.0
1987	26	0	2,982	0	0	3,008	1,648	1.8	0.7	11.9	61.7	5.5	1.1
1988	30	0	3,421	0	0	3,451	1,692	2.0	0.7	11.6	66.3	6.3	1.1
1989	30	0	3,509	0	0	3,539	1,737	2.0	0.7	10.7	66.5	6.8	1.1
1990	30	0	4,035	0	0	4,065	1,783	2.3	0.8	10.6	65.8	7.7	1.2
1991	25	0	4,123	0	0	4,148	1,830	2.3	0.8	10.2	59.9	7.9	1.4
1992	33	0	4,386	0	0	4,419	1,878	2.4	0.8	9.7	58.9	8.6	1.4
1993	40	0	4,912	0	0	4,952	1,927	2.6	0.9	10.5	59.4	8.7	1.5
1994	40	0	5,263	0	0	5,303	1,977	2.7	1.0	9.9	59.0	9.6	1.6
1995	40	0	5,702	0	0	5,742	2,027	2.8	1.0	10.0	68.5	10.1	1.5

Note: 1961-1978 data for fish apparent consumption not shown as trade statistics not available..

123 LIBERIA	- FOOD BALANCE SHEET OF FISH AND FISHERY PRODUCTS IN LIVE WEIGHT AND FISH CONTRIBUTION TO PROTEIN SUPPLY -	FAO-FIDI

YEAR	PRODUCTION	NON-FOOD USES	IMPORTS	EXPORTS	STOCKS VARIATIONS	TOTAL FOOD SUPPLY	POPULATION	PER CAPUT SUPPLY	FISH PROTEINS	ANIMAL PROTEINS	TOTAL PROTEINS	FISH/ANIMAL PROTEINS	FISH/TOTAL PROTEINS
	(.....................		metric tons in live weight			.....................)	(thousands)	(kilograms)		(grams per caput per day)		(%)	(%)
1961	3,700	0	6,050	...	0	9,750	1,068	9.1	2.7	9.8	41.8	27.2	6.4
1962	4,000	0	5,850	...	0	9,850	1,098	9.0	2.7	9.2	40.8	28.9	6.5
1963	4,000	0	4,550	...	0	8,550	1,129	7.6	2.2	8.1	40.6	27.8	5.5
1964	4,000	0	5,006	...	0	9,006	1,161	7.8	2.1	7.6	41.1	27.1	5.0
1965	4,500	0	5,800	...	0	10,300	1,195	8.6	2.3	8.1	40.0	28.7	5.8
1966	5,000	0	10,289	200	0	15,089	1,230	12.3	3.2	8.9	43.3	36.1	7.4
1967	6,000	0	8,685	200	0	14,485	1,267	11.4	3.2	9.2	41.7	34.1	7.6
1968	6,000	0	9,306	100	0	15,206	1,305	11.7	3.2	8.8	41.9	35.9	7.6
1969	6,000	0	8,503	400	0	14,103	1,345	10.5	2.9	8.4	39.2	34.8	7.4
1970	8,000	0	12,400	800	0	19,600	1,385	14.2	3.8	9.5	44.1	40.1	8.6
1971	8,200	0	13,601	600	0	21,201	1,427	14.9	4.0	10.3	47.2	38.7	8.5
1972	8,900	0	10,100	800	0	18,200	1,470	12.4	3.3	9.6	44.2	34.7	7.5
1973	9,300	0	15,100	1,365	0	23,035	1,515	15.2	4.1	10.2	45.1	40.7	9.2
1974	10,064	0	15,401	1,360	0	24,105	1,561	15.4	4.2	9.6	43.9	43.7	9.6
1975	10,046	0	15,901	1,312	-120	24,515	1,609	15.2	4.2	9.6	44.1	43.6	9.5
1976	11,163	0	14,960	1,582	120	24,660	1,659	14.9	4.1	9.9	43.7	41.9	9.5
1977	10,190	0	16,986	882	0	26,294	1,710	15.4	4.2	10.5	47.1	40.3	9.0
1978	10,812	0	16,912	1,524	0	26,200	1,763	14.9	4.1	10.5	47.7	38.7	8.5
1979	13,484	0	19,615	1,219	0	31,880	1,819	17.5	4.8	11.3	49.5	42.7	9.8
1980	11,791	0	14,914	971	0	25,734	1,876	13.7	3.8	11.6	48.8	32.6	7.7
1981	13,004	0	11,181	294	0	23,892	1,934	12.4	3.4	10.9	48.5	31.5	7.1
1982	13,553	0	14,193	374	0	27,371	1,991	13.8	3.8	11.6	50.1	32.5	7.5
1983	15,277	0	18,783	365	0	33,695	2,051	16.4	4.4	12.7	49.3	34.9	9.0
1984	14,657	0	23,616	573	0	37,699	2,120	17.8	4.8	11.6	48.1	41.1	9.9
1985	11,486	0	18,547	1,115	0	28,918	2,199	13.2	3.6	11.6	47.1	30.9	7.6
1986	16,059	0	17,825	1,319	0	32,566	2,297	14.2	3.9	11.4	46.6	33.8	8.3
1987	18,734	0	15,752	1,019	0	33,467	2,409	13.9	3.8	11.6	44.2	32.3	8.5
1988	16,076	21	16,736	688	-200	31,903	2,512	12.7	3.4	9.1	41.5	38.1	8.3
1989	14,805	0	16,382	879	-20	30,288	2,574	11.8	3.3	8.3	44.1	39.3	7.4
1990	6,463	0	13,227	372	0	19,318	2,575	7.5	2.1	5.9	34.0	35.0	6.0
1991	9,620	1	8,230	593	220	17,476	2,500	7.0	1.9	6.1	36.4	31.5	5.2
1992	8,891	3	3,929	173	0	12,644	2,364	5.4	1.5	6.3	36.0	23.7	4.1
1993	7,782	4	2,267	351	0	9,694	2,215	4.4	1.3	6.3	34.5	20.1	3.7
1994	7,721	0	2,333	345	0	9,709	2,119	4.6	1.3	6.2	31.8	21.0	4.1
1995	7,700	0	2,357	319	0	9,738	2,123	4.6	1.3	6.1	34.1	21.6	3.8

124 LIBYA	- FOOD BALANCE SHEET OF FISH AND FISHERY PRODUCTS IN LIVE WEIGHT AND FISH CONTRIBUTION TO PROTEIN SUPPLY -	FAO-FIDI

YEAR	PRODUCTION	NON-FOOD USES	IMPORTS	EXPORTS	STOCKS VARIATIONS	TOTAL FOOD SUPPLY	POPULATION	PER CAPUT SUPPLY	FISH PROTEINS	ANIMAL PROTEINS	TOTAL PROTEINS	FISH/ANIMAL PROTEINS	FISH/TOTAL PROTEINS
	(..		metric tons in live weight		....)		(thousands)	(kilograms)	(grams per caput per day)			(%)	(%)
1961	1,700	0	500	800	0	1,400	1,400	1.0	0.3	9.3	37.8	3.3	0.8
1962	1,500	0	200	600	0	1,100	1,452	0.8	0.2	9.4	39.5	2.4	0.6
1963	1,400	0	500	400	0	1,500	1,505	1.0	0.3	9.7	42.5	3.0	0.7
1964	1,600	0	400	0	0	2,000	1,562	1.3	0.4	10.1	43.1	3.7	0.9
1965	2,900	0	200	0	0	3,100	1,623	1.9	0.6	12.5	48.2	4.4	1.2
1966	3,300	0	700	0	0	4,000	1,688	2.4	0.7	14.2	51.2	4.9	1.4
1967	4,200	0	1,800	0	0	6,000	1,757	3.4	1.0	15.9	53.6	6.3	1.9
1968	5,500	0	4,200	0	0	9,700	1,830	5.3	1.6	17.6	56.2	8.8	2.8
1969	6,500	0	2,000	0	0	8,500	1,906	4.5	1.4	18.5	57.0	7.3	2.4
1970	5,500	0	6,600	0	0	12,100	1,986	6.1	1.8	19.9	59.7	9.1	3.0
1971	5,600	0	4,800	0	0	10,400	2,069	5.0	1.5	19.6	60.6	7.6	2.5
1972	2,549	0	7,600	0	0	10,149	2,156	4.7	1.4	21.2	64.5	6.5	2.1
1973	2,975	0	8,300	0	0	11,275	2,248	5.0	1.5	22.6	70.1	6.6	2.1
1974	4,769	0	18,000	0	0	22,769	2,344	9.7	2.8	24.2	77.1	11.7	3.7
1975	4,949	0	13,748	0	0	18,697	2,446	7.6	2.2	25.2	79.5	8.9	2.8
1976	4,059	1,500	6,041	0	0	8,600	2,553	3.4	1.0	24.5	79.9	4.1	1.2
1977	2,046	410	542	0	0	2,178	2,666	0.8	0.2	26.9	81.6	0.9	0.3
1978	4,355	925	13,994	0	0	17,424	2,785	6.3	1.8	29.3	83.8	6.2	2.2
1979	4,500	735	15,916	0	0	19,681	2,911	6.8	1.9	31.4	86.0	6.1	2.3
1980	12,752	800	13,525	0	0	25,477	3,043	8.4	3.3	35.5	89.8	9.3	3.7
1981	13,130	1,480	13,020	0	0	24,670	3,183	7.8	2.7	37.0	91.2	7.2	2.9
1982	14,706	1,305	17,657	430	-1,000	29,628	3,330	8.9	2.8	35.9	89.7	7.8	3.1
1983	10,122	1,220	5,616	856	1,000	14,662	3,481	4.2	1.2	34.9	87.7	3.5	1.4
1984	11,974	1,250	4,704	0	0	15,428	3,634	4.3	1.2	33.0	85.1	3.7	1.4
1985	15,895	1,300	3,996	0	0	18,591	3,786	4.9	1.5	30.1	81.2	4.8	1.8
1986	17,937	1,300	2,401	0	0	19,038	3,936	4.8	1.4	28.4	80.0	5.0	1.8
1987	18,614	1,300	2,881	112	0	20,083	4,085	4.9	1.4	27.7	80.4	5.1	1.8
1988	21,682	1,350	4,369	168	-430	24,103	4,235	5.7	1.6	27.0	79.4	5.9	2.0
1989	22,050	1,250	8,290	115	120	29,095	4,387	6.6	1.9	27.2	80.1	6.9	2.3
1990	24,796	1,000	4,838	169	170	28,635	4,545	6.3	1.8	26.1	79.7	6.8	2.2
1991	26,098	1,000	13,612	2,152	90	36,648	4,708	7.8	2.2	25.5	79.9	8.5	2.7
1992	28,900	1,250	13,200	2,150	50	38,750	4,875	8.0	2.2	23.7	77.1	9.3	2.9
1993	31,244	1,500	13,497	3,000	0	40,241	5,048	8.0	2.2	23.0	75.3	9.5	2.9
1994	34,059	1,650	5,225	5,500	0	32,134	5,225	6.2	1.6	20.3	72.4	8.1	2.3
1995	34,500	1,650	3,559	5,500	0	30,909	5,407	5.7	1.5	21.7	89.5	7.0	1.7

126 LITHUANIA	- FOOD BALANCE SHEET OF FISH AND FISHERY PRODUCTS IN LIVE WEIGHT AND FISH CONTRIBUTION TO PROTEIN SUPPLY -	FAO-FIDI

YEAR	PRODUCTION	NON-FOOD USES	IMPORTS	EXPORTS	STOCKS VARIATIONS	TOTAL FOOD SUPPLY	POPULATION	PER CAPUT SUPPLY	FISH PROTEINS	ANIMAL PROTEINS	TOTAL PROTEINS	FISH/ANIMAL PROTEINS	FISH/TOTAL PROTEINS
	(............ metric tons in live weight ....)						(thousands)	(kilograms)	(grams per caput per day)			(%)	(%)
1961	0	0	0	0	0	0	2,824	0.0	0.0	0.0	0.0	0.0	0.0
1962	0	0	0	0	0	0	2,864	0.0	0.0	0.0	0.0	0.0	0.0
1963	0	0	0	0	0	0	2,899	0.0	0.0	0.0	0.0	0.0	0.0
1964	0	0	0	0	0	0	2,935	0.0	0.0	0.0	0.0	0.0	0.0
1965	0	0	0	0	0	0	2,971	0.0	0.0	0.0	0.0	0.0	0.0
1966	0	0	0	0	0	0	3,008	0.0	0.0	0.0	0.0	0.0	0.0
1967	0	0	0	0	0	0	3,045	0.0	0.0	0.0	0.0	0.0	0.0
1968	0	0	0	0	0	0	3,079	0.0	0.0	0.0	0.0	0.0	0.0
1969	0	0	0	0	0	0	3,112	0.0	0.0	0.0	0.0	0.0	0.0
1970	0	0	0	0	0	0	3,148	0.0	0.0	0.0	0.0	0.0	0.0
1971	0	0	0	0	0	0	3,186	0.0	0.0	0.0	0.0	0.0	0.0
1972	0	0	0	0	0	0	3,220	0.0	0.0	0.0	0.0	0.0	0.0
1973	0	0	0	0	0	0	3,251	0.0	0.0	0.0	0.0	0.0	0.0
1974	0	0	0	0	0	0	3,281	0.0	0.0	0.0	0.0	0.0	0.0
1975	0	0	0	0	0	0	3,308	0.0	0.0	0.0	0.0	0.0	0.0
1976	0	0	0	0	0	0	3,335	0.0	0.0	0.0	0.0	0.0	0.0
1977	0	0	0	0	0	0	3,362	0.0	0.0	0.0	0.0	0.0	0.0
1978	0	0	0	0	0	0	3,386	0.0	0.0	0.0	0.0	0.0	0.0
1979	0	0	0	0	0	0	3,409	0.0	0.0	0.0	0.0	0.0	0.0
1980	0	0	0	0	0	0	3,433	0.0	0.0	0.0	0.0	0.0	0.0
1981	0	0	0	0	0	0	3,460	0.0	0.0	0.0	0.0	0.0	0.0
1982	0	0	0	0	0	0	3,490	0.0	0.0	0.0	0.0	0.0	0.0
1983	0	0	0	0	0	0	3,523	0.0	0.0	0.0	0.0	0.0	0.0
1984	0	0	0	0	0	0	3,555	0.0	0.0	0.0	0.0	0.0	0.0
1985	0	0	0	0	0	0	3,587	0.0	0.0	0.0	0.0	0.0	0.0
1986	0	0	0	0	0	0	3,622	0.0	0.0	0.0	0.0	0.0	0.0
1987	0	0	0	0	0	0	3,657	0.0	0.0	0.0	0.0	0.0	0.0
1988	0	0	0	0	0	0	3,682	0.0	0.0	0.0	0.0	0.0	0.0
1989	0	0	0	0	0	0	3,712	0.0	0.0	0.0	0.0	0.0	0.0
1990	0	0	0	0	0	0	3,737	0.0	0.0	0.0	0.0	0.0	0.0
1991	475,043	330,000	...	...	-20,000	125,043	3,751	33.3	10.3	41.1	91.0	25.0	11.3
1992	192,458	48,000	2,037	25,780	0	120,715	3,755	32.2	10.1	52.3	99.3	19.4	10.2
1993	120,078	0	10,071	8,414	0	121,736	3,751	32.5	10.5	49.2	92.1	21.3	11.4
1994	51,036	0	48,096	28,031	-134	70,967	3,744	19.0	5.7	45.7	91.7	12.6	6.3
1995	49,510	0	43,633	61,164	20,014	51,993	3,736	13.9	3.6	40.4	87.8	8.9	4.1

Note: for years before 1991 see USSR.

128 MACAU	- FOOD BALANCE SHEET OF FISH AND FISHERY PRODUCTS IN LIVE WEIGHT AND FISH CONTRIBUTION TO PROTEIN SUPPLY -	FAO-FIDI

YEAR	PRODUCTION	NON-FOOD USES	IMPORTS	EXPORTS	STOCKS VARIATIONS	TOTAL FOOD SUPPLY	POPULATION	PER CAPUT SUPPLY	FISH PROTEINS	ANIMAL PROTEINS	TOTAL PROTEINS	FISH/ANIMAL PROTEINS	FISH/TOTAL PROTEINS
	(............		metric tons in live weight			............)	(thousands)	(kilograms)		(grams per caput per day)		(%)	(%)
1961	5,800	0	2,500	7,400	0	900	177	5.1	1.1	14.2	50.1	7.7	2.2
1962	5,800	0	2,400	7,100	0	1,100	183	6.0	1.7	15.9	51.8	10.4	3.2
1963	5,500	0	2,700	6,600	0	1,600	191	8.4	2.4	18.8	54.3	12.6	4.4
1964	5,800	0	3,400	7,100	0	2,100	200	10.5	2.9	20.8	56.7	14.0	5.1
1965	7,000	0	4,100	8,500	0	2,600	209	12.4	3.5	20.9	54.9	16.8	6.4
1966	7,500	0	4,800	9,100	0	3,200	219	14.6	4.2	24.3	55.5	17.3	7.6
1967	7,500	0	3,600	8,500	0	2,600	230	11.3	3.1	19.7	51.1	15.7	6.1
1968	7,000	0	4,100	7,700	0	3,400	240	14.2	4.0	22.8	53.5	17.4	7.4
1969	7,000	0	3,400	7,200	0	3,200	248	12.9	3.7	21.9	52.4	16.7	7.0
1970	5,900	0	3,700	5,600	0	4,000	254	15.8	4.4	23.3	52.6	18.9	8.4
1971	6,100	0	3,900	5,900	0	4,100	257	16.0	4.6	24.3	53.0	18.9	8.6
1972	5,900	0	4,900	5,700	0	5,100	258	19.8	5.4	30.2	59.6	17.9	9.1
1973	6,000	0	4,400	5,500	0	4,900	257	19.1	4.9	30.5	58.5	16.1	8.4
1974	6,000	0	5,100	5,400	0	5,700	255	22.4	5.9	30.8	59.8	19.1	9.9
1975	6,000	0	4,700	6,800	0	3,900	253	15.4	3.8	30.4	61.5	12.6	6.2
1976	7,400	0	6,485	8,288	0	5,598	251	22.3	6.8	34.9	69.7	19.3	9.7
1977	7,160	0	5,630	6,162	0	6,628	249	26.6	8.6	38.4	73.0	22.4	11.8
1978	9,236	0	6,346	9,994	0	5,589	248	22.6	7.2	39.2	74.5	18.3	9.6
1979	7,881	0	7,063	6,390	0	8,554	248	34.4	9.5	45.8	84.1	20.6	11.2
1980	6,624	0	8,122	4,286	0	10,460	252	41.5	11.7	44.4	82.5	26.4	14.2
1981	7,545	0	6,236	2,520	0	11,261	259	43.5	12.9	46.5	85.3	27.8	15.1
1982	6,656	12	6,207	4,817	0	8,034	268	29.9	9.3	43.8	80.9	21.2	11.5
1983	6,250	2	7,568	5,554	0	8,262	280	29.5	9.2	44.9	80.1	20.4	11.5
1984	8,800	0	9,971	8,154	0	10,617	293	36.2	11.1	45.0	81.6	24.6	13.6
1985	9,100	3	9,743	8,486	0	10,353	306	33.8	10.4	45.5	81.1	22.8	12.8
1986	8,042	3	11,100	8,886	0	10,252	319	32.1	10.6	43.1	77.5	24.6	13.7
1987	3,517	3	13,886	9,542	0	7,857	333	23.6	7.0	37.2	72.0	18.7	9.7
1988	2,485	0	12,353	7,056	0	7,783	346	22.5	7.6	38.1	70.9	20.0	10.8
1989	3,464	0	13,622	6,768	0	10,318	359	28.7	9.3	40.4	75.9	23.1	12.3
1990	2,583	0	14,952	6,264	0	11,271	372	30.3	9.4	40.5	75.6	23.1	12.4
1991	2,322	0	17,261	5,339	0	14,244	385	37.0	10.7	42.6	76.4	25.0	13.9
1992	2,668	1	15,195	3,319	0	14,543	396	36.7	10.0	42.5	79.1	23.4	12.6
1993	1,898	47	16,608	3,905	0	14,555	408	35.7	10.0	42.3	76.3	23.6	13.1
1994	1,890	18	17,216	2,450	0	16,638	419	39.7	11.1	43.4	80.3	25.5	13.8
1995	1,604	49	16,185	1,083	0	16,657	430	38.8	11.1	43.8	80.3	25.4	13.9

154 MACEDONIA	- FOOD BALANCE SHEET OF FISH AND FISHERY PRODUCTS IN LIVE WEIGHT AND FISH CONTRIBUTION TO PROTEIN SUPPLY -	FAO-FIDI

YEAR	PRODUCTION	NON-FOOD USES	IMPORTS	EXPORTS	STOCKS VARIATIONS	TOTAL FOOD SUPPLY	POPULATION	PER CAPUT SUPPLY	FISH PROTEINS	ANIMAL PROTEINS	TOTAL PROTEINS	FISH/ANIMAL PROTEINS	FISH/TOTAL PROTEINS
	(.. metric tons in live weight ....)						(thousands)	(kilograms)	(grams per caput per day)			(%)	(%)
1961	0	0	0	0	0	0	1,405	0.0	0.0	0.0	0.0	0.0	0.0
1962	0	0	0	0	0	0	1,423	0.0	0.0	0.0	0.0	0.0	0.0
1963	0	0	0	0	0	0	1,442	0.0	0.0	0.0	0.0	0.0	0.0
1964	0	0	0	0	0	0	1,462	0.0	0.0	0.0	0.0	0.0	0.0
1965	0	0	0	0	0	0	1,481	0.0	0.0	0.0	0.0	0.0	0.0
1966	0	0	0	0	0	0	1,499	0.0	0.0	0.0	0.0	0.0	0.0
1967	0	0	0	0	0	0	1,516	0.0	0.0	0.0	0.0	0.0	0.0
1968	0	0	0	0	0	0	1,532	0.0	0.0	0.0	0.0	0.0	0.0
1969	0	0	0	0	0	0	1,550	0.0	0.0	0.0	0.0	0.0	0.0
1970	0	0	0	0	0	0	1,568	0.0	0.0	0.0	0.0	0.0	0.0
1971	0	0	0	0	0	0	1,588	0.0	0.0	0.0	0.0	0.0	0.0
1972	0	0	0	0	0	0	1,609	0.0	0.0	0.0	0.0	0.0	0.0
1973	0	0	0	0	0	0	1,630	0.0	0.0	0.0	0.0	0.0	0.0
1974	0	0	0	0	0	0	1,653	0.0	0.0	0.0	0.0	0.0	0.0
1975	0	0	0	0	0	0	1,676	0.0	0.0	0.0	0.0	0.0	0.0
1976	0	0	0	0	0	0	1,699	0.0	0.0	0.0	0.0	0.0	0.0
1977	0	0	0	0	0	0	1,722	0.0	0.0	0.0	0.0	0.0	0.0
1978	0	0	0	0	0	0	1,746	0.0	0.0	0.0	0.0	0.0	0.0
1979	0	0	0	0	0	0	1,770	0.0	0.0	0.0	0.0	0.0	0.0
1980	0	0	0	0	0	0	1,795	0.0	0.0	0.0	0.0	0.0	0.0
1981	0	0	0	0	0	0	1,820	0.0	0.0	0.0	0.0	0.0	0.0
1982	0	0	0	0	0	0	1,846	0.0	0.0	0.0	0.0	0.0	0.0
1983	0	0	0	0	0	0	1,872	0.0	0.0	0.0	0.0	0.0	0.0
1984	0	0	0	0	0	0	1,898	0.0	0.0	0.0	0.0	0.0	0.0
1985	0	0	0	0	0	0	1,923	0.0	0.0	0.0	0.0	0.0	0.0
1986	0	0	0	0	0	0	1,948	0.0	0.0	0.0	0.0	0.0	0.0
1987	0	0	0	0	0	0	1,973	0.0	0.0	0.0	0.0	0.0	0.0
1988	0	0	0	0	0	0	1,998	0.0	0.0	0.0	0.0	0.0	0.0
1989	0	0	0	0	0	0	2,022	0.0	0.0	0.0	0.0	0.0	0.0
1990	0	0	0	0	0	0	2,046	0.0	0.0	0.0	0.0	0.0	0.0
1991	0	0	0	0	0	0	2,070	0.0	0.0	0.0	0.0	0.0	0.0
1992	1,209	0	4,285	299	-75	5,120	2,093	2.5	0.7	22.0	68.4	3.2	1.0
1993	1,235	0	5,355	676	50	5,964	2,115	2.8	0.8	26.4	66.5	2.8	1.1
1994	1,230	0	7,561	883	25	7,932	2,137	3.7	1.0	28.8	70.7	3.4	1.4
1995	1,505	0	7,011	106	0	8,410	2,156	3.9	1.0	26.1	66.7	3.9	1.5

Note: for years before 1992 see Yugoslavia, S.F.R.

129 MADAGASCAR	- FOOD BALANCE SHEET OF FISH AND FISHERY PRODUCTS IN LIVE WEIGHT AND FISH CONTRIBUTION TO PROTEIN SUPPLY -	FAO-FIDI

YEAR	PRODUCTION	NON-FOOD USES	IMPORTS	EXPORTS	STOCKS VARIATIONS	TOTAL FOOD SUPPLY	POPULATION	PER CAPUT SUPPLY	FISH PROTEINS	ANIMAL PROTEINS	TOTAL PROTEINS	FISH/ANIMAL PROTEINS	FISH/TOTAL PROTEINS
	(..		metric tons in live weight			....)	(thousands)	(kilograms)	(grams per caput per day)			(%)	(%)
1961	29,000	0	1,250	253	0	29,997	5,506	5.5	1.7	18.4	60.4	9.1	2.8
1962	30,500	0	1,350	356	0	31,494	5,651	5.6	1.7	18.7	59.0	9.1	2.9
1963	35,900	0	1,450	352	0	36,998	5,800	6.4	2.0	18.8	61.0	10.4	3.2
1964	44,500	300	1,250	454	0	44,996	5,949	7.6	2.3	19.0	62.1	12.2	3.7
1965	43,800	300	1,450	354	0	44,596	6,099	7.3	2.2	17.2	60.1	13.1	3.7
1966	54,400	0	1,450	885	0	54,965	6,247	8.8	2.7	20.0	61.8	13.5	4.4
1967	57,500	300	1,200	1,650	0	56,750	6,396	8.9	2.7	19.9	62.7	13.8	4.4
1968	62,600	0	1,650	2,242	0	62,008	6,547	9.5	2.9	20.0	63.7	14.6	4.6
1969	42,600	0	1,450	2,311	0	41,739	6,701	6.2	1.9	18.3	62.7	10.5	3.1
1970	50,100	0	1,200	3,380	-300	47,620	6,860	6.9	2.1	18.0	61.0	11.8	3.5
1971	52,600	0	2,050	4,279	-300	50,071	7,025	7.1	2.2	17.0	61.4	12.8	3.5
1972	54,200	0	1,000	5,180	0	50,020	7,195	7.0	2.1	17.2	60.5	12.2	3.5
1973	61,000	0	1,600	8,434	0	54,166	7,375	7.3	2.2	17.0	59.6	13.1	3.7
1974	67,840	0	1,000	15,671	0	53,169	7,568	7.0	2.1	16.6	59.8	12.8	3.6
1975	60,620	0	1,540	7,967	-100	54,093	7,778	7.0	2.1	16.9	59.8	12.2	3.4
1976	59,950	0	324	4,194	-1,135	54,945	8,006	6.9	2.0	16.6	60.3	12.2	3.4
1977	60,970	0	641	5,829	232	56,014	8,252	6.8	2.0	14.3	57.0	13.9	3.5
1978	59,240	0	134	4,830	100	54,643	8,514	6.4	1.9	18.8	60.0	10.0	3.1
1979	58,375	0	228	4,964	1,099	54,738	8,788	6.2	1.8	16.8	57.6	10.8	3.2
1980	55,152	1	140	4,732	1,003	51,561	9,072	5.7	1.7	16.5	57.0	10.1	2.9
1981	55,175	50	279	3,385	0	52,019	9,366	5.6	1.7	16.6	57.7	10.1	2.9
1982	59,335	43	46	4,225	0	55,113	9,671	5.7	1.7	16.1	57.0	10.6	3.0
1983	63,352	42	274	3,858	0	59,726	9,988	6.0	1.8	15.7	57.7	11.6	3.2
1984	67,514	97	89	5,121	0	62,385	10,320	6.1	1.8	15.4	56.2	11.8	3.2
1985	69,216	53	260	5,601	0	63,822	10,669	6.0	1.8	15.1	55.4	11.9	3.2
1986	84,347	64	228	5,681	0	78,829	11,035	7.1	2.1	15.5	56.2	13.9	3.8
1987	92,428	0	175	8,277	0	84,326	11,418	7.4	2.2	15.1	54.1	14.7	4.1
1988	101,118	0	472	6,254	0	95,336	11,816	8.1	2.4	15.0	51.5	15.9	4.6
1989	97,661	0	460	6,623	0	91,498	12,224	7.5	2.3	14.4	50.3	15.6	4.5
1990	103,674	0	430	9,447	0	94,657	12,642	7.5	2.2	14.3	50.3	15.6	4.5
1991	101,100	0	411	10,679	0	90,832	13,069	7.0	2.1	13.9	49.3	15.0	4.2
1992	108,020	5,790	12,528	19,896	0	94,862	13,504	7.0	2.2	13.7	48.9	16.3	4.6
1993	116,550	5,000	14,189	24,766	0	100,973	13,950	7.2	2.2	13.3	48.1	16.8	4.7
1994	119,726	7,170	17,916	35,086	0	95,386	14,406	6.6	2.0	12.8	46.3	15.7	4.4
1995	119,352	10,580	11,903	38,176	0	82,499	14,874	5.6	1.7	12.3	46.1	13.8	3.7

130 MALAWI	- FOOD BALANCE SHEET OF FISH AND FISHERY PRODUCTS IN LIVE WEIGHT AND FISH CONTRIBUTION TO PROTEIN SUPPLY -	FAO-FIDI

YEAR	PRODUCTION	NON-FOOD USES	IMPORTS	EXPORTS	STOCKS VARIATIONS	TOTAL FOOD SUPPLY	POPULATION	PER CAPUT SUPPLY	FISH PROTEINS	ANIMAL PROTEINS	TOTAL PROTEINS	FISH/ANIMAL PROTEINS	FISH/TOTAL PROTEINS
	(... metric tons in live weight ...)						(thousands)	(kilograms)	(grams per caput per day)			(%)	(%)
1961	14,300	0	200	300	0	14,200	3,611	3.9	1.2	3.2	60.4	37.0	1.9
1962	19,500	0	200	300	0	19,400	3,697	5.3	1.6	3.6	60.4	43.4	2.6
1963	21,700	0	200	300	0	21,600	3,786	5.7	1.7	3.8	63.4	44.0	2.7
1964	21,200	0	200	300	0	21,100	3,879	5.4	1.6	3.8	66.3	42.5	2.4
1965	33,900	0	200	300	0	33,800	3,975	8.5	2.5	4.8	67.2	52.9	3.8
1966	32,500	0	200	300	0	32,400	4,073	8.0	2.4	4.6	67.6	51.3	3.5
1967	29,300	0	700	300	0	29,700	4,176	7.1	2.1	4.6	65.1	45.6	3.2
1968	32,900	0	700	400	0	33,200	4,283	7.8	2.3	4.9	68.4	47.2	3.3
1969	38,200	0	1,400	1,400	0	38,200	4,397	8.7	2.6	5.6	68.5	46.0	3.7
1970	47,000	0	1,400	2,200	0	46,200	4,518	10.2	3.0	6.0	71.5	50.3	4.2
1971	60,000	0	850	1,900	0	58,950	4,646	12.7	3.7	6.6	74.2	56.8	5.1
1972	70,000	0	600	2,500	0	68,100	4,783	14.2	4.2	7.0	73.0	59.8	5.8
1973	69,300	0	450	2,700	0	67,050	4,928	13.6	4.0	6.9	73.6	58.0	5.5
1974	70,277	0	800	2,100	0	68,977	5,081	13.6	4.0	6.9	72.7	57.8	5.5
1975	70,992	0	514	5,609	0	65,897	5,244	12.6	3.7	6.6	71.1	56.2	5.2
1976	74,900	0	763	2,387	0	73,276	5,419	13.5	4.0	7.0	71.0	57.0	5.6
1977	68,200	0	251	3,152	0	65,298	5,606	11.7	3.4	6.5	70.1	52.8	4.9
1978	67,800	0	584	2,312	0	66,072	5,800	11.4	3.4	6.7	69.5	50.3	4.8
1979	60,000	0	443	3,123	0	57,320	5,994	9.6	2.8	6.1	66.8	46.6	4.2
1980	65,757	0	489	7,301	0	58,945	6,183	9.5	2.8	6.2	65.6	45.8	4.3
1981	51,392	0	395	4,331	0	47,456	6,358	7.5	2.2	5.8	66.1	38.0	3.3
1982	58,416	0	462	4,214	0	54,664	6,524	8.4	2.5	6.1	66.7	40.4	3.7
1983	66,993	0	625	2,633	0	64,984	6,707	9.7	2.9	6.2	63.6	45.7	4.5
1984	65,068	0	284	892	0	64,460	6,940	9.3	2.7	5.9	61.6	46.0	4.4
1985	62,072	0	372	680	0	61,764	7,245	8.5	2.5	5.6	61.1	45.0	4.1
1986	72,858	0	169	465	0	72,562	7,641	9.5	2.8	6.2	59.3	45.4	4.7
1987	88,596	0	69	818	0	87,847	8,109	10.8	3.2	6.2	58.7	51.8	5.4
1988	78,817	0	328	537	0	78,608	8,593	9.2	2.7	5.5	58.2	48.8	4.6
1989	70,771	0	422	121	0	71,072	9,018	7.9	2.3	5.1	55.8	45.9	4.2
1990	74,100	0	281	122	-10	74,249	9,329	8.0	2.4	5.0	54.6	47.4	4.3
1991	63,729	0	212	88	0	63,853	9,505	6.7	2.0	4.8	53.8	41.7	3.7
1992	69,484	0	174	51	0	69,607	9,569	7.3	2.2	4.8	51.6	44.6	4.2
1993	68,207	0	191	235	5	68,168	9,571	7.1	2.1	4.8	55.7	43.5	3.8
1994	58,805	0	248	93	5	58,966	9,587	6.2	1.8	4.4	53.7	41.7	3.4
1995	53,890	0	240	3	0	54,127	9,673	5.6	1.7	4.2	55.0	39.8	3.0

131 MALAYSIA	- FOOD BALANCE SHEET OF FISH AND FISHERY PRODUCTS IN LIVE WEIGHT AND FISH CONTRIBUTION TO PROTEIN SUPPLY -	FAO-FIDI

YEAR	PRODUCTION	NON-FOOD USES	IMPORTS	EXPORTS	STOCKS VARIATIONS	TOTAL FOOD SUPPLY	POPULATION	PER CAPUT SUPPLY	FISH PROTEINS	ANIMAL PROTEINS	TOTAL PROTEINS	FISH/ANIMAL PROTEINS	FISH/TOTAL PROTEINS
	(....................................		metric tons in live weight			....................................)	(thousands)	(kilograms)	(grams per caput per day)			(%)	(%)
1961	166,400	24,500	59,832	46,401	0	155,331	8,401	18.5	5.9	13.6	46.7	43.5	12.7
1962	188,000	23,900	65,484	45,872	0	183,712	8,671	21.2	6.6	14.6	49.1	45.0	13.4
1963	219,000	28,700	68,329	48,880	-2,740	207,009	8,948	23.1	6.9	15.2	50.1	45.2	13.7
1964	212,000	31,500	81,655	61,912	-1,241	199,001	9,226	21.6	6.2	14.9	48.5	41.4	12.7
1965	228,200	34,600	71,126	75,229	1,761	191,259	9,502	20.1	5.8	14.9	48.4	39.1	12.0
1966	271,900	39,900	73,173	80,726	2,241	226,688	9,775	23.2	6.7	16.0	48.8	41.9	13.8
1967	342,700	54,400	65,955	80,490	0	273,765	10,044	27.3	8.0	17.2	49.6	46.5	16.1
1968	381,900	61,400	79,373	107,027	0	292,847	10,312	28.4	8.4	17.7	50.5	47.3	16.6
1969	347,600	54,400	78,785	101,330	0	270,655	10,581	25.6	7.6	17.3	52.0	43.9	14.6
1970	340,100	56,090	94,320	130,923	0	247,407	10,853	22.8	6.7	16.7	50.8	39.8	13.1
1971	367,400	52,543	83,125	132,624	0	265,358	11,128	23.9	7.0	16.8	50.7	41.5	13.8
1972	358,300	53,841	92,439	150,329	0	246,568	11,407	21.6	6.1	16.4	50.6	37.2	12.0
1973	444,100	75,819	114,047	161,194	0	321,135	11,690	27.5	7.7	18.2	53.2	42.5	14.5
1974	525,120	96,797	112,383	134,506	0	406,200	11,973	33.9	9.7	20.8	54.9	46.5	17.6
1975	473,382	112,486	128,252	128,754	0	360,394	12,258	29.4	8.9	21.6	54.5	41.4	16.4
1976	516,369	119,933	164,646	158,464	0	402,618	12,543	32.1	8.9	22.0	55.1	40.4	16.1
1977	618,895	87,080	151,950	127,709	0	556,055	12,831	43.3	11.7	24.8	57.1	47.1	20.5
1978	684,607	104,904	179,850	166,607	0	592,946	13,126	45.2	12.5	26.4	59.3	47.3	21.0
1979	696,029	137,415	245,925	220,926	0	583,614	13,435	43.4	11.6	25.6	58.7	45.2	19.7
1980	736,086	151,545	172,039	172,921	0	583,659	13,763	42.4	10.2	25.2	57.9	40.6	17.7
1981	803,594	211,466	192,910	224,930	-2,281	557,827	14,112	39.5	10.3	25.4	57.6	40.5	17.8
1982	743,866	113,469	211,137	189,289	-2,233	650,012	14,479	44.9	12.0	26.8	58.4	44.6	20.5
1983	803,983	177,499	197,321	182,366	-179	641,259	14,864	43.1	11.8	27.9	59.2	42.3	19.9
1984	804,957	120,753	244,239	173,178	-2,584	752,681	15,264	49.3	13.5	30.1	61.5	45.0	22.0
1985	800,338	104,106	226,385	210,907	-261	711,449	15,677	45.4	12.6	29.6	58.4	42.6	21.6
1986	826,825	105,814	261,903	235,696	-1,139	746,079	16,103	46.3	12.7	30.0	58.1	42.5	21.9
1987	905,951	198,021	309,456	254,716	1,262	763,932	16,541	46.2	12.3	30.9	59.6	39.8	20.7
1988	873,584	197,132	266,490	189,260	-21,919	731,763	16,987	43.1	11.7	30.6	60.1	38.3	19.5
1989	936,692	218,538	310,251	214,659	-2,609	811,137	17,439	46.5	12.4	31.5	60.1	39.4	20.7
1990	1,004,887	205,487	278,165	203,225	-1,282	873,058	17,891	48.8	13.1	34.2	63.1	38.3	20.7
1991	978,236	150,679	307,052	211,443	-7,313	915,853	18,344	49.9	14.1	37.6	68.9	37.5	20.5
1992	1,104,989	186,928	308,644	238,728	13,084	1,001,061	18,796	53.3	14.9	40.2	71.6	37.2	20.8
1993	1,154,562	221,524	320,375	239,131	23,468	1,037,750	19,247	53.9	14.4	41.0	72.2	35.2	20.0
1994	1,181,763	202,407	332,487	239,439	3,033	1,075,437	19,695	54.6	14.9	41.9	72.5	35.6	20.6
1995	1,239,755	209,502	301,905	226,447	2,067	1,107,778	20,140	55.0	15.3	43.1	73.4	35.4	20.8

132 MALDIVES	- FOOD BALANCE SHEET OF FISH AND FISHERY PRODUCTS IN LIVE WEIGHT AND FISH CONTRIBUTION TO PROTEIN SUPPLY -	FAO-FIDI

YEAR	PRODUCTION	NON-FOOD USES	IMPORTS	EXPORTS	STOCKS VARIATIONS	TOTAL FOOD SUPPLY	POPULATION	PER CAPUT SUPPLY	FISH PROTEINS	ANIMAL PROTEINS	TOTAL PROTEINS	FISH/ANIMAL PROTEINS	FISH/TOTAL PROTEINS
	(..		metric tons in live weight			....)	(thousands)	(kilograms)	(grams per caput per day)			(%)	(%)
1961	17,000	0	0	10,800	0	6,200	100	61.7	20.1	21.8	44.2	91.8	45.4
1962	17,000	0	0	10,400	0	6,600	102	64.4	21.0	22.8	45.9	92.1	45.8
1963	17,000	0	0	10,400	0	6,600	105	63.1	20.6	22.4	45.4	91.9	45.3
1964	17,000	0	0	10,000	0	7,000	107	65.6	21.5	23.3	47.0	92.2	45.7
1965	19,600	0	0	12,400	0	7,200	109	66.1	21.7	23.5	47.0	92.3	46.1
1966	22,400	0	0	15,000	0	7,400	111	66.6	22.4	24.2	48.7	92.5	46.0
1967	25,100	0	0	16,500	0	8,600	114	75.7	25.6	27.4	51.7	93.4	49.5
1968	23,700	0	0	15,500	0	8,200	116	70.6	23.8	25.7	49.9	92.9	47.8
1969	32,300	0	0	23,300	0	9,000	119	75.8	22.3	24.1	48.3	92.5	46.2
1970	33,200	0	0	23,830	0	9,370	121	77.2	25.9	27.7	51.9	93.5	49.9
1971	34,800	0	0	24,487	0	10,313	124	83.0	27.9	29.7	54.9	93.9	50.7
1972	31,400	0	0	21,212	0	10,188	127	80.1	23.5	25.4	51.8	92.8	45.4
1973	33,700	0	0	23,022	0	10,678	130	81.9	27.5	29.3	61.4	93.7	44.7
1974	35,424	0	0	23,908	0	11,516	134	86.1	28.3	30.2	62.7	93.9	45.2
1975	26,504	0	0	15,316	0	11,188	137	81.5	26.2	28.0	60.8	93.5	43.1
1976	32,544	5,000	0	15,941	0	11,603	141	82.3	26.4	28.2	62.3	93.5	42.3
1977	26,322	0	0	14,813	0	11,509	145	79.4	25.2	27.0	63.4	93.3	39.7
1978	25,436	0	0	13,155	0	12,281	149	82.3	25.4	27.2	65.3	93.4	38.9
1979	27,847	0	0	15,336	0	12,511	154	81.5	26.2	27.9	67.7	93.8	38.7
1980	34,648	0	0	21,982	0	12,666	158	80.1	25.0	26.7	67.0	93.7	37.4
1981	34,281	0	0	19,883	0	14,398	163	88.5	27.9	29.6	71.1	94.4	39.3
1982	30,533	0	0	16,736	0	13,797	168	82.3	25.3	27.2	69.2	92.8	36.5
1983	38,569	2,000	0	15,212	0	21,357	173	123.7	39.6	41.6	81.1	95.3	48.8
1984	55,076	10,500	0	22,200	0	22,376	178	125.7	39.2	41.1	80.5	95.2	48.7
1985	61,925	1,500	0	34,422	0	26,003	184	141.6	45.5	47.4	87.4	95.9	52.0
1986	59,281	0	0	34,775	0	24,506	189	129.3	42.1	47.0	86.2	89.6	48.8
1987	56,992	4,500	0	32,459	0	20,033	196	102.4	33.0	38.1	76.5	86.7	43.2
1988	71,483	4,500	0	37,610	0	29,373	202	145.4	48.8	53.4	91.8	91.4	53.2
1989	71,245	0	0	47,915	0	23,330	209	111.8	37.3	43.8	83.1	85.2	44.9
1990	78,733	0	0	58,731	0	20,002	216	92.7	29.9	36.7	75.4	81.5	39.7
1991	80,713	0	0	59,460	0	21,253	223	95.4	29.9	37.6	74.5	79.6	40.1
1992	82,154	0	0	50,816	0	31,338	230	136.1	43.0	50.5	85.3	85.1	50.4
1993	90,012	0	0	51,608	0	38,404	238	161.4	50.6	61.1	92.7	82.7	54.6
1994	104,110	0	0	60,623	-256	43,231	246	175.8	54.5	64.6	97.6	84.3	55.8
1995	104,566	0	0	56,672	-389	47,505	254	186.7	58.6	68.7	102.7	85.2	57.1

Note: prior to 1983 supply data are not consistent with results of national consumption survey, probably due to underestimated production.

133 MALI — FOOD BALANCE SHEET OF FISH AND FISHERY PRODUCTS IN LIVE WEIGHT AND FISH CONTRIBUTION TO PROTEIN SUPPLY — FAO-FIDI

YEAR	PRODUCTION	NON-FOOD USES	IMPORTS	EXPORTS	STOCKS VARIATIONS	TOTAL FOOD SUPPLY	POPULATION	PER CAPUT SUPPLY	FISH PROTEINS	ANIMAL PROTEINS	TOTAL PROTEINS	FISH/ANIMAL PROTEINS	FISH/TOTAL PROTEINS
	(...... metric tons in live weight					......)	(thousands)	(kilograms)	(grams per caput per day)			(%)	(%)
1961	80,000	0	0	34,483	0	45,517	4,480	10.2	2.6	16.3	64.8	16.2	4.1
1962	90,000	10,000	1,000	34,483	0	46,517	4,588	10.1	2.8	16.6	64.6	16.6	4.3
1963	90,000	5,003	200	35,200	0	49,998	4,699	10.6	2.9	17.3	65.0	16.5	4.4
1964	90,000	3,455	0	38,000	0	48,545	4,811	10.1	2.7	17.6	60.0	15.1	4.4
1965	90,000	3,109	0	38,000	0	48,891	4,922	9.9	2.6	18.3	59.4	14.2	4.4
1966	110,000	32,473	0	24,873	0	52,655	5,032	10.5	2.9	18.9	60.2	15.6	4.9
1967	100,000	29,364	0	17,964	0	52,673	5,140	10.3	2.8	19.1	62.6	14.6	4.4
1968	106,000	34,200	0	18,309	0	53,491	5,249	10.2	2.8	19.3	61.4	14.6	4.6
1969	108,000	36,964	0	15,891	0	55,145	5,364	10.3	2.9	19.9	62.9	14.4	4.5
1970	110,000	35,582	200	19,691	0	54,927	5,484	10.0	2.8	19.5	61.6	14.4	4.6
1971	90,000	19,345	0	15,200	0	55,455	5,614	9.9	2.6	18.7	56.4	13.8	4.6
1972	80,000	23,836	0	10,018	0	46,145	5,751	8.0	2.0	15.6	53.2	13.0	3.8
1973	85,000	32,818	459	5,586	0	47,055	5,892	8.0	2.1	13.7	49.8	15.0	4.1
1974	75,000	25,000	2,145	3,804	0	48,342	6,032	8.0	2.0	14.3	55.9	13.6	3.5
1975	75,000	22,403	598	8,251	0	44,943	6,169	7.3	2.0	15.7	58.2	12.7	3.4
1976	75,000	20,000	461	8,973	0	46,488	6,300	7.4	2.0	16.3	54.9	12.5	3.7
1977	75,000	19,000	415	6,956	0	49,459	6,428	7.7	2.2	16.6	54.1	13.2	4.1
1978	75,000	15,000	426	3,974	0	56,452	6,560	8.6	2.5	17.5	55.2	14.0	4.4
1979	83,586	0	309	3,939	0	79,956	6,703	11.9	3.4	17.8	55.5	19.1	6.1
1980	88,228	0	635	7,020	0	81,843	6,863	11.9	3.4	18.2	53.7	18.7	6.3
1981	75,564	0	0	4,268	0	71,296	7,043	10.1	2.9	17.9	51.2	16.2	5.7
1982	73,451	0	875	4,358	0	69,968	7,242	9.7	2.8	17.5	57.0	15.8	4.9
1983	61,289	0	99	879	0	60,508	7,457	8.1	2.3	15.8	56.1	14.8	4.2
1984	54,711	0	490	1,378	0	53,823	7,682	7.0	2.1	15.2	55.6	13.5	3.7
1985	54,184	0	2,435	3,371	0	53,249	7,915	6.7	2.0	14.3	54.1	13.9	3.7
1986	61,012	0	2,380	3,789	0	59,603	8,155	7.3	2.1	15.2	55.0	14.0	3.9
1987	55,713	0	4,187	1,257	0	58,643	8,403	7.0	2.0	14.6	53.9	13.9	3.8
1988	55,875	0	2,356	1,930	0	56,301	8,660	6.5	1.9	14.8	57.8	12.8	3.3
1989	71,838	0	2,220	2,450	0	71,608	8,929	8.0	2.3	15.4	56.2	15.2	4.2
1990	70,548	0	3,340	1,143	0	72,745	9,212	7.9	2.3	15.5	57.4	14.9	4.0
1991	68,780	0	2,914	1,215	0	70,479	9,507	7.4	2.2	15.4	59.4	14.1	3.7
1992	68,507	0	2,744	1,148	0	70,103	9,816	7.1	2.1	15.6	62.2	13.5	3.4
1993	64,352	0	3,216	999	0	66,569	10,135	6.6	1.9	15.3	57.9	12.6	3.3
1994	62,950	0	800	836	0	62,914	10,462	6.0	1.8	14.6	59.3	12.2	3.0
1995	133,000	0	1,934	1,000	0	133,934	10,795	12.4	3.6	16.3	65.0	21.9	5.5

134 MALTA	- FOOD BALANCE SHEET OF FISH AND FISHERY PRODUCTS IN LIVE WEIGHT AND FISH CONTRIBUTION TO PROTEIN SUPPLY -	FAO-FIDI

YEAR	PRODUCTION	NON-FOOD USES	IMPORTS	EXPORTS	STOCKS VARIATIONS	TOTAL FOOD SUPPLY	POPULATION	PER CAPUT SUPPLY	FISH PROTEINS	ANIMAL PROTEINS	TOTAL PROTEINS	FISH/ANIMAL PROTEINS	FISH/TOTAL PROTEINS
	(..		metric tons in live weight		....)		(thousands)	(kilograms)	(grams per caput per day)			(%)	(%)
1961	1,300	0	2,050	0	0	3,350	311	10.8	3.3	31.7	87.6	10.5	3.8
1962	1,300	0	1,900	0	0	3,200	309	10.4	3.3	31.3	86.8	10.5	3.8
1963	1,600	0	1,800	0	0	3,400	308	11.1	3.4	33.1	89.8	10.3	3.8
1964	1,400	0	1,800	0	0	3,200	306	10.5	3.2	32.1	89.9	9.8	3.5
1965	1,200	0	1,500	0	0	2,700	305	8.9	2.7	32.8	90.7	8.2	3.0
1966	1,300	0	1,500	0	0	2,800	304	9.2	2.8	32.8	88.5	8.6	3.2
1967	1,400	0	1,900	0	0	3,300	304	10.9	3.3	34.4	88.1	9.6	3.8
1968	1,200	0	1,500	0	0	2,700	303	8.9	2.7	36.9	91.2	7.3	3.0
1969	1,100	0	1,900	0	0	3,000	303	9.9	3.0	38.4	94.1	7.7	3.2
1970	1,212	0	2,283	0	0	3,495	303	11.6	3.4	41.6	94.5	8.1	3.6
1971	1,224	0	2,133	0	0	3,357	302	11.1	3.4	44.5	97.8	7.6	3.5
1972	1,224	0	2,533	0	0	3,757	302	12.4	3.7	43.8	98.1	8.4	3.8
1973	1,624	0	2,083	0	0	3,707	302	12.3	3.7	42.5	96.3	8.7	3.8
1974	1,546	0	3,194	0	0	4,740	303	15.7	4.4	43.0	94.8	10.2	4.6
1975	1,529	0	2,244	0	0	3,773	304	12.4	3.5	45.1	99.4	7.7	3.5
1976	1,571	0	2,520	5	0	4,086	307	13.3	4.1	51.4	103.6	7.9	3.9
1977	1,483	0	3,456	0	0	4,939	311	15.9	4.8	52.7	102.9	9.2	4.7
1978	1,079	0	4,043	0	0	5,122	315	16.3	4.8	54.2	101.6	8.9	4.7
1979	1,322	0	4,026	0	0	5,348	320	16.7	5.0	52.9	100.3	9.5	5.0
1980	1,078	0	7,351	45	0	8,384	324	25.9	7.8	55.4	105.4	14.0	7.4
1981	942	0	5,785	61	0	6,666	329	20.3	5.6	50.4	99.4	11.2	5.7
1982	1,216	0	3,051	76	0	4,191	333	12.6	3.8	48.8	94.4	7.7	4.0
1983	999	0	3,157	94	0	4,062	337	12.0	3.7	48.0	95.2	7.7	3.9
1984	1,226	0	3,539	628	0	4,137	341	12.1	3.9	46.9	93.5	8.2	4.1
1985	2,520	0	4,935	1,604	0	5,851	344	17.0	5.5	53.2	98.1	10.3	5.6
1986	1,121	0	4,120	46	0	5,195	347	15.0	4.7	47.6	92.1	9.9	5.1
1987	1,017	0	4,663	429	0	5,251	349	15.1	4.6	52.3	97.1	8.8	4.7
1988	869	0	4,423	77	0	5,215	351	14.9	4.5	51.0	99.5	8.7	4.5
1989	938	1,778	7,940	83	0	7,017	352	19.9	6.4	55.7	103.5	11.4	6.2
1990	6,365	2,600	4,440	1,222	-14	6,968	354	19.7	5.9	51.9	100.8	11.4	5.9
1991	5,169	4,000	6,419	669	-14	6,904	356	19.4	5.2	58.3	107.2	8.9	4.9
1992	5,270	4,000	7,860	457	-14	8,658	359	24.1	5.9	57.4	105.2	10.2	5.6
1993	1,373	0	6,313	1,273	0	6,412	361	17.7	4.3	53.6	104.7	8.0	4.1
1994	1,799	0	7,555	1,207	9	8,155	364	22.4	5.5	49.8	100.0	11.1	5.5
1995	1,758	0	8,560	822	34	9,530	367	26.0	6.7	59.7	107.9	11.2	6.2

Note: 1989-1992 non-food uses data refer to fish caught by vessels using flags of convenience.

127 MARSHALL ISLANDS	- FOOD BALANCE SHEET OF FISH AND FISHERY PRODUCTS IN LIVE WEIGHT AND FISH CONTRIBUTION TO PROTEIN SUPPLY -	FAO-FIDI

YEAR	PRODUCTION	NON-FOOD USES	IMPORTS	EXPORTS	STOCKS VARIATIONS	TOTAL FOOD SUPPLY	POPULATION	PER CAPUT SUPPLY	FISH PROTEINS	ANIMAL PROTEINS	TOTAL PROTEINS	FISH/ANIMAL PROTEINS	FISH/TOTAL PROTEINS
	(..		metric tons in live weight		....)		(thousands)	(kilograms)		(grams per caput per day)		(%)	(%)
1961	...	0	0	0	0	...	19	...	...	...	...	...	...
1962	...	0	0	0	0	...	19	...	...	...	...	...	...
1963	...	0	0	0	0	...	20	...	...	...	...	...	...
1964	...	0	0	0	0	...	20	...	...	...	...	...	...
1965	...	0	0	0	0	...	21	...	...	...	...	...	...
1966	100	0	0	0	0	100	22	4.6	1.3	...	...	...	...
1967	100	0	0	0	0	100	22	4.5	1.3	...	...	...	...
1968	100	0	0	0	0	100	23	4.3	1.2	...	...	...	...
1969	100	0	0	0	0	100	24	4.2	1.2	...	...	...	...
1970	100	0	0	0	0	100	25	4.0	1.1	...	...	...	...
1971	100	0	0	0	0	100	26	3.9	1.1	...	...	...	...
1972	100	0	0	0	0	100	27	3.7	1.1	...	...	...	...
1973	100	0	0	0	0	100	28	3.6	1.0	...	...	...	...
1974	100	0	0	0	0	100	29	3.5	1.0	...	...	...	...
1975	100	0	0	0	0	100	30	3.3	0.9	...	...	...	...
1976	100	0	0	0	0	100	31	3.2	0.9	...	...	...	...
1977	100	0	0	0	0	100	32	3.1	0.9	...	...	...	...
1978	100	0	0	0	0	100	33	3.0	0.9	...	...	...	...
1979	100	0	0	0	0	100	34	2.9	0.8	...	...	...	...
1980	150	0	0	0	0	150	35	4.3	1.2	...	...	...	...
1981	150	0	79	0	0	229	36	6.4	2.0	...	...	...	...
1982	160	0	333	0	0	493	37	13.4	4.5	...	...	...	...
1983	160	0	263	0	0	423	38	11.2	3.7	...	...	...	...
1984	170	0	104	0	0	274	39	7.0	2.2	...	...	...	...
1985	170	0	202	0	0	372	40	9.3	3.0	...	...	...	...
1986	180	0	196	0	0	376	41	9.2	3.0	...	...	...	...
1987	180	0	167	0	0	347	42	8.2	2.6	...	...	...	...
1988	190	0	118	0	0	308	43	7.1	2.2	...	...	...	...
1989	190	0	111	0	0	301	45	6.8	2.1	...	...	...	...
1990	200	0	102	0	0	302	46	6.6	2.0	...	...	...	...
1991	200	0	96	120	0	176	48	3.7	1.2	...	...	...	...
1992	204	0	105	110	0	199	49	4.0	1.3	...	...	...	...
1993	300	0	105	90	0	315	51	6.2	2.0	...	...	...	...
1994	260	0	114	100	0	274	53	5.2	1.7	...	...	...	...
1995	260	0	114	72	0	302	55	5.5	1.6	...	...	...	...

135 MARTINIQUE	- FOOD BALANCE SHEET OF FISH AND FISHERY PRODUCTS IN LIVE WEIGHT AND FISH CONTRIBUTION TO PROTEIN SUPPLY -												FAO-FIDI
YEAR	PRODUCTION	NON-FOOD USES	IMPORTS	EXPORTS	STOCKS VARIATIONS	TOTAL FOOD SUPPLY	POPULATION	PER CAPUT SUPPLY	FISH PROTEINS	ANIMAL PROTEINS	TOTAL PROTEINS	FISH/ANIMAL PROTEINS	FISH/TOTAL PROTEINS
	(.. metric tons in live weight ....)						(thousands)	(kilograms)	(grams per caput per day)			(%)	(%)
1961	3,100	0	6,764	0	0	9,864	289	34.2	11.5	32.2	63.9	35.8	18.0
1962	2,900	0	8,885	0	0	11,785	295	39.9	13.4	30.1	64.9	44.5	20.7
1963	2,500	0	9,188	0	0	11,688	301	38.8	13.1	31.1	66.1	42.0	19.7
1964	3,300	0	9,188	0	0	12,488	307	40.7	13.6	29.4	65.4	46.2	20.8
1965	3,400	0	8,782	0	0	12,182	311	39.1	13.0	32.0	65.4	40.6	19.9
1966	4,000	0	8,476	0	0	12,476	315	39.6	13.1	32.1	66.2	40.7	19.7
1967	4,800	0	9,079	0	0	13,879	319	43.5	14.3	35.2	70.3	40.5	20.3
1968	4,600	0	8,576	0	0	13,176	322	41.0	13.4	34.3	68.8	38.9	19.4
1969	4,000	0	8,676	0	0	12,676	324	39.2	12.8	35.6	69.0	36.1	18.6
1970	4,000	0	8,370	0	0	12,370	326	38.0	12.4	37.3	73.1	33.3	16.9
1971	3,700	0	8,064	0	0	11,764	327	36.0	11.7	36.8	71.5	31.8	16.4
1972	3,200	0	8,967	0	0	12,167	328	37.1	12.0	38.3	73.1	31.4	16.5
1973	3,100	0	8,564	0	0	11,664	329	35.5	11.4	39.0	74.6	29.3	15.3
1974	4,056	0	8,058	0	0	12,114	329	36.9	11.7	37.5	72.2	31.1	16.2
1975	3,665	0	7,352	0	0	11,017	329	33.5	10.7	36.9	70.8	28.9	15.1
1976	4,233	0	8,525	0	0	12,758	328	38.9	12.6	42.6	80.9	29.7	15.6
1977	2,432	0	8,370	33	0	10,769	327	32.9	10.8	41.3	78.7	26.2	13.7
1978	4,375	0	8,786	11	0	13,150	326	40.3	12.9	43.2	78.9	29.9	16.4
1979	4,664	0	9,437	64	0	14,036	326	43.1	13.8	46.0	82.0	29.9	16.8
1980	4,891	0	8,432	28	0	13,295	326	40.8	13.0	44.9	81.3	29.0	16.0
1981	4,700	0	9,953	48	0	14,605	328	44.6	14.2	47.1	84.4	30.2	16.8
1982	5,506	0	7,983	32	0	13,457	330	40.8	13.0	46.0	83.7	28.2	15.5
1983	5,120	0	9,848	46	0	14,922	333	44.8	14.3	45.9	86.1	31.1	16.6
1984	5,204	0	8,420	24	0	13,600	337	40.4	12.9	45.3	83.1	28.4	15.5
1985	4,639	0	9,732	76	0	14,295	341	42.0	13.4	47.1	88.3	28.4	15.2
1986	4,088	0	9,986	23	0	14,051	344	40.8	12.9	46.4	86.9	27.8	14.9
1987	3,184	0	9,581	51	0	12,715	348	36.5	11.5	44.6	80.9	25.9	14.3
1988	3,053	2	11,828	163	0	14,716	352	41.8	12.3	48.3	86.2	25.5	14.3
1989	3,314	2	12,318	112	0	15,518	356	43.6	12.6	47.9	86.0	26.4	14.7
1990	3,585	2	11,678	344	0	14,918	360	41.4	12.3	49.0	85.4	25.1	14.4
1991	6,391	4	13,347	100	0	19,633	364	53.9	16.1	54.7	92.1	29.4	17.5
1992	4,641	2	13,153	35	0	17,758	368	48.2	14.0	53.6	91.6	26.1	15.3
1993	5,957	2	13,540	68	0	19,428	372	52.2	15.3	56.3	95.4	27.2	16.0
1994	5,905	2	14,204	40	0	20,067	376	53.3	15.1	55.9	95.4	27.0	15.8
1995	5,377	2	14,319	41	0	19,654	380	51.7	15.0	55.9	91.6	26.9	16.4

136 MAURITANIA	- FOOD BALANCE SHEET OF FISH AND FISHERY PRODUCTS IN LIVE WEIGHT AND FISH CONTRIBUTION TO PROTEIN SUPPLY -												FAO-FIDI
YEAR	PRODUCTION	NON-FOOD USES	IMPORTS	EXPORTS	STOCKS VARIATIONS	TOTAL FOOD SUPPLY	POPULATION	PER CAPUT SUPPLY	FISH PROTEINS	ANIMAL PROTEINS	TOTAL PROTEINS	FISH/ANIMAL PROTEINS	FISH/TOTAL PROTEINS
	(............ metric tons in live weight ....)						(thousands)	(kilograms)	(grams per caput per day)			(%)	(%)
1961	22,000	0	100	794	0	21,306	1,010	21.1	7.8	42.9	79.8	18.1	9.7
1962	23,000	0	100	1,935	0	21,165	1,030	20.5	7.4	42.9	79.3	17.1	9.3
1963	23,000	0	100	1,200	0	21,900	1,051	20.8	7.6	43.0	79.3	17.6	9.6
1964	23,000	0	0	1,218	0	21,782	1,073	20.3	7.6	43.1	78.8	17.5	9.6
1965	25,000	0	0	3,450	0	21,550	1,096	19.7	7.1	43.2	78.1	16.4	9.1
1966	27,000	0	100	7,000	0	20,100	1,119	18.0	5.7	42.0	78.4	13.5	7.2
1967	30,700	0	0	8,000	0	22,700	1,143	19.9	6.3	42.9	79.4	14.7	8.0
1968	30,000	0	0	8,000	0	22,000	1,168	18.8	6.0	43.4	79.4	13.7	7.5
1969	30,000	0	0	8,000	0	22,000	1,194	18.4	5.8	41.6	78.3	14.0	7.4
1970	51,850	5,700	3,600	23,214	0	26,536	1,221	21.7	6.8	42.2	75.4	16.1	9.0
1971	62,400	15,000	800	14,532	-15,696	17,971	1,249	14.4	4.3	39.9	73.3	10.9	5.9
1972	37,450	0	0	15,429	0	22,021	1,278	17.2	5.3	37.7	69.6	14.2	7.7
1973	34,450	0	0	20,429	6,000	20,021	1,308	15.3	4.7	34.1	69.1	13.7	6.7
1974	41,597	0	0	22,000	10,000	29,597	1,339	22.1	7.6	34.2	69.9	22.1	10.8
1975	33,921	0	0	15,514	0	18,407	1,371	13.4	4.6	33.2	71.0	13.9	6.5
1976	33,834	0	129	17,615	0	16,348	1,405	11.6	3.9	33.0	69.0	11.8	5.6
1977	37,897	0	3	21,103	0	16,797	1,439	11.7	3.9	35.5	72.1	11.0	5.4
1978	41,435	0	0	25,590	0	15,845	1,475	10.7	3.6	35.9	73.2	9.9	4.9
1979	24,541	0	0	10,293	-2,000	12,248	1,512	8.1	2.6	34.5	71.6	7.6	3.7
1980	21,632	0	13	8,472	0	13,173	1,551	8.5	2.5	35.1	72.8	7.2	3.5
1981	58,779	18,000	0	23,337	2,000	19,442	1,591	12.2	3.8	36.9	75.0	10.2	5.0
1982	56,288	8,000	193	26,122	0	22,359	1,633	13.7	4.1	36.7	75.8	11.1	5.4
1983	81,600	0	111	56,293	0	25,418	1,677	15.2	4.4	36.6	81.1	12.0	5.4
1984	93,800	0	348	73,674	0	20,474	1,721	11.9	3.6	35.5	78.2	10.2	4.6
1985	103,300	0	261	81,854	0	21,707	1,766	12.3	3.8	36.4	81.2	10.5	4.7
1986	98,200	0	188	76,538	0	21,850	1,812	12.1	3.5	34.1	78.2	10.4	4.5
1987	99,400	0	205	83,930	0	15,675	1,858	8.4	2.5	32.8	77.9	7.5	3.2
1988	97,600	0	268	80,850	0	17,018	1,905	8.9	2.7	33.4	78.6	8.0	3.4
1989	92,612	0	516	76,115	0	17,013	1,953	8.7	2.6	33.1	80.5	7.9	3.3
1990	80,530	0	453	55,963	0	25,020	2,003	12.5	3.5	33.3	77.1	10.4	4.5
1991	86,660	0	561	50,571	0	36,650	2,054	17.8	4.7	34.0	80.8	13.7	5.8
1992	93,900	0	531	56,353	0	38,078	2,107	18.1	4.7	33.6	81.5	14.1	5.8
1993	100,353	0	647	58,451	-1,000	41,549	2,161	19.2	5.5	32.2	77.7	17.0	7.1
1994	85,000	0	664	49,907	0	35,757	2,217	16.1	4.2	30.9	81.1	13.7	5.2
1995	90,000	0	3,654	56,367	1,000	38,287	2,274	16.8	4.6	30.4	76.4	15.1	6.0

137 MAURITIUS	- FOOD BALANCE SHEET OF FISH AND FISHERY PRODUCTS IN LIVE WEIGHT AND FISH CONTRIBUTION TO PROTEIN SUPPLY -	FAO-FIDI

YEAR	PRODUCTION	NON-FOOD USES	IMPORTS	EXPORTS	STOCKS VARIATIONS	TOTAL FOOD SUPPLY	POPULATION	PER CAPUT SUPPLY	FISH PROTEINS	ANIMAL PROTEINS	TOTAL PROTEINS	FISH/ANIMAL PROTEINS	FISH/TOTAL PROTEINS
	(.. metric tons in live weight ....)						(thousands)	(kilograms)	(grams per caput per day)			(%)	(%)
1961	2,500	0	4,900	0	0	7,400	679	10.9	3.4	12.7	49.0	26.4	6.8
1962	2,800	0	5,700	0	0	8,500	698	12.2	3.7	12.5	50.1	29.5	7.3
1963	2,800	0	4,950	0	0	7,750	717	10.8	3.3	12.5	48.5	26.1	6.7
1964	2,800	0	6,400	0	0	9,200	736	12.5	3.8	13.4	50.8	28.1	7.4
1965	3,000	0	5,900	0	0	8,900	753	11.8	3.5	12.5	47.7	28.0	7.4
1966	3,000	0	5,500	0	0	8,500	769	11.1	3.7	12.8	48.5	28.9	7.7
1967	3,000	0	5,867	0	0	8,867	785	11.3	3.5	12.0	48.3	29.0	7.2
1968	3,300	0	4,933	0	0	8,233	799	10.3	3.5	12.0	47.4	28.8	7.3
1969	4,000	0	4,460	200	0	8,260	813	10.2	3.3	11.8	46.7	28.4	7.2
1970	5,400	0	3,450	275	0	8,575	826	10.4	3.5	13.2	50.0	26.5	7.0
1971	5,200	0	3,600	167	0	8,633	839	10.3	3.4	12.5	49.6	26.8	6.8
1972	6,600	0	4,571	267	0	10,903	852	12.8	4.4	14.4	50.8	30.3	8.6
1973	6,400	0	7,316	1,727	0	11,989	865	13.9	4.8	14.0	49.9	34.2	9.6
1974	7,679	0	10,704	2,567	0	15,816	878	18.0	5.6	16.3	53.4	34.0	10.4
1975	7,038	0	8,097	1,956	0	13,179	892	14.8	5.1	16.2	55.4	31.1	9.1
1976	6,660	0	10,723	1,894	0	15,489	907	17.1	5.4	18.5	55.6	29.3	9.8
1977	7,668	0	13,502	2,695	0	18,474	922	20.0	6.2	22.4	60.9	27.8	10.2
1978	7,111	0	14,031	2,686	0	18,456	938	19.7	6.2	22.8	62.6	27.0	9.8
1979	6,564	0	14,040	2,125	0	18,479	953	19.4	6.3	23.2	60.2	27.4	10.5
1980	6,381	0	13,108	2,058	0	17,430	966	18.0	5.9	22.2	61.7	26.6	9.6
1981	7,129	0	14,207	2,559	0	18,777	978	19.2	6.3	22.7	62.5	27.6	10.0
1982	9,569	0	9,382	2,518	0	16,433	989	16.6	5.2	20.9	60.4	24.7	8.5
1983	9,763	0	8,514	3,189	0	15,088	998	15.1	4.8	20.8	61.7	23.2	7.8
1984	10,602	0	9,968	4,317	0	16,253	1,007	16.1	5.1	21.9	61.6	23.2	8.3
1985	12,351	0	8,699	5,349	0	15,701	1,016	15.5	4.7	21.5	63.9	21.7	7.3
1986	12,886	0	9,877	5,432	0	17,331	1,024	16.9	5.3	22.3	63.4	23.9	8.4
1987	18,188	0	7,613	5,449	0	20,352	1,031	19.7	6.2	24.9	65.5	25.0	9.5
1988	17,154	0	8,368	7,181	-25	18,316	1,039	17.6	5.5	25.9	66.5	21.2	8.3
1989	17,194	0	10,335	7,918	25	19,636	1,047	18.8	5.9	25.5	65.2	23.2	9.1
1990	14,659	0	14,217	5,755	0	23,121	1,057	21.9	7.4	27.7	69.0	26.5	10.7
1991	18,875	0	13,525	8,847	-175	23,377	1,067	21.9	7.9	29.6	70.9	26.9	11.2
1992	19,207	13	28,144	11,686	-6,136	29,516	1,079	27.4	8.2	30.6	72.9	26.9	11.3
1993	21,160	6	16,041	12,923	4,615	28,887	1,091	26.5	8.4	31.9	78.9	26.4	10.7
1994	18,969	7	24,896	15,026	0	28,831	1,104	26.1	8.3	34.7	81.5	23.8	10.2
1995	17,048	10	34,045	20,763	1,694	32,014	1,117	28.7	9.0	34.3	77.5	26.3	11.6

138 MEXICO	- FOOD BALANCE SHEET OF FISH AND FISHERY PRODUCTS IN LIVE WEIGHT AND FISH CONTRIBUTION TO PROTEIN SUPPLY -	FAO-FIDI

YEAR	PRODUCTION	NON-FOOD USES	IMPORTS	EXPORTS	STOCKS VARIATIONS	TOTAL FOOD SUPPLY	POPULATION	PER CAPUT SUPPLY	FISH PROTEINS	ANIMAL PROTEINS	TOTAL PROTEINS	FISH/ANIMAL PROTEINS	FISH/TOTAL PROTEINS
	(............ metric tons in live weight ....)						(thousands)	(kilograms)	(grams per caput per day)			(%)	(%)
1961	206,700	35,800	2,897	77,359	0	96,438	38,102	2.5	0.6	17.0	65.7	3.6	0.9
1962	197,300	25,800	2,874	76,697	0	97,677	39,295	2.5	0.6	17.5	64.8	3.3	0.9
1963	224,900	27,300	3,040	76,916	0	123,724	40,529	3.1	0.8	17.6	65.2	4.6	1.2
1964	225,300	28,200	3,065	70,381	0	129,784	41,812	3.1	0.8	17.5	67.4	4.6	1.2
1965	239,300	36,200	3,533	62,166	0	144,467	43,148	3.4	0.9	17.3	67.1	4.9	1.3
1966	263,800	52,100	2,948	66,962	0	147,687	44,536	3.3	0.9	18.9	68.6	4.5	1.2
1967	327,600	82,900	3,656	64,098	0	184,258	45,975	4.0	1.0	19.6	69.8	5.1	1.4
1968	332,800	93,400	3,621	56,223	0	186,797	47,464	3.9	1.0	19.1	69.9	5.0	1.4
1969	323,500	89,800	2,717	56,390	0	180,027	49,004	3.7	1.0	19.3	69.9	5.2	1.4
1970	348,485	99,000	12,191	57,926	0	203,750	50,596	4.0	1.2	19.6	70.0	6.3	1.8
1971	395,100	108,700	3,344	58,127	0	231,617	52,238	4.4	1.1	19.6	70.7	5.8	1.6
1972	424,687	122,900	3,519	66,432	0	238,874	53,924	4.4	1.2	21.6	71.7	5.3	1.6
1973	447,006	127,900	3,065	65,081	0	257,091	55,641	4.6	1.2	21.9	72.0	5.5	1.7
1974	399,571	83,270	2,880	57,376	-6,407	255,398	57,370	4.5	1.1	23.1	72.6	4.6	1.5
1975	465,900	120,500	8,437	64,117	-5,858	283,863	59,099	4.8	1.3	23.5	73.3	5.6	1.8
1976	493,677	165,600	3,994	66,827	0	265,244	60,822	4.4	1.0	24.9	72.9	4.1	1.4
1977	531,382	204,200	3,335	64,786	0	265,732	62,538	4.3	1.1	26.1	74.7	4.4	1.5
1978	794,222	266,300	12,700	73,003	0	467,619	64,238	7.3	1.9	27.2	76.6	7.0	2.5
1979	976,990	342,600	18,399	64,107	0	588,683	65,917	8.9	2.4	27.8	76.6	8.6	3.1
1980	1,250,073	501,200	16,862	67,461	10,630	708,904	67,570	10.5	2.8	31.7	84.8	8.9	3.3
1981	1,610,515	534,617	15,826	60,889	-60,940	969,894	69,192	14.0	3.8	32.2	88.9	11.7	4.2
1982	1,399,536	493,500	3,964	57,607	-11	852,382	70,784	12.0	3.3	31.3	85.2	10.6	3.9
1983	1,064,612	321,452	601	52,018	11	691,755	72,354	9.6	2.6	30.8	84.3	8.5	3.1
1984	1,107,842	284,693	941	44,521	-24,365	755,204	73,912	10.2	2.7	30.3	81.2	9.0	3.4
1985	1,226,166	394,535	3,018	75,916	-982	757,751	75,465	10.0	2.7	31.3	83.8	8.5	3.2
1986	1,315,360	421,484	1,908	118,048	900	778,636	77,015	10.1	2.7	31.2	82.4	8.6	3.3
1987	1,418,903	442,209	2,711	136,035	80	843,450	78,561	10.7	2.9	30.4	79.7	9.5	3.6
1988	1,372,185	381,017	15,738	211,279	81,222	876,849	80,109	11.0	3.0	30.7	80.5	9.7	3.7
1989	1,469,518	462,391	17,360	207,508	69,609	886,588	81,663	10.9	3.0	29.2	77.6	10.1	3.8
1990	1,400,684	281,674	33,356	152,699	-1,792	997,817	83,226	12.0	3.2	30.3	80.2	10.5	4.0
1991	1,453,137	302,135	17,681	125,675	3,348	1,046,357	84,801	12.3	3.4	30.8	80.9	11.0	4.2
1992	1,247,455	169,779	24,001	76,934	0	1,024,742	86,386	11.9	3.2	32.7	82.9	9.8	3.8
1993	1,200,517	130,870	38,073	84,378	0	1,023,342	87,976	11.6	3.2	33.9	82.9	9.4	3.8
1994	1,264,192	183,034	61,916	71,241	-50	1,071,783	89,564	12.0	3.2	34.3	83.3	9.4	3.9
1995	1,358,353	276,351	27,750	242,130	50	867,672	91,145	9.5	2.7	32.6	83.7	8.3	3.2

145 MICRONESIA | - FOOD BALANCE SHEET OF FISH AND FISHERY PRODUCTS IN LIVE WEIGHT AND FISH CONTRIBUTION TO PROTEIN SUPPLY - | FAO-FIDI

YEAR	PRODUCTION	NON-FOOD USES	IMPORTS	EXPORTS	STOCKS VARIATIONS	TOTAL FOOD SUPPLY	POPULATION	PER CAPUT SUPPLY	FISH PROTEINS	ANIMAL PROTEINS	TOTAL PROTEINS	FISH/ANIMAL PROTEINS	FISH/TOTAL PROTEINS
	(.. metric tons in live weight					....)	(thousands)	(kilograms)	(grams per caput per day)			(%)	(%)
1961	500	0	0	0	0	500	45	11.0	...	...	...	...	...
1962	500	0	0	0	0	500	47	10.6	...	...	...	...	...
1963	600	0	0	0	0	600	49	12.3	...	...	...	...	...
1964	600	0	0	0	0	600	50	11.9	...	...	...	...	...
1965	700	0	0	0	0	700	52	13.5	...	...	...	...	...
1966	800	0	0	0	0	800	54	14.9	...	...	...	...	...
1967	1,000	0	0	0	0	1,000	56	18.0	...	...	...	...	...
1968	1,000	0	0	0	0	1,000	57	17.5	...	...	...	...	...
1969	1,000	0	0	0	0	1,000	59	16.9	...	...	...	...	...
1970	1,500	0	0	0	0	1,500	61	24.6	...	...	...	...	...
1971	1,500	0	0	0	0	1,500	63	23.9	...	...	...	...	...
1972	1,500	0	0	0	0	1,500	65	23.1	...	...	...	...	...
1973	1,500	0	0	0	0	1,500	67	22.5	...	...	...	...	...
1974	1,500	0	0	0	0	1,500	69	21.8	...	...	...	...	...
1975	1,500	0	0	0	0	1,500	71	21.1	...	...	...	...	...
1976	1,500	0	0	0	0	1,500	73	20.5	...	...	...	...	...
1977	1,503	0	0	3	0	1,500	76	19.8	...	...	...	...	...
1978	1,505	0	0	150	0	1,355	78	17.4	...	...	...	...	...
1979	1,505	0	0	5	0	1,500	81	18.6	...	...	...	...	...
1980	1,720	0	0	3	0	1,717	83	20.7	...	...	...	...	...
1981	1,720	0	0	0	0	1,720	85	20.1	...	...	...	...	...
1982	1,777	0	0	0	0	1,777	88	20.2	...	...	...	...	...
1983	1,777	0	0	5	0	1,772	90	19.7	...	...	...	...	...
1984	1,777	0	0	10	0	1,767	93	19.1	...	...	...	...	...
1985	3,432	0	0	17	0	3,415	95	36.0	...	...	...	...	...
1986	3,532	0	0	19	0	3,513	97	36.0	...	...	...	...	...
1987	3,634	0	0	22	0	3,612	100	36.1	...	...	...	...	...
1988	2,634	0	562	14	0	3,182	103	31.0	...	...	...	...	...
1989	2,134	0	1,090	17	0	3,207	105	30.5	...	...	...	...	...
1990	2,140	0	896	62	0	2,974	108	27.5	...	...	...	...	...
1991	12,335	9,000	657	237	0	3,755	111	33.9	...	...	...	...	...
1992	16,518	13,500	1,169	166	0	4,021	114	35.4	...	...	...	...	...
1993	17,201	14,000	983	26	0	4,158	117	35.6	...	...	...	...	...
1994	23,105	21,000	2,543	20	0	4,628	120	38.6	...	...	...	...	...
1995	8,513	7,000	3,683	14	-56	5,127	123	41.7	13.4	28.7	59.5	46.7	22.5

Note: from 1991 non-food uses data refer to fish caught by vessels using flags of convenience.

146 MOLDOVA, REPUBLIC	- FOOD BALANCE SHEET OF FISH AND FISHERY PRODUCTS IN LIVE WEIGHT AND FISH CONTRIBUTION TO PROTEIN SUPPLY -	FAO-FIDI

YEAR	PRODUCTION	NON-FOOD USES	IMPORTS	EXPORTS	STOCKS VARIATIONS	TOTAL FOOD SUPPLY	POPULATION	PER CAPUT SUPPLY	FISH PROTEINS	ANIMAL PROTEINS	TOTAL PROTEINS	FISH/ANIMAL PROTEINS	FISH/TOTAL PROTEINS
	(..		metric tons in live weight		....	)	(thousands)	(kilograms)	(grams per caput per day)			(%)	(%)
1961	0	0	0	0	0	0	3,074	0.0	0.0	0.0	0.0	0.0	0.0
1962	0	0	0	0	0	0	3,141	0.0	0.0	0.0	0.0	0.0	0.0
1963	0	0	0	0	0	0	3,209	0.0	0.0	0.0	0.0	0.0	0.0
1964	0	0	0	0	0	0	3,274	0.0	0.0	0.0	0.0	0.0	0.0
1965	0	0	0	0	0	0	3,339	0.0	0.0	0.0	0.0	0.0	0.0
1966	0	0	0	0	0	0	3,399	0.0	0.0	0.0	0.0	0.0	0.0
1967	0	0	0	0	0	0	3,457	0.0	0.0	0.0	0.0	0.0	0.0
1968	0	0	0	0	0	0	3,510	0.0	0.0	0.0	0.0	0.0	0.0
1969	0	0	0	0	0	0	3,553	0.0	0.0	0.0	0.0	0.0	0.0
1970	0	0	0	0	0	0	3,595	0.0	0.0	0.0	0.0	0.0	0.0
1971	0	0	0	0	0	0	3,647	0.0	0.0	0.0	0.0	0.0	0.0
1972	0	0	0	0	0	0	3,700	0.0	0.0	0.0	0.0	0.0	0.0
1973	0	0	0	0	0	0	3,748	0.0	0.0	0.0	0.0	0.0	0.0
1974	0	0	0	0	0	0	3,795	0.0	0.0	0.0	0.0	0.0	0.0
1975	0	0	0	0	0	0	3,839	0.0	0.0	0.0	0.0	0.0	0.0
1976	0	0	0	0	0	0	3,877	0.0	0.0	0.0	0.0	0.0	0.0
1977	0	0	0	0	0	0	3,910	0.0	0.0	0.0	0.0	0.0	0.0
1978	0	0	0	0	0	0	3,943	0.0	0.0	0.0	0.0	0.0	0.0
1979	0	0	0	0	0	0	3,976	0.0	0.0	0.0	0.0	0.0	0.0
1980	0	0	0	0	0	0	4,011	0.0	0.0	0.0	0.0	0.0	0.0
1981	0	0	0	0	0	0	4,050	0.0	0.0	0.0	0.0	0.0	0.0
1982	0	0	0	0	0	0	4,097	0.0	0.0	0.0	0.0	0.0	0.0
1983	0	0	0	0	0	0	4,137	0.0	0.0	0.0	0.0	0.0	0.0
1984	0	0	0	0	0	0	4,176	0.0	0.0	0.0	0.0	0.0	0.0
1985	0	0	0	0	0	0	4,214	0.0	0.0	0.0	0.0	0.0	0.0
1986	0	0	0	0	0	0	4,255	0.0	0.0	0.0	0.0	0.0	0.0
1987	0	0	0	0	0	0	4,290	0.0	0.0	0.0	0.0	0.0	0.0
1988	0	0	0	0	0	0	4,321	0.0	0.0	0.0	0.0	0.0	0.0
1989	0	0	0	0	0	0	4,340	0.0	0.0	0.0	0.0	0.0	0.0
1990	0	0	0	0	0	0	4,364	0.0	0.0	0.0	0.0	0.0	0.0
1991	5,171	0	...	...	0	5,171	4,385	1.2	0.4	37.7	96.5	0.9	0.4
1992	6,500	0	...	...	0	6,500	4,402	1.5	0.4	30.9	87.1	1.4	0.5
1993	9,500	0	...	...	0	9,500	4,417	2.2	0.6	27.0	84.5	2.4	0.8
1994	2,100	4	9,062	451	0	10,707	4,428	2.4	0.7	25.1	74.5	2.9	1.0
1995	500	0	8,170	212	-111	8,347	4,437	1.9	0.5	22.8	68.1	2.4	0.8

Note: for years before 1991 see USSR.

140 MONACO	- FOOD BALANCE SHEET OF FISH AND FISHERY PRODUCTS IN LIVE WEIGHT AND FISH CONTRIBUTION TO PROTEIN SUPPLY -												FAO-FIDI
YEAR	PRODUCTION	NON-FOOD USES	IMPORTS	EXPORTS	STOCKS VARIATIONS	TOTAL FOOD SUPPLY	POPULATION	PER CAPUT SUPPLY	FISH PROTEINS	ANIMAL PROTEINS	TOTAL PROTEINS	FISH/ANIMAL PROTEINS	FISH/TOTAL PROTEINS
	(........ metric tons in live weight ....)						(thousands)	(kilograms)	(grams per caput per day)			(%)	(%)
1961	0	0	0	0	0	0	21	0.0	0.0	...	...	...	...
1962	0	0	0	0	0	0	21	0.0	0.0	...	...	...	...
1963	0	0	0	0	0	0	22	0.0	0.0	...	...	...	...
1964	0	0	0	0	0	0	22	0.0	0.0	...	...	...	...
1965	0	0	0	0	0	0	22	0.0	0.0	...	...	...	...
1966	0	0	0	0	0	0	22	0.0	0.0	...	...	...	...
1967	0	0	0	0	0	0	23	0.0	0.0	...	...	...	...
1968	0	0	0	0	0	0	23	0.0	0.0	...	...	...	...
1969	0	0	0	0	0	0	24	0.0	0.0	...	...	...	...
1970	0	0	0	0	0	0	24	0.0	0.0	...	...	...	...
1971	0	0	0	0	0	0	24	0.0	0.0	...	...	...	...
1972	0	0	0	0	0	0	24	0.0	0.0	...	...	...	...
1973	1	0	0	0	0	1	25	0.0	0.0	...	...	...	...
1974	1	0	0	0	0	1	25	0.0	0.0	...	...	...	...
1975	1	0	0	0	0	1	25	0.0	0.0	...	...	...	...
1976	1	0	0	0	0	1	25	0.0	0.0	...	...	...	...
1977	1	0	0	0	0	1	26	0.0	0.0	...	...	...	...
1978	1	0	0	0	0	1	26	0.0	0.0	...	...	...	...
1979	1	0	0	0	0	1	27	0.0	0.0	...	...	...	...
1980	1	0	0	0	0	1	27	0.0	0.0	...	...	...	...
1981	2	0	0	0	0	2	27	0.1	0.0	...	...	...	...
1982	2	0	0	0	0	2	27	0.1	0.0	...	...	...	...
1983	2	0	0	0	0	2	28	0.1	0.0	...	...	...	...
1984	2	0	0	0	0	2	28	0.1	0.0	...	...	...	...
1985	2	0	0	0	0	2	28	0.1	0.0	...	...	...	...
1986	2	0	0	0	0	2	28	0.1	0.0	...	...	...	...
1987	2	0	0	0	0	2	29	0.1	0.0	...	...	...	...
1988	2	0	0	0	0	2	29	0.1	0.0	...	...	...	...
1989	2	0	0	0	0	2	30	0.1	0.0	...	...	...	...
1990	2	0	0	0	0	2	30	0.1	0.0	...	...	...	...
1991	2	0	0	0	0	2	30	0.1	0.0	...	...	...	...
1992	3	0	0	0	0	3	31	0.1	0.0	...	...	...	...
1993	3	0	0	0	0	3	31	0.1	0.0	...	...	...	...
1994	3	0	0	0	0	3	32	0.1	0.0	...	...	...	...
1995	3	0	0	0	0	3	32	0.1	0.0	...	...	...	...

Note: trade included with France. Per caput supply apparent consumption underestimated.

141 MONGOLIA	- FOOD BALANCE SHEET OF FISH AND FISHERY PRODUCTS IN LIVE WEIGHT AND FISH CONTRIBUTION TO PROTEIN SUPPLY -	FAO-FIDI

YEAR	PRODUCTION	NON-FOOD USES	IMPORTS	EXPORTS	STOCKS VARIATIONS	TOTAL FOOD SUPPLY	POPULATION	PER CAPUT SUPPLY	FISH PROTEINS	ANIMAL PROTEINS	TOTAL PROTEINS	FISH/ANIMAL PROTEINS	FISH/TOTAL PROTEINS
	(............ metric tons in live weight					............)	(thousands)	(kilograms)	(grams per caput per day)			(%)	(%)
1961	250	0	...	...	0	250	984	0.3	0.1	63.8	87.7	0.1	0.1
1962	250	0	...	...	0	250	1,010	0.3	0.1	55.7	79.9	0.1	0.1
1963	250	0	...	...	0	250	1,037	0.2	0.1	57.2	81.9	0.1	0.1
1964	260	0	...	...	0	260	1,065	0.2	0.1	57.5	84.1	0.1	0.1
1965	260	0	...	...	0	260	1,094	0.2	0.1	58.6	87.3	0.1	0.1
1966	280	0	360	240	0	400	1,124	0.4	0.1	59.4	88.3	0.2	0.1
1967	280	0	370	240	0	410	1,155	0.4	0.1	62.1	88.3	0.2	0.1
1968	280	0	370	240	0	410	1,187	0.4	0.1	59.0	83.5	0.2	0.1
1969	350	0	350	200	0	500	1,221	0.4	0.1	58.0	83.0	0.2	0.2
1970	300	0	450	240	0	510	1,256	0.4	0.1	55.3	83.5	0.2	0.2
1971	300	0	400	200	0	500	1,292	0.4	0.1	52.2	80.3	0.2	0.1
1972	300	0	240	100	0	440	1,329	0.3	0.1	53.2	80.2	0.2	0.1
1973	300	0	440	200	0	540	1,368	0.4	0.1	52.6	81.6	0.2	0.1
1974	300	0	300	0	0	600	1,407	0.4	0.1	54.3	82.5	0.2	0.2
1975	300	0	700	0	0	1,000	1,447	0.7	0.2	56.0	83.6	0.4	0.3
1976	300	0	2,632	0	0	2,932	1,489	2.0	0.7	56.0	85.9	1.2	0.8
1977	265	0	1,053	0	0	1,318	1,531	0.9	0.3	52.6	81.6	0.6	0.4
1978	494	0	1,754	0	0	2,248	1,574	1.4	0.5	56.4	85.2	0.9	0.6
1979	537	0	1,789	0	0	2,326	1,618	1.4	0.5	54.8	84.1	0.9	0.6
1980	346	0	1,682	0	0	2,028	1,663	1.2	0.4	51.7	81.2	0.8	0.5
1981	262	0	1,446	0	0	1,708	1,709	1.0	0.4	49.8	79.3	0.7	0.4
1982	184	0	1,184	0	0	1,368	1,755	0.8	0.3	47.5	77.8	0.6	0.4
1983	347	0	1,326	0	0	1,673	1,803	0.9	0.3	47.9	79.9	0.7	0.4
1984	370	0	1,847	0	0	2,217	1,854	1.2	0.4	46.6	78.2	0.9	0.5
1985	374	0	1,449	0	0	1,823	1,909	1.0	0.3	45.2	75.3	0.7	0.4
1986	412	0	2,349	0	0	2,761	1,968	1.4	0.5	47.2	78.1	1.0	0.6
1987	383	0	2,026	0	0	2,409	2,031	1.2	0.4	46.0	75.8	0.9	0.6
1988	281	0	2,028	0	0	2,309	2,095	1.1	0.4	43.5	74.1	0.9	0.5
1989	254	0	1,844	0	0	2,098	2,158	1.0	0.3	45.0	74.3	0.8	0.5
1990	124	0	1,842	0	0	1,966	2,216	0.9	0.3	46.4	73.3	0.7	0.4
1991	100	0	0	0	0	100	2,270	0.0	0.0	51.4	75.8	0.0	0.0
1992	120	0	0	0	0	120	2,319	0.1	0.0	48.6	69.1	0.0	0.0
1993	165	0	60	0	0	225	2,366	0.1	0.0	46.6	66.0	0.1	0.0
1994	184	0	0	0	0	184	2,414	0.1	0.0	44.2	65.5	0.1	0.0
1995	158	0	59	89	0	128	2,463	0.1	0.0	45.8	70.2	0.0	0.0

142 MONTSERRAT	- FOOD BALANCE SHEET OF FISH AND FISHERY PRODUCTS IN LIVE WEIGHT AND FISH CONTRIBUTION TO PROTEIN SUPPLY -												FAO-FIDI
YEAR	PRODUCTION	NON-FOOD USES	IMPORTS	EXPORTS	STOCKS VARIATIONS	TOTAL FOOD SUPPLY	POPULATION	PER CAPUT SUPPLY	FISH PROTEINS	ANIMAL PROTEINS	TOTAL PROTEINS	FISH/ANIMAL PROTEINS	FISH/TOTAL PROTEINS
	(....................... metric tons in live weight ....)						(thousands)	(kilograms)	(grams per caput per day)			(%)	(%)
1961	100	0	0	0	0	100	12	8.4	2.4	25.9	47.6	9.1	5.0
1962	100	0	0	0	0	100	12	8.3	2.4	30.7	53.7	7.7	4.4
1963	100	0	0	0	0	100	12	8.3	2.3	30.9	51.9	7.6	4.5
1964	100	0	0	0	0	100	12	8.3	2.3	32.0	57.5	7.3	4.1
1965	0	0	0	0	0	0	12	0.0	0.0	32.0	58.9	0.0	0.0
1966	0	0	0	0	0	0	12	0.0	0.0	35.0	57.5	0.0	0.0
1967	0	0	0	0	0	0	12	0.0	0.0	39.3	61.5	0.0	0.0
1968	100	0	0	0	0	100	11	8.9	2.5	47.2	71.0	5.3	3.5
1969	100	0	0	0	0	100	11	9.0	2.6	46.2	70.0	5.5	3.6
1970	0	0	0	0	0	0	11	0.0	0.0	46.5	67.6	0.0	0.0
1971	0	0	0	0	0	0	11	0.0	0.0	49.6	66.8	0.0	0.0
1972	100	0	0	0	0	100	11	8.9	2.5	52.9	73.0	4.7	3.4
1973	100	0	0	0	0	100	12	8.7	2.5	48.0	67.1	5.1	3.7
1974	77	0	0	0	0	77	12	6.5	1.8	50.3	79.3	3.7	2.3
1975	89	0	0	0	0	89	12	7.4	2.1	50.2	75.7	4.2	2.8
1976	95	0	0	0	0	95	12	7.8	2.2	50.3	74.0	4.4	3.0
1977	98	0	0	0	0	98	12	8.1	2.3	52.3	78.1	4.3	2.9
1978	100	0	0	0	0	100	12	8.2	2.3	57.6	80.3	4.0	2.9
1979	102	0	0	0	0	102	12	8.4	2.4	59.8	81.4	4.0	2.9
1980	109	0	0	0	0	109	12	9.1	2.6	66.0	87.9	3.9	2.9
1981	104	0	0	0	0	104	12	8.8	2.5	75.4	101.0	3.3	2.5
1982	111	0	0	0	0	111	12	9.6	2.7	78.6	102.6	3.4	2.6
1983	110	0	0	0	0	110	11	9.7	2.7	75.4	100.4	3.6	2.7
1984	110	0	0	0	0	110	11	9.9	2.8	74.8	95.4	3.7	2.9
1985	110	0	0	0	0	110	11	10.0	2.8	77.0	96.5	3.7	2.9
1986	110	0	0	0	0	110	11	10.1	2.8	76.7	104.0	3.7	2.7
1987	117	0	0	0	0	117	11	10.7	3.0	86.1	117.4	3.5	2.6
1988	58	0	0	0	0	58	11	5.3	1.5	83.1	111.8	1.8	1.3
1989	28	0	0	0	0	28	11	2.6	0.7	82.0	116.7	0.9	0.6
1990	150	0	0	0	0	150	11	13.6	3.9	134.3	188.1	2.9	2.1
1991	101	0	0	0	0	101	11	9.2	2.6	134.2	197.7	1.9	1.3
1992	115	0	0	0	0	115	11	10.4	3.0	106.4	171.3	2.8	1.7
1993	160	0	0	0	0	160	11	14.5	4.1	115.9	181.5	3.5	2.3
1994	188	0	0	0	0	188	11	17.1	4.8	116.4	181.4	4.1	2.7
1995	175	0	0	0	0	175	11	15.9	4.5	104.6	169.6	4.3	2.7

143 MOROCCO	- FOOD BALANCE SHEET OF FISH AND FISHERY PRODUCTS IN LIVE WEIGHT AND FISH CONTRIBUTION TO PROTEIN SUPPLY -	FAO-FIDI

YEAR	PRODUCTION	NON-FOOD USES	IMPORTS	EXPORTS	STOCKS VARIATIONS	TOTAL FOOD SUPPLY	POPULATION	PER CAPUT SUPPLY	FISH PROTEINS	ANIMAL PROTEINS	TOTAL PROTEINS	FISH/ANIMAL PROTEINS	FISH/TOTAL PROTEINS
	(..		metric tons in live weight			....)	(thousands)	(kilograms)		(grams per caput per day)		(%)	(%)
1961	159,200	78,900	1,025	66,747	0	14,679	11,948	1.2	0.4	9.8	70.7	3.9	0.6
1962	163,000	72,600	400	71,949	0	18,852	12,275	1.5	0.5	9.4	62.6	5.3	0.8
1963	171,500	84,700	500	59,892	0	27,408	12,611	2.2	0.7	9.0	58.1	7.3	1.1
1964	197,600	86,000	200	82,110	0	29,690	12,959	2.3	0.7	9.1	60.8	7.6	1.1
1965	212,490	117,300	100	44,386	-16,789	34,114	13,323	2.6	0.8	9.4	56.8	8.3	1.4
1966	300,577	196,000	600	68,249	-6,409	30,519	13,703	2.2	0.7	9.5	66.0	7.4	1.1
1967	255,359	151,700	300	63,193	-13,663	27,103	14,097	1.9	0.6	9.5	60.7	6.1	1.0
1968	216,473	120,700	100	74,282	5,318	26,909	14,500	1.9	0.6	9.8	60.2	5.9	1.0
1969	225,647	126,600	100	70,513	7,632	36,266	14,906	2.4	0.8	10.0	66.3	7.8	1.2
1970	247,437	137,000	0	62,144	-13,994	34,299	15,310	2.2	0.8	9.9	64.4	7.6	1.2
1971	224,712	72,100	100	91,161	14,905	76,456	15,710	4.9	1.6	10.6	65.3	14.8	2.4
1972	244,178	92,900	100	86,005	13,606	78,979	16,108	4.9	1.5	10.7	67.9	13.9	2.2
1973	397,301	206,300	200	115,352	8,489	84,339	16,505	5.1	1.7	10.6	71.3	15.7	2.3
1974	285,473	110,400	100	108,169	9,060	76,064	16,903	4.5	1.4	10.1	67.5	13.9	2.1
1975	225,339	75,000	3,400	80,367	-9,236	64,136	17,305	3.7	1.2	9.5	70.5	12.7	1.7
1976	283,803	111,000	2,728	98,871	9,372	86,032	17,710	4.9	1.6	9.5	67.9	17.0	2.4
1977	260,239	70,500	92	95,149	0	94,683	18,119	5.2	1.7	10.3	70.4	16.2	2.4
1978	292,949	113,900	36	59,744	0	119,340	18,533	6.4	2.3	11.3	70.8	20.1	3.2
1979	285,182	87,000	30	80,968	0	117,244	18,953	6.2	2.1	11.5	72.1	17.9	2.9
1980	330,203	121,752	22	84,994	0	123,479	19,382	6.4	2.0	11.3	71.4	17.9	2.8
1981	390,505	108,130	21	155,111	-4,683	122,601	19,820	6.2	2.0	11.9	73.1	16.7	2.7
1982	363,613	87,500	27	133,642	-7,660	134,839	20,267	6.7	2.1	12.0	74.9	17.2	2.8
1983	453,885	144,600	24	181,505	6,593	134,397	20,721	6.5	2.0	11.0	76.2	18.0	2.6
1984	467,525	145,200	352	191,593	6,471	137,556	21,182	6.5	2.0	11.1	76.7	18.3	2.7
1985	473,191	100,300	60	202,144	0	170,807	21,648	7.9	2.4	11.2	79.4	21.7	3.1
1986	595,414	196,000	408	205,081	-1,590	193,152	22,118	8.7	2.6	12.4	81.8	20.7	3.1
1987	494,157	122,339	1,502	228,204	1,837	146,953	22,592	6.5	1.9	13.2	83.6	14.5	2.3
1988	551,722	176,742	2,768	223,914	0	153,836	23,071	6.7	2.0	14.1	85.3	14.2	2.3
1989	520,359	159,189	4,272	214,658	-2,000	148,784	23,555	6.3	1.9	15.0	84.4	12.9	2.3
1990	565,741	157,053	2,464	222,177	-19,000	169,975	24,043	7.1	2.2	15.0	87.7	14.6	2.5
1991	593,217	160,554	1,386	246,609	-1,350	186,091	24,535	7.6	2.4	15.0	86.9	15.8	2.7
1992	548,052	119,454	3,575	217,872	4,300	218,602	25,030	8.7	2.7	15.7	85.1	17.3	3.2
1993	622,844	201,562	1,283	223,042	-2,000	197,525	25,527	7.7	2.4	14.7	80.6	16.4	3.0
1994	751,994	315,240	3,159	241,879	17,500	215,534	26,025	8.3	2.6	14.6	85.5	17.6	3.0
1995	846,201	388,865	3,881	241,179	2,550	222,589	26,524	8.4	2.7	15.1	83.1	17.7	3.2

144 MOZAMBIQUE	- FOOD BALANCE SHEET OF FISH AND FISHERY PRODUCTS IN LIVE WEIGHT AND FISH CONTRIBUTION TO PROTEIN SUPPLY -	FAO-FIDI

YEAR	PRODUCTION	NON-FOOD USES	IMPORTS	EXPORTS	STOCKS VARIATIONS	TOTAL FOOD SUPPLY	POPULATION	PER CAPUT SUPPLY	FISH PROTEINS	ANIMAL PROTEINS	TOTAL PROTEINS	FISH/ANIMAL PROTEINS	FISH/TOTAL PROTEINS
	(.........................		metric tons in live weight			..........................)	(thousands)	(kilograms)	(grams per caput per day)			(%)	(%)
1961	12,800	0	18,975	650	0	31,125	7,623	4.1	1.4	4.3	36.1	31.5	3.8
1962	12,800	0	19,600	975	0	31,425	7,790	4.0	1.3	4.4	35.6	30.1	3.8
1963	13,500	0	18,642	1,100	0	31,042	7,965	3.9	1.2	4.3	35.9	27.4	3.3
1964	13,900	0	19,300	1,400	0	31,800	8,147	3.9	1.2	4.4	36.0	28.1	3.4
1965	16,200	0	20,585	1,645	0	35,140	8,338	4.2	1.4	4.7	37.5	29.1	3.6
1966	17,300	0	22,375	1,965	0	37,710	8,539	4.4	1.4	4.8	36.8	28.8	3.7
1967	17,000	0	25,060	2,570	0	39,490	8,749	4.5	1.4	4.7	35.9	29.3	3.9
1968	17,700	0	28,558	625	0	45,633	8,965	5.1	1.6	5.0	34.8	31.5	4.5
1969	19,000	0	30,539	850	0	48,689	9,182	5.3	1.6	5.0	36.4	31.9	4.4
1970	20,100	0	27,450	900	0	46,650	9,395	5.0	1.6	5.1	34.9	30.4	4.4
1971	22,900	0	26,750	1,500	0	48,150	9,601	5.0	1.6	5.1	35.4	30.5	4.4
1972	22,900	0	20,440	2,000	0	41,340	9,804	4.2	1.3	4.9	33.9	26.9	3.9
1973	26,300	0	17,523	2,700	0	41,123	10,013	4.1	1.2	4.9	36.3	25.5	3.4
1974	28,660	0	11,285	5,000	0	34,945	10,241	3.4	1.0	4.6	34.8	21.9	2.9
1975	25,490	0	9,458	5,168	0	29,780	10,498	2.8	0.8	4.4	33.7	18.9	2.5
1976	27,900	0	6,824	3,803	0	30,921	10,786	2.9	0.9	4.0	34.9	23.2	2.6
1977	26,950	0	...	3,000	0	23,950	11,099	2.2	0.7	4.1	33.0	17.4	2.2
1978	25,940	0	13,343	3,300	0	35,983	11,429	3.2	1.0	4.8	33.0	20.4	3.0
1979	28,630	0	18,241	3,800	0	43,071	11,764	3.7	1.1	4.3	32.3	26.1	3.5
1980	33,850	0	15,739	5,000	0	44,589	12,095	3.7	1.2	4.5	33.3	27.2	3.7
1981	40,630	0	9,625	7,600	0	42,655	12,428	3.4	1.1	4.4	31.8	25.5	3.5
1982	38,180	0	15,221	5,900	0	47,501	12,762	3.7	1.2	4.5	30.5	26.6	3.9
1983	41,016	0	18,388	4,800	0	54,604	13,076	4.2	1.3	4.4	30.7	30.2	4.3
1984	35,336	0	14,381	4,400	0	45,317	13,341	3.4	1.1	4.5	32.7	23.6	3.3
1985	36,806	0	11,221	5,400	0	42,627	13,542	3.2	1.0	4.4	35.6	22.7	2.8
1986	42,443	0	14,010	4,754	-300	51,399	13,659	3.8	1.1	4.4	31.5	25.5	3.6
1987	39,609	0	15,478	4,761	0	50,326	13,710	3.7	1.1	4.9	32.0	22.2	3.4
1988	35,457	0	17,193	4,395	0	48,255	13,760	3.5	1.0	4.2	34.2	24.7	3.0
1989	31,084	0	10,590	4,198	300	37,776	13,897	2.7	0.8	4.0	31.5	20.7	2.6
1990	36,439	0	10,157	6,336	0	40,260	14,182	2.8	0.9	4.1	32.9	21.4	2.7
1991	29,355	5	11,328	7,842	0	32,836	14,641	2.2	0.7	4.3	31.4	16.0	2.2
1992	31,622	0	11,850	8,604	0	34,868	15,249	2.3	0.7	3.7	30.5	19.7	2.4
1993	30,205	0	11,389	8,978	0	32,616	15,944	2.1	0.7	3.5	31.8	18.9	2.1
1994	27,572	0	11,304	9,016	0	29,860	16,636	1.8	0.6	3.5	30.5	17.0	1.9
1995	26,870	6	11,325	8,047	0	30,142	17,260	1.8	0.6	3.3	33.9	17.3	1.7

28 MYANMAR	- FOOD BALANCE SHEET OF FISH AND FISHERY PRODUCTS IN LIVE WEIGHT AND FISH CONTRIBUTION TO PROTEIN SUPPLY -												FAO-FIDI
YEAR	PRODUCTION	NON-FOOD USES	IMPORTS	EXPORTS	STOCKS VARIATIONS	TOTAL FOOD SUPPLY	POPULATION	PER CAPUT SUPPLY	FISH PROTEINS	ANIMAL PROTEINS	TOTAL PROTEINS	FISH/ANIMAL PROTEINS	FISH/TOTAL PROTEINS
	(...................................... metric tons in live weight					......................................)	(thousands)	(kilograms)	(grams per caput per day)			(%)	(%)
1961	360,000	48,000	20,050	0	0	332,050	22,207	15.0	5.1	8.3	44.3	61.7	11.5
1962	360,000	48,000	17,533	0	0	329,533	22,674	14.5	4.9	8.1	45.1	61.0	11.0
1963	360,000	48,000	13,306	0	0	325,306	23,152	14.1	4.7	8.2	46.7	57.8	10.1
1964	360,000	48,000	10,033	0	0	322,033	23,648	13.6	4.5	7.9	48.1	56.8	9.4
1965	360,000	48,000	8,033	0	0	320,033	24,167	13.2	4.4	8.1	47.2	54.2	9.3
1966	360,000	48,000	0	0	0	312,000	24,711	12.6	4.2	7.8	47.5	53.5	8.8
1967	380,700	49,500	0	0	0	331,200	25,278	13.1	4.1	8.0	47.8	51.6	8.6
1968	396,100	51,500	0	0	0	344,600	25,866	13.3	4.2	8.0	49.2	52.3	8.5
1969	413,900	53,800	200	0	0	360,300	26,475	13.6	4.2	8.0	52.2	52.9	8.1
1970	432,400	56,200	0	0	0	376,200	27,102	13.9	3.7	7.5	52.0	49.3	7.1
1971	442,700	57,600	0	0	0	385,100	27,748	13.9	3.7	7.5	52.0	49.2	7.1
1972	453,300	58,900	0	0	0	394,400	28,412	13.9	3.7	7.3	52.7	50.7	7.0
1973	463,400	60,300	0	0	0	403,100	29,087	13.9	3.7	7.7	52.5	48.3	7.0
1974	433,840	56,400	0	0	0	377,440	29,766	12.7	3.4	7.2	53.1	46.8	6.4
1975	485,140	63,100	12	121	0	421,931	30,441	13.9	3.7	7.6	54.0	48.8	6.9
1976	501,560	65,210	12	1,554	0	434,809	31,111	14.0	3.8	7.8	53.5	48.6	7.1
1977	518,700	67,400	38	2,481	0	448,857	31,776	14.1	3.8	7.9	55.2	48.3	6.9
1978	539,690	70,400	0	7,000	0	462,290	32,443	14.3	3.9	8.0	56.7	48.5	6.8
1979	564,070	73,500	0	7,750	-375	482,445	33,122	14.6	3.9	8.2	58.4	47.7	6.7
1980	580,010	75,400	0	10,026	0	494,584	33,821	14.6	4.0	8.4	59.8	47.5	6.6
1981	594,540	77,290	0	12,672	371	504,949	34,540	14.6	4.1	8.8	61.3	46.4	6.7
1982	584,410	75,980	0	13,650	0	494,780	35,276	14.0	3.5	8.4	63.1	41.7	5.6
1983	587,550	76,380	0	7,285	0	503,885	36,026	14.0	3.5	8.7	65.4	40.4	5.4
1984	613,691	79,270	0	6,678	0	527,743	36,784	14.4	3.6	9.2	68.0	39.5	5.3
1985	648,804	83,680	0	6,676	-1,860	556,588	37,544	14.8	3.6	9.3	67.8	39.3	5.4
1986	686,523	89,247	0	6,694	0	590,582	38,307	15.4	4.0	9.5	69.3	42.0	5.7
1987	685,864	89,162	0	5,450	0	591,252	39,073	15.1	3.9	9.4	68.6	41.3	5.7
1988	704,547	91,590	0	7,040	-872	605,044	39,839	15.2	3.7	9.3	68.5	39.7	5.4
1989	733,763	95,389	0	13,436	873	625,811	40,601	15.4	3.8	8.1	63.9	46.5	5.9
1990	743,818	96,696	0	17,098	0	630,024	41,354	15.2	3.9	8.2	63.7	47.5	6.1
1991	769,236	100,001	0	17,757	1,864	653,342	42,097	15.5	3.9	8.3	63.9	46.3	6.0
1992	800,000	100,000	0	20,043	-2,469	677,488	42,832	15.8	3.9	8.3	64.5	46.7	6.0
1993	836,878	108,794	0	17,182	4,417	715,320	43,569	16.4	4.1	8.5	66.9	47.6	6.1
1994	824,468	107,181	672	19,000	0	698,959	44,323	15.8	4.0	8.6	67.4	46.7	6.0
1995	832,469	108,000	607	20,264	-100	704,712	45,107	15.6	3.9	8.6	67.3	45.5	5.8

147 NAMIBIA	- FOOD BALANCE SHEET OF FISH AND FISHERY PRODUCTS IN LIVE WEIGHT AND FISH CONTRIBUTION TO PROTEIN SUPPLY -	FAO-FIDI

YEAR	PRODUCTION	NON-FOOD USES	IMPORTS	EXPORTS	STOCKS VARIATIONS	TOTAL FOOD SUPPLY	POPULATION	PER CAPUT SUPPLY	FISH PROTEINS	ANIMAL PROTEINS	TOTAL PROTEINS	FISH/ANIMAL PROTEINS	FISH/TOTAL PROTEINS
	(.. metric tons in live weight					....)	(thousands)	(kilograms)	(grams per caput per day)			(%)	(%)
1961	5,800	0	0	0	0	5,800	639	9.1	2.6	24.3	61.8	10.6	4.2
1962	7,200	0	0	0	0	7,200	653	11.0	3.1	26.2	64.8	11.9	4.8
1963	5,900	0	0	0	0	5,900	669	8.8	2.5	26.0	64.2	9.6	3.9
1964	5,000	0	0	0	0	5,000	684	7.3	2.1	26.0	64.2	7.9	3.2
1965	5,000	0	0	0	0	5,000	701	7.1	2.0	26.0	64.6	7.8	3.1
1966	5,000	0	0	0	0	5,000	718	7.0	2.0	27.8	65.8	7.1	3.0
1967	5,000	0	0	0	0	5,000	735	6.8	1.9	27.7	65.3	6.9	2.9
1968	5,000	0	0	0	0	5,000	753	6.6	1.9	27.0	64.7	6.9	2.9
1969	5,000	0	0	0	0	5,000	772	6.5	1.8	26.9	64.5	6.8	2.8
1970	6,500	0	0	0	0	6,500	792	8.2	2.3	27.1	65.3	8.4	3.5
1971	6,500	0	0	0	0	6,500	812	8.0	2.2	27.4	66.1	8.1	3.3
1972	6,500	0	0	0	0	6,500	833	7.8	2.2	27.0	65.5	8.0	3.3
1973	6,500	0	0	0	0	6,500	855	7.6	2.1	27.1	65.9	7.7	3.2
1974	6,550	0	0	0	0	6,550	877	7.5	2.1	27.5	66.4	7.5	3.1
1975	7,050	0	0	0	0	7,050	900	7.8	2.2	23.7	63.4	9.2	3.4
1976	7,850	0	0	0	0	7,850	925	8.5	2.4	25.3	64.7	9.3	3.6
1977	8,750	0	0	0	0	8,750	950	9.2	2.6	27.2	66.6	9.4	3.8
1978	9,050	0	0	0	0	9,050	976	9.3	2.6	27.8	67.2	9.3	3.8
1979	9,750	0	0	0	0	9,750	1,002	9.7	2.7	28.6	68.1	9.5	4.0
1980	10,250	250	0	0	0	10,000	1,030	9.7	2.7	28.6	68.5	9.5	3.9
1981	10,850	500	0	0	0	10,350	1,058	9.8	2.7	23.5	64.4	11.6	4.2
1982	11,650	1,000	0	0	0	10,650	1,087	9.8	2.7	21.9	63.1	12.5	4.3
1983	12,100	750	0	0	0	11,350	1,116	10.2	2.8	19.5	61.1	14.5	4.6
1984	12,900	900	0	0	0	12,000	1,147	10.5	2.9	20.3	62.3	14.4	4.7
1985	13,700	1,500	0	0	0	12,200	1,179	10.4	2.9	20.5	61.4	14.1	4.7
1986	14,923	2,200	0	0	0	12,723	1,212	10.5	2.9	21.0	61.5	14.0	4.8
1987	32,437	19,400	0	0	0	13,037	1,246	10.5	2.8	20.6	62.6	13.4	4.4
1988	33,395	19,800	0	0	0	13,595	1,280	10.6	2.8	20.8	62.7	13.6	4.5
1989	21,225	7,300	0	0	0	13,925	1,316	10.6	2.8	19.5	61.4	14.1	4.5
1990	262,452	174,050	0	74,000	0	14,402	1,352	10.7	2.5	20.7	62.1	12.3	4.1
1991	206,190	128,130	0	63,200	0	14,860	1,388	10.7	2.5	21.9	63.8	11.5	3.9
1992	294,505	189,480	0	89,600	0	15,425	1,424	10.8	2.6	21.9	61.3	11.9	4.3
1993	330,026	192,800	0	121,400	0	15,826	1,461	10.8	2.6	18.5	57.1	13.9	4.5
1994	300,928	139,828	0	144,472	0	16,627	1,499	11.1	2.8	19.4	60.8	14.6	4.7
1995	285,980	103,833	0	164,367	0	17,780	1,536	11.6	3.4	18.6	58.4	18.2	5.8

Note: from 1990 production data include quantities caught in Namibia waters by licensed foreign vessels and processed in Walvis Bay and Luderitz mostly for exports. Exports data estimated.

148 NAURU	- FOOD BALANCE SHEET OF FISH AND FISHERY PRODUCTS IN LIVE WEIGHT AND FISH CONTRIBUTION TO PROTEIN SUPPLY -	FAO-FIDI

YEAR	PRODUCTION	NON-FOOD USES	IMPORTS	EXPORTS	STOCKS VARIATIONS	TOTAL FOOD SUPPLY	POPULATION	PER CAPUT SUPPLY	FISH PROTEINS	ANIMAL PROTEINS	TOTAL PROTEINS	FISH/ANIMAL PROTEINS	FISH/TOTAL PROTEINS
	(......... metric tons in live weight					.........)	(thousands)	(kilograms)	(grams per caput per day)			(%)	(%)
1961	...	0	0	0	0	...	4	...	...	30.1	62.2	...	...
1962	...	0	0	0	0	...	4	...	...	32.9	64.0	...	...
1963	100	0	0	0	0	100	5	22.1	6.3	42.8	73.6	14.6	8.5
1964	100	0	0	0	0	100	5	21.0	5.9	44.3	79.2	13.4	7.5
1965	100	0	0	0	0	100	5	20.0	5.6	42.2	72.4	13.4	7.8
1966	100	0	0	0	0	100	5	19.1	5.4	42.4	71.3	12.7	7.6
1967	100	0	0	0	0	100	5	18.2	5.2	44.9	77.4	11.5	6.7
1968	100	0	0	0	0	100	6	17.5	5.0	40.5	72.4	12.2	6.8
1969	100	0	0	0	0	100	6	17.0	4.8	36.2	66.7	13.3	7.2
1970	100	0	0	0	0	100	6	16.7	4.7	37.4	66.0	12.6	7.1
1971	100	0	0	0	0	100	6	16.6	4.7	31.9	64.4	14.7	7.3
1972	100	0	0	0	0	100	6	16.6	4.7	40.0	70.3	11.7	6.7
1973	100	0	0	0	0	100	6	16.7	4.7	38.9	69.2	12.1	6.8
1974	100	0	0	0	0	100	6	16.8	4.7	39.2	70.6	12.1	6.7
1975	120	0	0	0	0	120	6	20.0	5.6	42.2	70.0	13.4	8.1
1976	120	0	0	0	0	120	6	19.6	5.5	39.0	64.0	14.2	8.7
1977	120	0	0	0	0	120	6	19.0	5.4	38.5	63.7	13.9	8.4
1978	130	0	0	0	0	130	7	19.9	5.6	40.1	67.5	14.0	8.3
1979	130	0	0	0	0	130	7	19.2	5.4	39.0	65.9	13.9	8.2
1980	140	0	0	0	0	140	7	20.0	5.6	36.7	63.2	15.4	8.9
1981	140	0	0	0	0	140	7	19.5	5.5	39.7	66.0	13.9	8.3
1982	140	0	0	0	0	140	7	19.1	5.4	39.1	64.8	13.8	8.3
1983	140	0	0	0	0	140	8	18.6	5.3	38.9	63.9	13.5	8.2
1984	150	0	0	0	0	150	8	19.4	5.5	38.8	62.5	14.1	8.8
1985	150	0	0	0	0	150	8	18.8	5.3	37.1	60.3	14.3	8.8
1986	150	0	0	0	0	150	8	18.0	5.1	37.4	59.7	13.6	8.5
1987	170	0	0	0	0	170	9	19.4	5.5	37.1	58.4	14.8	9.4
1988	170	0	0	0	0	170	9	18.4	5.2	39.6	59.9	13.2	8.7
1989	180	0	0	0	0	180	10	18.7	5.3	35.1	54.7	15.0	9.6
1990	180	0	0	0	0	180	10	18.0	5.1	35.7	54.6	14.2	9.3
1991	190	0	0	0	0	190	10	18.5	5.2	24.2	30.3	21.5	17.2
1992	377	0	0	0	0	377	11	35.9	10.1	30.6	37.1	33.1	27.3
1993	500	0	0	0	0	500	11	46.9	13.2	36.0	42.0	36.8	31.5
1994	500	0	0	0	0	500	11	46.2	13.0	35.5	41.7	36.7	31.3
1995	450	0	0	0	0	450	11	40.9	11.5	34.4	40.5	33.6	28.5

149 NEPAL	- FOOD BALANCE SHEET OF FISH AND FISHERY PRODUCTS IN LIVE WEIGHT AND FISH CONTRIBUTION TO PROTEIN SUPPLY -												FAO-FIDI
YEAR	PRODUCTION	NON-FOOD USES	IMPORTS	EXPORTS	STOCKS VARIATIONS	TOTAL FOOD SUPPLY	POPULATION	PER CAPUT SUPPLY	FISH PROTEINS	ANIMAL PROTEINS	TOTAL PROTEINS	FISH/ANIMAL PROTEINS	FISH/TOTAL PROTEINS
	(.. metric tons in live weight ....)						(thousands)	(kilograms)	(grams per caput per day)			(%)	(%)
1961	1,500	0	0	0	0	1,500	9,440	0.2	0.1	8.1	50.9	0.6	0.1
1962	1,500	0	0	0	0	1,500	9,624	0.2	0.1	8.1	51.1	0.6	0.1
1963	1,500	0	0	0	0	1,500	9,815	0.2	0.1	8.2	51.3	0.6	0.1
1964	1,600	0	0	0	0	1,600	10,010	0.2	0.1	8.1	51.5	0.6	0.1
1965	1,600	0	0	0	0	1,600	10,211	0.2	0.1	8.1	51.5	0.6	0.1
1966	1,600	0	0	0	0	1,600	10,416	0.2	0.1	8.1	50.8	0.6	0.1
1967	1,800	0	0	0	0	1,800	10,626	0.2	0.1	8.2	51.4	0.6	0.1
1968	1,800	0	0	0	0	1,800	10,845	0.2	0.1	8.2	51.0	0.6	0.1
1969	1,900	0	0	0	0	1,900	11,078	0.2	0.1	8.3	51.4	0.6	0.1
1970	1,900	0	0	0	0	1,900	11,327	0.2	0.1	8.2	51.7	0.6	0.1
1971	2,100	0	0	0	0	2,100	11,593	0.2	0.1	8.3	51.2	0.7	0.1
1972	2,200	0	0	0	0	2,200	11,876	0.2	0.1	8.4	50.4	0.7	0.1
1973	2,200	0	0	0	0	2,200	12,174	0.2	0.1	8.4	51.3	0.6	0.1
1974	2,200	0	0	0	0	2,200	12,482	0.2	0.1	8.5	51.2	0.6	0.1
1975	2,500	0	0	0	0	2,500	12,797	0.2	0.1	8.7	49.9	0.7	0.1
1976	2,800	0	0	0	0	2,800	13,118	0.2	0.1	8.7	50.7	0.7	0.1
1977	3,000	0	0	0	0	3,000	13,447	0.2	0.1	8.7	50.8	0.8	0.1
1978	3,300	0	0	0	0	3,300	13,785	0.2	0.1	8.7	51.3	0.8	0.1
1979	3,500	0	0	0	0	3,500	14,135	0.3	0.1	8.5	49.2	0.9	0.2
1980	3,654	0	0	0	0	3,654	14,498	0.3	0.1	8.7	50.4	0.9	0.2
1981	3,868	0	0	0	0	3,868	14,875	0.3	0.1	8.8	51.8	0.9	0.2
1982	4,400	0	0	0	0	4,400	15,266	0.3	0.1	9.3	52.1	0.9	0.2
1983	4,700	0	0	0	0	4,700	15,669	0.3	0.1	9.3	55.1	1.0	0.2
1984	4,860	0	0	0	0	4,860	16,081	0.3	0.1	9.1	54.5	1.0	0.2
1985	9,076	0	852	291	0	9,637	16,503	0.6	0.2	9.0	53.2	1.9	0.3
1986	9,443	0	260	413	0	9,290	16,932	0.6	0.2	9.0	52.6	1.8	0.3
1987	10,716	0	226	353	0	10,589	17,370	0.6	0.2	8.7	56.1	2.1	0.3
1988	12,100	0	390	246	0	12,244	17,820	0.7	0.2	8.8	58.4	2.3	0.4
1989	12,522	0	323	107	0	12,738	18,286	0.7	0.2	8.8	61.8	2.4	0.3
1990	14,546	0	152	76	0	14,622	18,772	0.8	0.2	8.8	61.6	2.7	0.4
1991	15,595	0	...	...	0	15,595	19,279	0.8	0.2	8.6	60.1	2.8	0.4
1992	16,516	0	...	...	0	16,516	19,804	0.8	0.3	8.5	57.1	2.9	0.4
1993	16,852	0	...	...	0	16,852	20,346	0.8	0.3	8.3	59.9	3.0	0.4
1994	17,003	0	...	...	0	17,003	20,898	0.8	0.2	8.3	58.6	2.9	0.4
1995	21,148	0	150	...	0	21,298	21,456	1.0	0.3	8.4	58.1	3.5	0.5

150 NETHERLANDS	- FOOD BALANCE SHEET OF FISH AND FISHERY PRODUCTS IN LIVE WEIGHT AND FISH CONTRIBUTION TO PROTEIN SUPPLY -	FAO-FIDI

YEAR	PRODUCTION	NON-FOOD USES	IMPORTS	EXPORTS	STOCKS VARIATIONS	TOTAL FOOD SUPPLY	POPULATION	PER CAPUT SUPPLY	FISH PROTEINS	ANIMAL PROTEINS	TOTAL PROTEINS	FISH/ANIMAL PROTEINS	FISH/TOTAL PROTEINS
	(.....		metric tons in live weight			.....)	(thousands)	(kilograms)	(grams per caput per day)			(%)	(%)
1961	346,100	31,100	47,338	238,043	0	124,295	11,639	10.7	3.4	48.8	86.6	6.9	3.9
1962	321,200	25,740	46,452	223,298	0	118,614	11,802	10.1	2.9	51.3	87.3	5.6	3.3
1963	361,000	43,100	55,667	218,039	0	155,528	11,967	13.0	4.2	51.6	87.2	8.2	4.8
1964	388,100	41,200	59,943	245,939	0	160,903	12,131	13.3	4.0	51.7	85.8	7.8	4.7
1965	377,200	43,300	63,370	246,687	0	150,583	12,292	12.3	3.5	51.3	83.7	6.7	4.1
1966	353,000	32,700	74,138	230,356	0	164,082	12,449	13.2	3.9	53.5	86.7	7.2	4.5
1967	314,500	32,500	91,402	216,190	0	157,213	12,601	12.5	3.8	55.8	89.2	6.8	4.3
1968	323,300	34,100	97,960	229,786	0	157,374	12,749	12.3	3.5	54.1	86.9	6.5	4.1
1969	323,200	24,431	113,511	239,262	0	173,019	12,893	13.4	3.9	53.6	86.1	7.3	4.5
1970	300,800	9,082	126,702	244,483	0	173,938	13,032	13.4	4.0	55.1	86.7	7.3	4.6
1971	321,200	6,653	111,276	254,356	0	171,467	13,167	13.0	3.8	56.0	87.1	6.8	4.4
1972	348,300	9,637	117,650	285,963	0	170,349	13,296	12.8	3.8	56.8	88.8	6.6	4.2
1973	343,700	10,968	115,104	283,113	0	164,723	13,420	12.3	3.7	56.0	89.8	6.6	4.1
1974	325,861	12,594	130,352	270,005	0	173,613	13,539	12.8	3.7	57.7	91.3	6.4	4.1
1975	350,548	20,631	146,960	278,820	-20,000	178,057	13,653	13.0	3.9	56.9	89.5	6.8	4.3
1976	284,972	14,554	198,789	278,075	-35,000	156,132	13,763	11.3	3.6	55.9	89.8	6.5	4.0
1977	313,041	16,443	164,223	281,964	-15,000	163,858	13,868	11.8	3.1	57.3	90.4	5.5	3.5
1378	324,435	12,551	172,548	320,973	-5,000	158,459	13,968	11.3	3.0	60.9	93.5	4.9	3.2
1979	323,693	2,101	184,736	367,700	0	138,629	14,060	9.9	2.7	60.3	93.7	4.5	2.9
1980	340,429	721	207,465	389,170	-20,000	138,003	14,144	9.8	2.9	62.3	93.4	4.6	3.1
1981	479,665	1,064	207,665	491,197	-1,111	193,959	14,219	13.6	3.4	61.2	92.9	5.5	3.6
1982	544,753	67	193,315	556,344	3,400	185,057	14,287	13.0	3.4	63.8	95.6	5.3	3.5
1983	561,821	225	182,085	554,538	14,700	203,844	14,352	14.2	3.3	59.7	92.0	5.5	3.6
1984	496,061	59	229,431	556,146	12,022	181,309	14,419	12.6	3.1	62.2	95.3	5.0	3.3
1985	554,911	59	240,154	610,488	3,245	187,763	14,492	13.0	3.2	63.7	96.7	5.0	3.3
1986	502,877	60	243,562	642,117	50,389	154,652	14,572	10.6	2.9	59.8	93.2	4.8	3.1
1987	511,773	149	276,326	616,168	2,833	174,615	14,659	11.9	3.0	60.1	93.4	5.0	3.2
1988	476,392	2,135	327,957	595,112	-4,116	202,986	14,752	13.8	3.7	63.3	97.8	5.9	3.8
1989	530,226	7,643	355,581	670,902	-638	206,623	14,850	13.9	3.6	60.8	92.9	5.9	3.9
1990	507,237	11,619	443,130	775,569	-697	162,482	14,952	10.9	2.9	61.9	94.7	4.7	3.1
1991	458,814	17,301	522,507	770,694	-37,996	155,330	15,057	10.3	3.3	63.2	97.0	5.2	3.4
1992	487,075	21,646	539,876	835,019	32,505	202,792	15,166	13.4	3.7	64.4	100.2	5.8	3.7
1993	532,881	17,324	486,800	807,176	14,808	209,989	15,276	13.8	4.4	67.2	102.3	6.5	4.3
1994	529,635	21,591	535,972	829,295	569	215,291	15,382	14.0	4.5	64.2	99.4	7.0	4.5
1995	521,377	37,661	568,055	786,242	-42,790	222,739	15,482	14.4	4.7	68.8	101.7	6.8	4.6

151 NETHERLANDS ANTILLES	- FOOD BALANCE SHEET OF FISH AND FISHERY PRODUCTS IN LIVE WEIGHT AND FISH CONTRIBUTION TO PROTEIN SUPPLY -	FAO-FIDI

YEAR	PRODUCTION	NON-FOOD USES	IMPORTS	EXPORTS	STOCKS VARIATIONS	TOTAL FOOD SUPPLY	POPULATION	PER CAPUT SUPPLY	FISH PROTEINS	ANIMAL PROTEINS	TOTAL PROTEINS	FISH/ANIMAL PROTEINS	FISH/TOTAL PROTEINS
	(.. metric tons in live weight ....)						(thousands)	(kilograms)	(grams per caput per day)			(%)	(%)
1961	600	0	1,827	0	0	2,427	137	17.7	4.2	34.0	61.0	12.2	6.8
1962	600	0	2,278	0	0	2,878	139	20.7	4.9	36.9	64.3	13.2	7.6
1963	700	0	2,980	0	0	3,680	141	26.0	6.2	40.0	68.0	15.6	9.2
1964	700	0	2,103	0	0	2,803	144	19.5	4.6	37.7	66.2	12.1	6.9
1965	700	0	2,403	0	0	3,103	146	21.2	5.0	37.4	64.8	13.2	7.6
1966	700	0	2,503	0	0	3,203	149	21.5	5.0	37.0	64.6	13.5	7.8
1967	600	0	2,903	0	0	3,503	152	23.1	5.0	38.9	66.7	12.9	7.5
1968	600	0	3,178	0	0	3,778	154	24.5	5.4	40.5	68.5	13.2	7.8
1969	600	0	2,707	0	0	3,307	157	21.1	4.7	39.8	66.6	11.8	7.1
1970	657	0	3,378	0	0	4,035	159	25.4	5.5	44.0	70.2	12.6	7.9
1971	757	0	4,101	0	0	4,858	161	30.2	5.2	44.8	70.5	11.7	7.4
1972	857	0	2,953	0	0	3,810	162	23.5	5.1	44.5	70.4	11.4	7.2
1973	857	0	3,053	0	0	3,910	164	23.9	5.2	45.1	71.1	11.4	7.3
1974	857	0	3,880	0	0	4,737	165	28.7	6.3	44.9	75.5	14.1	8.4
1975	887	0	3,153	0	0	4,040	166	24.3	5.2	44.7	75.7	11.7	6.9
1976	937	0	3,439	77	0	4,299	168	25.7	5.4	50.2	83.5	10.9	6.5
1977	987	0	3,332	72	0	4,247	169	25.1	5.3	52.9	86.7	10.0	6.1
1978	1,010	0	3,320	0	0	4,330	171	25.4	5.6	52.1	87.0	10.8	6.4
1979	1,010	0	3,756	114	0	4,653	172	27.0	5.9	60.2	97.9	9.8	6.0
1980	1,090	0	3,728	0	0	4,818	174	27.7	6.1	56.5	92.8	10.7	6.5
1981	990	0	3,828	21	0	4,797	175	27.4	6.0	66.4	100.7	9.0	5.9
1982	1,030	0	4,167	37	0	5,161	177	29.2	6.3	65.9	101.2	9.5	6.2
1983	1,030	0	4,428	23	0	5,435	179	30.4	6.4	60.8	94.0	10.6	6.8
1984	1,030	0	4,246	42	0	5,234	180	29.0	6.2	57.0	87.0	10.9	7.1
1985	1,030	0	2,953	81	0	3,902	182	21.5	6.4	54.0	82.2	11.9	7.8
1986	1,060	0	3,172	44	0	4,188	183	22.9	6.9	52.1	79.6	13.2	8.7
1987	1,130	0	3,768	46	0	4,852	184	26.3	7.7	57.4	87.1	13.5	8.9
1988	1,220	0	3,166	16	0	4,370	185	23.6	7.1	52.6	82.7	13.5	8.6
1989	1,220	0	3,767	34	0	4,954	187	26.6	7.7	52.0	80.4	14.9	9.6
1990	1,220	0	4,291	258	0	5,252	188	28.0	8.6	50.0	79.2	17.3	10.9
1991	1,120	0	2,650	158	0	3,612	189	19.1	5.8	51.1	80.0	11.4	7.3
1992	1,160	0	2,958	37	0	4,081	190	21.5	6.3	51.4	83.1	12.3	7.6
1993	1,210	0	2,934	40	0	4,105	191	21.4	6.5	54.6	87.1	11.9	7.4
1994	1,110	0	2,723	28	0	3,806	193	19.8	5.6	53.2	85.0	10.6	6.6
1995	1,010	0	2,650	48	0	3,613	194	18.6	5.5	56.1	86.2	9.8	6.4

Note: up to 1983 trade include those for Aruba.

153 NEW CALEDONIA	- FOOD BALANCE SHEET OF FISH AND FISHERY PRODUCTS IN LIVE WEIGHT AND FISH CONTRIBUTION TO PROTEIN SUPPLY -												FAO-FIDI
YEAR	PRODUCTION	NON-FOOD USES	IMPORTS	EXPORTS	STOCKS VARIATIONS	TOTAL FOOD SUPPLY	POPULATION	PER CAPUT SUPPLY	FISH PROTEINS	ANIMAL PROTEINS	TOTAL PROTEINS	FISH/ANIMAL PROTEINS	FISH/TOTAL PROTEINS
	(........ metric tons in live weight ....)						(thousands)	(kilograms)	(grams per caput per day)			(%)	(%)
1961	400	0	...	...	0	...	81	...	...	38.6	72.3	...	...
1962	800	0	...	...	0	...	84	...	...	38.6	72.7	...	...
1963	1,200	0	...	...	0	...	87	...	...	40.6	73.3	...	...
1964	900	0	...	...	0	...	89	...	...	40.8	74.1	...	...
1965	900	0	...	...	0	...	92	...	...	39.8	74.8	...	...
1966	800	0	...	...	0	...	95	...	...	38.7	74.6	...	...
1967	500	0	...	...	0	...	98	...	...	40.2	75.5	...	...
1968	500	0	...	...	0	...	101	...	...	40.9	76.5	...	...
1969	500	0	...	...	0	...	104	...	...	43.6	79.7	...	...
1970	500	0	...	...	0	...	108	...	...	45.0	82.2	...	...
1971	500	0	...	...	0	...	113	...	...	46.6	82.5	...	...
1972	500	0	...	...	0	...	118	...	...	50.7	85.7	...	...
1973	800	0	...	...	0	...	124	...	...	48.3	83.6	...	...
1974	868	0	...	...	0	...	129	...	...	47.5	81.4	...	...
1975	843	0	...	...	0	...	133	...	...	43.5	77.9	...	...
1976	871	0	1,821	19	0	2,673	136	19.6	6.1	46.3	80.7	13.2	7.6
1977	2,094	1,176	2,001	154	0	2,765	138	20.0	6.3	46.9	79.6	13.4	7.9
1978	1,047	574	1,781	55	-300	1,898	140	13.6	4.4	42.7	76.5	10.2	5.7
1979	1,382	532	1,890	85	-300	2,355	141	16.7	5.3	43.0	75.4	12.4	7.1
1980	2,099	1,372	2,818	98	-325	3,122	143	21.8	5.5	45.7	79.4	12.1	7.0
1981	1,723	826	1,962	266	-13	2,580	145	17.8	5.7	44.0	78.0	13.0	7.3
1982	2,904	1,582	1,855	741	588	3,024	147	20.5	6.8	46.3	79.8	14.7	8.5
1983	1,276	42	1,796	622	300	2,708	150	18.1	6.0	45.0	77.9	13.3	7.7
1984	3,349	1,620	1,612	660	13	2,693	152	17.7	5.8	44.9	78.9	12.8	7.3
1985	2,694	1,260	1,704	392	0	2,746	155	17.7	6.0	42.9	79.7	14.0	7.5
1986	4,006	2,521	2,241	666	-8	3,053	157	19.4	5.4	44.2	77.7	12.3	7.0
1987	4,798	1,841	2,251	1,508	20	3,720	160	23.3	6.7	46.6	80.6	14.3	8.3
1988	3,684	1,240	1,934	1,519	0	2,860	162	17.6	5.2	39.6	75.1	13.1	6.9
1989	3,319	548	2,095	1,487	13	3,392	165	20.6	6.3	40.6	76.6	15.5	8.2
1990	5,306	1,294	2,067	1,829	-24	4,227	168	25.2	7.8	46.4	79.6	16.8	9.8
1991	4,869	1,240	2,066	1,861	36	3,870	170	22.8	7.1	47.6	80.1	15.0	8.9
1992	3,799	1,090	2,181	1,651	0	3,239	173	18.8	5.8	48.3	80.0	11.9	7.2
1993	3,331	490	2,497	1,953	0	3,385	175	19.3	5.1	45.5	79.2	11.1	6.4
1994	3,910	481	2,780	2,206	0	4,003	178	22.5	6.0	49.5	83.2	12.1	7.2
1995	3,483	263	2,552	2,001	0	3,772	181	20.9	5.5	46.7	81.8	11.9	6.8

Note: 1961-1975 data for fish apparent consumption not shown as trade statistics not available.

156 NEW ZEALAND	- FOOD BALANCE SHEET OF FISH AND FISHERY PRODUCTS IN LIVE WEIGHT AND FISH CONTRIBUTION TO PROTEIN SUPPLY -	FAO-FIDI

YEAR	PRODUCTION	NON-FOOD USES	IMPORTS	EXPORTS	STOCKS VARIATIONS	TOTAL FOOD SUPPLY	POPULATION	PER CAPUT SUPPLY	FISH PROTEINS	ANIMAL PROTEINS	TOTAL PROTEINS	FISH/ANIMAL PROTEINS	FISH/TOTAL PROTEINS
	(............		metric tons in live weight			............)	(thousands)	(kilograms)		(grams per caput per day)		(%)	(%)
1961	43,000	0	3,503	4,300	0	42,203	2,424	17.4	3.2	60.5	95.6	5.4	3.4
1962	41,300	0	5,200	4,900	0	41,600	2,477	16.8	3.6	62.7	95.2	5.7	3.7
1963	40,800	0	3,538	4,600	0	39,738	2,531	15.7	3.3	63.8	97.7	5.1	3.3
1964	44,200	0	6,379	5,200	0	45,379	2,582	17.6	3.5	61.2	96.1	5.8	3.7
1965	48,400	0	5,779	6,267	0	47,913	2,628	18.2	3.5	61.3	94.0	5.8	3.8
1966	55,700	0	6,026	8,500	0	53,226	2,669	20.0	3.8	62.5	95.3	6.2	4.0
1967	59,600	0	4,344	11,321	0	52,623	2,704	19.5	3.8	61.4	92.3	6.2	4.1
1968	59,100	0	3,405	15,028	0	47,477	2,738	17.3	3.5	63.7	94.6	5.4	3.7
1969	49,400	0	3,605	14,128	0	38,877	2,775	14.0	3.0	63.6	97.6	4.8	3.1
1970	59,700	3,500	6,700	17,454	0	45,446	2,820	16.1	3.2	65.7	96.7	4.8	3.3
1971	66,300	1,000	9,667	24,554	0	50,413	2,873	17.6	3.8	67.1	97.1	5.6	3.9
1972	58,500	3,000	8,285	28,154	0	35,631	2,933	12.2	2.6	66.9	98.7	3.8	2.6
1973	66,300	4,000	11,608	25,805	0	48,102	2,994	16.1	3.1	69.9	101.1	4.4	3.1
1974	69,432	16,000	15,983	22,552	0	46,863	3,046	15.4	3.0	70.6	101.2	4.3	3.0
1975	64,776	2,083	7,503	23,522	0	46,674	3,083	15.1	2.6	72.5	103.6	3.6	2.5
1976	71,337	149	8,380	32,866	0	46,702	3,103	15.1	3.0	72.5	102.8	4.1	2.9
1977	78,929	131	7,465	39,463	0	46,799	3,109	15.1	2.9	71.0	103.2	4.2	2.9
1978	99,513	209	7,685	51,756	0	55,234	3,107	17.8	4.0	70.8	104.4	5.7	3.9
1979	120,146	125	11,156	82,091	0	49,086	3,106	15.8	3.4	67.5	99.2	5.0	3.4
1980	155,859	152	9,149	116,979	0	47,876	3,113	15.4	3.2	65.0	98.1	4.9	3.3
1981	168,615	82	12,146	133,832	0	46,847	3,131	15.0	3.2	65.7	96.8	4.8	3.3
1982	175,675	308	15,179	132,774	0	57,772	3,157	18.3	3.6	67.5	100.6	5.3	3.5
1983	207,603	280	18,945	150,137	-9,129	67,002	3,188	21.0	4.1	66.8	98.3	6.2	4.2
1984	218,342	325	14,536	164,201	-7,160	61,192	3,219	19.0	3.8	61.8	93.1	6.1	4.0
1985	213,544	286	14,537	174,499	0	53,296	3,247	16.4	3.2	68.3	99.5	4.6	3.2
1986	217,865	465	16,222	185,469	9,146	57,298	3,270	17.5	3.6	64.2	96.1	5.6	3.7
1987	247,252	314	18,374	205,357	0	59,955	3,290	18.2	4.2	66.9	99.2	6.3	4.2
1988	291,866	418	20,038	261,904	8,120	57,702	3,310	17.4	3.6	59.0	90.6	6.1	4.0
1989	333,649	801	23,774	289,811	0	66,812	3,333	20.1	4.6	61.8	93.5	7.4	4.9
1990	372,451	567	23,290	291,435	-35,147	68,592	3,360	20.4	4.8	64.0	98.1	7.6	4.9
1991	417,065	332	25,538	377,434	-78	64,759	3,394	19.1	4.8	61.0	96.5	7.8	4.9
1992	503,198	899	23,222	465,305	10,001	70,216	3,433	20.5	5.8	64.5	100.3	9.0	5.8
1993	470,383	869	20,439	434,473	19,268	74,748	3,476	21.5	5.8	68.5	106.3	8.5	5.5
1994	491,495	958	19,354	435,116	59	74,835	3,519	21.3	6.1	68.4	106.6	8.9	5.7
1995	612,223	813	22,867	539,346	-10,184	84,747	3,561	23.8	6.8	71.2	110.0	9.5	6.2

Note: trade with the Cook Islands, Niue and Tokelau is excluded.

157 NICARAGUA	- FOOD BALANCE SHEET OF FISH AND FISHERY PRODUCTS IN LIVE WEIGHT AND FISH CONTRIBUTION TO PROTEIN SUPPLY -	FAO-FIDI

YEAR	PRODUCTION	NON-FOOD USES	IMPORTS	EXPORTS	STOCKS VARIATIONS	TOTAL FOOD SUPPLY	POPULATION	PER CAPUT SUPPLY	FISH PROTEINS	ANIMAL PROTEINS	TOTAL PROTEINS	FISH/ANIMAL PROTEINS	FISH/TOTAL PROTEINS
	(.. metric tons in live weight ....)						(thousands)	(kilograms)	(grams per caput per day)			(%)	(%)
1961	4,000	0	702	400	0	4,302	1,541	2.8	0.8	23.7	66.8	3.4	1.2
1962	4,000	0	877	1,500	0	3,377	1,591	2.1	0.6	22.8	64.8	2.8	1.0
1963	5,000	0	877	1,500	0	4,377	1,642	2.7	0.8	24.1	65.2	3.3	1.2
1964	5,000	0	1,080	2,000	0	4,080	1,696	2.4	0.8	23.9	67.1	3.3	1.2
1965	5,600	0	1,467	1,600	0	5,467	1,751	3.1	1.0	23.9	68.7	4.4	1.5
1966	7,000	0	1,200	2,550	0	5,650	1,808	3.1	0.9	25.0	71.1	3.8	1.3
1967	8,000	0	1,300	3,076	0	6,224	1,866	3.3	1.0	24.7	72.2	4.2	1.4
1968	8,700	0	929	3,251	0	6,378	1,926	3.3	1.0	26.4	74.3	3.8	1.4
1969	9,500	0	600	5,288	0	4,812	1,989	2.4	0.8	26.9	74.2	3.0	1.1
1970	9,800	0	743	5,511	0	5,032	2,054	2.5	0.8	27.4	73.7	2.9	1.1
1971	9,400	0	743	5,475	0	4,667	2,123	2.2	0.7	27.3	71.0	2.5	1.0
1972	11,200	0	733	6,830	0	5,103	2,195	2.3	0.8	26.6	68.6	2.9	1.1
1973	13,900	0	929	6,825	0	8,003	2,268	3.5	1.1	25.8	68.4	4.4	1.7
1974	8,689	0	1,800	7,187	0	3,302	2,342	1.4	0.4	25.1	66.0	1.7	0.7
1975	10,778	0	720	7,070	0	4,428	2,417	1.8	0.6	28.4	68.6	2.0	0.8
1976	9,663	1	1,631	7,835	0	3,457	2,490	1.4	0.4	29.3	68.3	1.3	0.6
1977	10,554	2	2,036	8,900	0	3,688	2,563	1.4	0.4	29.3	69.3	1.4	0.6
1978	10,160	1	1,125	7,225	0	4,059	2,636	1.5	0.4	27.8	66.9	1.5	0.6
1979	7,241	1	777	6,053	0	1,964	2,711	0.7	0.2	25.1	63.9	0.8	0.3
1980	6,996	2	1,236	5,476	0	2,754	2,790	1.0	0.3	19.6	61.6	1.4	0.5
1981	5,944	0	173	3,997	0	2,120	2,871	0.7	0.2	19.2	68.3	1.1	0.3
1982	5,000	0	1,327	3,071	0	3,255	2,957	1.1	0.3	19.7	59.2	1.7	0.6
1983	4,548	0	877	1,902	0	3,523	3,042	1.2	0.3	18.6	59.9	1.8	0.6
1984	4,356	0	81	1,470	0	2,967	3,125	1.0	0.3	20.6	64.9	1.3	0.4
1985	4,191	0	0	1,746	0	2,445	3,203	0.8	0.2	19.5	63.1	1.1	0.3
1986	2,501	0	18	1,206	0	1,312	3,275	0.4	0.1	18.5	63.0	0.6	0.2
1987	5,031	0	0	1,417	0	3,614	3,343	1.1	0.3	16.6	57.9	1.8	0.5
1988	4,728	0	0	1,581	0	3,147	3,410	0.9	0.3	14.8	61.3	1.7	0.4
1989	4,656	0	0	2,211	0	2,445	3,483	0.7	0.2	13.7	57.3	1.4	0.3
1990	3,162	0	0	1,681	0	1,481	3,568	0.4	0.1	15.1	59.1	0.8	0.2
1991	5,783	0	1,406	2,907	0	4,282	3,665	1.2	0.4	14.5	57.8	2.4	0.6
1992	6,778	0	2,142	3,190	0	5,730	3,773	1.5	0.5	15.4	56.0	3.0	0.8
1993	8,773	3	1,961	5,281	-5	5,446	3,889	1.4	0.4	14.2	56.9	2.8	0.7
1994	12,338	3	1,577	9,591	5	4,326	4,007	1.1	0.3	13.4	52.5	2.2	0.6
1995	13,503	0	2,692	12,231	0	3,963	4,123	1.0	0.3	12.7	53.5	2.2	0.5

158 NIGER	- FOOD BALANCE SHEET OF FISH AND FISHERY PRODUCTS IN LIVE WEIGHT AND FISH CONTRIBUTION TO PROTEIN SUPPLY -	FAO-FIDI

YEAR	PRODUCTION	NON-FOOD USES	IMPORTS	EXPORTS	STOCKS VARIATIONS	TOTAL FOOD SUPPLY	POPULATION	PER CAPUT SUPPLY	FISH PROTEINS	ANIMAL PROTEINS	TOTAL PROTEINS	FISH/ANIMAL PROTEINS	FISH/TOTAL PROTEINS
	(..		metric tons in live weight		....)		(thousands)	(kilograms)	(grams per caput per day)			(%)	(%)
1961	4,200	0	0	4,000	0	200	3,138	0.1	0.0	13.4	48.6	0.1	0.0
1962	6,000	0	400	4,750	0	1,650	3,266	0.5	0.1	13.0	48.4	1.1	0.3
1963	9,000	0	0	2,750	0	6,250	3,402	1.8	0.6	13.8	50.0	4.5	1.2
1964	6,300	0	0	1,263	0	5,037	3,536	1.4	0.5	13.6	45.9	3.4	1.0
1965	5,000	0	250	1,000	0	4,250	3,660	1.2	0.4	13.5	55.2	2.9	0.7
1966	5,000	0	200	1,750	0	3,450	3,771	0.9	0.3	14.0	54.2	2.1	0.6
1967	3,700	0	200	1,750	0	2,150	3,872	0.6	0.2	14.6	56.7	1.2	0.3
1968	4,300	0	200	1,750	0	2,750	3,968	0.7	0.2	15.8	58.4	1.4	0.4
1969	3,000	0	200	2,250	0	950	4,064	0.2	0.1	14.3	55.5	0.7	0.2
1970	2,400	0	495	2,400	0	495	4,165	0.1	0.1	13.4	55.2	0.4	0.1
1971	5,400	0	175	4,909	0	666	4,273	0.2	0.1	12.7	56.7	0.4	0.1
1972	16,400	0	175	6,226	0	10,350	4,386	2.4	0.8	12.4	55.9	6.2	1.4
1973	16,200	0	0	6,400	0	9,800	4,507	2.2	0.7	10.0	55.1	7.0	1.3
1974	15,050	0	0	7,250	0	7,800	4,635	1.7	0.5	9.7	56.9	5.6	1.0
1975	9,142	0	0	3,265	0	5,877	4,771	1.2	0.4	10.4	58.8	3.8	0.7
1976	4,715	0	64	19	0	4,760	4,915	1.0	0.3	10.1	62.8	3.1	0.5
1977	7,372	0	119	3,718	0	3,774	5,069	0.7	0.2	11.8	63.4	2.0	0.4
1978	8,783	0	174	868	0	8,090	5,232	1.6	0.5	12.4	64.9	4.1	0.8
1979	8,934	0	534	7,100	0	2,368	5,404	0.4	0.1	12.4	65.8	1.1	0.2
1980	8,892	0	424	30	0	9,286	5,586	1.7	0.6	12.6	66.4	4.4	0.8
1981	8,208	0	1,238	234	0	9,212	5,777	1.6	0.5	12.9	66.6	4.1	0.8
1982	6,840	0	767	100	0	7,507	5,977	1.3	0.4	11.9	64.4	3.5	0.7
1983	3,251	0	318	49	0	3,520	6,184	0.6	0.3	12.1	61.8	2.7	0.5
1984	3,000	0	306	23	0	3,283	6,396	0.5	0.2	9.3	57.2	2.6	0.4
1985	2,000	0	370	80	0	2,290	6,608	0.4	0.1	8.0	53.5	1.7	0.3
1986	2,350	0	812	57	0	3,105	6,822	0.5	0.2	8.5	54.6	1.8	0.3
1987	2,288	0	986	23	0	3,252	7,036	0.5	0.2	8.7	56.7	1.8	0.3
1988	2,504	0	712	0	0	3,216	7,257	0.4	0.1	9.1	58.9	1.6	0.3
1989	4,751	0	2,090	12	0	6,828	7,487	0.9	0.3	9.1	57.0	3.3	0.5
1990	3,362	0	1,294	0	-70	4,586	7,731	0.6	0.2	8.9	55.6	2.2	0.4
1991	3,150	0	1,640	0	0	4,790	7,991	0.6	0.2	8.6	56.4	2.2	0.3
1992	2,054	0	1,018	0	0	3,072	8,264	0.4	0.1	8.9	53.7	1.2	0.2
1993	2,172	0	1,118	132	70	3,228	8,550	0.4	0.1	8.9	47.7	1.3	0.3
1994	2,516	0	1,383	62	0	3,837	8,846	0.4	0.1	8.7	65.6	1.6	0.2
1995	3,586	0	1,716	15	0	5,288	9,151	0.6	0.2	8.8	48.5	2.0	0.4

159 NIGERIA	- FOOD BALANCE SHEET OF FISH AND FISHERY PRODUCTS IN LIVE WEIGHT AND FISH CONTRIBUTION TO PROTEIN SUPPLY -												FAO-FIDI
YEAR	PRODUCTION	NON-FOOD USES	IMPORTS	EXPORTS	STOCKS VARIATIONS	TOTAL FOOD SUPPLY	POPULATION	PER CAPUT SUPPLY	FISH PROTEINS	ANIMAL PROTEINS	TOTAL PROTEINS	FISH/ANIMAL PROTEINS	FISH/TOTAL PROTEINS
	(............ metric tons in live weight ....)						(thousands)	(kilograms)	(grams per caput per day)			(%)	(%)
1961	55,000	0	103,707	0	0	158,707	43,515	3.7	1.2	5.0	50.1	23.1	2.3
1962	65,200	0	85,561	100	0	150,661	44,782	3.4	1.0	4.8	51.4	21.9	2.0
1963	66,200	0	139,275	0	0	205,475	46,084	4.5	1.4	5.1	49.7	27.3	2.8
1964	85,400	0	162,644	100	0	247,944	47,389	5.2	1.6	5.5	50.6	28.6	3.1
1965	97,200	0	136,570	100	0	233,670	48,676	4.8	1.4	5.4	49.3	26.2	2.9
1966	110,000	0	126,492	100	0	236,392	49,937	4.7	1.4	5.3	43.7	26.5	3.2
1967	119,300	0	107,295	200	0	226,395	51,183	4.4	1.3	5.2	40.9	24.5	3.1
1968	120,000	0	59,238	700	0	178,538	52,435	3.4	0.9	4.9	42.4	19.1	2.2
1969	115,700	0	57,764	800	0	172,664	53,723	3.2	0.9	5.3	49.1	16.4	1.8
1970	155,800	0	59,187	300	0	214,687	55,070	3.9	1.1	5.6	50.6	19.6	2.2
1971	245,300	0	83,732	300	0	328,732	56,486	5.8	1.6	6.1	45.4	25.7	3.4
1972	246,400	0	110,192	801	0	355,791	57,965	6.1	1.6	5.8	38.5	27.8	4.2
1973	250,600	0	118,393	1,700	0	367,293	59,508	6.2	1.6	6.0	39.5	27.1	4.1
1974	257,333	0	138,946	2,800	0	393,479	61,111	6.4	1.7	6.0	46.5	27.6	3.6
1975	251,513	0	231,525	2,200	0	480,838	62,770	7.7	2.0	6.5	48.4	30.9	4.1
1976	260,586	0	345,773	837	0	605,522	64,488	9.4	2.5	7.5	44.8	33.0	5.5
1977	262,180	0	423,407	179	0	685,408	66,272	10.3	2.7	7.9	42.1	33.5	6.3
1978	264,072	0	612,517	487	0	876,102	68,121	12.9	3.3	9.1	41.6	36.4	8.0
1979	268,755	0	730,033	779	0	998,009	70,037	14.3	3.8	9.4	41.1	40.0	9.1
1980	261,291	0	723,150	530	0	983,911	72,024	13.7	3.7	10.2	43.4	36.4	8.6
1981	260,131	0	998,521	1,307	0	1,257,345	74,082	17.0	4.7	11.0	44.1	42.8	10.7
1982	269,712	0	1,101,619	861	0	1,370,470	76,214	18.0	4.6	11.2	45.4	41.4	10.2
1983	272,309	0	713,123	1,389	0	984,043	78,422	12.6	3.3	9.5	40.0	34.6	8.2
1984	263,933	0	455,615	732	0	718,816	80,706	8.9	2.2	8.1	40.8	26.7	5.3
1985	244,871	0	268,204	727	0	512,348	83,068	6.2	1.6	7.5	43.3	21.5	3.7
1986	271,573	0	238,438	704	0	509,307	85,506	6.0	1.6	6.6	43.9	24.3	3.6
1987	260,808	0	524,605	2,119	0	783,294	88,023	8.9	2.2	7.1	45.2	30.3	4.8
1988	279,514	0	424,361	3,152	0	700,723	90,629	7.7	1.9	6.6	47.5	28.5	3.9
1989	299,833	0	602,313	8,115	0	894,030	93,336	9.6	2.2	6.7	49.9	32.3	4.4
1990	316,328	0	719,881	7,155	0	1,029,054	96,154	10.7	2.3	7.1	50.9	32.7	4.5
1991	267,216	0	672,801	6,030	0	933,988	99,087	9.4	2.1	7.0	53.1	29.4	3.9
1992	318,384	0	557,065	8,224	-1,719	865,506	102,129	8.5	2.0	6.9	53.6	28.5	3.7
1993	255,499	0	378,770	8,161	156	626,264	105,264	6.0	1.6	6.9	55.4	23.5	2.9
1994	282,089	0	290,425	5,420	625	567,720	108,467	5.2	1.5	6.5	53.4	23.2	2.8
1995	366,101	0	311,348	8,503	938	669,883	111,721	6.0	1.6	7.3	54.2	22.5	3.0

160 NIUE	- FOOD BALANCE SHEET OF FISH AND FISHERY PRODUCTS IN LIVE WEIGHT AND FISH CONTRIBUTION TO PROTEIN SUPPLY -												FAO-FIDI
YEAR	PRODUCTION	NON-FOOD USES	IMPORTS	EXPORTS	STOCKS VARIATIONS	TOTAL FOOD SUPPLY	POPULATION	PER CAPUT SUPPLY	FISH PROTEINS	ANIMAL PROTEINS	TOTAL PROTEINS	FISH/ANIMAL PROTEINS	FISH/TOTAL PROTEINS
	(..................................... metric tons in live weight ....)						(thousands)	(kilograms)	(grams per caput per day)			(%)	(%)
1961	...	0	...	...	0	...	5	...	...	19.6	37.9	...	...
1962	...	0	...	...	0	...	5	...	...	17.7	35.9	...	...
1963	...	0	...	...	0	...	5	...	...	17.2	35.3	...	...
1964	...	0	...	...	0	...	5	...	...	17.2	35.3	...	...
1965	...	0	...	...	0	...	5	...	...	19.8	39.1	...	...
1966	...	0	...	...	0	...	5	...	...	20.0	39.8	...	...
1967	...	0	...	...	0	...	5	...	...	20.0	40.1	...	...
1968	...	0	...	...	0	...	5	...	...	17.2	35.5	...	...
1969	...	0	...	...	0	...	5	...	...	17.1	36.4	...	...
1970	...	0	...	...	0	...	5	...	...	17.4	37.3	...	...
1971	...	0	...	...	0	...	5	...	...	19.1	39.5	...	...
1972	...	0	...	...	0	...	5	...	...	20.1	41.3	...	...
1973	...	0	...	...	0	...	4	...	...	21.3	44.4	...	...
1974	20	0	...	...	0	20	4	4.7	1.3	28.2	54.0	4.7	2.5
1975	20	0	...	...	0	20	4	5.0	1.4	30.2	55.0	4.7	2.6
1976	30	0	...	...	0	30	4	8.0	2.3	33.7	55.3	6.7	4.1
1977	40	0	...	...	0	40	4	11.4	3.2	38.7	61.1	8.3	5.3
1978	60	0	...	...	0	60	3	18.2	5.1	42.9	64.2	12.0	8.0
1979	30	0	...	...	0	30	3	9.6	2.7	42.1	64.8	6.5	4.2
1980	45	0	...	...	0	45	3	15.0	4.2	40.7	62.5	10.4	6.8
1981	45	0	...	...	0	45	3	15.2	4.3	33.1	52.4	13.0	8.2
1982	54	0	...	...	0	54	3	18.1	5.1	32.9	51.2	15.5	10.0
1983	60	0	...	...	0	60	3	19.8	5.6	30.1	48.1	18.5	11.6
1984	72	0	...	...	0	72	3	23.6	6.7	27.6	45.8	24.1	14.5
1985	90	0	...	...	0	90	3	30.0	8.5	39.4	66.8	21.5	12.7
1986	105	0	...	...	0	105	3	36.8	10.4	41.4	68.6	25.1	15.1
1987	115	0	...	...	0	115	3	43.6	12.3	50.2	81.8	24.5	15.0
1988	115	0	...	...	0	115	2	48.2	13.6	49.9	86.3	27.3	15.8
1989	115	0	...	...	0	115	2	53.2	15.0	54.5	100.0	27.6	15.0
1990	115	0	...	...	0	115	2	57.5	16.2	60.2	99.8	26.9	16.3
1991	120	0	...	...	0	120	2	62.5	17.6	49.6	92.6	35.5	19.1
1992	115	0	...	...	0	115	2	60.4	17.0	54.2	98.8	31.4	17.3
1993	120	0	...	...	0	120	2	62.1	17.5	57.5	100.8	30.5	17.4
1994	110	0	...	...	0	110	2	55.8	15.7	57.4	100.3	27.4	15.7
1995	115	0	...	...	0	115	2	57.5	16.2	56.5	99.2	28.7	16.4

163 NORTHERN MARIANA ISLANDS	- FOOD BALANCE SHEET OF FISH AND FISHERY PRODUCTS IN LIVE WEIGHT AND FISH CONTRIBUTION TO PROTEIN SUPPLY -												FAO-FIDI
YEAR	PRODUCTION	NON-FOOD USES	IMPORTS	EXPORTS	STOCKS VARIATIONS	TOTAL FOOD SUPPLY	POPULATION	PER CAPUT SUPPLY	FISH PROTEINS	ANIMAL PROTEINS	TOTAL PROTEINS	FISH/ANIMAL PROTEINS	FISH/TOTAL PROTEINS
	(....................................... metric tons in live weight ....)						(thousands)	(kilograms)	(grams per caput per day)			(%)	(%)
1961	150	0	0	0	0	150	10	14.6	4.1	...	...	...	...
1962	150	0	0	0	0	150	10	14.4	4.1	...	...	...	...
1963	150	0	0	0	0	150	11	14.2	4.0	...	...	...	...
1964	150	0	0	0	0	150	11	13.9	3.9	...	...	...	...
1965	200	0	0	0	0	200	11	18.2	5.4	...	...	...	...
1966	200	0	0	0	0	200	11	17.7	5.3	...	...	...	...
1967	200	0	0	0	0	200	12	17.1	5.1	...	...	...	...
1968	200	0	0	0	0	200	12	16.5	4.9	...	...	...	...
1969	200	0	0	0	0	200	13	15.9	4.7	...	...	...	...
1970	150	0	0	0	0	150	13	11.5	3.5	...	...	...	...
1971	150	0	0	0	0	150	13	11.2	3.4	...	...	...	...
1972	150	0	0	0	0	150	14	10.9	3.3	...	...	...	...
1973	150	0	0	0	0	150	14	10.6	3.2	...	...	...	...
1974	150	0	0	0	0	150	15	10.3	3.1	...	...	...	...
1975	150	0	0	0	0	150	15	10.0	3.0	...	...	...	...
1976	150	0	0	0	0	150	15	9.7	2.9	...	...	...	...
1977	150	0	0	0	0	150	16	9.4	2.8	...	...	...	...
1978	150	0	0	0	0	150	17	9.1	2.8	...	...	...	...
1979	145	0	0	0	0	145	17	8.6	2.6	...	...	...	...
1980	143	0	0	0	0	143	17	8.4	2.5	...	...	...	...
1981	143	0	0	0	0	143	17	8.5	2.5	...	...	...	...
1982	98	0	0	0	0	98	16	6.0	1.8	...	...	...	...
1983	217	0	0	0	0	217	16	13.5	4.1	...	...	...	...
1984	308	0	0	0	0	308	17	18.3	5.6	...	...	...	...
1985	218	0	0	0	0	218	19	11.5	3.5	...	...	...	...
1986	279	0	0	0	0	279	23	12.3	3.7	...	...	...	...
1987	201	0	0	0	0	201	28	7.2	2.2	...	...	...	...
1988	201	0	0	0	0	201	34	6.0	1.9	...	...	...	...
1989	229	0	0	25	0	204	39	5.2	1.6	...	...	...	...
1990	160	0	0	30	0	130	43	3.0	0.9	...	...	...	...
1991	120	0	0	17	0	103	46	2.3	0.7	...	...	...	...
1992	140	0	0	20	0	120	47	2.6	0.8	...	...	...	...
1993	136	0	0	14	0	122	47	2.6	0.8	...	...	...	...
1994	176	0	0	0	0	176	48	3.7	1.1	...	...	...	...
1995	192	0	0	0	0	192	48	4.0	1.2	...	...	...	...

162 NORWAY	- FOOD BALANCE SHEET OF FISH AND FISHERY PRODUCTS IN LIVE WEIGHT AND FISH CONTRIBUTION TO PROTEIN SUPPLY -	FAO-FIDI

YEAR	PRODUCTION	NON-FOOD USES	IMPORTS	EXPORTS	STOCKS VARIATIONS	TOTAL FOOD SUPPLY	POPULATION	PER CAPUT SUPPLY	FISH PROTEINS	ANIMAL PROTEINS	TOTAL PROTEINS	FISH/ANIMAL PROTEINS	FISH/TOTAL PROTEINS
	(..............................		metric tons in live weight		....	)	(thousands)	(kilograms)		(grams per caput per day)		(%)	(%)
1961	1,454,900	712,200	19,565	522,781	-94,574	144,911	3,610	40.1	11.3	52.3	87.4	21.7	13.0
1962	1,250,100	541,000	28,609	602,195	8,782	144,295	3,638	39.7	11.4	52.7	88.7	21.7	12.9
1963	1,331,000	596,700	37,741	627,804	7,719	151,957	3,665	41.5	11.3	51.3	85.4	21.9	13.2
1964	1,544,200	875,500	35,011	513,533	-40,132	150,046	3,694	40.6	11.3	51.7	86.7	21.8	13.0
1965	2,222,100	1,471,800	36,606	648,656	18,943	157,193	3,723	42.2	11.1	52.5	86.9	21.1	12.7
1966	2,810,900	1,984,000	54,759	682,745	-45,164	153,751	3,753	41.0	11.4	53.1	87.9	21.6	13.0
1967	3,189,300	2,391,600	49,444	644,002	-68,101	135,041	3,785	35.7	10.3	51.6	85.7	19.9	12.0
1968	2,731,700	1,973,800	50,692	668,710	14,601	154,483	3,816	40.5	11.9	53.3	86.9	22.4	13.7
1969	2,418,400	1,607,300	57,614	774.017	63,359	158,057	3,847	41.1	11.5	53.3	88.2	21.6	13.1
1970	2,908,770	2,118,300	43,117	745,029	77,140	165,698	3,877	42.7	12.2	54.3	87.4	22.5	14.0
1971	3,000,471	2,130,300	75,600	792,580	-15,928	137,262	3,906	35.1	10.7	52.8	85.6	20.2	12.5
1972	3,121,929	2,225,800	54,353	795,935	13,827	168,374	3,935	42.8	11.5	55.2	87.9	20.9	13.1
1973	2,949,991	2,110,100	56,436	766,495	41,160	170,992	3,961	43.2	12.9	56.6	89.9	22.7	14.3
1974	2,581,978	1,795,373	75,631	648,765	-43,131	170,340	3,985	42.7	10.5	56.7	90.2	18.5	11.6
1975	2,495,425	1,694,749	33,257	702,221	38,814	170,527	4,007	42.6	12.6	59.0	92.1	21.3	13.6
1976	3,364,996	2,548,007	47,228	687,885	-18,033	158,299	4,027	39.3	12.7	59.3	93.9	21.3	13.5
1977	3,407,414	2,570,178	48,518	692,590	-29,116	164,049	4,043	40.6	13.7	60.1	94.8	22.8	14.4
1978	2,593,798	1,808,721	39,805	646,637	-10,911	167,334	4,058	41.2	14.9	63.3	98.1	23.5	15.1
1979	2,658,418	1,862,952	54,463	701,396	21,570	170,103	4,072	41.8	14.5	64.4	100.1	22.5	14.5
1980	2,408,960	1,619,109	68,573	635,206	-45,191	178,028	4,086	43.6	15.2	67.1	104.0	22.6	14.6
1981	2,551,975	1,661,416	89,357	798,003	-7,412	174,501	4,099	42.6	13.9	64.2	103.0	21.6	13.5
1982	2,500,479	1,586,947	67,508	784,742	-16,341	179,958	4,111	43.8	14.2	64.5	98.8	22.0	14.3
1983	2,835,777	1,872,416	56,870	877,952	34,430	176,708	4,124	42.9	14.1	61.9	98.2	22.8	14.4
1984	2,465,828	1,540,842	45,386	802,629	6,559	174,301	4,138	42.1	14.6	63.2	98.9	23.2	14.8
1985	2,119,012	1,214,689	87,308	842,457	31,609	180,783	4,153	43.5	14.5	64.6	100.1	22.5	14.5
1986	1,915,121	979,789	67,659	853,676	31,531	180,847	4,169	43.4	14.7	64.3	101.1	22.8	14.5
1987	1,949,458	907,428	65,622	903,350	-20,043	184,260	4,186	44.0	15.5	64.2	102.4	24.2	15.2
1988	1,839,893	904,908	86,819	839,641	4,436	186,599	4,204	44.4	15.0	59.8	97.7	25.0	15.3
1989	1,908,357	929,457	93,607	884,508	-5,885	182,115	4,223	43.1	14.2	60.2	98.6	23.6	14.4
1990	1,746,700	770,920	121,909	906,310	0	191,379	4,241	45.1	14.3	60.0	97.8	23.8	14.6
1991	2,172,815	1,079,769	198,391	1,098,332	0	193,105	4,260	45.3	13.9	59.5	98.3	23.3	14.1
1992	2,561,481	1,436,740	214,248	1,162,790	20,000	196,194	4,278	45.9	14.6	59.9	98.8	24.3	14.8
1993	2,561,701	1,222,101	252,105	1,386,121	-5,000	200,583	4,297	46.7	15.5	62.2	101.4	25.0	15.3
1994	2,551,182	901,896	225,887	1,684,469	5,000	195,704	4,314	45.4	14.6	62.6	102.6	23.3	14.2
1995	2,807,549	1,002,317	283,979	1,848,197	-22,556	218,459	4,332	50.4	15.5	63.9	102.7	24.2	15.1

221 OMAN	- FOOD BALANCE SHEET OF FISH AND FISHERY PRODUCTS IN LIVE WEIGHT AND FISH CONTRIBUTION TO PROTEIN SUPPLY -												FAO-FIDI
YEAR	PRODUCTION	NON-FOOD USES	IMPORTS	EXPORTS	STOCKS VARIATIONS	TOTAL FOOD SUPPLY	POPULATION	PER CAPUT SUPPLY	FISH PROTEINS	ANIMAL PROTEINS	TOTAL PROTEINS	FISH/ANIMAL PROTEINS	FISH/TOTAL PROTEINS
	(.. metric tons in live weight ....)						(thousands)	(kilograms)	(grams per caput per day)			(%)	(%)
1961	55,000	47,300	0	...	0	7,700	571	13.5	3.2	9.1	19.1	35.4	16.8
1962	60,000	52,000	0	...	0	8,000	586	13.7	3.3	9.2	19.5	36.1	17.0
1963	60,000	51,400	0	...	0	8,600	600	14.3	3.6	9.4	20.0	38.0	17.9
1964	60,000	51,000	0	...	0	9,000	616	14.6	3.7	9.5	20.4	38.9	18.2
1965	65,000	56,000	0	...	0	9,000	631	14.3	3.4	9.3	20.2	37.1	17.0
1966	70,000	60,200	0	...	0	9,800	647	15.2	3.8	9.5	20.8	39.6	18.1
1967	70,000	59,700	0	...	0	10,300	663	15.5	3.9	9.8	21.5	40.4	18.4
1968	70,000	59,000	0	...	0	11,000	681	16.2	4.2	10.0	21.9	41.9	19.2
1969	70,000	58,200	104	...	0	11,904	701	17.0	4.5	10.4	22.7	43.4	19.9
1970	180,000	167,032	0	...	0	12,968	723	17.9	3.8	9.8	23.4	39.0	16.4
1971	180,000	166,524	154	...	0	13,629	748	18.2	4.1	10.4	24.1	39.7	17.0
1972	180,000	165,803	14	...	0	14,212	776	18.3	4.0	10.6	26.0	38.1	15.5
1973	180,000	165,803	564	...	0	14,762	807	18.3	4.1	11.9	27.4	34.1	14.8
1974	180,000	165,803	910	...	0	15,108	841	18.0	4.0	13.3	31.0	30.2	13.0
1975	198,850	183,822	0	...	0	15,028	880	17.1	3.8	14.4	31.8	26.3	11.9
1976	197,984	182,844	1,754	...	0	16,894	924	18.3	4.4	16.3	34.4	27.0	12.8
1977	70,000	53,500	1,818	300	0	18,018	971	18.6	4.3	16.2	36.4	26.6	11.8
1978	73,000	53,628	1,654	2,505	0	18,521	1,022	18.1	4.1	16.7	38.5	24.8	10.8
1979	76,000	53,848	1,562	5,782	0	17,932	1,076	16.7	3.8	18.3	40.9	20.5	9.2
1980	79,000	51,922	1,317	8,132	0	20,263	1,130	17.9	4.1	21.0	49.4	19.4	8.2
1981	83,650	54,560	2,423	9,568	0	21,945	1,186	18.5	4.3	22.6	53.7	18.9	7.9
1982	89,376	55,020	2,327	13,422	0	23,261	1,242	18.7	4.1	25.6	54.9	16.1	7.5
1983	108,766	70,448	1,501	15,211	0	24,608	1,301	18.9	4.0	28.0	57.5	14.4	7.0
1984	105,200	63,236	2,534	18,876	0	25,623	1,361	18.8	4.3	28.4	55.6	15.2	7.8
1985	101,277	55,228	2,640	20,874	0	27,815	1,425	19.5	4.6	28.7	58.8	15.9	7.8
1986	96,464	47,932	2,476	20,524	0	30,484	1,492	20.4	4.8	29.3	68.2	16.4	7.1
1987	136,157	83,540	1,810	23,291	0	31,136	1,561	19.9	4.8	26.8	63.3	17.9	7.6
1988	165,676	92,006	1,793	42,100	0	33,363	1,634	20.4	4.9	27.4	61.2	18.0	8.0
1989	117,703	51,125	1,405	33,226	0	34,757	1,708	20.4	4.5	26.7	56.0	16.8	8.0
1990	123,092	47,053	1,906	38,841	0	39,104	1,785	21.9	5.1	25.9	56.0	19.6	9.1
1991	117,780	46,726	4,201	35,019	0	40,236	1,864	21.6	4.9	26.2	54.0	18.7	9.1
1992	112,313	37,765	4,575	34,713	-10	44,400	1,945	22.8	5.3	26.7	57.1	19.8	9.3
1993	116,470	26,761	6,934	51,780	0	44,863	2,029	22.1	5.0	26.6	57.4	18.9	8.8
1994	118,568	25,965	6,135	51,355	1	47,384	2,116	22.4	5.3	26.2	55.5	20.3	9.6
1995	139,864	24,221	5,467	69,065	9	52,054	2,207	23.6	5.4	26.5	53.4	20.4	10.1

165 PAKISTAN	- FOOD BALANCE SHEET OF FISH AND FISHERY PRODUCTS IN LIVE WEIGHT AND FISH CONTRIBUTION TO PROTEIN SUPPLY -	FAO-FIDI

YEAR	PRODUCTION	NON-FOOD USES	IMPORTS	EXPORTS	STOCKS VARIATIONS	TOTAL FOOD SUPPLY	POPULATION	PER CAPUT SUPPLY	FISH PROTEINS	ANIMAL PROTEINS	TOTAL PROTEINS	FISH/ANIMAL PROTEINS	FISH/TOTAL PROTEINS
	(.................. metric tons in live weight ....)						(thousands)	(kilograms)	(grams per caput per day)			(%)	(%)
1961	80,100	6,300	0	20,507	0	53,293	51,273	1.0	0.3	13.4	49.6	2.3	0.6
1962	87,300	3,300	0	32,030	0	51,970	52,650	1.0	0.3	13.3	49.9	2.2	0.6
1963	93,500	2,600	0	28,307	0	62,593	54,086	1.2	0.4	13.4	49.6	2.9	0.8
1964	109,000	2,700	0	29,597	0	76,703	55,584	1.4	0.5	13.6	53.3	3.4	0.9
1965	118,600	3,500	0	35,583	0	79,517	57,145	1.4	0.5	13.5	54.8	3.4	0.9
1966	151,400	14,400	0	44,218	0	92,782	58,772	1.6	0.5	13.3	51.3	4.1	1.1
1967	150,000	17,300	0	42,332	0	90,368	60,461	1.5	0.5	13.2	51.4	3.9	1.0
1968	156,700	38,500	0	31,018	0	87,182	62,195	1.4	0.5	13.2	49.2	3.7	1.0
1969	176,200	39,800	0	37,381	0	99,019	63,948	1.6	0.5	13.3	55.7	4.1	1.0
1970	176,000	55,000	794	45,879	0	75,915	65,706	1.2	0.4	13.1	55.5	2.8	0.7
1971	174,600	59,700	400	39,126	0	76,174	67,471	1.1	0.4	13.1	53.7	2.7	0.7
1972	211,300	98,500	153	44,785	0	68,168	69,252	1.0	0.3	13.0	55.7	2.3	0.5
1973	235,700	116,400	411	47,200	0	72,510	71,053	1.0	0.3	13.0	56.1	2.4	0.6
1974	189,502	75,500	608	37,094	0	77,516	72,877	1.1	0.3	13.1	56.0	2.5	0.6
1975	195,039	68,876	0	40,507	0	85,656	74,734	1.2	0.4	13.1	53.5	2.7	0.7
1976	205,659	55,000	0	34,906	0	115,753	76,616	1.5	0.5	13.4	54.4	3.4	0.8
1977	267,954	86,500	0	28,737	0	152,717	78,539	1.9	0.6	13.6	53.4	4.3	1.1
1978	293,029	100,200	0	25,255	0	167,574	80,573	2.1	0.6	13.9	57.4	4.6	1.1
1979	300,390	138,500	4	27,228	0	134,666	82,805	1.6	0.5	13.6	55.0	3.8	0.9
1980	279,263	129,850	3	23,901	0	125,515	85,299	1.5	0.5	13.7	51.7	3.3	0.9
1981	317,849	131,605	43	24,859	0	161,428	88,075	1.8	0.6	13.8	51.2	4.2	1.1
1982	337,289	148,225	16	27,008	0	162,072	91,108	1.8	0.6	13.7	52.6	4.1	1.1
1983	343,402	152,472	3	37,126	0	153,807	94,351	1.6	0.5	13.8	52.9	3.8	1.0
1984	372,264	163,278	24	46,504	0	162,506	97,733	1.7	0.6	14.0	53.5	3.9	1.0
1985	408,404	165,165	10	57,026	0	186,223	101,196	1.8	0.6	14.4	52.4	4.0	1.1
1986	415,732	172,115	0	58,280	0	185,337	104,735	1.8	0.6	15.0	52.7	3.7	1.1
1987	427,736	170,090	25	63,545	0	194,126	108,350	1.8	0.6	15.3	52.2	3.6	1.1
1988	445,495	164,642	9	80,916	0	199,946	111,992	1.8	0.5	15.8	52.1	3.4	1.0
1989	446,289	166,377	122	81,258	0	198,776	115,601	1.7	0.5	16.1	58.5	3.2	0.9
1990	479,093	172,857	57	83,986	0	222,307	119,141	1.9	0.6	16.4	59.0	3.4	1.0
1991	515,554	177,391	5	91,110	-2,300	244,758	122,582	2.0	0.6	16.7	57.5	3.5	1.0
1992	553,176	185,711	40	103,171	-5,320	259,014	125,943	2.1	0.6	17.1	59.4	3.4	1.0
1993	621,753	188,405	30	120,401	-1,528	311,449	129,285	2.4	0.7	17.7	60.7	3.9	1.1
1994	551,961	189,202	13	105,890	9,318	266,200	132,701	2.0	0.6	18.3	59.7	3.1	1.0
1995	541,917	185,488	2	88,319	354	268,466	136,257	2.0	0.6	18.7	57.7	3.1	1.0

180 PALAU	- FOOD BALANCE SHEET OF FISH AND FISHERY PRODUCTS IN LIVE WEIGHT AND FISH CONTRIBUTION TO PROTEIN SUPPLY -													FAO-FIDI
YEAR	PRODUCTION	NON-FOOD USES	IMPORTS	EXPORTS	STOCKS VARIATIONS	TOTAL FOOD SUPPLY	POPULATION	PER CAPUT SUPPLY	FISH PROTEINS	ANIMAL PROTEINS	TOTAL PROTEINS	FISH/ANIMAL PROTEINS	FISH/TOTAL PROTEINS
	(.. metric tons in live weight ....)						(thousands)	(kilograms)	(grams per caput per day)			(%)	(%)
1961	500	0	0	0	0	500	7	69.6	19.6	...	...	...	...
1962	600	0	0	0	0	600	7	81.7	23.1	...	...	...	...
1963	600	0	0	0	0	600	8	79.9	22.5	...	...	...	...
1964	800	0	0	0	0	800	8	103.6	29.2	...	...	...	...
1965	800	0	0	0	0	800	8	100.0	28.2	...	...	...	...
1966	800	0	0	0	0	800	8	95.7	27.0	...	...	...	...
1967	1,000	0	0	0	0	1,000	9	113.9	32.9	...	...	...	...
1968	1,000	0	0	0	0	1,000	9	108.4	31.3	...	...	...	...
1969	1,000	0	0	0	0	1,000	10	103.7	29.9	...	...	...	...
1970	1,100	0	0	0	0	1,100	10	110.0	31.7	...	...	...	...
1971	1,100	0	0	0	0	1,100	10	107.0	30.8	...	...	...	...
1972	1,100	0	0	0	0	1,100	11	104.7	30.2	...	...	...	...
1973	1,100	0	0	0	0	1,100	11	103.0	29.7	...	...	...	...
1974	1,100	0	0	0	0	1,100	11	101.5	29.2	...	...	...	...
1975	1,100	0	0	0	0	1,100	11	100.0	28.8	...	...	...	...
1976	1,100	0	0	0	0	1,100	11	98.6	28.4	...	...	...	...
1977	1,100	0	0	0	0	1,100	11	97.2	28.0	...	...	...	...
1978	1,100	0	0	0	0	1,100	11	95.7	27.6	...	...	...	...
1979	1,100	0	0	0	0	1,100	12	93.9	27.0	...	...	...	...
1980	1,050	0	0	0	0	1,050	12	87.5	25.0	...	...	...	...
1981	1,050	0	0	415	0	635	12	51.4	14.8	...	...	...	...
1982	1,050	0	0	250	0	800	13	62.6	17.9	...	...	...	...
1983	1,291	0	0	0	0	1,291	13	97.5	27.9	...	...	...	...
1984	1,287	0	0	0	0	1,287	14	94.3	26.7	...	...	...	...
1985	1,407	0	0	0	0	1,407	14	100.5	28.8	...	...	...	...
1986	1,441	0	0	89	0	1,352	14	94.9	27.3	...	...	...	...
1987	1,511	0	0	50	0	1,461	14	101.3	29.3	...	...	...	...
1988	1,507	0	0	60	0	1,447	15	99.3	28.7	...	...	...	...
1989	1,427	0	0	156	0	1,271	15	86.2	24.6	...	...	...	...
1990	1,438	0	0	161	0	1,277	15	85.1	24.4	...	...	...	...
1991	1,362	2	0	0	0	1,360	15	88.6	25.0	...	...	...	...
1992	1,437	2	0	0	0	1,435	16	91.0	26.0	...	...	...	...
1993	1,490	2	123	0	0	1,611	16	99.3	28.8	...	...	...	...
1994	1,500	3	125	0	0	1,622	17	97.4	28.3	...	...	...	...
1995	1,450	2	116	32	0	1,532	17	90.1	25.9	...	...	...	...

166 PANAMA	- FOOD BALANCE SHEET OF FISH AND FISHERY PRODUCTS IN LIVE WEIGHT AND FISH CONTRIBUTION TO PROTEIN SUPPLY -												FAO-FIDI
YEAR	PRODUCTION	NON-FOOD USES	IMPORTS	EXPORTS	STOCKS VARIATIONS	TOTAL FOOD SUPPLY	POPULATION	PER CAPUT SUPPLY	FISH PROTEINS	ANIMAL PROTEINS	TOTAL PROTEINS	FISH/ANIMAL PROTEINS	FISH/TOTAL PROTEINS
	(...... metric tons in live weight ......)						(thousands)	(kilograms)	(grams per caput per day)			(%)	(%)
1961	11,400	6,200	9,018	6,535	0	7,683	1,159	6.6	2.5	21.2	53.4	11.6	4.6
1962	15,200	8,200	10,121	7,655	0	9,466	1,193	7.9	2.8	22.1	53.0	12.5	5.2
1963	14,000	7,100	9,418	7,228	0	9,090	1,229	7.4	2.5	22.4	55.3	11.1	4.5
1964	25,800	18,500	12,321	10,913	0	8,709	1,265	6.9	2.3	21.9	54.3	10.6	4.3
1965	39,600	33,300	12,221	8,445	0	10,076	1,303	7.7	2.7	22.9	52.7	11.6	5.1
1966	73,000	62,500	7,018	5,600	0	11,918	1,341	8.9	2.8	23.5	56.5	11.8	4.9
1967	72,200	62,000	5,321	6,500	0	9,021	1,381	6.5	2.1	23.0	56.3	9.1	3.7
1968	71,800	63,000	6,036	6,000	0	8,836	1,422	6.2	2.0	23.2	56.3	8.5	3.5
1969	36,700	24,100	6,546	5,600	0	13,546	1,464	9.3	2.8	25.2	59.3	11.2	4.7
1970	58,060	39,900	5,658	8,333	0	15,484	1,506	10.3	3.3	25.7	57.3	12.8	5.7
1971	78,640	60,800	6,297	8,481	0	15,656	1,549	10.1	3.2	27.7	60.1	11.6	5.4
1972	70,056	49,400	5,264	8,100	0	17,820	1,592	11.2	3.5	28.4	54.8	12.3	6.4
1973	114,829	89,300	3,966	8,663	0	20,832	1,635	12.7	4.2	28.7	56.4	14.7	7.5
1974	89,412	58,659	3,582	7,889	0	26,446	1,679	15.8	5.3	30.6	61.2	17.2	8.6
1975	117,177	83,530	5,027	8,055	0	30,619	1,723	17.8	5.9	31.9	63.1	18.7	9.4
1976	185,303	148,806	4,030	10,088	0	30,439	1,768	17.2	5.8	33.0	57.9	17.7	10.1
1977	240,053	202,258	3,798	10,363	0	31,230	1,814	17.2	5.9	32.6	62.9	18.0	9.3
1978	140,019	103,041	4,854	10,309	0	31,522	1,860	17.0	5.8	31.3	56.9	18.4	10.1
1979	166,060	130,563	6,332	10,092	-443	31,294	1,905	16.4	5.4	30.7	55.9	17.6	9.7
1980	217,747	183,569	7,056	10,105	376	31,505	1,950	16.2	5.4	33.1	57.1	16.3	9.4
1981	143,848	116,586	6,773	13,095	0	20,940	1,994	10.5	3.0	29.9	58.2	10.2	5.2
1982	113,530	76,944	7,147	14,017	0	29,716	2,037	14.6	4.8	33.2	61.4	14.4	7.8
1983	168,298	139,118	8,559	14,283	0	23,456	2,080	11.3	3.6	33.5	62.1	10.8	5.8
1984	138,381	107,625	8,879	13,074	0	26,561	2,123	12.5	3.9	33.3	61.8	11.6	6.3
1985	291,563	245,180	8,724	18,270	-35	36,801	2,167	17.0	5.4	34.1	66.6	15.8	8.1
1986	134,374	90,874	7,603	19,663	0	31,440	2,212	14.2	3.8	33.6	65.9	11.2	5.7
1987	154,163	131,452	9,401	13,195	0	18,917	2,258	8.4	1.9	30.7	58.6	6.3	3.3
1988	123,682	87,511	7,439	13,518	0	30,090	2,304	13.1	4.0	30.2	58.6	13.1	6.8
1989	185,214	144,218	10,385	15,824	0	35,597	2,351	15.1	4.6	31.2	60.7	14.9	7.6
1990	137,718	97,835	9,660	13,160	0	36,383	2,398	15.2	4.4	31.8	61.8	14.0	7.2
1991	150,236	102,190	9,393	14,396	-5	43,038	2,444	17.6	5.3	33.7	61.6	15.7	8.6
1992	156,329	104,863	10,900	13,300	40	49,106	2,491	19.7	5.8	34.7	66.8	16.7	8.7
1993	165,299	118,747	10,180	16,388	0	40,344	2,538	15.9	4.8	33.9	65.0	14.1	7.3
1994	169,844	116,058	10,419	17,751	-100	46,353	2,585	17.9	5.4	35.0	67.0	15.4	8.0
1995	181,781	130,783	10,625	21,285	100	40,438	2,631	15.4	4.5	34.4	68.1	13.2	6.7

168 PAPUA NEW GUINEA	- FOOD BALANCE SHEET OF FISH AND FISHERY PRODUCTS IN LIVE WEIGHT AND FISH CONTRIBUTION TO PROTEIN SUPPLY -												FAO-FIDI
YEAR	PRODUCTION	NON-FOOD USES	IMPORTS	EXPORTS	STOCKS VARIATIONS	TOTAL FOOD SUPPLY	POPULATION	PER CAPUT SUPPLY	FISH PROTEINS	ANIMAL PROTEINS	TOTAL PROTEINS	FISH/ANIMAL PROTEINS	FISH/TOTAL PROTEINS
	(..		metric tons in live weight		....	)	(thousands)	(kilograms)		(grams per caput per day)		(%)	(%)
1961	12,000	0	7,400	0	0	19,400	1,962	9.9	2.9	10.2	31.8	27.8	9.0
1962	13,000	0	10,400	0	0	23,400	2,005	11.7	3.4	10.7	32.5	31.3	10.3
1963	15,000	0	8,600	0	0	23,600	2,051	11.5	3.3	10.8	32.6	30.7	10.2
1964	15,000	0	11,000	0	0	26,000	2,099	12.4	3.6	11.3	33.4	31.5	10.7
1965	15,500	0	12,500	0	0	28,000	2,148	13.0	3.7	11.8	34.7	31.6	10.8
1966	15,000	0	13,500	0	0	28,500	2,200	13.0	3.7	12.1	35.4	31.0	10.6
1967	15,000	0	17,100	0	0	32,100	2,253	14.3	4.1	12.3	35.6	33.2	11.5
1968	15,400	0	18,200	300	0	33,300	2,308	14.4	4.2	12.8	36.6	32.5	11.4
1969	16,100	0	20,400	200	0	36,300	2,365	15.4	4.4	13.2	37.0	33.3	11.9
1970	15,700	0	25,000	1,500	0	39,200	2,422	16.2	4.7	13.8	38.7	33.6	12.0
1971	30,700	0	30,400	9,500	0	51,600	2,480	20.8	6.1	15.0	40.3	40.8	15.2
1972	27,200	0	30,500	14,150	0	43,550	2,540	17.2	4.9	15.2	40.3	32.2	12.2
1973	45,400	0	30,800	12,850	-10,000	53,350	2,600	20.5	6.0	16.0	41.2	37.5	14.6
1974	49,963	0	44,156	38,100	0	56,019	2,663	21.0	6.1	15.9	40.1	38.6	15.3
1975	34,938	0	22,485	16,750	5,000	45,673	2,729	16.7	4.9	15.2	40.5	32.4	12.1
1976	50,640	0	44,531	20,412	-28,000	46,759	2,797	16.7	4.8	15.1	39.9	31.6	12.0
1977	26,648	0	24,000	25,455	16,667	41,860	2,868	14.6	4.0	14.8	40.8	26.9	9.8
1978	52,199	0	34,000	46,937	5,000	44,262	2,941	15.1	4.5	14.3	40.9	31.5	11.0
1979	34,562	0	55,800	28,521	0	61,842	3,014	20.5	5.9	16.0	43.2	36.9	13.6
1980	43,617	0	56,122	34,474	-303	64,962	3,086	21.1	6.0	16.5	44.9	36.3	13.3
1981	44,886	0	67,932	31,132	0	81,687	3,157	25.9	7.4	17.7	45.1	41.6	16.3
1982	17,694	0	53,556	4,189	3,333	70,395	3,227	21.8	6.2	16.3	44.0	38.0	14.1
1983	15,532	0	56,483	1,779	1,111	71,347	3,298	21.6	6.3	17.0	43.9	37.1	14.4
1984	17,764	0	58,168	3,959	3,235	75,208	3,369	22.3	6.4	17.1	43.9	37.4	14.6
1985	25,470	0	59,882	9,806	5,917	81,463	3,442	23.7	7.1	17.9	47.3	39.7	15.1
1986	25,311	0	69,127	2,220	303	92,522	3,517	26.3	7.5	19.1	46.8	39.2	16.0
1987	25,533	0	69,444	2,024	82	93,034	3,594	25.9	7.4	18.8	46.4	39.2	15.9
1988	25,971	0	69,031	1,931	-344	92,727	3,673	25.3	7.2	18.5	46.7	38.8	15.4
1989	26,039	0	67,094	2,018	-238	90,877	3,755	24.2	6.9	18.2	46.2	37.9	14.9
1990	26,188	0	65,067	2,707	-100	88,448	3,839	23.0	6.6	17.7	45.6	37.1	14.4
1991	24,992	0	43,197	3,298	482	65,373	3,926	16.7	5.0	16.0	44.1	31.3	11.4
1992	25,413	0	40,467	2,961	300	63,218	4,017	15.7	4.7	16.3	44.9	28.9	10.5
1993	25,747	0	45,897	1,938	-289	69,417	4,109	16.9	5.0	16.9	46.3	29.4	10.8
1994	27,017	0	41,516	3,770	222	64,985	4,204	15.5	4.6	17.1	47.1	26.8	9.7
1995	26,021	0	26,883	4,392	67	47,939	4,301	11.2	3.2	14.6	45.2	22.2	7.2

169 PARAGUAY	- FOOD BALANCE SHEET OF FISH AND FISHERY PRODUCTS IN LIVE WEIGHT AND FISH CONTRIBUTION TO PROTEIN SUPPLY -												FAO-FIDI
YEAR	PRODUCTION	NON-FOOD USES	IMPORTS	EXPORTS	STOCKS VARIATIONS	TOTAL FOOD SUPPLY	POPULATION	PER CAPUT SUPPLY	FISH PROTEINS	ANIMAL PROTEINS	TOTAL PROTEINS	FISH/ANIMAL PROTEINS	FISH/TOTAL PROTEINS
	(.......................... metric tons in live weight ....)						(thousands)	(kilograms)	(grams per caput per day)			(%)	(%)
1961	600	0	0	0	0	600	1,886	0.3	0.1	38.2	69.5	0.3	0.1
1962	600	0	0	0	0	600	1,932	0.3	0.1	35.4	67.6	0.3	0.1
1963	700	0	0	0	0	700	1,980	0.4	0.1	32.6	65.5	0.3	0.2
1964	700	0	0	0	0	700	2,030	0.3	0.1	33.2	66.5	0.3	0.2
1965	700	0	0	0	0	700	2,081	0.3	0.1	33.5	68.9	0.3	0.2
1966	700	0	0	0	0	700	2,133	0.3	0.1	35.2	70.5	0.3	0.1
1967	800	0	0	0	0	800	2,186	0.4	0.1	33.8	70.4	0.3	0.2
1968	1,000	0	0	0	0	1,000	2,240	0.5	0.1	35.0	73.0	0.4	0.2
1969	1,400	0	0	0	0	1,400	2,295	0.6	0.2	33.7	71.8	0.5	0.3
1970	1,800	0	0	0	0	1,800	2,350	0.8	0.2	34.1	73.5	0.7	0.3
1971	2,200	0	0	0	0	2,200	2,406	0.9	0.3	33.4	73.0	0.8	0.4
1972	2,500	0	0	0	0	2,500	2,463	1.0	0.3	32.3	70.6	0.9	0.4
1973	2,700	0	0	0	0	2,700	2,522	1.1	0.3	29.2	68.8	1.1	0.5
1974	2,700	0	0	0	0	2,700	2,587	1.0	0.3	31.3	70.4	1.0	0.4
1975	2,800	0	0	0	0	2,800	2,659	1.1	0.3	28.3	67.0	1.1	0.5
1976	2,900	0	0	0	0	2,900	2,740	1.1	0.3	30.2	70.7	1.1	0.5
1977	3,000	0	0	0	0	3,000	2,828	1.1	0.3	32.8	73.7	1.0	0.4
1978	3,100	0	0	0	0	3,100	2,922	1.1	0.3	31.9	74.0	1.0	0.4
1979	3,200	0	70	153	0	3,118	3,018	1.0	0.3	34.1	73.0	0.9	0.4
1980	3,300	0	432	26	0	3,706	3,114	1.2	0.4	34.6	72.3	1.0	0.5
1981	3,350	0	434	26	0	3,757	3,209	1.2	0.4	34.7	72.9	1.0	0.5
1982	3,400	0	294	23	0	3,671	3,304	1.1	0.3	34.0	73.3	1.0	0.5
1983	3,500	0	387	11	0	3,876	3,400	1.1	0.3	33.3	72.6	1.0	0.5
1984	5,000	0	495	0	0	5,495	3,501	1.6	0.5	33.9	71.8	1.4	0.7
1985	7,500	0	361	0	0	7,861	3,609	2.2	0.7	34.1	70.2	1.9	0.9
1986	13,000	0	173	5	0	13,168	3,723	3.5	1.1	30.3	67.2	3.5	1.6
1987	10,022	0	284	26	0	10,280	3,844	2.7	0.8	30.7	67.6	2.6	1.2
1988	10,054	0	412	28	0	10,438	3,969	2.6	0.8	33.2	67.8	2.4	1.2
1989	11,051	0	673	16	0	11,708	4,095	2.9	0.8	31.7	68.5	2.6	1.2
1990	12,551	0	602	16	0	13,137	4,219	3.1	0.9	31.1	66.1	3.0	1.4
1991	13,050	0	1,221	73	0	14,198	4,341	3.3	1.0	38.7	69.9	2.5	1.4
1992	18,060	0	1,039	14	0	19,085	4,461	4.3	1.3	42.0	73.3	3.0	1.7
1993	16,060	0	1,917	25	0	17,952	4,581	3.9	1.2	40.3	72.8	2.9	1.6
1994	17,000	0	2,576	19	0	19,557	4,703	4.2	1.2	41.6	75.2	2.9	1.6
1995	18,000	0	1,011	10	0	19,001	4,828	3.9	1.2	42.8	76.2	2.7	1.5

170 PERU	- FOOD BALANCE SHEET OF FISH AND FISHERY PRODUCTS IN LIVE WEIGHT AND FISH CONTRIBUTION TO PROTEIN SUPPLY -												FAO-FIDI
YEAR	PRODUCTION	NON-FOOD USES	IMPORTS	EXPORTS	STOCKS VARIATIONS	TOTAL FOOD SUPPLY	POPULATION	PER CAPUT SUPPLY	FISH PROTEINS	ANIMAL PROTEINS	TOTAL PROTEINS	FISH/ANIMAL PROTEINS	FISH/TOTAL PROTEINS
	(.. metric tons in live weight ....)						(thousands)	(kilograms)	(grams per caput per day)			(%)	(%)
1961	5,216,000	5,012,100	1,123	73,361	3,897	135,559	10,218	13.3	4.0	18.1	58.4	22.2	6.9
1962	6,883,800	6,691,200	1,203	70,969	13,575	136,409	10,516	13.0	3.9	18.1	58.5	21.5	6.7
1963	6,821,500	6,635,300	1,577	62,646	-4,494	120,638	10,826	11.1	3.2	17.5	54.6	18.3	5.9
1964	9,037,200	8,894,500	940	35,823	4,853	112,670	11,143	10.1	3.2	18.6	58.3	17.1	5.4
1965	7,384,900	7,260,700	1,352	33,867	1,402	93,087	11,467	8.1	2.3	18.0	54.6	13.0	4.3
1966	8,712,000	8,555,100	2,218	46,549	13,836	126,406	11,797	10.7	3.1	19.3	55.0	16.2	5.7
1967	10,056,700	9,858,800	1,846	31,594	0	168,153	12,133	13.9	3.7	19.4	56.9	18.9	6.4
1968	10,443,400	10,284,900	190	23,031	0	135,659	12,476	10.9	3.0	18.5	55.4	16.3	5.4
1969	9,146,500	8,976,900	0	28,504	0	141,096	12,829	11.0	3.0	18.6	61.2	16.0	4.9
1970	12,483,200	12,295,700	133	28,078	-3,000	156,555	13,193	11.9	3.2	19.3	58.2	16.7	5.5
1971	10,506,100	10,298,800	100	34,407	-1,813	171,181	13,567	12.6	3.3	19.5	58.9	16.7	5.5
1972	4,678,200	4,462,200	240	31,593	0	184,647	13,951	13.2	3.5	19.0	58.2	18.3	6.0
1973	2,295,500	1,899,420	691	52,895	-1,500	342,375	14,345	23.9	6.1	21.7	58.7	28.2	10.4
1974	4,125,603	3,791,477	1,229	76,504	-2,100	256,751	14,748	17.4	4.2	19.8	56.8	21.2	7.4
1975	3,414,975	3,117,000	790	65,322	-7,500	225,943	15,161	14.9	3.5	19.3	57.9	17.9	6.0
1976	4,343,096	4,003,982	788	96,226	1,441	245,117	15,583	15.7	3.4	19.3	56.0	17.7	6.1
1977	2,503,157	2,017,413	811	183,889	-2,180	300,486	16,012	18.8	4.3	19.9	55.2	21.5	7.7
1978	3,442,829	2,811,490	75	285,385	681	346,710	16,448	21.1	4.5	18.8	55.0	23.9	8.2
1979	3,652,096	2,881,722	1,600	242,282	-71,408	458,284	16,886	27.1	5.7	19.5	56.1	29.1	10.1
1980	2,708,612	1,726,208	2,513	527,874	8,994	466,037	17,324	26.9	5.2	19.3	57.6	26.7	9.0
1981	2,717,160	1,850,316	1,384	524,430	20,463	364,261	17,762	20.5	4.3	19.6	57.1	22.1	7.6
1982	3,513,245	2,946,026	887	292,629	12,225	287,702	18,198	15.8	3.6	19.5	56.8	18.2	6.3
1983	1,569,496	1,227,729	626	140,455	2,518	204,457	18,632	11.0	2.8	18.3	51.2	15.2	5.4
1984	3,319,884	2,740,782	2,322	214,043	-19,685	347,696	19,064	18.2	4.5	19.1	55.4	23.5	8.1
1985	4,138,053	3,597,806	832	243,156	39,375	337,297	19,492	17.3	4.1	18.6	50.4	22.3	8.2
1986	5,616,326	4,995,724	44	80,539	-3,630	536,477	19,918	26.9	6.4	23.7	58.7	26.9	10.9
1987	4,587,586	3,969,385	12	128,018	9,729	499,925	20,339	24.6	5.8	23.9	58.6	24.3	9.9
1988	6,641,844	6,059,998	40	116,701	5,843	471,028	20,756	22.7	5.5	23.0	59.1	24.1	9.4
1989	6,853,937	6,149,684	0	111,751	-6,825	585,677	21,166	27.7	6.4	20.5	52.0	31.4	12.4
1990	6,874,131	6,135,451	166	78,064	-64,493	596,288	21,569	27.7	6.5	20.7	50.2	31.4	13.0
1991	6,904,972	6,479,577	2,028	97,566	39,963	369,825	21,964	16.8	3.9	18.8	50.1	20.5	7.7
1992	7,507,570	7,072,806	1,790	65,368	14,137	385,352	22,353	17.2	4.3	19.3	47.8	22.5	9.1
1993	9,008,371	8,497,528	2,552	76,519	235	437,111	22,739	19.2	4.5	19.4	51.5	23.4	8.8
1994	12,005,120	11,399,050	2,457	150,496	28,696	486,730	23,131	21.0	5.0	20.9	56.3	24.1	9.0
1995	8,943,111	8,204,100	4,087	88,408	-11,971	642,718	23,532	27.3	5.9	22.8	59.1	25.9	10.0

171 PHILIPPINES	- FOOD BALANCE SHEET OF FISH AND FISHERY PRODUCTS IN LIVE WEIGHT AND FISH CONTRIBUTION TO PROTEIN SUPPLY -												FAO-FIDI
YEAR	PRODUCTION	NON-FOOD USES	IMPORTS	EXPORTS	STOCKS VARIATIONS	TOTAL FOOD SUPPLY	POPULATION	PER CAPUT SUPPLY	FISH PROTEINS	ANIMAL PROTEINS	TOTAL PROTEINS	FISH/ANIMAL PROTEINS	FISH/TOTAL PROTEINS
	(............ metric tons in live weight ....)						(thousands)	(kilograms)	(grams per caput per day)			(%)	(%)
1961	475,000	0	129,582	607	0	603,975	28,380	21.3	8.1	14.7	40.2	55.4	20.3
1962	504,300	0	63,829	175	0	567,954	29,228	19.4	7.5	14.6	38.9	51.6	19.3
1963	564,800	0	91,208	254	0	655,754	30,113	21.8	8.0	15.1	40.4	52.8	19.7
1964	622,600	0	79,320	654	0	701,267	31,045	22.6	8.2	15.6	40.8	52.8	20.2
1965	684,800	0	84,103	654	0	768,249	32,030	24.0	8.7	16.5	42.8	52.5	20.3
1966	722,700	0	82,939	1,453	0	804,186	33,071	24.3	8.7	16.9	42.0	51.3	20.7
1967	763,300	0	106,741	1,554	0	868,487	34,160	25.4	9.1	17.2	42.3	52.8	21.5
1968	940,100	0	133,138	1,401	0	1,071,836	35,282	30.4	10.7	19.6	44.0	54.5	24.3
1969	974,500	0	119,769	1,938	0	1,092,332	36,414	30.0	10.5	19.7	43.9	53.3	23.9
1970	1,030,800	0	105,500	2,939	-50	1,133,311	37,540	30.2	11.2	19.4	43.3	57.9	25.9
1971	1,115,400	0	116,900	6,533	-100	1,225,667	38,656	31.7	11.8	20.0	44.6	59.3	26.5
1972	1,220,100	0	104,100	10,233	-100	1,313,867	39,764	33.0	12.4	20.3	46.3	61.3	26.9
1973	1,303,400	0	79,200	15,310	-100	1,367,190	40,861	33.5	11.7	20.0	44.8	58.5	26.2
1974	1,370,958	0	93,700	17,500	-130	1,447,028	41,943	34.5	12.1	21.7	47.3	55.7	25.6
1975	1,442,981	0	115,467	14,432	-130	1,543,886	43,010	35.9	12.7	20.7	47.9	61.3	26.5
1976	1,393,420	0	64,890	17,306	30	1,441,033	44,054	32.7	11.5	19.7	47.8	58.2	24.1
1977	1,508,700	0	38,142	29,379	-50	1,517,413	45,081	33.7	12.2	20.9	48.8	58.3	25.0
1978	1,494,564	0	42,974	46,810	400	1,491,128	46,113	32.3	11.4	20.0	48.2	57.0	23.7
1979	1,475,032	0	32,646	60,930	230	1,446,978	47,185	30.7	11.1	20.8	49.5	53.5	22.5
1980	1,555,656	0	41,396	84,466	-450	1,512,136	48,317	31.3	11.2	21.0	51.6	53.7	21.8
1981	1,686,054	0	43,775	79,591	-10,371	1,639,867	49,522	33.1	11.9	21.3	51.8	56.1	23.0
1982	1,786,790	0	59,203	62,564	-14,186	1,769,243	50,791	34.8	12.2	21.6	52.9	56.7	23.1
1983	1,976,141	0	11,805	72,172	850	1,916,624	52,097	36.8	13.0	22.2	51.3	58.4	25.3
1984	1,934,206	0	1,452	70,442	8,143	1,873,359	53,398	35.1	12.4	21.0	50.5	59.2	24.6
1985	1,865,490	0	6,088	83,283	3,157	1,791,452	54,668	32.8	11.8	19.2	49.3	61.3	23.9
1986	1,916,347	0	36,928	88,277	2,870	1,867,868	55,895	33.4	12.0	20.0	50.7	60.1	23.7
1987	1,988,718	0	75,469	102,974	1,445	1,962,658	57,091	34.4	12.2	21.0	50.9	58.3	24.0
1988	2,010,363	0	131,073	128,632	0	2,012,805	58,282	34.5	12.4	21.5	53.0	57.6	23.4
1989	2,098,787	0	158,258	145,909	-432	2,110,704	59,502	35.5	13.0	22.8	55.2	56.9	23.5
1990	2,209,602	0	153,651	143,604	436	2,220,085	60,779	36.5	13.3	23.8	56.2	55.7	23.6
1991	2,316,495	0	169,926	146,148	2,160	2,342,433	62,119	37.7	13.7	23.9	56.4	57.2	24.3
1992	2,271,953	0	167,488	133,938	2,162	2,307,666	63,512	36.3	13.1	23.5	55.5	55.9	23.6
1993	2,263,792	0	133,847	173,021	4,405	2,229,022	64,943	34.3	12.7	24.8	56.4	51.3	22.6
1994	2,276,202	0	141,266	178,505	0	2,238,963	66,391	33.7	12.4	25.1	57.6	49.4	21.6
1995	2,269,231	0	158,040	164,868	0	2,262,404	67,839	33.4	12.5	25.2	56.7	49.4	22.0

173 POLAND	- FOOD BALANCE SHEET OF FISH AND FISHERY PRODUCTS IN LIVE WEIGHT AND FISH CONTRIBUTION TO PROTEIN SUPPLY -	FAO-FIDI

YEAR	PRODUCTION	NON-FOOD USES	IMPORTS	EXPORTS	STOCKS VARIATIONS	TOTAL FOOD SUPPLY	POPULATION	PER CAPUT SUPPLY	FISH PROTEINS	ANIMAL PROTEINS	TOTAL PROTEINS	FISH/ANIMAL PROTEINS	FISH/TOTAL PROTEINS
	(..		metric tons in live weight	....		)	(thousands)	(kilograms)	(grams per caput per day)			(%)	(%)
1961	185,500	0	21,467	7,767	0	199,200	29,994	6.6	3.0	42.6	96.9	7.0	3.1
1962	180,600	0	17,580	8,200	0	189,980	30,419	6.3	2.7	42.6	96.9	6.2	2.7
1963	226,700	0	18,817	11,420	0	234,097	30,822	7.6	3.2	44.3	98.0	7.1	3.2
1964	264,300	0	22,030	13,540	0	272,790	31,185	8.8	3.6	44.3	98.6	8.1	3.6
1965	297,500	0	24,280	41,460	0	280,320	31,496	8.9	3.7	45.5	100.8	8.2	3.7
1966	334,900	0	15,806	40,188	0	310,518	31,748	9.8	3.7	47.4	101.8	7.9	3.7
1967	338,900	0	15,206	56,738	0	297,368	31,951	9.3	3.4	48.4	101.0	7.1	3.4
1968	406,700	0	16,723	65,161	0	358,262	32,127	11.2	3.9	49.4	100.5	7.9	3.9
1969	408,100	0	11,567	71,835	0	347,832	32,311	10.8	3.7	50.3	101.3	7.3	3.6
1970	469,200	30,000	18,002	83,060	0	374,142	32,526	11.5	4.2	51.7	102.9	8.2	4.1
1971	517,500	66,600	16,962	71,040	0	396,822	32,781	12.1	4.6	54.4	104.7	8.5	4.4
1972	543,900	66,600	31,902	98,427	0	410,775	33,071	12.4	4.6	56.2	106.4	8.2	4.4
1973	579,700	66,600	36,298	108,834	0	440,564	33,384	13.2	4.8	57.9	107.8	8.3	4.5
1974	678,944	120,000	44,807	117,873	0	485,878	33,706	14.4	5.0	59.7	108.6	8.4	4.6
1975	800,737	170,000	70,467	134,379	0	566,826	34,022	16.7	5.6	63.4	111.9	8.9	5.0
1976	750,072	168,000	106,225	159,498	-2,000	526,800	34,330	15.4	6.3	63.9	112.9	9.9	5.6
1977	660,482	194,000	103,524	127,617	2,000	444,390	34,634	12.8	5.5	63.5	112.0	8.6	4.9
1978	571,398	163,000	139,308	76,203	0	471,503	34,938	13.5	5.6	63.8	113.1	8.8	4.9
1979	601,153	196,000	130,237	112,170	0	423,220	35,250	12.0	5.3	63.0	112.5	8.4	4.7
1980	640,588	195,000	36,898	77,719	0	404,767	35,574	11.4	5.3	62.2	112.9	8.5	4.7
1981	630,093	168,400	71,482	81,953	0	451,222	35,912	12.6	5.3	58.7	107.9	9.1	5.0
1982	608,147	127,000	39,637	100,376	0	420,408	36,259	11.6	4.8	51.1	98.9	9.4	4.9
1983	735,097	181,300	73,604	140,492	-4,000	482,909	36,600	13.2	5.3	53.9	100.2	9.9	5.3
1984	719,174	176,000	100,545	124,943	-4,000	514,776	36,919	13.9	5.3	52.9	99.0	10.0	5.4
1985	683,455	176,200	140,883	119,176	4,000	532,962	37,203	14.3	5.4	54.9	101.4	9.9	5.4
1986	645,220	186,900	153,708	115,052	4,000	500,977	37,448	13.4	5.0	56.0	103.5	8.8	4.8
1987	670,906	201,000	209,662	138,805	0	540,763	37,656	14.4	5.3	56.0	103.3	9.4	5.1
1988	654,890	185,200	222,681	147,159	0	545,212	37,832	14.4	5.1	56.8	104.6	9.0	4.9
1989	564,805	161,500	212,995	134,243	0	482,057	37,984	12.7	4.8	55.5	103.6	8.6	4.6
1990	473,011	131,000	166,359	125,331	0	383,038	38,119	10.1	3.8	53.9	100.7	7.0	3.7
1991	457,389	84,000	97,753	122,649	0	348,493	38,238	9.1	3.5	54.3	101.6	6.4	3.4
1992	505,897	168,300	207,343	78,379	0	466,556	38,341	12.2	4.8	53.3	102.1	9.0	4.7
1993	423,031	112,000	247,494	60,807	0	497,719	38,427	13.0	4.8	51.2	101.1	9.4	4.8
1994	460,229	110,000	321,665	80,480	0	591,414	38,499	15.4	5.1	48.4	97.1	10.5	5.2
1995	451,346	90,004	295,716	105,840	0	551,217	38,557	14.3	4.8	48.6	98.1	9.9	4.9

174 PORTUGAL	- FOOD BALANCE SHEET OF FISH AND FISHERY PRODUCTS IN LIVE WEIGHT AND FISH CONTRIBUTION TO PROTEIN SUPPLY -	FAO-FIDI

YEAR	PRODUCTION	NON-FOOD USES	IMPORTS	EXPORTS	STOCKS VARIATIONS	TOTAL FOOD SUPPLY	POPULATION	PER CAPUT SUPPLY	FISH PROTEINS	ANIMAL PROTEINS	TOTAL PROTEINS	FISH/ANIMAL PROTEINS	FISH/TOTAL PROTEINS
	(.......... metric tons in live weight ....)						(thousands)	(kilograms)	(grams per caput per day)			(%)	(%)
1961	502,393	0	117,151	120,930	-317	498,297	8,891	56.0	12.6	27.9	70.7	45.0	17.8
1962	517,870	0	100,672	127,519	2,380	493,403	8,965	55.0	11.9	27.6	73.7	43.0	16.1
1963	537,370	0	80,099	119,860	2,382	499,990	9,037	55.3	12.0	28.2	75.0	42.6	16.0
1964	577,895	0	104,971	129,582	-19,838	533,445	9,095	58.7	13.2	29.4	73.2	44.9	18.0
1965	552,685	0	88,056	139,629	476	501,588	9,129	54.9	12.5	30.0	75.4	41.7	16.5
1966	507,880	0	96,994	125,307	9,998	489,566	9,138	53.6	12.0	29.5	73.7	40.7	16.3
1967	563,178	0	190,517	122,369	-4,128	627,198	9,126	68.7	14.7	33.2	80.2	44.4	18.4
1968	507,590	0	115,601	121,909	4,603	505,885	9,100	55.6	11.7	30.1	78.1	38.7	14.9
1969	460,338	0	192,318	96,158	8,891	565,390	9,070	62.3	12.3	32.5	80.3	37.7	15.3
1970	462,826	0	197,680	86,181	0	574,325	9,044	63.5	13.0	33.8	82.2	38.5	15.8
1971	439,199	0	206,133	77,478	0	567,855	9,023	62.9	14.0	35.6	81.6	39.3	17.2
1972	446,533	20,700	223,649	80,985	0	568,497	9,006	63.1	14.0	36.8	84.1	38.1	16.6
1973	478,506	38,900	111,977	83,350	0	468,233	9,003	52.0	12.2	35.7	83.6	34.3	14.7
1974	431,657	37,100	131,161	54,243	0	471,476	9,030	52.2	12.2	38.1	86.1	32.0	14.2
1975	377,411	39,600	125,300	53,748	-7,936	401,427	9,093	44.1	10.4	36.9	85.2	28.1	12.2
1976	345,709	26,124	129,204	61,999	7,934	394,723	9,200	42.9	11.0	37.9	83.5	29.1	13.2
1977	306,771	6,612	126,081	67,544	-2,522	356,174	9,343	38.1	10.5	39.2	85.1	26.9	12.4
1978	252,598	7,074	90,142	75,249	2,476	262,893	9,503	27.7	7.6	34.6	78.7	22.1	9.7
1979	243,648	6,976	92,963	73,455	-8,865	247,315	9,651	25.6	7.0	33.9	75.4	20.7	9.3
1980	270,579	9,373	98,620	77,745	22	282,103	9,766	28.9	8.1	36.1	77.7	22.3	10.4
1981	263,233	7,675	101,665	63,707	-333	293,182	9,842	29.8	8.3	37.6	76.9	22.1	10.8
1982	254,947	9,040	125,587	72,056	-2,561	296,877	9,885	30.0	8.2	38.1	79.3	21.4	10.3
1983	249,900	7,000	226,004	77,168	2,325	394,062	9,901	39.8	10.4	39.9	81.8	26.1	12.7
1984	303,896	10,900	226,386	83,669	3,889	439,602	9,904	44.4	11.7	39.2	83.5	29.8	14.0
1985	318,892	13,200	328,355	90,731	4,117	547,434	9,904	55.3	14.3	43.2	88.9	33.1	16.1
1986	410,230	37,400	285,011	105,701	1,456	553,595	9,903	55.9	14.4	45.0	95.3	32.0	15.1
1987	391,361	35,600	350,213	95,075	-41,499	569,400	9,898	57.5	15.1	48.5	99.4	31.1	15.2
1988	348,308	35,513	389,879	93,515	-23,538	585,620	9,890	59.2	15.1	49.9	99.7	30.2	15.1
1989	333,241	34,847	356,858	126,080	52,324	581,497	9,880	58.9	14.9	52.7	101.7	28.2	14.6
1990	323,141	39,326	412,427	126,491	12,682	582,433	9,869	59.0	15.0	55.4	104.7	27.1	14.4
1991	325,901	38,538	452,043	132,004	-1,627	605,775	9,856	61.5	15.5	56.7	106.3	27.4	14.6
1992	299,569	32,788	422,821	118,617	-324	570,661	9,845	58.0	15.0	57.7	106.5	25.9	14.0
1993	293,743	34,038	431,970	106,607	-290	584,777	9,834	59.5	15.4	60.5	109.7	25.5	14.0
1994	269,780	34,051	446,908	108,285	870	575,223	9,824	58.6	15.1	62.4	112.4	24.2	13.5
1995	265,508	32,465	468,141	129,888	1,411	572,708	9,815	58.4	14.9	61.6	110.6	24.2	13.5

177 PUERTO RICO	- FOOD BALANCE SHEET OF FISH AND FISHERY PRODUCTS IN LIVE WEIGHT AND FISH CONTRIBUTION TO PROTEIN SUPPLY -												FAO-FIDI
YEAR	PRODUCTION	NON-FOOD USES	IMPORTS	EXPORTS	STOCKS VARIATIONS	TOTAL FOOD SUPPLY	POPULATION	PER CAPUT SUPPLY	FISH PROTEINS	ANIMAL PROTEINS	TOTAL PROTEINS	FISH/ANIMAL PROTEINS	FISH/TOTAL PROTEINS
	(........................ metric tons in live weight ....)						(thousands)	(kilograms)	(grams per caput per day)			(%)	(%)
1961	3,000	0	...	...	0	3,000	2,400	1.3	0.3	17.6	23.1	1.7	1.3
1962	3,300	0	...	...	0	3,300	2,447	1.4	0.3	18.5	24.2	1.8	1.3
1963	3,300	0	...	...	0	3,300	2,496	1.3	0.3	19.5	24.9	1.6	1.3
1964	2,900	0	...	...	0	2,900	2,543	1.1	0.3	19.5	25.0	1.4	1.1
1965	2,800	0	...	...	0	2,800	2,583	1.1	0.3	20.2	25.4	1.3	1.0
1966	2,500	0	...	...	0	2,500	2,615	1.0	0.2	20.4	25.7	1.1	0.9
1967	2,500	0	...	...	0	2,500	2,640	1.0	0.2	20.4	25.5	1.1	0.9
1968	2,200	0	...	...	0	2,200	2,662	0.8	0.2	20.5	24.7	0.9	0.8
1969	2,200	0	...	...	0	2,200	2,687	0.8	0.2	20.3	24.4	0.9	0.8
1970	2,100	0	...	...	0	2,100	2,716	0.8	0.2	20.1	24.4	1.0	0.8
1971	2,200	0	...	...	0	2,200	2,753	0.8	0.2	20.6	24.5	1.0	0.8
1972	2,100	0	...	...	0	2,100	2,794	0.8	0.2	20.8	24.6	0.9	0.8
1973	2,000	0	...	...	0	2,000	2,840	0.7	0.2	21.0	24.7	0.8	0.7
1974	2,000	0	...	...	0	2,100	2,888	0.7	0.2	20.6	24.2	0.8	0.7
1975	2,275	0	...	...	0	2,275	2,939	0.8	0.2	20.5	24.3	0.9	0.8
1976	2,663	0	...	...	0	2,663	2,991	0.9	0.2	21.7	25.3	1.0	0.8
1977	2,881	0	...	...	0	2,881	3,044	1.0	0.2	22.6	26.0	1.0	0.9
1978	3,197	0	...	...	0	3,197	3,098	1.0	0.2	23.8	27.5	1.0	0.9
1979	3,597	0	...	...	0	3,597	3,149	1.1	0.3	24.3	28.0	1.1	1.0
1980	2,557	0	...	...	0	2,557	3,197	0.8	0.2	21.6	25.3	0.9	0.8
1981	1,784	0	...	...	0	1,784	3,239	0.6	0.1	21.0	24.8	0.7	0.6
1982	2,202	0	...	...	0	2,202	3,278	0.7	0.2	20.0	23.6	0.8	0.7
1983	2,655	0	...	...	0	2,655	3,313	0.8	0.2	19.4	23.1	1.0	0.8
1984	2,354	0	...	...	0	2,354	3,346	0.7	0.2	18.9	22.2	0.9	0.7
1985	1,496	0	...	...	0	1,496	3,378	0.4	0.1	18.6	21.8	0.5	0.5
1986	1,327	0	...	...	0	1,327	3,409	0.4	0.1	20.2	23.0	0.5	0.4
1987	1,288	0	...	...	0	1,288	3,438	0.4	0.1	21.8	25.0	0.4	0.4
1988	1,697	0	...	...	0	1,697	3,467	0.5	0.1	21.2	24.2	0.6	0.5
1989	2,107	0	...	...	0	2,107	3,497	0.6	0.2	21.9	24.8	0.7	0.6
1990	2,198	0	...	...	0	2,198	3,528	0.6	0.2	21.9	24.3	0.7	0.6
1991	2,649	0	...	...	0	2,649	3,560	0.7	0.2	22.0	24.6	0.9	0.8
1992	2,172	0	...	...	0	2,172	3,594	0.6	0.2	20.6	23.1	0.8	0.7
1993	1,903	0	...	...	0	1,903	3,629	0.5	0.1	19.9	22.2	0.7	0.6
1994	2,240	0	...	...	0	2,240	3,665	0.6	0.2	19.5	21.5	0.8	0.7
1995	2,629	0	...	...	0	2,629	3,701	0.7	0.2	19.5	21.3	1.0	0.9

Note: trade statistics included with those of USA. Food supply data nor reliable.

179 QATAR	- FOOD BALANCE SHEET OF FISH AND FISHERY PRODUCTS IN LIVE WEIGHT AND FISH CONTRIBUTION TO PROTEIN SUPPLY -												FAO-FIDI
YEAR	PRODUCTION	NON-FOOD USES	IMPORTS	EXPORTS	STOCKS VARIATIONS	TOTAL FOOD SUPPLY	POPULATION	PER CAPUT SUPPLY	FISH PROTEINS	ANIMAL PROTEINS	TOTAL PROTEINS	FISH/ANIMAL PROTEINS	FISH/TOTAL PROTEINS
	(.. metric tons in live weight ....)						(thousands)	(kilograms)	(grams per caput per day)			(%)	(%)
1961	800	0	...	0	0	800	49	16.5	4.7	45.1	118.7	10.3	3.9
1962	1,000	0	...	0	0	1,000	53	18.9	5.3	44.3	115.9	12.0	4.6
1963	1,000	0	...	0	0	1,000	58	17.2	4.9	42.7	111.5	11.4	4.4
1964	1,000	0	...	0	0	1,000	64	15.7	4.4	40.5	106.7	11.0	4.2
1965	1,300	0	...	0	0	1,300	70	18.6	5.1	40.7	105.0	12.6	4.9
1966	1,300	0	1,250	0	0	2,550	77	33.2	10.4	43.3	104.4	24.0	10.0
1967	1,500	0	1,250	0	0	2,750	84	32.6	10.1	42.5	102.1	23.8	9.9
1968	1,500	0	1,250	45	0	2,705	92	29.3	9.1	40.9	98.7	22.2	9.2
1969	1,500	0	2,478	36	0	3,942	101	38.9	11.8	43.1	95.4	27.3	12.3
1970	1,500	0	2,284	0	0	3,784	111	34.0	10.3	41.2	92.9	25.0	11.1
1971	1,500	0	2,896	187	0	4,209	122	34.4	10.4	49.6	105.2	20.9	9.9
1972	2,000	0	1,482	433	0	3,049	135	22.6	7.0	39.0	89.2	18.0	7.9
1973	2,200	0	1,654	382	0	3,472	147	23.6	7.2	37.8	82.9	19.0	8.7
1974	2,047	0	1,328	0	0	3,375	160	21.1	6.4	36.6	75.1	17.5	8.5
1975	1,989	0	200	0	0	2,189	171	12.8	3.5	37.3	83.3	9.5	4.2
1976	2,400	0	0	311	0	2,089	181	11.5	3.2	40.8	85.1	7.7	3.7
1977	2,433	0	0	0	0	2,433	191	12.8	3.5	50.5	103.5	6.9	3.4
1978	2,200	0	0	0	0	2,200	200	11.0	3.0	56.2	111.6	5.4	2.7
1979	2,200	0	389	0	0	2,589	213	12.2	3.5	60.4	121.6	5.8	2.9
1980	2,178	0	778	0	0	2,956	229	12.9	3.8	69.6	127.2	5.5	3.0
1981	2,604	0	801	0	0	3,405	250	13.6	4.0	62.1	120.1	6.4	3.3
1982	2,331	0	1,307	0	0	3,638	274	13.3	4.0	58.0	113.5	6.8	3.5
1983	2,114	0	585	0	0	2,699	301	9.0	2.5	66.8	117.5	3.7	2.1
1984	3,173	0	666	0	0	3,839	330	11.7	3.2	60.4	113.4	5.3	2.8
1985	2,484	0	866	0	0	3,350	358	9.4	2.7	54.7	107.4	5.0	2.6
1986	1,980	0	1,120	0	0	3,100	386	8.0	2.4	64.2	113.2	3.7	2.1
1987	2,678	0	1,346	0	0	4,024	414	9.7	2.8	53.6	98.4	5.3	2.9
1988	3,086	0	1,312	0	0	4,398	440	10.0	2.9	52.2	110.2	5.5	2.6
1989	4,374	0	1,182	0	0	5,556	464	12.0	3.3	47.9	93.9	7.0	3.6
1990	5,702	0	1,530	0	0	7,232	485	14.9	4.2	48.7	102.0	8.6	4.1
1991	8,136	0	2,102	54	0	10,185	503	20.3	5.7	41.8	81.0	13.5	7.0
1992	7,845	0	2,943	11	0	10,777	516	20.9	5.9	40.9	85.7	14.4	6.9
1993	6,994	0	2,780	46	0	9,728	528	18.4	5.2	40.7	88.4	12.7	5.9
1994	5,086	0	3,510	46	0	8,550	538	15.9	4.5	41.9	82.1	10.8	5.5
1995	4,271	0	2,906	20	0	7,157	548	13.1	3.8	41.6	84.1	9.0	4.5

182 REUNION	- FOOD BALANCE SHEET OF FISH AND FISHERY PRODUCTS IN LIVE WEIGHT AND FISH CONTRIBUTION TO PROTEIN SUPPLY -	FAO-FIDI

YEAR	PRODUCTION	NON-FOOD USES	IMPORTS	EXPORTS	STOCKS VARIATIONS	TOTAL FOOD SUPPLY	POPULATION	PER CAPUT SUPPLY	FISH PROTEINS	ANIMAL PROTEINS	TOTAL PROTEINS	FISH/ANIMAL PROTEINS	FISH/TOTAL PROTEINS
	(.......... metric tons in live weight ....)						(thousands)	(kilograms)	(grams per caput per day)			(%)	(%)
1961	1,600	0	4,100	0	0	5,700	345	16.5	5.2	18.1	59.2	28.6	8.8
1962	1,500	0	4,100	0	0	5,600	355	15.8	5.0	19.1	60.4	26.0	8.2
1963	1,700	0	5,050	0	0	6,750	365	18.5	5.9	21.2	62.6	27.7	9.4
1964	2,200	0	5,050	0	0	7,250	377	19.3	6.0	21.8	63.0	27.5	9.5
1965	2,200	0	5,550	0	0	7,750	390	19.9	6.2	21.2	61.6	29.2	10.1
1966	2,600	0	6,667	10	0	9,257	404	22.9	7.1	22.2	62.5	31.9	11.4
1967	2,800	0	6,283	200	0	8,883	420	21.2	6.6	22.0	61.5	29.8	10.7
1968	2,900	0	7,583	8	0	10,475	435	24.1	7.5	24.1	63.9	30.9	11.7
1969	2,600	0	7,100	0	0	9,700	449	21.6	6.7	23.5	62.2	28.6	10.8
1970	2,200	0	7,350	48	0	9,502	461	20.6	6.3	24.4	62.7	25.9	10.1
1971	2,800	0	8,036	295	0	10,542	469	22.5	6.7	24.3	62.0	27.7	10.9
1972	2,700	0	7,550	2	0	10,248	474	21.6	6.6	25.0	63.3	26.6	10.5
1973	2,900	0	7,300	0	0	10,200	477	21.4	6.5	25.3	60.6	25.6	10.7
1974	3,489	0	8,346	124	0	11,711	480	24.4	7.2	27.4	65.0	26.4	11.1
1975	3,284	0	7,650	233	0	10,701	483	22.2	6.9	29.0	67.8	23.8	10.2
1976	2,789	0	7,729	0	0	10,518	486	21.6	7.3	31.2	70.5	23.3	10.3
1977	1,188	0	8,454	99	0	9,543	490	19.5	6.7	30.1	72.0	22.2	9.3
1978	2,633	0	8,744	181	0	11,195	494	22.6	7.6	32.3	75.1	23.5	10.1
1979	2,482	0	9,859	171	0	12,170	500	24.4	8.1	35.0	75.7	23.2	10.7
1980	2,120	0	10,655	130	0	12,646	506	25.0	8.4	35.1	76.4	23.8	10.9
1981	3,581	0	9,561	623	0	12,519	514	24.3	8.0	34.1	77.4	23.3	10.3
1982	5,290	0	8,605	984	-250	12,661	524	24.2	7.7	35.1	76.1	22.0	10.2
1983	2,607	0	10,542	1,969	250	11,430	534	21.4	7.4	36.1	77.4	20.4	9.5
1984	2,648	0	8,931	720	0	10,860	545	19.9	6.7	35.8	75.8	18.7	8.8
1985	2,171	0	10,403	878	0	11,697	555	21.1	7.2	36.5	79.0	19.8	9.2
1986	1,744	0	11,433	603	0	12,574	565	22.3	7.5	39.5	84.4	19.0	8.9
1987	1,798	0	11,625	563	0	12,860	575	22.4	7.3	39.8	85.3	18.4	8.6
1988	2,053	0	12,527	425	0	14,156	584	24.2	7.6	41.6	88.9	18.1	8.5
1989	1,727	0	13,524	599	-337	14,316	594	24.1	7.6	44.0	92.1	17.3	8.3
1990	1,733	0	12,447	896	0	13,284	604	22.0	6.9	43.9	93.1	15.8	7.4
1991	2,284	0	12,853	1,444	282	13,975	614	22.8	6.9	45.4	91.4	15.3	7.6
1992	2,498	0	14,365	930	-333	15,600	625	25.0	7.7	50.6	98.9	15.3	7.8
1993	2,775	0	14,642	1,450	222	16,189	635	25.5	7.9	49.1	98.9	16.1	8.0
1994	4,491	0	15,751	2,221	111	18,132	645	28.1	8.5	49.6	99.3	17.1	8.5
1995	4,825	0	15,823	2,161	18	18,505	655	28.3	8.5	49.9	98.8	17.0	8.6

183 ROMANIA	- FOOD BALANCE SHEET OF FISH AND FISHERY PRODUCTS IN LIVE WEIGHT AND FISH CONTRIBUTION TO PROTEIN SUPPLY -												FAO-FIDI
YEAR	PRODUCTION	NON-FOOD USES	IMPORTS	EXPORTS	STOCKS VARIATIONS	TOTAL FOOD SUPPLY	POPULATION	PER CAPUT SUPPLY	FISH PROTEINS	ANIMAL PROTEINS	TOTAL PROTEINS	FISH/ANIMAL PROTEINS	FISH/TOTAL PROTEINS
	(........ metric tons in live weight					........)	(thousands)	(kilograms)	(grams per caput per day)			(%)	(%)
1961	27,600	0	8,206	0	0	35,806	18,533	1.9	0.6	23.5	87.8	2.4	0.7
1962	33,400	0	11,328	0	0	44,728	18,640	2.4	0.7	23.7	87.7	3.0	0.8
1963	36,100	0	31,422	0	0	67,522	18,745	3.6	1.1	22.6	86.6	4.7	1.2
1964	34,000	0	25,306	0	0	59,306	18,870	3.1	0.9	23.1	86.8	4.0	1.1
1965	37,500	0	20,786	0	0	58,286	19,032	3.1	0.9	23.7	87.1	3.6	1.0
1966	36,600	0	10,493	0	0	47,093	19,237	2.5	0.8	25.1	89.0	3.0	0.8
1967	48,300	0	9,710	0	0	58,010	19,480	3.0	0.9	26.5	89.2	3.5	1.0
1968	40,600	0	26,269	0	0	66,869	19,745	3.4	1.0	26.7	89.2	3.7	1.1
1969	46,000	0	23,769	0	0	69,769	20,008	3.5	1.0	26.9	88.7	3.8	1.2
1970	58,600	8,400	74,375	0	0	124,575	20,253	6.2	1.7	28.5	85.2	5.9	2.0
1971	67,700	11,100	63,500	0	0	120,100	20,473	5.9	1.7	29.1	87.8	5.8	1.9
1972	84,400	13,700	56,375	0	0	127,075	20,675	6.2	1.8	31.9	89.8	5.8	2.1
1973	102,300	24,300	45,125	0	0	123,125	20,864	5.9	2.0	33.5	91.0	5.9	2.2
1974	129,459	43,782	34,500	0	0	120,177	21,051	5.7	2.0	35.1	95.4	5.7	2.1
1975	136,624	48,466	45,477	0	0	133,635	21,245	6.3	2.2	36.6	92.1	5.9	2.4
1976	127,197	49,771	44,390	0	0	121,816	21,447	5.7	2.2	37.3	94.0	5.8	2.3
1977	150,701	52,078	30,711	0	0	129,334	21,653	6.0	2.1	38.7	94.3	5.4	2.2
1978	137,676	50,678	35,112	0	0	122,110	21,854	5.6	1.9	40.7	96.5	4.8	2.0
1979	179,087	63,856	33,763	0	0	148,994	22,040	6.8	2.5	45.3	97.5	5.5	2.6
1980	173,598	47,060	51,068	0	0	177,606	22,201	8.0	3.0	43.8	97.2	6.9	3.1
1981	192,013	48,097	36,838	0	0	180,754	22,334	8.1	3.3	43.0	97.8	7.6	3.3
1982	235,653	60,258	33,991	0	0	209,386	22,440	9.3	3.6	40.4	90.7	8.9	4.0
1983	242,644	63,838	29,772	0	0	208,578	22,531	9.3	3.3	40.2	90.8	8.2	3.6
1984	232,449	60,288	25,110	0	0	197,271	22,623	8.7	3.4	42.3	93.2	7.9	3.6
1985	237,637	74,162	23,653	0	0	187,128	22,725	8.2	3.3	45.0	95.2	7.4	3.5
1986	271,126	93,542	19,471	0	0	197,055	22,844	8.6	3.7	42.2	91.2	8.9	4.1
1987	264,371	80,210	31,018	0	0	215,179	22,975	9.4	4.0	43.1	91.9	9.3	4.4
1988	267,618	82,728	40,218	0	0	225,108.	23,096	9.8	4.2	41.9	91.8	10.1	4.6
1989	224,810	69,455	11,900	9,655	0	157,600	23,179	6.8	2.9	38.9	90.1	7.5	3.2
1990	127,734	26,144	27,368	148	-434	128,376	23,207	5.5	2.2	41.4	91.2	5.3	2.4
1991	125,003	34,390	2,995	535	159	93,231	23,170	4.0	1.5	41.1	85.9	3.7	1.8
1992	95,381	28,455	1,755	3,408	275	65,548	23,078	2.8	1.0	37.9	82.3	2.6	1.2
1993	34,919	1,044	8,549	270	0	42,155	22,955	1.8	0.7	39.4	89.3	1.7	0.7
1994	42,651	866	23,647	782	-150	64,500	22,830	2.8	1.0	39.5	89.0	2.4	1.1
1995	69,105	1,478	10,043	22	0	77,648	22,728	3.4	1.3	40.3	91.6	3.1	1.4

185 RUSSIAN FEDERATION	- FOOD BALANCE SHEET OF FISH AND FISHERY PRODUCTS IN LIVE WEIGHT AND FISH CONTRIBUTION TO PROTEIN SUPPLY -												FAO-FIDI
YEAR	PRODUCTION	NON-FOOD USES	IMPORTS	EXPORTS	STOCKS VARIATIONS	TOTAL FOOD SUPPLY	POPULATION	PER CAPUT SUPPLY	FISH PROTEINS	ANIMAL PROTEINS	TOTAL PROTEINS	FISH/ANIMAL PROTEINS	FISH/TOTAL PROTEINS
	(... metric tons in live weight ....)						(thousands)	(kilograms)	(grams per caput per day)			(%)	(%)
1961	0	0	0	0	0	0	121,809	0.0	0.0	0.0	0.0	0.0	0.0
1962	0	0	0	0	0	0	123,353	0.0	0.0	0.0	0.0	0.0	0.0
1963	0	0	0	0	0	0	124,741	0.0	0.0	0.0	0.0	0.0	0.0
1964	0	0	0	0	0	0	125,974	0.0	0.0	0.0	0.0	0.0	0.0
1965	0	0	0	0	0	0	126,876	0.0	0.0	0.0	0.0	0.0	0.0
1966	0	0	0	0	0	0	127,735	0.0	0.0	0.0	0.0	0.0	0.0
1967	0	0	0	0	0	0	128,489	0.0	0.0	0.0	0.0	0.0	0.0
1968	0	0	0	0	0	0	129,167	0.0	0.0	0.0	0.0	0.0	0.0
1969	0	0	0	0	0	0	129,859	0.0	0.0	0.0	0.0	0.0	0.0
1970	0	0	0	0	0	0	130,392	0.0	0.0	0.0	0.0	0.0	0.0
1971	0	0	0	0	0	0	131,075	0.0	0.0	0.0	0.0	0.0	0.0
1972	0	0	0	0	0	0	131,828	0.0	0.0	0.0	0.0	0.0	0.0
1973	0	0	0	0	0	0	132,576	0.0	0.0	0.0	0.0	0.0	0.0
1974	0	0	0	0	0	0	133,358	0.0	0.0	0.0	0.0	0.0	0.0
1975	0	0	0	0	0	0	134,233	0.0	0.0	0.0	0.0	0.0	0.0
1976	0	0	0	0	0	0	135,168	0.0	0.0	0.0	0.0	0.0	0.0
1977	0	0	0	0	0	0	136,121	0.0	0.0	0.0	0.0	0.0	0.0
1978	0	0	0	0	0	0	137,074	0.0	0.0	0.0	0.0	0.0	0.0
1979	0	0	0	0	0	0	137,896	0.0	0.0	0.0	0.0	0.0	0.0
1980	0	0	0	0	0	0	138,483	0.0	0.0	0.0	0.0	0.0	0.0
1981	0	0	0	0	0	0	139,307	0.0	0.0	0.0	0.0	0.0	0.0
1982	0	0	0	0	0	0	140,229	0.0	0.0	0.0	0.0	0.0	0.0
1983	0	0	0	0	0	0	141,122	0.0	0.0	0.0	0.0	0.0	0.0
1984	0	0	0	0	0	0	142,053	0.0	0.0	0.0	0.0	0.0	0.0
1985	0	0	0	0	0	0	143,033	0.0	0.0	0.0	0.0	0.0	0.0
1986	0	0	0	0	0	0	144,140	0.0	0.0	0.0	0.0	0.0	0.0
1987	0	0	0	0	0	0	145,386	0.0	0.0	0.0	0.0	0.0	0.0
1988	0	0	0	0	0	0	146,494	0.0	0.0	0.0	0.0	0.0	0.0
1989	0	0	0	0	0	0	147,593	0.0	0.0	0.0	0.0	0.0	0.0
1990	0	0	0	0	0	0	148,292	0.0	0.0	0.0	0.0	0.0	0.0
1991	7,006,834	3,035,789	226,516	601,149	-42,852	3,553,558	148,720	23.9	9.5	55.4	98.9	17.2	9.7
1992	5,611,164	2,138,118	52,436	613,726	42,827	2,954,582	148,902	19.8	6.9	46.3	91.2	14.9	7.6
1993	4,461,375	1,051,060	50,653	1,220,430	-833	2,240,305	148,878	15.1	5.3	45.6	91.0	11.5	5.8
1994	3,780,538	893,369	398,910	1,554,622	385	1,731,841	148,714	11.7	4.5	46.0	89.3	9.7	5.0
1995	4,373,827	895,544	528,045	1,466,776	385	2,539,936	148,460	17.1	5.6	42.5	86.7	13.1	6.4

Note: for years before 1991 see USSR.

184 RWANDA | **- FOOD BALANCE SHEET OF FISH AND FISHERY PRODUCTS IN LIVE WEIGHT AND FISH CONTRIBUTION TO PROTEIN SUPPLY -** | **FAO-FIDI**

YEAR	PRODUCTION	NON-FOOD USES	IMPORTS	EXPORTS	STOCKS VARIATIONS	TOTAL FOOD SUPPLY	POPULATION	PER CAPUT SUPPLY	FISH PROTEINS	ANIMAL PROTEINS	TOTAL PROTEINS	FISH/ANIMAL PROTEINS	FISH/TOTAL PROTEINS
	(............ metric tons in live weight ....)						(thousands)	(kilograms)	(grams per caput per day)			(%)	(%)
1961	500	0	0	0	0	500	2,822	0.2	0.1	2.8	48.1	1.9	0.1
1962	500	0	0	0	0	500	2,907	0.2	0.1	2.7	46.6	1.9	0.1
1963	500	0	0	0	0	500	2,995	0.2	0.1	2.8	51.3	1.8	0.1
1964	700	0	200	0	0	900	3,087	0.3	0.1	2.7	48.8	3.2	0.2
1965	700	0	200	0	0	900	3,183	0.3	0.1	2.6	48.8	3.2	0.2
1966	700	0	200	0	0	900	3,284	0.3	0.1	2.7	58.0	3.0	0.1
1967	800	0	0	0	0	800	3,389	0.2	0.1	2.9	59.9	2.5	0.1
1968	800	0	0	0	0	800	3,498	0.2	0.1	2.7	56.1	2.5	0.1
1969	800	0	0	0	0	800	3,611	0.2	0.1	2.7	58.4	2.4	0.1
1970	900	0	0	0	0	900	3,728	0.2	0.1	2.8	56.7	2.6	0.1
1971	900	0	0	0	0	900	3,850	0.2	0.1	2.7	55.9	2.6	0.1
1972	800	0	0	0	0	800	3,977	0.2	0.1	2.7	54.7	2.3	0.1
1973	1,000	0	200	0	0	1,200	4,108	0.3	0.1	2.8	53.7	3.1	0.2
1974	1,368	0	200	0	0	1,568	4,243	0.4	0.1	2.8	53.5	3.9	0.2
1975	1,198	0	120	0	0	1,318	4,384	0.3	0.1	3.0	55.2	3.0	0.2
1976	1,008	0	69	0	0	1,077	4,530	0.2	0.1	3.1	56.2	2.3	0.1
1977	1,325	0	108	0	0	1,433	4,681	0.3	0.1	3.4	57.3	2.8	0.2
1978	715	0	89	0	0	804	4,836	0.2	0.1	3.7	55.3	1.4	0.1
1979	954	0	158	0	0	1,112	4,997	0.2	0.1	3.5	54.8	1.9	0.1
1980	1,200	0	635	0	0	1,835	5,163	0.4	0.1	4.0	53.9	2.9	0.2
1981	955	0	300	0	0	1,255	5,327	0.2	0.1	3.9	54.4	1.9	0.1
1982	1,210	0	160	0	0	1,370	5,487	0.3	0.1	4.0	53.7	1.9	0.1
1983	1,212	0	215	0	0	1,427	5,653	0.3	0.1	3.8	53.7	2.1	0.1
1984	825	0	429	0	0	1,254	5,840	0.2	0.1	4.0	53.1	1.7	0.1
1985	945	0	931	0	0	1,876	6,052	0.3	0.1	4.3	56.0	2.4	0.2
1986	1,546	0	444	0	0	1,990	6,313	0.3	0.1	4.2	54.5	2.4	0.2
1987	1,694	0	1,497	0	0	3,191	6,610	0.5	0.2	3.8	50.8	4.1	0.3
1988	1,328	0	425	0	0	1,753	6,873	0.3	0.1	3.5	45.8	2.3	0.2
1989	1,516	0	286	0	0	1,802	7,007	0.3	0.1	3.5	45.4	2.3	0.2
1990	2,514	0	291	0	0	2,805	6,954	0.4	0.1	3.5	45.7	3.5	0.3
1991	3,561	0	825	0	0	4,386	6,670	0.7	0.2	4.0	48.3	5.1	0.4
1992	3,697	0	1,793	0	0	5,490	6,200	0.9	0.3	4.1	52.4	6.9	0.5
1993	3,553	0	473	0	0	4,026	5,678	0.7	0.2	4.4	49.4	4.9	0.4
1994	3,450	0	546	0	0	3,996	5,296	0.8	0.2	4.2	49.8	5.5	0.5
1995	3,349	0	546	0	0	3,895	5,184	0.8	0.2	4.4	47.6	5.2	0.5

187 SAINT HELENA	- FOOD BALANCE SHEET OF FISH AND FISHERY PRODUCTS IN LIVE WEIGHT AND FISH CONTRIBUTION TO PROTEIN SUPPLY -												FAO-FIDI
YEAR	PRODUCTION	NON-FOOD USES	IMPORTS	EXPORTS	STOCKS VARIATIONS	TOTAL FOOD SUPPLY	POPULATION	PER CAPUT SUPPLY	FISH PROTEINS	ANIMAL PROTEINS	TOTAL PROTEINS	FISH/ANIMAL PROTEINS	FISH/TOTAL PROTEINS
	(.................................... metric tons in live weight ....)						(thousands)	(kilograms)	(grams per caput per day)			(%)	(%)
1961	600	0	0	0	0	600	5	120.0	32.4	32.4	56.7	100.0	57.1
1962	600	0	0	0	0	600	5	120.0	32.4	35.2	59.6	92.1	54.3
1963	600	0	0	0	0	600	5	120.0	32.4	41.5	65.9	78.1	49.1
1964	600	0	0	0	0	600	5	120.0	32.4	40.2	71.0	80.5	45.6
1965	700	0	0	0	0	700	5	140.0	37.5	44.0	62.9	85.3	59.6
1966	700	0	0	0	0	700	5	140.0	37.5	40.6	80.6	92.4	46.5
1967	700	0	0	0	0	700	5	140.0	37.5	41.5	61.5	90.3	61.0
1968	700	0	0	0	0	700	5	140.0	37.5	40.8	60.6	91.9	61.8
1969	700	0	0	0	0	700	5	140.0	37.5	41.5	65.2	90.3	57.5
1970	800	200	0	0	0	600	5	120.0	34.2	46.7	68.3	73.2	50.1
1971	800	200	0	0	0	600	5	119.9	34.2	39.4	60.8	86.7	56.2
1972	1,100	500	0	0	0	600	5	119.8	34.1	40.7	67.9	83.8	50.3
1973	800	200	0	0	0	600	5	119.7	34.1	53.3	73.8	64.0	46.2
1974	754	150	0	0	0	604	5	120.5	33.1	56.5	78.2	58.5	42.3
1975	912	310	0	0	0	602	5	120.4	32.8	48.0	85.3	68.3	38.4
1976	1,000	400	0	0	0	600	5	120.7	33.7	49.4	263.0	68.2	12.8
1977	719	120	0	0	0	599	5	121.5	32.3	49.3	85.6	65.6	37.8
1978	700	110	0	0	0	590	5	120.3	33.2	50.1	76.1	66.2	43.6
1979	632	50	0	0	0	582	5	118.3	33.7	53.8	98.1	62.7	34.4
1980	637	0	0	52	0	585	5	117.0	33.0	51.7	81.3	63.8	40.6
1981	687	0	9	105	0	591	5	114.6	32.3	53.0	110.9	60.9	29.1
1982	928	320	7	9	0	606	5	112.7	30.9	43.7	68.5	70.8	45.2
1983	619	0	7	20	0	606	6	107.8	31.0	47.9	76.8	64.7	40.4
1984	652	0	5	78	0	579	6	99.1	28.5	45.9	78.2	62.1	36.5
1985	572	0	12	163	0	421	6	70.2	20.1	46.9	75.9	42.9	26.5
1986	632	0	11	284	0	359	6	59.0	17.6	41.4	69.4	42.6	25.4
1987	741	0	0	252	0	489	6	80.2	23.4	44.9	73.2	52.1	31.9
1988	786	0	11	220	0	577	6	95.0	27.9	52.6	79.4	53.1	35.2
1989	1,013	0	0	228	0	785	6	130.2	40.0	76.8	107.5	52.1	37.2
1990	802	0	0	366	0	436	6	72.7	20.7	78.0	107.9	26.6	19.2
1991	626	0	0	333	0	293	6	48.9	12.4	42.1	82.2	29.5	15.1
1992	651	0	0	337	0	314	6	52.4	14.7	30.2	62.1	48.7	23.7
1993	726	0	0	309	0	417	6	69.4	20.1	36.8	66.1	54.7	30.4
1994	627	0	0	290	0	337	6	56.1	15.5	31.5	53.4	49.2	28.9
1995	818	0	150	466	0	502	6	83.7	23.9	43.1	74.5	55.4	32.0

188 SAINT KITTS and NEVIS	- FOOD BALANCE SHEET OF FISH AND FISHERY PRODUCTS IN LIVE WEIGHT AND FISH CONTRIBUTION TO PROTEIN SUPPLY -	FAO-FIDI

YEAR	PRODUCTION	NON-FOOD USES	IMPORTS	EXPORTS	STOCKS VARIATIONS	TOTAL FOOD SUPPLY	POPULATION	PER CAPUT SUPPLY	FISH PROTEINS	ANIMAL PROTEINS	TOTAL PROTEINS	FISH/ANIMAL PROTEINS	FISH/TOTAL PROTEINS
	(............ metric tons in live weight ....)						(thousands)	(kilograms)	(grams per caput per day)			(%)	(%)
1961	600	0	...	0	0	600	57	10.6	3.0	11.4	31.3	26.2	9.5
1962	600	0	...	0	0	600	56	10.6	3.0	11.7	32.7	25.7	9.2
1963	700	0	...	0	0	700	56	12.5	3.5	12.7	34.7	27.8	10.2
1964	700	0	...	0	0	700	55	12.6	3.6	13.3	35.6	26.9	10.0
1965	700	0	...	0	0	700	55	12.7	3.6	13.9	35.7	25.9	10.1
1966	800	0	...	0	0	800	55	14.7	4.1	15.6	36.3	26.5	11.4
1967	800	0	...	0	0	800	54	14.8	4.2	17.6	38.2	23.7	10.9
1968	800	0	...	0	0	800	54	14.9	4.2	18.4	38.6	22.8	10.9
1969	800	0	...	0	0	800	53	15.0	4.2	18.3	38.1	23.1	11.1
1970	1,000	0	...	0	0	1,000	53	18.9	5.3	19.9	40.0	26.8	13.3
1971	1,000	0	...	0	0	1,000	53	19.0	5.4	21.2	41.2	25.3	13.0
1972	1,000	0	725	0	0	1,725	52	33.1	10.3	25.9	46.2	39.9	22.4
1973	1,000	0	725	13	0	1,713	52	33.2	10.3	24.2	45.5	42.8	22.7
1974	1,000	0	780	8	0	1,773	51	34.6	10.8	24.6	45.6	44.1	23.8
1975	1,218	0	661	25	0	1,855	51	36.4	11.1	25.2	45.8	44.0	24.3
1976	1,684	0	709	95	0	2,298	51	45.2	13.7	29.5	49.3	46.4	27.7
1977	1,695	0	514	105	0	2,104	51	41.4	12.4	28.2	49.6	43.9	24.9
1978	1,790	0	495	100	0	2,185	51	42.9	12.8	30.1	52.9	42.4	24.1
1979	1,777	0	514	100	0	2,191	51	43.0	12.8	31.1	52.8	41.2	24.3
1980	1,935	0	568	90	0	2,413	51	47.3	14.1	33.1	55.8	42.7	25.3
1981	1,900	0	541	24	0	2,417	51	47.5	14.2	34.3	55.2	41.2	25.7
1982	1,930	0	642	51	0	2,521	51	49.7	14.9	34.4	54.9	43.4	27.2
1983	1,177	0	567	80	0	1,664	50	33.0	10.1	32.8	54.2	30.8	18.6
1984	1,115	0	727	127	0	1,715	50	34.1	10.6	32.0	55.7	33.2	19.1
1985	995	0	714	102	0	1,607	50	32.1	10.1	33.7	58.6	29.9	17.2
1986	901	0	638	105	0	1,434	50	28.8	9.0	35.1	57.2	25.8	15.8
1987	924	0	435	227	0	1,132	50	22.9	7.0	34.9	60.5	20.2	11.6
1988	800	0	628	130	0	1,298	49	26.3	8.2	35.6	61.8	23.2	13.3
1989	710	0	525	100	0	1,135	49	23.1	7.2	31.4	59.4	22.9	12.1
1990	620	0	574	120	0	1,074	49	21.9	6.9	36.3	64.3	19.1	10.8
1991	453	0	340	110	0	683	49	14.0	4.4	33.7	61.3	13.1	7.2
1992	303	0	264	60	0	507	49	10.4	3.3	29.9	57.5	10.9	5.7
1993	253	0	260	9	0	504	49	10.3	3.2	29.1	55.5	11.1	5.8
1994	216	0	163	10	0	369	49	7.5	2.2	27.5	54.3	8.1	4.1
1995	222	0	478	4	0	696	49	14.2	4.3	30.6	54.9	14.2	7.9

Note: 1961-1971 food supply data not reliable as imports statistics not available.

189 SAINT LUCIA	- FOOD BALANCE SHEET OF FISH AND FISHERY PRODUCTS IN LIVE WEIGHT AND FISH CONTRIBUTION TO PROTEIN SUPPLY -													FAO-FIDI
YEAR	PRODUCTION	NON-FOOD USES	IMPORTS	EXPORTS	STOCKS VARIATIONS	TOTAL FOOD SUPPLY	POPULATION	PER CAPUT SUPPLY	FISH PROTEINS	ANIMAL PROTEINS	TOTAL PROTEINS	FISH/ANIMAL PROTEINS	FISH/TOTAL PROTEINS
	(........................ metric tons in live weight ....)						(thousands)	(kilograms)	(grams per caput per day)			(%)	(%)
1961	600	0	1,175	0	0	1,775	87	20.3	6.7	18.0	39.9	37.0	16.7
1962	600	0	1,175	0	0	1,775	89	20.0	6.5	17.8	39.7	36.8	16.5
1963	600	0	1,175	0	0	1,775	91	19.6	6.4	17.7	38.0	36.2	16.9
1964	800	0	1,175	0	0	1,975	92	21.4	6.9	18.4	40.5	37.4	17.0
1965	800	0	1,175	0	0	1,975	94	21.0	6.8	19.1	39.4	35.3	17.1
1966	800	0	1,175	0	0	1,975	96	20.7	6.7	20.3	41.8	32.7	15.9
1967	800	0	1,175	0	0	1,975	97	20.4	6.6	20.5	41.7	32.0	15.7
1968	900	0	1,175	0	0	2,075	98	21.1	6.7	21.7	43.9	31.1	15.4
1969	900	0	1,175	0	0	2,075	100	20.8	6.7	22.5	45.3	29.6	14.7
1970	1,719	0	1,235	0	0	2,954	101	29.3	9.1	27.3	50.4	33.3	18.0
1971	1,719	0	1,551	0	0	3,270	102	31.9	9.9	28.2	53.6	35.3	18.6
1972	1,719	0	1,751	0	0	3,470	104	33.4	10.4	28.9	52.4	35.8	19.7
1973	1,627	0	1,475	0	0	3,102	105	29.5	9.1	26.4	49.5	34.5	18.4
1974	1,700	0	1,225	0	0	2,925	107	27.4	8.4	24.7	46.7	33.9	17.9
1975	2,000	0	1,059	18	0	3,041	108	28.2	8.5	26.0	48.4	32.6	17.5
1976	2,200	0	1,118	9	0	3,309	109	30.3	9.0	27.9	51.0	32.3	17.6
1977	2,500	0	865	14	0	3,351	111	30.3	9.0	31.0	53.1	29.0	16.9
1978	2,600	0	926	1	0	3,525	112	31.5	9.2	33.9	64.9	27.2	14.2
1979	1,024	0	967	1	0	1,990	113	17.5	5.4	31.6	63.8	17.2	8.5
1980	969	0	971	0	0	1,940	115	16.9	5.3	28.9	51.5	18.2	10.2
1981	891	0	1,161	15	0	2,037	117	17.5	5.5	32.2	62.1	17.0	8.8
1982	921	0	1,113	3	0	2,032	118	17.2	5.2	32.3	59.6	16.1	8.7
1983	910	0	1,202	0	0	2,112	120	17.6	5.6	33.6	62.6	16.5	8.9
1984	946	0	1,371	0	0	2,317	122	19.0	5.9	36.9	67.4	15.9	8.7
1985	1,052	0	1,265	5	0	2,312	124	18.7	6.0	37.0	68.1	16.3	8.8
1986	840	0	1,061	27	0	1,874	126	14.9	4.9	39.9	73.1	12.1	6.6
1987	651	0	1,265	40	0	1,876	128	14.7	4.8	40.0	74.5	12.1	6.5
1988	782	0	1,359	37	0	2,104	129	16.3	5.1	40.0	74.6	12.8	6.9
1989	794	0	1,765	4	0	2,555	131	19.5	6.3	45.6	82.1	13.7	7.6
1990	931	0	1,637	23	0	2,545	133	19.1	6.1	43.8	78.0	13.8	7.8
1991	947	0	1,884	15	0	2,815	135	20.9	6.7	49.5	85.5	13.6	7.9
1992	1,010	0	1,897	2	0	2,905	137	21.3	6.5	49.7	84.1	13.1	7.7
1993	1,195	0	2,001	5	0	3,191	138	23.1	7.1	51.4	85.4	13.8	8.3
1994	1,155	0	1,938	3	0	3,090	140	22.1	7.0	53.8	91.4	13.0	7.6
1995	1,177	0	1,532	0	0	2,709	142	19.1	5.9	49.4	85.5	11.9	6.9

190 SAINT PIERRE and MIQUELON	- FOOD BALANCE SHEET OF FISH AND FISHERY PRODUCTS IN LIVE WEIGHT AND FISH CONTRIBUTION TO PROTEIN SUPPLY -	FAO-FIDI

YEAR	PRODUCTION	NON-FOOD USES	IMPORTS	EXPORTS	STOCKS VARIATIONS	TOTAL FOOD SUPPLY	POPULATION	PER CAPUT SUPPLY	FISH PROTEINS	ANIMAL PROTEINS	TOTAL PROTEINS	FISH/ANIMAL PROTEINS	FISH/TOTAL PROTEINS
	(.......... metric tons in live weight ....)						(thousands)	(kilograms)	(grams per caput per day)			(%)	(%)
1961	13,600	3,100	100	10,300	0	300	5	60.0	16.0	47.5	69.3	33.7	23.1
1962	8,100	2,605	100	5,300	0	295	5	58.9	15.7	41.6	59.4	37.9	26.5
1963	8,200	3,236	100	4,774	0	290	5	57.8	15.5	47.4	65.6	32.7	23.6
1964	8,700	2,550	150	6,000	0	300	5	59.9	20.5	51.9	69.8	39.6	29.4
1965	10,000	[illegible]	150	4,782	0	310	5	62.0	21.0	56.1	74.3	37.5	28.3
1966	10,500	2,040	150	8,300	0	310	5	62.4	21.2	66.2	108.5	31.9	19.5
1967	8,300	3,630	150	4,500	0	320	5	64.9	21.8	69.3	113.0	31.4	19.3
1968	6,500	3,220	50	3,000	0	330	5	67.3	20.0	72.3	117.6	27.6	17.0
1969	7,700	3,195	75	4,250	0	330	5	67.1	22.2	72.6	119.6	30.6	18.6
1970	6,700	2,500	75	3,930	0	345	5	69.0	24.8	100.3	146.7	24.7	16.9
1971	5,700	1,800	75	3,615	0	360	5	69.8	23.6	91.9	139.9	25.7	16.9
1972	4,700	1,000	75	3,395	0	380	5	70.7	24.5	100.3	148.2	24.4	16.5
1973	5,500	2,000	100	3,200	0	400	6	71.1	28.5	86.4	133.1	33.0	21.4
1974	4,515	2,000	100	2,190	0	425	6	72.7	26.9	80.0	123.0	33.7	21.9
1975	6,397	2,500	0	3,476	0	421	6	70.2	21.7	74.3	104.6	29.2	20.7
1976	8,327	3,442	0	4,460	0	425	6	69.9	15.9	41.6	69.9	38.2	22.8
1977	7,360	3,084	0	3,850	0	426	6	69.9	15.9	43.8	65.7	36.3	24.2
1978	8,905	5,419	0	3,060	0	426	6	70.2	16.0	51.0	78.3	31.3	20.4
1979	10,542	3,556	39	5,900	-700	425	6	70.5	16.3	61.5	95.3	26.5	17.1
1980	8,617	3,235	47	5,105	100	424	6	70.7	16.4	60.7	88.2	27.0	18.5
1981	11,111	5,456	41	5,868	600	428	6	71.5	16.5	59.5	86.2	27.7	19.1
1982	10,474	4,802	39	5,290	0	421	6	70.3	16.9	81.4	111.2	20.8	15.2
1983	9,561	4,050	285	4,860	-510	426	6	71.2	18.1	119.7	149.7	15.1	12.1
1984	12,026	5,830	149	5,920	0	425	6	70.8	16.0	96.0	125.3	16.7	12.8
1985	12,304	5,500	131	6,790	280	425	6	70.8	16.3	95.5	123.4	17.1	13.2
1986	23,771	5,629	311	18,010	0	443	6	73.9	20.1	82.6	113.1	24.4	17.8
1987	23,679	4,815	203	18,630	0	437	6	72.8	16.2	86.8	122.5	18.6	13.2
1988	13,970	2,404	224	11,370	0	420	6	69.8	15.4	74.8	107.3	20.6	14.4
1989	18,520	2,677	200	15,613	0	430	6	71.5	16.6	58.7	90.2	28.2	18.4
1990	19,138	3,187	208	15,730	0	429	6	71.4	16.6	67.1	101.9	24.7	16.3
1991	19,164	3,972	121	14,940	0	373	6	62.5	16.5	89.3	130.6	18.5	12.7
1992	13,847	2,743	102	10,770	-120	316	6	53.3	14.1	86.5	122.4	16.3	11.5
1993	155	5	120	100	120	290	6	49.2	22.1	87.7	123.3	25.2	17.9
1994	293	35	181	158	0	281	6	47.4	13.0	83.7	116.2	15.5	11.2
1995	901	34	2,081	2,643	0	305	6	50.9	10.6	90.2	125.3	11.7	8.4

Note: the absence of fishing quota agreements between France and Canada has caused the cessation of almost all industrial fishing activities since August 1992.

191 SAINT VINCENT and the GRENADINES	- FOOD BALANCE SHEET OF FISH AND FISHERY PRODUCTS IN LIVE WEIGHT AND FISH CONTRIBUTION TO PROTEIN SUPPLY -	FAO-FIDI

YEAR	PRODUCTION	NON-FOOD USES	IMPORTS	EXPORTS	STOCKS VARIATIONS	TOTAL FOOD SUPPLY	POPULATION	PER CAPUT SUPPLY	FISH PROTEINS	ANIMAL PROTEINS	TOTAL PROTEINS	FISH/ANIMAL PROTEINS	FISH/TOTAL PROTEINS
	(...................... metric tons in live weight					......................)	(thousands)	(kilograms)	(grams per caput per day)			(%)	(%)
1961	400	0	1,212	0	0	1,612	81	19.9	6.5	16.7	42.6	39.1	15.3
1962	400	0	1,212	0	0	1,612	82	19.7	6.5	16.7	45.0	38.6	14.3
1963	500	0	909	0	0	1,409	83	17.0	5.5	16.0	42.3	34.2	13.0
1964	500	0	909	0	0	1,409	83	16.9	5.4	15.2	42.4	35.8	12.8
1965	500	0	909	0	0	1,409	84	16.8	5.4	15.6	44.9	34.5	12.0
1966	500	0	1,229	0	0	1,729	85	20.5	6.6	18.4	48.3	35.8	13.6
1967	500	0	815	0	0	1,315	85	15.5	4.9	16.6	44.8	29.4	10.9
1968	400	0	909	0	0	1,309	86	15.3	5.0	18.3	48.2	27.1	10.3
1969	500	0	909	0	0	1,409	86	16.4	5.3	18.9	49.1	27.8	10.7
1970	600	0	1,109	0	0	1,709	87	19.6	6.2	21.9	53.1	28.2	11.7
1971	300	0	1,109	0	0	1,409	88	16.0	5.2	22.9	54.0	22.5	9.5
1972	300	0	1,109	0	0	1,409	89	15.8	5.1	21.7	54.2	23.4	9.4
1973	200	0	1,109	0	0	1,309	90	14.5	4.7	20.9	52.6	22.4	8.9
1974	283	0	1,412	0	0	1,695	92	18.5	6.0	21.0	52.1	28.6	11.6
1975	549	0	930	0	0	1,479	93	15.9	4.9	20.0	49.6	24.5	9.9
1976	379	0	941	0	0	1,320	94	14.0	4.7	21.1	50.8	22.0	9.2
1977	581	0	603	59	0	1,125	95	11.8	3.8	20.8	52.0	18.2	7.3
1978	698	0	627	295	0	1,030	96	10.7	3.5	20.9	52.5	16.7	6.6
1979	547	0	596	1	0	1,142	97	11.8	3.7	20.2	56.0	18.5	6.7
1980	500	0	596	0	0	1,096	98	11.2	3.6	19.5	55.2	18.4	6.5
1981	500	0	637	1	-70	1,066	99	10.8	3.5	19.1	56.7	18.1	6.1
1982	500	0	607	231	70	947	100	9.5	3.0	22.7	57.6	13.2	5.2
1983	480	0	647	56	0	1,071	100	10.7	3.4	21.6	57.2	15.7	5.9
1984	480	0	760	36	0	1,204	101	11.9	3.8	25.5	59.7	15.1	6.4
1985	483	0	760	119	0	1,124	102	11.0	3.6	24.7	59.5	14.5	6.0
1986	599	0	464	105	0	958	103	9.3	2.9	24.9	57.9	11.8	5.0
1987	703	0	523	149	0	1,077	104	10.4	3.3	29.9	61.6	10.9	5.3
1988	4,647	0	551	4,060	0	1,138	105	10.8	3.4	28.3	60.4	12.1	5.7
1989	5,820	0	627	5,351	0	1,096	106	10.3	3.3	29.3	58.6	11.3	5.7
1990	8,971	0	662	7,786	0	1,847	107	17.3	5.4	34.0	61.7	15.7	8.7
1991	8,449	0	667	7,287	-20	1,809	108	16.7	5.3	33.5	59.7	15.8	8.8
1992	2,060	0	689	920	0	1,829	109	16.8	5.2	36.0	62.9	14.5	8.3
1993	1,944	0	691	682	0	1,953	110	17.8	5.6	34.7	64.2	16.1	8.7
1994	1,517	0	453	265	0	1,705	111	15.4	4.6	32.2	64.0	14.3	7.2
1995	1,320	0	767	221	20	1,886	112	16.8	5.2	34.3	66.0	15.1	7.8

244 SAMOA	- FOOD BALANCE SHEET OF FISH AND FISHERY PRODUCTS IN LIVE WEIGHT AND FISH CONTRIBUTION TO PROTEIN SUPPLY -	FAO-FIDI

YEAR	PRODUCTION	NON-FOOD USES	IMPORTS	EXPORTS	STOCKS VARIATIONS	TOTAL FOOD SUPPLY	POPULATION	PER CAPUT SUPPLY	FISH PROTEINS	ANIMAL PROTEINS	TOTAL PROTEINS	FISH/ANIMAL PROTEINS	FISH/TOTAL PROTEINS
	(........ metric tons in live weight ....)						(thousands)	(kilograms)	(grams per caput per day)			(%)	(%)
1961	600	0	3,000	0	0	3,600	116	31.1	8.9	20.8	44.2	42.6	20.0
1962	600	0	3,100	0	0	3,700	119	31.1	8.9	20.7	44.6	42.7	19.8
1963	600	0	3,100	0	0	3,700	122	30.3	8.6	22.0	45.9	39.2	18.8
1964	600	0	3,100	0	0	3,700	125	29.5	8.4	22.7	48.1	37.1	17.5
1965	600	0	3,100	0	0	3,700	129	28.8	8.2	21.3	46.9	38.5	17.4
1966	700	0	3,100	0	0	3,800	132	28.8	8.2	20.4	45.5	40.2	18.0
1967	700	0	3,100	0	0	3,800	136	28.0	8.0	18.8	45.0	42.3	17.7
1968	700	0	3,315	0	0	4,015	139	28.9	8.2	19.4	46.7	42.4	17.6
1969	700	0	2,980	0	0	3,680	142	25.9	7.4	19.1	46.7	38.7	15.8
1970	900	0	3,150	0	0	4,050	145	28.0	8.0	20.9	47.8	38.1	16.7
1971	900	0	4,384	0	0	5,284	147	36.1	10.3	23.1	50.4	44.4	20.3
1972	900	0	3,542	0	0	4,442	148	30.0	8.5	22.6	50.5	37.7	16.9
1973	900	0	2,814	0	0	3,714	149	24.9	7.1	19.8	48.7	35.8	14.5
1974	900	0	3,200	0	0	4,100	150	27.3	7.8	20.5	50.0	37.9	15.5
1975	1,000	0	3,200	0	0	4,200	151	27.8	7.9	24.2	56.8	32.6	13.9
1976	1,100	0	3,961	0	0	5,061	152	33.3	11.2	24.8	57.5	45.1	19.5
1977	1,250	0	3,897	0	0	5,147	153	33.7	11.3	27.6	58.1	40.9	19.4
1978	1,090	0	2,488	0	0	3,578	154	23.3	7.7	27.4	58.0	28.0	13.3
1979	1,890	0	4,315	0	0	6,205	155	40.2	13.7	29.3	59.7	46.7	22.9
1980	1,990	0	3,578	0	0	5,568	155	35.9	12.2	29.8	60.5	40.9	20.1
1981	3,095	0	4,618	0	0	7,713	156	49.6	17.0	33.1	63.4	51.3	26.8
1982	4,020	0	2,999	0	0	7,019	156	45.0	15.0	33.4	65.7	44.8	22.8
1983	3,820	0	2,018	0	0	5,838	156	37.4	12.6	29.1	62.9	43.4	20.1
1984	3,720	0	2,698	0	0	6,418	157	41.0	13.8	34.3	67.4	40.3	20.5
1985	3,641	0	3,447	0	0	7,088	157	45.2	15.2	38.9	72.8	39.1	20.9
1986	3,186	0	2,519	0	0	5,705	157	36.3	12.1	36.9	71.0	32.9	17.1
1987	3,076	0	3,939	5	0	7,010	158	44.4	15.1	42.0	75.3	35.8	20.0
1988	2,500	0	4,386	8	0	6,878	159	43.4	14.9	39.7	74.1	37.4	20.1
1989	1,890	0	4,912	3	0	6,799	159	42.7	15.0	41.2	74.8	36.4	20.0
1990	595	0	5,375	76	0	5,894	160	36.8	13.0	41.0	74.6	31.8	17.5
1991	565	0	5,439	60	0	5,944	161	37.0	13.2	41.1	74.2	32.0	17.8
1992	1,298	0	5,614	70	0	6,842	162	42.4	15.0	43.2	76.0	34.7	19.7
1993	1,608	0	5,873	11	0	7,470	162	46.0	16.2	45.5	75.5	35.5	21.4
1994	1,500	0	4,837	15	0	6,322	163	38.7	13.4	42.5	79.6	31.6	16.9
1995	1,100	0	4,980	44	0	6,036	165	36.7	12.4	40.0	77.4	31.0	16.0

193 SAO TOME and PRINCIPE	- FOOD BALANCE SHEET OF FISH AND FISHERY PRODUCTS IN LIVE WEIGHT AND FISH CONTRIBUTION TO PROTEIN SUPPLY -												FAO-FIDI
YEAR	PRODUCTION	NON-FOOD USES	IMPORTS	EXPORTS	STOCKS VARIATIONS	TOTAL FOOD SUPPLY	POPULATION	PER CAPUT SUPPLY	FISH PROTEINS	ANIMAL PROTEINS	TOTAL PROTEINS	FISH/ANIMAL PROTEINS	FISH/TOTAL PROTEINS
	(............................... metric tons in live weight ....)						(thousands)	(kilograms)	(grams per caput per day)			(%)	(%)
1961	700	0	0	0	0	700	65	10.8	3.1	7.1	50.3	43.2	6.1
1962	600	0	0	0	0	600	65	9.2	2.6	6.7	46.1	38.4	5.6
1963	500	0	0	0	0	500	66	7.6	2.1	6.4	47.9	33.2	4.5
1964	800	0	0	0	0	800	67	11.9	3.4	7.6	44.9	44.2	7.5
1965	900	0	0	0	0	900	68	13.2	3.7	7.8	47.6	47.8	7.9
1966	800	0	1,000	0	0	1,800	69	26.1	8.4	12.8	50.3	65.3	16.7
1967	900	0	750	0	0	1,650	70	23.6	7.4	12.1	52.7	61.1	14.1
1968	800	0	425	0	0	1,225	71	17.3	5.3	10.4	52.3	51.0	10.2
1969	800	0	425	0	0	1,225	72	17.1	5.2	10.0	48.3	52.6	10.8
1970	800	0	675	0	0	1,475	73	20.2	6.4	10.8	46.6	59.0	13.7
1971	900	0	250	0	0	1,150	74	15.5	4.6	10.2	46.4	45.3	9.9
1972	800	0	250	0	0	1,050	76	13.9	4.2	8.5	45.0	48.8	9.2
1973	800	0	0	0	0	800	77	10.4	2.9	7.4	43.1	39.3	6.8
1974	600	0	0	0	0	600	79	7.6	2.1	5.6	37.5	38.5	5.7
1975	600	0	0	0	0	600	81	7.4	2.0	5.7	35.1	35.0	5.7
1976	646	0	0	0	0	646	83	7.8	1.8	6.1	39.3	29.0	4.5
1977	889	0	0	0	0	889	86	10.4	2.9	7.0	42.8	40.7	6.7
1978	1,175	0	0	0	0	1,175	89	13.3	3.4	7.6	44.0	44.5	7.7
1979	1,618	0	84	0	0	1,702	91	18.6	5.5	9.5	43.9	57.3	12.4
1980	1,847	0	0	0	0	1,847	94	19.7	5.6	11.2	47.1	50.2	12.0
1981	2,158	0	0	0	0	2,158	97	22.4	4.6	8.6	45.6	53.7	10.1
1982	2,688	0	0	0	0	2,688	99	27.2	6.6	12.2	44.2	54.1	14.9
1983	4,012	0	0	761	0	3,251	101	32.1	9.1	12.9	44.9	70.7	20.3
1984	4,444	0	0	0	0	4,444	104	42.9	12.1	16.0	50.7	75.4	23.8
1985	3,988	0	614	0	0	4,602	106	43.4	12.9	16.9	46.9	76.3	27.5
1986	2,833	0	333	0	0	3,166	108	29.2	8.8	12.8	41.9	69.1	21.0
1987	2,798	0	217	0	0	3,015	111	27.2	8.2	14.0	42.1	58.2	19.4
1988	2,900	0	465	0	0	3,365	114	29.6	9.0	14.5	44.4	62.0	20.3
1989	3,116	0	500	0	0	3,616	116	31.1	9.5	14.3	49.5	65.9	19.1
1990	3,583	0	668	0	0	4,251	119	35.7	10.9	13.6	46.8	80.3	23.3
1991	2,221	0	389	0	0	2,610	122	21.4	6.5	11.6	47.1	56.4	13.9
1992	2,094	0	795	0	0	2,889	125	23.2	7.2	10.7	47.9	67.8	15.1
1993	2,600	0	224	0	0	2,824	127	22.2	6.7	10.6	47.2	62.7	14.1
1994	3,000	0	153	0	0	3,153	130	24.2	7.2	11.3	47.8	64.4	15.1
1995	2,800	0	439	7	0	3,232	133	24.3	7.2	11.7	46.3	61.6	15.5

194 SAUDI ARABIA	- FOOD BALANCE SHEET OF FISH AND FISHERY PRODUCTS IN LIVE WEIGHT AND FISH CONTRIBUTION TO PROTEIN SUPPLY -												FAO-FIDI
YEAR	PRODUCTION	NON-FOOD USES	IMPORTS	EXPORTS	STOCKS VARIATIONS	TOTAL FOOD SUPPLY	POPULATION	PER CAPUT SUPPLY	FISH PROTEINS	ANIMAL PROTEINS	TOTAL PROTEINS	FISH/ANIMAL PROTEINS	FISH/TOTAL PROTEINS
	(............ metric tons in live weight ....)						(thousands)	(kilograms)	(grams per caput per day)			(%)	(%)
1961	16,200	0	3,700	200	0	19,700	4,199	4.7	1.3	7.9	45.4	15.9	2.8
1962	18,300	0	400	200	0	18,500	4,335	4.3	1.1	7.8	46.2	14.6	2.5
1963	19,600	0	1,000	300	0	20,300	4,481	4.5	1.2	8.0	47.0	14.8	2.5
1964	17,700	0	600	400	0	17,900	4,634	3.9	1.0	8.0	47.1	12.6	2.1
1965	18,600	0	600	1,000	0	18,200	4,793	3.8	1.0	8.1	47.7	12.4	2.1
1966	20,200	0	2,100	2,400	0	19,900	4,958	4.0	1.1	8.7	48.9	12.3	2.2
1967	22,100	0	600	1,400	0	21,300	5,132	4.2	1.1	9.0	48.6	12.3	2.3
1968	21,100	0	400	600	0	20,900	5,318	3.9	1.0	9.8	48.8	10.6	2.1
1969	22,000	0	200	1,000	0	21,200	5,521	3.8	1.0	10.0	49.0	10.3	2.1
1970	21,700	0	2,700	134	0	24,266	5,745	4.2	1.1	9.9	48.5	11.5	2.3
1971	22,300	0	2,000	1,000	0	23,300	5,991	3.9	1.1	10.5	48.0	10.0	2.2
1972	23,800	0	3,600	300	0	27,100	6,260	4.3	1.2	11.0	44.3	10.6	2.6
1973	26,400	0	10,500	1,400	0	35,500	6,557	5.4	1.5	11.7	45.2	12.6	3.3
1974	23,600	0	9,200	400	0	32,400	6,886	4.7	1.3	13.6	46.1	9.5	2.8
1975	23,000	0	11,466	1,683	0	32,783	7,251	4.5	1.2	14.1	47.8	8.8	2.6
1976	23,300	0	18,669	2,526	0	39,443	7,654	5.2	1.4	17.6	53.8	8.1	2.6
1977	22,800	0	30,785	2,722	0	50,863	8,092	6.3	1.7	21.9	61.7	7.9	2.8
1978	24,950	210	32,236	2,486	0	54,490	8,565	6.4	1.7	23.9	66.3	7.3	2.6
1979	24,460	546	47,882	2,482	0	69,314	9,070	7.6	2.1	28.2	74.3	7.4	2.8
1980	24,775	0	63,511	7,916	0	80,370	9,604	8.4	2.3	31.4	74.6	7.3	3.1
1981	28,570	0	64,207	3,441	0	89,336	10,163	8.8	2.4	30.6	72.9	7.8	3.3
1982	31,820	0	74,068	4,671	0	101,217	10,745	9.4	2.6	32.8	77.5	7.9	3.4
1983	35,300	0	76,680	2,632	0	109,348	11,352	9.6	2.7	35.7	79.1	7.4	3.4
1984	38,490	0	69,077	1,735	0	105,832	11,987	8.8	2.4	34.5	78.9	7.1	3.1
1985	43,696	0	73,923	5,374	0	112,245	12,649	8.9	2.5	32.5	75.8	7.8	3.3
1986	45,526	0	56,444	2,321	-800	98,849	13,347	7.4	2.1	33.3	75.7	6.2	2.7
1987	47,888	0	58,250	4,011	0	102,127	14,070	7.3	2.1	32.6	79.6	6.3	2.6
1988	46,064	0	58,525	2,860	0	101,729	14,788	6.9	1.9	32.0	80.7	6.0	2.4
1989	48,390	0	63,758	4,355	800	108,593	15,457	7.0	2.0	29.4	77.4	6.8	2.6
1990	42,619	450	49,170	967	0	90,371	16,048	5.6	1.6	30.5	80.1	5.1	2.0
1991	42,443	1,280	63,116	2,087	0	102,191	16,545	6.2	1.7	30.4	78.9	5.5	2.1
1992	48,290	1,510	43,276	2,479	0	87,576	16,964	5.2	1.3	25.1	73.0	5.2	1.8
1993	50,524	1,230	42,803	1,487	0	90,609	17,350	5.2	1.3	23.1	71.3	5.8	1.9
1994	56,847	900	49,670	435	0	105,182	17,765	5.9	1.5	23.0	71.6	6.7	2.2
1995	48,387	835	73,169	301	0	120,420	18,255	6.6	1.7	29.0	78.5	5.9	2.2

195 SENEGAL	- FOOD BALANCE SHEET OF FISH AND FISHERY PRODUCTS IN LIVE WEIGHT AND FISH CONTRIBUTION TO PROTEIN SUPPLY -	FAO-FIDI

YEAR	PRODUCTION	NON-FOOD USES	IMPORTS	EXPORTS	STOCKS VARIATIONS	TOTAL FOOD SUPPLY	POPULATION	PER CAPUT SUPPLY	FISH PROTEINS	ANIMAL PROTEINS	TOTAL PROTEINS	FISH/ANIMAL PROTEINS	FISH/TOTAL PROTEINS
	(.. metric tons in live weight					....)	(thousands)	(kilograms)	(grams per caput per day)			(%)	(%)
1961	82,600	3,000	1,300	18,600	0	62,300	3,269	19.1	5.6	14.8	60.5	38.0	9.3
1962	85,900	3,000	0	18,600	0	64,300	3,354	19.2	5.6	15.0	63.1	37.2	8.9
1963	87,500	1,000	400	23,700	0	63,200	3,441	18.4	5.5	15.5	62.7	35.6	8.8
1964	97,600	3,000	200	20,900	0	73,900	3,532	20.9	6.2	16.4	70.4	37.7	8.8
1965	106,600	10,000	200	17,400	0	79,400	3,626	21.9	6.6	16.8	71.8	39.0	9.1
1966	111,700	10,000	200	20,400	0	81,500	3,725	21.9	6.5	17.3	71.0	37.2	9.1
1967	118,000	10,000	200	24,700	0	83,500	3,826	21.8	6.6	17.8	69.3	37.3	9.6
1968	114,900	10,000	600	30,500	0	75,000	3,932	19.1	5.7	16.9	73.0	33.9	7.8
1969	118,900	12,000	0	30,627	0	76,273	4,043	18.9	5.5	16.7	66.2	33.1	8.4
1970	115,087	14,500	913	25,313	0	76,187	4,158	18.3	5.2	15.8	67.2	33.0	7.8
1971	134,612	16,500	153	32,453	0	85,812	4,279	20.1	5.7	15.7	61.9	36.6	9.3
1972	153,022	30,500	3,200	35,608	0	90,114	4,405	20.5	5.8	14.9	64.1	38.7	9.0
1973	193,044	49,000	1,570	45,459	0	100,156	4,535	22.1	6.2	15.7	58.5	39.3	10.6
1974	217,950	57,500	2,160	54,354	-712	107,544	4,669	23.0	6.7	15.1	60.7	44.1	11.0
1975	226,232	56,500	3,248	53,744	0	119,235	4,806	24.8	7.3	16.1	64.1	45.0	11.3
1976	217,641	49,845	11,295	58,166	500	121,426	4,945	24.6	7.4	16.2	65.4	45.6	11.3
1977	216,701	41,900	19,141	70,707	0	123,236	5,088	24.2	7.2	17.4	65.4	41.5	11.1
1978	221,175	48,400	24,384	75,840	-200	121,120	5,234	23.1	6.9	16.9	60.4	40.7	11.4
1979	225,185	36,400	13,929	76,833	-2,019	123,862	5,384	23.0	6.8	16.7	66.4	40.7	10.2
1980	232,752	20,600	16,767	98,696	2,220	132,442	5,538	23.9	7.0	15.9	65.4	43.7	10.6
1981	225,509	17,900	20,833	94,934	0	133,508	5,697	23.4	6.8	16.5	68.1	41.0	10.0
1982	229,343	...	15,129	118,658	0	125,814	5,859	21.5	6.3	16.2	69.3	39.2	9.1
1983	263,534	23,000	23,278	126,234	0	137,577	6,026	22.8	7.1	17.8	66.3	39.9	10.7
1984	247,383	...	22,021	137,568	0	131,836	6,198	21.3	6.4	17.3	63.1	36.8	10.1
1985	246,045	...	24,572	132,079	0	138,538	6,375	21.7	6.8	16.8	67.7	40.2	10.0
1986	255,602	...	21,426	131,324	0	145,704	6,558	22.2	7.0	18.2	70.5	38.4	9.9
1987	256,067	...	20,414	106,788	-11,110	158,582	6,747	23.5	7.3	19.2	67.4	37.8	10.8
1988	260,769	14,000	35,663	96,479	-11,320	174,633	6,940	25.2	7.9	19.3	70.9	40.7	11.1
1989	287,107	10,000	51,335	129,782	-16,850	181,810	7,134	25.5	8.0	19.1	66.6	41.9	12.0
1990	314,520	15,000	55,840	148,334	-16,612	190,415	7,327	26.0	8.0	19.4	70.4	41.0	11.3
1991	324,390	22,000	32,496	183,017	55,658	207,526	7,518	27.6	7.6	19.8	69.1	38.1	10.9
1992	370,253	35,000	26,312	113,326	-670	247,569	7,709	32.1	9.1	22.3	67.4	40.8	13.5
1993	382,238	40,000	18,078	73,237	10	287,089	7,902	36.3	10.5	24.2	68.9	43.5	15.3
1994	350,463	30,000	17,731	75,451	599	263,343	8,102	32.5	9.2	21.4	67.0	43.1	13.8
1995	358,677	20,000	18,948	83,604	130	274,151	8,312	33.0	9.5	21.3	69.1	44.6	13.8

196 SEYCHELLES	- FOOD BALANCE SHEET OF FISH AND FISHERY PRODUCTS IN LIVE WEIGHT AND FISH CONTRIBUTION TO PROTEIN SUPPLY -												FAO-FIDI
YEAR	PRODUCTION	NON-FOOD USES	IMPORTS	EXPORTS	STOCKS VARIATIONS	TOTAL FOOD SUPPLY	POPULATION	PER CAPUT SUPPLY	FISH PROTEINS	ANIMAL PROTEINS	TOTAL PROTEINS	FISH/ANIMAL PROTEINS	FISH/TOTAL PROTEINS
	(.................................... metric tons in live weight ....)						(thousands)	(kilograms)	(grams per caput per day)			(%)	(%)
1961	2,000	0	0	0	0	2,000	43	46.6	13.9	18.3	46.1	76.2	30.2
1962	2,000	0	0	0	0	2,000	44	45.6	13.6	17.9	47.5	76.1	28.6
1963	2,000	0	0	0	0	2,000	45	44.6	13.3	18.2	47.1	73.0	28.3
1964	2,000	0	0	0	0	2,000	46	43.6	13.0	17.7	45.3	73.4	28.7
1965	2,000	0	0	0	0	2,000	47	42.6	12.7	18.4	44.2	69.1	28.7
1966	2,500	0	0	0	0	2,500	48	51.9	15.3	21.3	46.7	72.0	32.8
1967	2,500	0	0	0	0	2,500	49	50.7	15.0	21.3	44.1	70.3	33.9
1968	2,500	0	0	0	0	2,500	51	49.5	14.7	21.1	47.1	69.8	31.3
1969	2,500	0	0	0	0	2,500	52	48.3	14.4	20.8	44.7	69.2	32.2
1970	3,000	0	0	0	0	3,000	53	56.6	16.7	24.4	51.6	68.3	32.3
1971	3,000	0	0	0	0	3,000	54	55.3	16.3	24.6	56.2	66.2	29.0
1972	3,000	0	0	0	0	3,000	56	54.0	16.0	25.2	55.6	63.3	28.7
1973	3,000	0	0	0	0	3,000	57	52.9	16.2	27.3	58.1	59.1	27.8
1974	3,500	0	0	220	0	3,280	58	56.6	16.9	28.3	59.2	59.8	28.6
1975	3,950	0	0	110	0	3,840	59	65.1	19.3	29.6	60.0	65.1	32.2
1976	4,010	0	86	481	0	3,615	60	60.3	17.5	29.0	59.4	60.1	29.4
1977	4,600	0	114	461	0	4,253	61	69.8	19.4	31.0	62.5	62.5	31.0
1978	5,400	630	62	414	0	4,418	62	71.6	21.0	33.5	66.8	62.6	31.4
1979	4,957	52	87	504	-26	4,462	62	71.5	21.7	34.9	66.5	62.1	32.6
1980	4,676	52	44	767	26	3,927	63	62.3	19.4	33.0	65.2	58.8	29.8
1981	5,221	52	45	374	0	4,840	63	76.3	24.9	36.9	71.1	67.4	35.0
1982	4,045	52	20	693	0	3,320	64	52.1	17.3	30.9	64.7	56.0	26.7
1983	3,853	52	20	833	0	2,988	64	46.7	14.4	28.5	60.3	50.5	23.8
1984	3,831	50	43	741	0	3,082	64	47.9	15.6	31.2	62.9	50.0	24.8
1985	4,090	10	44	849	0	3,275	65	50.4	16.1	33.4	65.2	48.2	24.7
1986	4,542	10	19	424	0	4,126	66	62.7	18.9	35.1	68.3	53.8	27.6
1987	3,900	10	2,811	3,246	0	3,455	67	51.7	14.9	30.5	62.9	48.8	23.6
1988	4,343	10	8,305	8,421	0	4,217	68	62.0	18.6	34.3	68.8	54.2	27.0
1989	4,419	10	7,479	8,418	0	3,469	69	50.2	14.9	31.7	66.4	46.8	22.4
1990	5,452	10	8,341	9,085	0	4,698	70	67.1	19.7	36.0	69.5	54.8	28.4
1991	8,165	2,110	10,888	12,045	0	4,897	71	69.2	19.7	35.6	72.1	55.5	27.4
1992	6,713	1,430	8,670	9,024	0	4,929	71	69.1	19.9	38.3	72.1	51.9	27.6
1993	5,278	800	11,425	11,064	0	4,839	72	67.3	19.6	36.4	71.1	53.9	27.6
1994	4,633	10	11,093	10,97[illegible]	0	4,741	72	65.5	20.0	44.8	76.8	44.7	26.0
1995	4,177	0	13,304	12,944	0	4,537	73	62.2	19.1	36.8	75.2	51.9	25.4

197 SIERRA LEONE	- FOOD BALANCE SHEET OF FISH AND FISHERY PRODUCTS IN LIVE WEIGHT AND FISH CONTRIBUTION TO PROTEIN SUPPLY -	FAO-FIDI

YEAR	PRODUCTION	NON-FOOD USES	IMPORTS	EXPORTS	STOCKS VARIATIONS	TOTAL FOOD SUPPLY	POPULATION	PER CAPUT SUPPLY	FISH PROTEINS	ANIMAL PROTEINS	TOTAL PROTEINS	FISH/ANIMAL PROTEINS	FISH/TOTAL PROTEINS
	(...... metric tons in live weight ......)						(thousands)	(kilograms)	(grams per caput per day)			(%)	(%)
1961	22,800	0	2,754	0	0	25,554	2,276	11.2	3.5	6.5	37.8	54.5	9.3
1962	26,000	0	5,644	200	0	31,444	2,313	13.6	4.3	7.3	39.3	58.0	10.8
1963	28,100	0	4,644	0	0	32,744	2,350	13.9	4.2	7.5	41.5	56.4	10.2
1964	29,900	0	10,061	0	0	39,961	2,388	16.7	4.9	8.3	41.3	58.6	11.8
1965	32,600	0	9,771	400	0	41,971	2,429	17.3	5.1	8.7	44.3	58.4	11.5
1966	32,200	0	12,000	0	0	44,200	2,471	17.9	5.3	9.1	46.0	58.9	11.6
1967	33,600	0	7,321	0	0	40,921	2,514	16.3	4.9	8.5	46.7	57.6	10.5
1968	23,500	0	5,750	0	0	29,250	2,559	11.4	3.5	7.1	44.8	49.6	7.8
1969	25,500	0	15,369	400	0	40,469	2,606	15.5	4.5	8.2	43.5	54.7	10.3
1970	30,600	0	15,975	0	0	46,575	2,656	17.5	4.8	8.6	49.1	56.5	9.9
1971	30,600	0	15,095	0	0	45,695	2,707	16.9	4.8	8.3	43.8	57.2	10.9
1972	51,100	0	9,972	0	0	61,072	2,761	22.1	6.4	9.8	43.3	65.1	14.7
1973	66,700	0	19,028	0	0	85,728	2,816	30.4	8.7	12.1	47.7	71.7	18.2
1974	67,739	0	11,606	0	0	79,345	2,873	27.6	8.1	11.3	47.0	71.3	17.2
1975	68,597	0	8,490	26	-500	76,561	2,931	26.1	7.5	10.8	44.6	69.7	16.9
1976	53,792	0	6,236	1	-500	59,527	2,989	19.9	6.3	9.6	44.5	64.9	14.1
1977	52,652	0	8,766	4	0	61,414	3,049	20.1	6.7	10.0	46.2	67.5	14.6
1978	50,080	0	10,589	0	0	60,668	3,110	19.5	6.0	9.5	46.4	63.0	12.9
1979	57,592	0	12,308	69	0	69,831	3,172	22.0	6.8	10.6	51.9	64.0	13.0
1980	49,187	0	25,990	448	0	74,729	3,236	23.1	6.9	10.9	45.0	63.2	15.4
1981	50,010	0	31,889	850	200	81,249	3,301	24.6	7.5	11.5	44.9	65.2	16.7
1982	52,038	0	7,266	940	300	58,663	3,367	17.4	5.9	9.6	45.5	61.0	12.9
1983	50,184	0	13,434	1,811	500	62,307	3,435	18.1	5.9	9.2	41.3	63.9	14.2
1984	51,767	0	8,893	748	0	59,912	3,507	17.1	5.7	9.0	38.8	63.1	14.7
1985	52,521	0	9,674	1,645	0	60,550	3,585	16.9	5.6	8.6	43.1	65.1	13.1
1986	52,313	0	8,108	1,558	0	58,862	3,670	16.0	5.4	8.5	41.4	63.4	13.0
1987	52,206	0	2,793	1,455	0	53,544	3,762	14.2	5.2	8.6	44.7	60.8	11.7
1988	52,273	0	2,872	1,569	-444	53,132	3,853	13.8	5.1	8.3	39.7	62.1	12.9
1989	52,197	0	3,246	1,889	-1,211	52,343	3,933	13.3	4.9	8.0	41.5	61.6	11.9
1990	50,070	0	3,069	3,138	-344	49,656	3,997	12.4	4.6	7.8	42.5	59.6	10.9
1991	60,215	0	2,592	2,097	-20	60,690	4,040	15.0	5.5	8.8	44.3	63.0	12.5
1992	60,648	0	2,101	2,189	464	61,024	4,066	15.0	5.5	8.6	41.7	64.0	13.2
1993	61,017	0	3,682	4,056	1,111	61,755	4,089	15.1	5.5	8.8	40.8	62.8	13.6
1994	61,888	0	2,976	2,160	0	62,704	4,127	15.2	5.6	8.8	44.4	63.1	12.5
1995	62,313	0	2,960	5,655	444	60,062	4,195	14.3	5.3	8.4	43.6	63.4	12.2

200 SINGAPORE	- FOOD BALANCE SHEET OF FISH AND FISHERY PRODUCTS IN LIVE WEIGHT AND FISH CONTRIBUTION TO PROTEIN SUPPLY -	FAO-FIDI

YEAR	PRODUCTION	NON-FOOD USES	IMPORTS	EXPORTS	STOCKS VARIATIONS	TOTAL FOOD SUPPLY	POPULATION	PER CAPUT SUPPLY	FISH PROTEINS	ANIMAL PROTEINS	TOTAL PROTEINS	FISH/ANIMAL PROTEINS	FISH/TOTAL PROTEINS
	(.......... metric tons in live weight					)	(thousands)	(kilograms)	(grams per caput per day)			(%)	(%)
1961	9,700	2,000	113,630	61,473	0	59,857	1,691	35.4	10.5	27.3	75.5	38.4	13.9
1962	11,500	2,000	98,110	53,539	0	54,071	1,743	31.0	9.2	26.1	66.0	35.3	14.0
1963	12,500	2,000	96,414	45,463	0	61,451	1,791	34.3	10.0	28.1	72.7	35.6	13.7
1964	10,400	2,000	88,693	37,906	0	59,187	1,837	32.2	9.6	26.9	61.2	35.7	15.7
1965	11,000	2,000	96,299	36,526	0	68,773	1,880	36.6	11.1	28.6	66.0	38.7	16.7
1966	18,500	2,000	102,169	38,426	-3,200	77,043	1,922	40.1	11.7	30.2	69.2	38.8	16.9
1967	18,200	2,000	110,037	37,518	0	88,719	1,961	45.2	12.9	32.4	75.4	39.8	17.1
1968	17,300	2,000	105,947	38,679	0	82,568	1,999	41.3	12.4	34.1	86.5	36.4	14.4
1969	17,000	2,000	112,073	37,552	0	89,521	2,037	44.0	13.1	38.3	85.9	34.1	15.2
1970	18,300	1,400	124,206	39,185	-200	101,721	2,075	49.0	14.0	43.7	114.6	32.1	12.2
1971	15,200	5,000	139,510	41,474	-200	108,036	2,113	51.1	15.3	45.3	98.2	33.8	15.6
1972	15,700	3,900	132,486	46,820	-1,133	96,333	2,153	44.7	13.2	43.0	111.9	30.7	11.8
1973	18,600	26,500	154,220	51,903	-1,267	93,150	2,192	42.5	12.1	41.3	96.1	29.3	12.6
1974	19,236	20,000	139,490	47,983	-2,067	88,676	2,229	39.8	10.6	42.6	91.2	25.0	11.7
1975	17,560	16,000	135,403	45,847	-300	90,816	2,263	40.1	11.2	42.7	86.1	26.3	13.0
1976	16,429	23,570	154,427	55,028	100	92,358	2,292	40.3	8.9	46.6	98.4	19.0	9.0
1977	15,105	17,430	173,325	79,213	-650	91,136	2,318	39.3	8.8	47.4	112.0	18.6	7.9
1978	16,172	15,545	172,545	83,361	-100	89,711	2,344	38.3	7.8	46.9	89.1	16.5	8.7
1979	16,932	16,595	130,267	78,411	3,393	55,586	2,375	23.4	5.9	44.6	104.1	13.3	5.7
1980	16,044	5,985	147,380	93,118	600	64,921	2,415	26.9	7.2	44.7	103.7	16.1	6.9
1981	16,112	12,085	163,988	89,321	-271	78,423	2,463	31.8	9.0	46.3	83.8	19.5	10.8
1982	19,346	13,365	153,525	90,938	-1,714	66,854	2,519	26.5	7.0	46.1	94.0	15.3	7.5
1983	19,549	16,115	180,058	101,656	3,886	85,722	2,581	33.2	8.5	46.7	92.3	18.2	9.2
1984	26,261	33,790	198,389	93,454	-1,710	95,697	2,645	36.2	9.4	50.0	97.4	18.9	9.7
1985	23,906	13,240	180,037	93,974	-1,629	95,101	2,709	35.1	9.5	47.1	88.9	20.1	10.7
1986	21,487	9,630	206,897	116,327	-1,992	100,434	2,771	36.3	10.6	48.7	99.6	21.8	10.6
1987	16,908	8,010	206,647	138,173	273	77,645	2,832	27.4	8.0	49.2	92.7	16.3	8.7
1988	15,250	10,600	234,416	152,265	-1,939	84,515	2,893	29.2	9.6	54.6	109.6	17.6	8.8
1989	12,633	8,000	210,742	145,714	-218	69,088	2,954	23.4	7.6	48.5	99.5	15.7	7.7
1990	13,348	4,200	230,649	164,164	-559	74,567	3,016	24.7	8.8	49.7	102.2	17.6	8.6
1991	13,099	...	258,910	178,148	1,103	94,225	3,080	30.6	9.2	54.3	109.3	16.9	8.4
1992	11,551	...	263,865	168,904	1,433	106,666	3,143	33.9	10.5	54.5	106.4	19.3	9.9
1993	11,654	...	245,673	154,369	1,630	103,388	3,206	32.3	10.4	54.5	111.1	19.0	9.3
1994	13,661	...	272,249	185,596	2,741	101,805	3,268	31.2	9.2	55.1	106.0	16.7	8.7
1995	13,727	...	268,944	177,412	1,966	105,976	3,327	31.9	9.3	56.3	101.1	16.6	9.2

199 SLOVAKIA	- FOOD BALANCE SHEET OF FISH AND FISHERY PRODUCTS IN LIVE WEIGHT AND FISH CONTRIBUTION TO PROTEIN SUPPLY -	FAO-FIDI

YEAR	PRODUCTION	NON-FOOD USES	IMPORTS	EXPORTS	STOCKS VARIATIONS	TOTAL FOOD SUPPLY	POPULATION	PER CAPUT SUPPLY	FISH PROTEINS	ANIMAL PROTEINS	TOTAL PROTEINS	FISH/ANIMAL PROTEINS	FISH/TOTAL PROTEINS
	(.................................... metric tons in live weight					....................................)	(thousands)	(kilograms)	(grams per caput per day)			(%)	(%)
1961	0	0	0	0	0	0	4,199	0.0	0.0	0.0	0.0	0.0	0.0
1962	0	0	0	0	0	0	4,246	0.0	0.0	0.0	0.0	0.0	0.0
1963	0	0	0	0	0	0	4,288	0.0	0.0	0.0	0.0	0.0	0.0
1964	0	0	0	0	0	0	4,326	0.0	0.0	0.0	0.0	0.0	0.0
1965	0	0	0	0	0	0	4,362	0.0	0.0	0.0	0.0	0.0	0.0
1966	0	0	0	0	0	0	4,396	0.0	0.0	0.0	0.0	0.0	0.0
1967	0	0	0	0	0	0	4,429	0.0	0.0	0.0	0.0	0.0	0.0
1968	0	0	0	0	0	0	4,461	0.0	0.0	0.0	0.0	0.0	0.0
1969	0	0	0	0	0	0	4,494	0.0	0.0	0.0	0.0	0.0	0.0
1970	0	0	0	0	0	0	4,528	0.0	0.0	0.0	0.0	0.0	0.0
1971	0	0	0	0	0	0	4,559	0.0	0.0	0.0	0.0	0.0	0.0
1972	0	0	0	0	0	0	4,596	0.0	0.0	0.0	0.0	0.0	0.0
1973	0	0	0	0	0	0	4,641	0.0	0.0	0.0	0.0	0.0	0.0
1974	0	0	0	0	0	0	4,691	0.0	0.0	0.0	0.0	0.0	0.0
1975	0	0	0	0	0	0	4,736	0.0	0.0	0.0	0.0	0.0	0.0
1976	0	0	0	0	0	0	4,786	0.0	0.0	0.0	0.0	0.0	0.0
1977	0	0	0	0	0	0	4,837	0.0	0.0	0.0	0.0	0.0	0.0
1978	0	0	0	0	0	0	4,888	0.0	0.0	0.0	0.0	0.0	0.0
1979	0	0	0	0	0	0	4,936	0.0	0.0	0.0	0.0	0.0	0.0
1980	0	0	0	0	0	0	4,976	0.0	0.0	0.0	0.0	0.0	0.0
1981	0	0	0	0	0	0	5,016	0.0	0.0	0.0	0.0	0.0	0.0
1982	0	0	0	0	0	0	5,052	0.0	0.0	0.0	0.0	0.0	0.0
1983	0	0	0	0	0	0	5,083	0.0	0.0	0.0	0.0	0.0	0.0
1984	0	0	0	0	0	0	5,113	0.0	0.0	0.0	0.0	0.0	0.0
1985	0	0	0	0	0	0	5,140	0.0	0.0	0.0	0.0	0.0	0.0
1986	0	0	0	0	0	0	5,167	0.0	0.0	0.0	0.0	0.0	0.0
1987	0	0	0	0	0	0	5,191	0.0	0.0	0.0	0.0	0.0	0.0
1988	0	0	0	0	0	0	5,214	0.0	0.0	0.0	0.0	0.0	0.0
1989	0	0	0	0	0	0	5,236	0.0	0.0	0.0	0.0	0.0	0.0
1990	0	0	0	0	0	0	5,256	0.0	0.0	0.0	0.0	0.0	0.0
1991	0	0	0	0	0	0	5,275	0.0	0.0	0.0	0.0	0.0	0.0
1992	0	0	0	0	0	0	5,295	0.0	0.0	0.0	0.0	0.0	0.0
1993	2,773	0	...	516	0	...	5,311	...	...	42.9	84.3	...	...
1994	3,488	0	35,629	732	0	38,386	5,325	7.2	2.0	38.9	87.4	5.0	2.2
1995	3,565	0	34,789	883	0	37,471	5,338	7.0	1.9	38.2	77.5	4.9	2.4

Note: 1993 food supply data not shown as imports statistics not available. For years before 1993 see Czechoslovakia.

198 SLOVENIA	- FOOD BALANCE SHEET OF FISH AND FISHERY PRODUCTS IN LIVE WEIGHT AND FISH CONTRIBUTION TO PROTEIN SUPPLY -												FAO-FIDI
YEAR	PRODUCTION	NON-FOOD USES	IMPORTS	EXPORTS	STOCKS VARIATIONS	TOTAL FOOD SUPPLY	POPULATION	PER CAPUT SUPPLY	FISH PROTEINS	ANIMAL PROTEINS	TOTAL PROTEINS	FISH/ANIMAL PROTEINS	FISH/TOTAL PROTEINS
	(...... metric tons in live weight ......)						(thousands)	(kilograms)	(grams per caput per day)			(%)	(%)
1961	0	0	0	0	0	0	1,590	0.0	0.0	0.0	0.0	0.0	0.0
1962	0	0	0	0	0	0	1,601	0.0	0.0	0.0	0.0	0.0	0.0
1963	0	0	0	0	0	0	1,611	0.0	0.0	0.0	0.0	0.0	0.0
1964	0	0	0	0	0	0	1,621	0.0	0.0	0.0	0.0	0.0	0.0
1965	0	0	0	0	0	0	1,630	0.0	0.0	0.0	0.0	0.0	0.0
1966	0	0	0	0	0	0	1,638	0.0	0.0	0.0	0.0	0.0	0.0
1967	0	0	0	0	0	0	1,645	0.0	0.0	0.0	0.0	0.0	0.0
1968	0	0	0	0	0	0	1,652	0.0	0.0	0.0	0.0	0.0	0.0
1969	0	0	0	0	0	0	1,660	0.0	0.0	0.0	0.0	0.0	0.0
1970	0	0	0	0	0	0	1,670	0.0	0.0	0.0	0.0	0.0	0.0
1971	0	0	0	0	0	0	1,681	0.0	0.0	0.0	0.0	0.0	0.0
1972	0	0	0	0	0	0	1,694	0.0	0.0	0.0	0.0	0.0	0.0
1973	0	0	0	0	0	0	1,709	0.0	0.0	0.0	0.0	0.0	0.0
1974	0	0	0	0	0	0	1,725	0.0	0.0	0.0	0.0	0.0	0.0
1975	0	0	0	0	0	0	1,742	0.0	0.0	0.0	0.0	0.0	0.0
1976	0	0	0	0	0	0	1,760	0.0	0.0	0.0	0.0	0.0	0.0
1977	0	0	0	0	0	0	1,779	0.0	0.0	0.0	0.0	0.0	0.0
1978	0	0	0	0	0	0	1,798	0.0	0.0	0.0	0.0	0.0	0.0
1979	0	0	0	0	0	0	1,816	0.0	0.0	0.0	0.0	0.0	0.0
1980	0	0	0	0	0	0	1,832	0.0	0.0	0.0	0.0	0.0	0.0
1981	0	0	0	0	0	0	1,845	0.0	0.0	0.0	0.0	0.0	0.0
1982	0	0	0	0	0	0	1,856	0.0	0.0	0.0	0.0	0.0	0.0
1983	0	0	0	0	0	0	1,864	0.0	0.0	0.0	0.0	0.0	0.0
1984	0	0	0	0	0	0	1,873	0.0	0.0	0.0	0.0	0.0	0.0
1985	0	0	0	0	0	0	1,881	0.0	0.0	0.0	0.0	0.0	0.0
1986	0	0	0	0	0	0	1,888	0.0	0.0	0.0	0.0	0.0	0.0
1987	0	0	0	0	0	0	1,896	0.0	0.0	0.0	0.0	0.0	0.0
1988	0	0	0	0	0	0	1,906	0.0	0.0	0.0	0.0	0.0	0.0
1989	0	0	0	0	0	0	1,912	0.0	0.0	0.0	0.0	0.0	0.0
1990	0	0	0	0	0	0	1,918	0.0	0.0	0.0	0.0	0.0	0.0
1991	0	0	0	0	0	0	1,922	0.0	0.0	0.0	0.0	0.0	0.0
1992	4,738	0	9,144	5,643	0	8,240	1,924	4.3	1.3	40.7	82.4	3.1	1.5
1993	2,969	0	12,048	4,100	0	10,917	1,925	5.7	1.6	51.8	93.3	3.1	1.7
1994	3,111	0	12,772	4,144	0	11,739	1,925	6.1	1.7	53.8	97.1	3.2	1.8
1995	2,930	0	13,826	2,547	0	14,209	1,925	7.4	2.1	57.5	102.2	3.7	2.1

Note: for years before 1992 see Yugoslavia, S.F.R.

25 SOLOMON ISLANDS	- FOOD BALANCE SHEET OF FISH AND FISHERY PRODUCTS IN LIVE WEIGHT AND FISH CONTRIBUTION TO PROTEIN SUPPLY -	FAO-FIDI

YEAR	PRODUCTION	NON-FOOD USES	IMPORTS	EXPORTS	STOCKS VARIATIONS	TOTAL FOOD SUPPLY	POPULATION	PER CAPUT SUPPLY	FISH PROTEINS	ANIMAL PROTEINS	TOTAL PROTEINS	FISH/ANIMAL PROTEINS	FISH/TOTAL PROTEINS
	(.........	....	metric tons in live weight	....	....	.........)	(thousands)	(kilograms)	(grams per caput per day)			(%)	(%)
1961	4,000	0	...	...	0	4,000	122	32.8	9.3	15.4	47.0	60.2	19.7
1962	5,000	0	...	...	0	5,000	126	39.8	11.2	17.1	48.4	65.9	23.2
1963	5,000	0	...	...	0	5,000	129	38.7	10.9	17.3	48.7	62.9	22.4
1964	5,000	0	...	...	0	5,000	133	37.5	10.6	18.5	51.3	57.2	20.6
1965	5,000	0	...	...	0	5,000	137	36.4	10.3	17.0	49.1	60.3	20.9
1966	7,000	0	...	...	0	7,000	142	49.4	13.9	21.2	52.9	65.9	26.4
1967	7,000	0	600	...	0	7,600	146	52.0	14.7	21.7	55.2	67.8	26.6
1968	7,000	0	600	...	0	7,600	151	50.4	14.2	22.4	54.5	63.5	26.2
1969	7,000	0	800	...	0	7,800	156	50.1	14.2	22.1	54.0	64.0	26.2
1970	8,000	0	800	...	0	8,800	161	54.8	15.5	24.2	56.6	64.1	27.3
1971	13,000	0	1,000	4,100	0	9,900	166	59.6	17.1	25.3	56.1	67.4	30.4
1972	16,400	0	1,000	6,800	-1,000	9,600	172	55.9	15.8	24.5	55.8	64.7	28.4
1973	15,300	0	800	5,100	-1,000	10,000	178	56.3	16.0	24.3	55.4	66.0	28.9
1974	19,390	18	1,000	9,850	50	10,572	184	57.5	16.4	25.4	55.9	64.8	29.4
1975	16,483	9	595	5,834	50	11,285	190	59.3	17.2	24.9	55.2	69.0	31.1
1976	25,693	0	475	13,412	-400	12,356	197	62.7	18.3	26.1	56.6	69.9	32.3
1977	22,294	0	569	11,407	500	11,956	204	58.6	17.0	24.8	55.9	68.6	30.4
1978	28,231	0	558	12,225	-4,000	12,563	211	59.5	17.3	25.4	56.8	68.1	30.4
1979	35,329	0	592	25,157	2,700	13,464	219	61.5	18.1	26.2	57.3	68.9	31.6
1980	34,782	0	875	23,328	1,800	14,130	227	62.4	18.5	27.0	58.4	68.5	31.7
1981	37,534	0	665	25,068	1,200	14,331	235	61.1	18.1	26.0	56.9	69.8	31.9
1982	32,788	0	841	17,640	-1,100	14,889	243	61.3	18.0	26.9	58.2	66.9	30.9
1983	47,338	0	474	32,697	550	15,664	252	62.2	18.4	26.0	57.2	70.8	32.2
1984	48,935	0	728	35,327	650	14,985	261	57.5	16.6	25.0	57.6	66.3	28.8
1985	44,049	0	1,208	29,529	0	15,728	270	58.3	17.1	25.5	58.0	67.0	29.4
1986	55,456	0	1,401	41,821	0	15,036	279	53.8	15.6	23.2	56.3	67.3	27.8
1987	44,594	0	386	29,515	0	15,464	289	53.5	15.5	22.6	55.5	68.5	27.9
1988	55,141	0	777	38,705	0	17,213	299	57.5	17.6	24.2	56.4	72.9	31.2
1989	51,594	0	1,285	31,307	-3,500	18,072	310	58.4	17.7	25.4	57.0	69.6	31.0
1990	44,022	0	187	23,613	-3,600	16,996	320	53.1	15.6	20.5	51.9	76.2	30.1
1991	63,466	0	202	47,392	0	16,275	331	49.1	14.3	18.5	50.5	77.1	28.3
1992	48,704	0	102	38,721	7,100	17,185	343	50.2	15.4	19.6	52.1	78.7	29.6
1993	46,653	405	18	27,023	0	19,243	354	54.4	18.3	22.7	55.3	80.5	33.1
1994	50,541	417	26	38,262	0	11,888	366	32.5	11.3	15.7	47.7	71.7	23.6
1995	70,610	481	61	54,580	0	15,610	378	41.3	14.2	18.7	51.2	76.2	27.8

201 SOMALIA	- FOOD BALANCE SHEET OF FISH AND FISHERY PRODUCTS IN LIVE WEIGHT AND FISH CONTRIBUTION TO PROTEIN SUPPLY -	FAO-FIDI

YEAR	PRODUCTION	NON-FOOD USES	IMPORTS	EXPORTS	STOCKS VARIATIONS	TOTAL FOOD SUPPLY	POPULATION	PER CAPUT SUPPLY	FISH PROTEINS	ANIMAL PROTEINS	TOTAL PROTEINS	FISH/ANIMAL PROTEINS	FISH/TOTAL PROTEINS
	(..		metric tons in live weight		....)		(thousands)	(kilograms)		(grams per caput per day)		(%)	(%)
1961	4,500	0	0	3,700	0	800	3,869	0.2	0.1	33.7	59.3	0.3	0.1
1962	4,500	0	0	4,300	0	200	3,957	0.1	0.0	33.8	59.3	0.1	0.0
1963	4,500	0	0	3,000	0	1,500	4,048	0.4	0.2	34.8	60.1	0.5	0.3
1964	4,500	0	0	2,820	0	1,680	4,142	0.4	0.2	35.0	59.4	0.5	0.3
1965	4,600	0	0	3,200	0	1,400	4,239	0.3	0.1	35.2	61.4	0.4	0.2
1966	4,600	0	0	2,100	0	2,500	4,342	0.6	0.2	35.8	58.8	0.7	0.4
1967	4,700	0	0	1,200	0	3,500	4,450	0.8	0.3	36.7	58.9	0.9	0.5
1968	5,000	0	0	900	0	4,100	4,562	0.9	0.4	38.1	61.0	1.0	0.6
1969	5,000	0	0	700	0	4,300	4,676	0.9	0.4	37.8	61.0	1.0	0.7
1970	5,600	0	0	1,300	0	4,300	4,791	0.9	0.3	37.7	61.5	0.9	0.5
1971	5,700	0	0	3,700	0	2,000	4,905	0.4	0.2	36.1	62.1	0.4	0.2
1972	5,800	0	0	3,100	0	2,700	5,020	0.5	0.2	36.2	59.3	0.5	0.3
1973	5,900	0	0	3,500	0	2,400	5,145	0.5	0.2	34.4	57.3	0.5	0.3
1974	5,980	0	0	3,581	0	2,399	5,293	0.5	0.2	31.8	50.4	0.5	0.3
1975	10,350	600	0	4,766	0	4,984	5,471	0.9	0.3	34.7	57.4	1.0	0.6
1976	8,268	150	0	5,756	0	2,362	5,683	0.4	0.2	34.3	57.7	0.5	0.3
1977	9,830	550	229	1,432	0	8,077	5,924	1.4	0.5	36.8	60.8	1.4	0.8
1978	8,384	0	0	3,607	0	4,777	6,185	0.8	0.3	39.2	57.2	0.7	0.5
1979	10,984	0	0	3,175	0	7,809	6,451	1.2	0.4	38.8	58.6	1.1	0.7
1980	14,330	0	199	1,977	0	12,552	6,713	1.9	0.7	36.7	59.0	1.8	1.1
1981	9,823	0	0	2,207	0	7,616	6,968	1.1	0.4	35.2	66.4	1.1	0.6
1982	9,030	0	20	4,892	0	4,158	7,218	0.6	0.2	34.5	67.7	0.6	0.3
1983	11,495	0	111	1,986	0	9,620	7,456	1.3	0.5	33.5	60.3	1.4	0.8
1984	19,939	0	0	1,206	0	18,733	7,676	2.4	0.9	33.3	59.8	2.6	1.4
1985	16,767	0	0	1,509	0	15,258	7,875	1.9	0.7	32.0	60.0	2.1	1.1
1986	16,900	0	0	1,219	0	15,681	8,053	2.0	0.7	31.4	61.8	2.2	1.1
1987	17,450	0	0	1,811	0	15,639	8,211	1.9	0.7	33.0	64.4	2.0	1.0
1988	18,227	0	0	3,016	0	15,211	8,354	1.8	0.6	32.1	59.4	1.9	1.0
1989	18,196	0	4	4,703	0	13,497	8,489	1.6	0.5	33.3	60.2	1.6	0.9
1990	17,495	0	9	5,552	0	11,952	8,623	1.4	0.5	29.6	57.1	1.6	0.8
1991	16,610	0	9	3,354	0	13,265	8,753	1.5	0.6	29.5	51.2	1.9	1.1
1992	15,500	0	57	3,999	0	11,558	8,884	1.3	0.5	26.1	49.3	1.9	1.0
1993	15,160	0	135	3,598	0	11,697	9,035	1.3	0.5	31.7	51.2	1.6	1.0
1994	16,300	0	54	2,912	0	13,442	9,232	1.5	0.6	36.6	53.3	1.6	1.1
1995	15,500	0	0	3,579	0	11,921	9,491	1.3	0.5	37.1	54.2	1.4	0.9

202 SOUTH AFRICA	- FOOD BALANCE SHEET OF FISH AND FISHERY PRODUCTS IN LIVE WEIGHT AND FISH CONTRIBUTION TO PROTEIN SUPPLY -	FAO-FIDI

YEAR	PRODUCTION	NON-FOOD USES	IMPORTS	EXPORTS	STOCKS VARIATIONS	TOTAL FOOD SUPPLY	POPULATION	PER CAPUT SUPPLY	FISH PROTEINS	ANIMAL PROTEINS	TOTAL PROTEINS	FISH/ANIMAL PROTEINS	FISH/TOTAL PROTEINS
	(....................................		metric tons in live weight		....................................)		(thousands)	(kilograms)		(grams per caput per day)		(%)	(%)
1961	1,010,800	767,600	5,356	150,678	0	97,878	17,855	5.5	2.6	25.0	71.1	10.5	3.7
1962	1,061,100	822,600	4,879	126,272	-17,147	99,960	18,333	5.5	2.6	24.5	68.4	10.4	3.7
1963	1,169,214	972,600	6,036	103,082	7,146	106,715	18,827	5.7	2.5	25.8	71.2	9.8	3.5
1964	1,224,206	990,500	7,121	140,183	10,000	110,644	19,328	5.7	2.4	23.5	68.4	10.4	3.6
1965	1,322,309	1,069,500	7,179	113,066	0	146,922	19,832	7.4	3.2	25.9	72.0	12.3	4.4
1966	1,335,692	1,050,000	7,265	131,906	0	161,052	20,335	7.9	3.3	27.3	73.2	12.0	4.5
1967	1,658,900	1,367,600	7,867	113,489	0	185,678	20,840	8.9	3.6	27.2	73.1	13.3	5.0
1968	2,108,900	1,850,200	9,701	115,408	0	152,993	21,356	7.2	2.8	26.9	72.5	10.4	3.9
1969	1,858,600	1,624,800	10,127	96,818	0	147,109	21,893	6.7	2.7	27.0	73.1	10.2	3.8
1970	1,218,950	984,400	12,242	53,954	0	192,838	22,458	8.6	3.4	27.3	74.4	12.4	4.5
1971	1,178,550	972,000	17,537	41,509	0	182,578	23,056	7.9	3.1	27.1	74.1	11.4	4.2
1972	1,144,004	869,800	7,544	92,359	0	189,389	23,683	8.0	3.2	27.9	76.4	11.3	4.1
1973	1,364,825	998,600	14,731	127,571	0	253,385	24,333	10.4	4.1	29.0	76.2	14.0	5.3
1974	1,430,496	1,007,309	18,554	183,997	0	257,744	24,997	10.3	4.0	29.1	76.6	13.6	5.2
1975	1,368,326	917,519	13,019	131,144	-31,060	301,622	25,669	11.8	4.5	29.7	76.9	15.0	5.8
1976	1,167,103	790,120	12,193	106,656	-18,533	263,988	26,347	10.0	3.3	28.3	75.5	11.8	4.4
1977	966,060	680,517	10,757	79,887	47,313	263,728	27,032	9.8	3.1	27.7	75.4	11.3	4.2
1978	1,013,576	729,113	31,947	50,395	0	266,015	27,728	9.6	3.0	27.6	75.3	10.9	4.0
1979	996,460	690,218	26,319	71,184	-3,748	257,628	28,440	9.1	2.8	27.7	74.6	10.2	3.8
1980	854,280	604,716	64,369	64,549	4,545	253,930	29,170	8.7	2.7	27.5	75.1	9.9	3.6
1981	862,971	651,516	131,016	63,385	2,424	281,510	29,919	9.4	3.0	27.3	73.9	11.1	4.1
1982	824,207	555,212	39,631	52,413	1,212	257,426	30,686	8.4	2.4	27.0	76.1	8.8	3.1
1983	935,869	669,911	63,109	60,275	303	269,095	31,467	8.6	2.6	27.1	73.9	9.4	3.5
1984	735,512	465,112	81,763	57,276	-3,241	291,645	32,254	9.0	2.7	27.3	73.7	9.8	3.6
1985	777,328	456,211	42,757	77,752	424	286,546	33,043	8.7	2.4	26.6	73.2	9.2	3.3
1986	821,412	496,510	39,372	91,180	826	273,920	33,833	8.1	2.4	25.3	70.5	9.4	3.4
1987	1,427,422	1,065,910	55,150	80,870	1,990	337,781	34,626	9.8	3.0	27.3	74.7	10.9	4.0
1988	1,302,361	935,905	63,129	65,634	-3,250	360,702	35,425	10.2	3.1	27.3	75.1	11.5	4.2
1989	878,512	500,010	58,098	65,574	-2,499	368,527	36,237	10.2	3.0	26.3	72.4	11.5	4.2
1990	537,627	125,500	69,754	54,255	-25,626	402,000	37,066	10.9	3.3	26.4	75.8	12.5	4.3
1991	501,041	113,835	50,276	71,764	2,501	368,219	37,912	9.7	2.8	25.2	73.5	11.2	3.9
1992	696,471	279,330	53,656	152,316	26,249	344,731	38,776	8.9	2.8	26.2	72.4	10.7	3.9
1993	563,081	153,900	63,396	104,388	1,873	370,065	39,656	9.3	2.9	24.6	73.0	12.0	4.0
1994	521,067	170,000	73,071	171,982	1,249	253,405	40,552	6.3	2.0	23.5	70.8	8.5	2.8
1995	575,177	108,400	68,443	135,772	0	399,448	41,465	9.6	3.1	25.1	71.7	12.5	4.4

Note: from 1990 trade excludes estimates of Namibian exports.

203 SPAIN	- FOOD BALANCE SHEET OF FISH AND FISHERY PRODUCTS IN LIVE WEIGHT AND FISH CONTRIBUTION TO PROTEIN SUPPLY -	FAO-FIDI

YEAR	PRODUCTION	NON-FOOD USES	IMPORTS	EXPORTS	STOCKS VARIATIONS	TOTAL FOOD SUPPLY	POPULATION	PER CAPUT SUPPLY	FISH PROTEINS	ANIMAL PROTEINS	TOTAL PROTEINS	FISH/ANIMAL PROTEINS	FISH/TOTAL PROTEINS
	(............		metric tons in live weight		....	............)	(thousands)	(kilograms)		(grams per caput per day)		(%)	(%)
1961	957,195	71,900	34,076	124,346	0	795,025	30,750	25.9	7.7	26.4	79.0	29.2	9.8
1962	1,025,321	126,900	41,465	125,433	0	814,453	31,062	26.2	7.3	26.9	80.8	27.1	9.0
1963	1,036,467	120,000	60,735	97,388	0	879,814	31,387	28.0	7.7	31.3	81.1	24.7	9.5
1964	1,107,190	126,500	54,364	150,744	0	884,310	31,720	27.9	7.4	30.6	78.5	24.0	9.4
1965	1,238,994	237,862	75,642	137,235	0	939,539	32,056	29.3	8.0	31.0	77.6	25.9	10.4
1966	1,241,082	163,500	67,102	185,078	0	959,606	32,394	29.6	7.3	33.7	81.7	21.7	8.9
1967	1,324,234	288,434	79,199	185,377	0	929,622	32,735	28.4	7.6	35.5	82.2	21.4	9.2
1968	1,464,567	347,960	95,045	181,388	0	1,030,264	33,079	31.2	8.3	37.1	83.8	22.4	9.9
1969	1,431,980	302,990	53,389	255,475	0	926,905	33,427	27.7	6.9	37.2	80.6	18.6	8.6
1970	1,532,017	253,289	73,921	323,191	0	1,029,458	33,779	30.5	7.2	39.8	83.4	18.0	8.6
1971	1,490,013	305,251	129,253	285,121	0	1,028,894	34,131	30.2	7.6	40.9	83.8	18.6	9.1
1972	1,528,091	344,672	219,765	353,300	0	1,049,884	34,483	30.5	8.0	42.6	86.9	18.7	9.2
1973	1,563,401	406,720	202,089	303,576	0	1,055,195	34,840	30.3	8.1	43.3	88.4	18.8	9.2
1974	1,496,719	358,596	277,870	286,788	0	1,129,205	35,210	32.1	9.0	46.8	92.7	19.2	9.7
1975	1,507,990	326,667	197,851	264,634	0	1,114,539	35,596	31.3	8.6	46.9	92.8	18.4	9.3
1976	1,518,959	193,361	162,601	314,957	0	1,173,242	36,003	32.6	8.8	49.7	93.1	17.6	9.4
1977	1,438,662	189,123	175,970	267,168	0	1,158,341	36,424	31.8	8.5	49.4	92.6	17.3	9.2
1978	1,423,163	233,518	259,950	264,602	-55	1,184,939	36,839	32.2	8.8	51.3	94.1	17.1	9.3
1979	1,282,545	236,513	352,337	272,677	-27	1,125,664	37,218	30.3	8.2	51.8	94.6	15.7	8.6
1980	1,314,727	329,532	399,519	202,676	-10,013	1,172,026	37,542	31.2	8.7	53.8	97.1	16.1	9.0
1981	1,356,363	240,704	378,136	334,390	9,853	1,169,260	37,802	30.9	8.0	54.5	94.3	14.6	8.5
1982	1,474,421	410,539	460,636	298,887	-826	1,224,805	38,005	32.2	8.0	56.4	97.8	14.2	8.2
1983	1,412,764	260,445	370,115	260,941	-964	1,260,530	38,168	33.0	8.3	56.6	97.6	14.7	8.5
1984	1,441,395	310,458	391,785	259,402	991	1,264,311	38,318	33.0	8.2	56.8	99.2	14.4	8.2
1985	1,484,764	400,780	494,544	301,467	30	1,277,092	38,474	33.2	7.7	55.8	98.7	13.8	7.8
1986	1,493,493	341,605	464,758	289,075	255	1,327,825	38,642	34.4	8.6	56.1	98.0	15.4	8.8
1987	1,524,084	688,234	805,949	302,585	0	1,339,214	38,816	34.5	9.7	57.9	100.9	16.7	9.6
1988	1,593,486	798,581	1,000,902	419,460	114	1,376,462	38,987	35.3	9.2	59.2	102.6	15.6	9.0
1989	1,522,433	707,351	962,553	423,999	600	1,354,236	39,142	34.6	9.9	59.5	103.2	16.7	9.6
1990	1,300,047	661,504	1,120,445	411,853	0	1,347,143	39,272	34.3	9.9	60.4	103.9	16.4	9.5
1991	1,271,984	681,534	1,176,759	428,974	0	1,338,228	39,376	34.0	9.9	62.4	106.0	15.9	9.4
1992	1,254,981	599,534	1,079,099	346,090	0	1,388,459	39,458	35.2	10.6	62.6	105.8	17.0	10.1
1993	1,219,981	702,094	1,342,245	461,108	0	1,399,025	39,522	35.4	10.5	64.2	106.1	16.3	9.9
1994	1,310,000	586,764	1,230,543	518,567	0	1,435,209	39,577	36.3	10.7	65.2	108.5	16.4	9.8
1995	1,369,985	575,909	1,266,391	579,820	0	1,480,647	39,627	37.4	10.6	64.7	107.0	16.4	9.9

38 SRI LANKA	- FOOD BALANCE SHEET OF FISH AND FISHERY PRODUCTS IN LIVE WEIGHT AND FISH CONTRIBUTION TO PROTEIN SUPPLY -	FAO-FIDI

YEAR	PRODUCTION	NON-FOOD USES	IMPORTS	EXPORTS	STOCKS VARIATIONS	TOTAL FOOD SUPPLY	POPULATION	PER CAPUT SUPPLY	FISH PROTEINS	ANIMAL PROTEINS	TOTAL PROTEINS	FISH/ANIMAL PROTEINS	FISH/TOTAL PROTEINS
	(.. metric tons in live weight					....)	(thousands)	(kilograms)	(grams per caput per day)			(%)	(%)
1961	74,000	0	103,586	277	0	177,309	10,136	17.5	4.9	8.7	42.7	56.4	11.5
1962	83,900	0	82,909	271	0	166,538	10,385	16.0	4.6	8.5	42.3	54.7	11.0
1963	92,600	0	99,871	373	0	192,093	10,639	18.1	5.1	9.5	43.0	53.9	11.9
1964	101,600	0	111,063	370	0	212,292	10,899	19.5	5.6	10.4	47.7	53.8	11.7
1965	86,000	0	100,887	403	0	186,484	11,164	16.7	4.5	9.3	45.2	48.1	9.9
1966	89,400	0	175,809	500	0	264,709	11,437	23.1	6.0	10.5	48.1	57.6	12.6
1967	99,400	0	100,306	900	0	198,806	11,716	17.0	4.6	9.0	47.3	51.0	9.7
1968	119,500	100	113,506	700	0	232,206	11,994	19.4	5.3	9.8	47.0	54.5	11.3
1969	105,500	300	118,400	800	0	222,800	12,262	18.2	4.9	9.1	46.0	53.6	10.6
1970	98,100	100	84,800	700	0	182,100	12,514	14.6	4.8	9.2	45.3	52.1	10.5
1971	85,200	300	66,090	760	0	150,230	12,747	11.8	3.8	8.1	42.6	47.2	9.0
1972	101,800	200	88,126	1,301	0	188,425	12,965	14.5	4.6	9.3	45.1	49.3	10.2
1973	100,700	400	46,524	2,670	0	144,154	13,174	10.9	3.4	8.0	43.1	43.3	8.0
1974	110,683	600	41,306	2,315	0	149,075	13,384	11.1	3.4	7.3	43.4	45.7	7.7
1975	124,681	600	28,548	1,604	0	151,025	13,603	11.1	3.7	7.7	43.5	48.3	8.6
1976	135,853	1,000	12,975	3,007	0	144,821	13,833	10.5	3.5	7.5	42.9	46.3	8.1
1977	138,747	700	10,481	3,137	0	145,391	14,072	10.3	3.4	7.6	45.2	44.7	7.5
1978	156,567	821	9,304	4,134	0	160,916	14,317	11.2	3.7	8.5	45.9	43.8	8.1
1979	166,254	943	35,056	4,060	0	196,308	14,567	13.5	4.4	9.1	45.6	48.2	9.6
1980	185,506	347	41,343	2,394	-1,129	222,979	14,819	15.1	5.0	9.8	45.1	51.4	11.2
1981	206,816	542	12,072	3,528	0	214,818	15,073	14.3	4.7	9.0	44.7	52.0	10.4
1982	216,304	465	28,052	4,416	-1,432	238,043	15,328	15.5	5.1	9.2	46.0	55.3	11.0
1983	218,842	1,931	26,537	3,479	2,905	242,874	15,581	15.6	5.1	10.1	46.9	50.6	10.9
1984	169,261	580	78,968	4,722	0	242,927	15,827	15.4	5.3	10.1	47.6	52.5	11.1
1985	179,413	560	63,813	4,072	0	238,594	16,060	14.9	5.0	9.9	46.0	50.4	10.8
1986	181,911	0	57,000	4,402	15	234,525	16,280	14.4	4.8	9.1	45.7	53.3	10.5
1987	186,243	0	66,308	3,256	15	249,310	16,487	15.1	5.1	10.5	47.4	48.9	10.8
1988	198,205	0	63,654	4,322	45	257,582	16,684	15.4	5.2	10.3	47.1	50.6	11.1
1989	205,986	0	42,916	4,533	30	244,399	16,873	14.5	4.9	10.2	45.1	47.4	10.8
1990	166,397	0	82,066	4,070	-13	244,380	17,057	14.3	5.1	9.8	46.2	51.4	10.9
1991	199,563	0	92,105	2,936	13	288,745	17,237	16.8	5.9	11.1	47.4	52.6	12.3
1992	208,168	0	98,166	4,977	-5	301,351	17,413	17.3	6.0	11.1	48.2	53.8	12.4
1993	223,400	0	59,877	5,663	-138	277,477	17,585	15.8	5.4	10.4	49.0	51.5	10.9
1994	226,800	0	53,960	5,818	143	275,085	17,757	15.5	5.2	10.5	47.2	49.4	10.9
1995	235,829	0	95,641	6,765	0	324,705	17,928	18.1	6.1	11.9	50.4	51.6	12.1

206 SUDAN	- FOOD BALANCE SHEET OF FISH AND FISHERY PRODUCTS IN LIVE WEIGHT AND FISH CONTRIBUTION TO PROTEIN SUPPLY -												FAO-FIDI
YEAR	PRODUCTION	NON-FOOD USES	IMPORTS	EXPORTS	STOCKS VARIATIONS	TOTAL FOOD SUPPLY	POPULATION	PER CAPUT SUPPLY	FISH PROTEINS	ANIMAL PROTEINS	TOTAL PROTEINS	FISH/ANIMAL PROTEINS	FISH/TOTAL PROTEINS
	(........ metric tons in live weight ....)						(thousands)	(kilograms)	(grams per caput per day)			(%)	(%)
1961	17,300	0	400	2,655	0	15,045	11,386	1.3	0.4	18.8	54.4	2.1	0.7
1962	18,600	0	400	2,700	0	16,300	11,616	1.4	0.4	18.9	59.4	2.2	0.7
1963	19,000	0	800	1,829	0	17,971	11,855	1.5	0.4	19.0	58.3	2.3	0.8
1964	19,300	0	800	2,100	0	18,000	12,102	1.5	0.4	19.3	60.3	2.3	0.7
1965	19,200	0	400	300	0	19,300	12,359	1.6	0.5	19.2	55.5	2.4	0.8
1966	20,000	0	800	0	0	20,800	12,625	1.7	0.5	19.4	55.9	2.5	0.9
1967	20,000	0	800	0	0	20,800	12,902	1.6	0.5	19.8	57.5	2.4	0.8
1968	21,500	0	1,000	0	0	22,500	13,195	1.7	0.5	20.2	59.8	2.5	0.8
1969	21,800	0	800	0	0	22,600	13,513	1.7	0.5	19.6	59.3	2.5	0.8
1970	22,200	0	500	300	0	22,400	13,859	1.6	0.5	19.2	61.3	2.5	0.8
1971	22,200	0	600	600	0	22,200	14,235	1.6	0.5	19.0	61.0	2.4	0.8
1972	22,200	0	600	300	0	22,500	14,641	1.5	0.5	18.7	60.4	2.4	0.7
1973	22,600	0	1,400	0	0	24,000	15,073	1.6	0.5	18.7	60.3	2.5	0.8
1974	22,600	0	200	0	0	22,800	15,531	1.5	0.4	18.5	59.6	2.3	0.7
1975	22,600	0	206	75	0	22,731	16,012	1.4	0.4	19.1	60.2	2.2	0.7
1976	24,700	0	48	54	0	24,694	16,515	1.5	0.4	19.9	60.5	2.2	0.7
1977	23,570	0	152	0	0	23,722	17,037	1.4	0.4	20.6	60.9	2.0	0.7
1978	26,610	0	88	261	0	26,437	17,576	1.5	0.4	21.9	63.4	2.0	0.7
1979	28,520	0	66	710	0	27,876	18,125	1.5	0.5	23.3	63.5	2.0	0.7
1980	26,060	0	1,146	1,590	0	25,616	18,681	1.4	0.4	25.1	64.9	1.6	0.6
1981	28,530	0	1,130	1,914	0	27,746	19,240	1.4	0.4	26.5	65.2	1.6	0.7
1982	29,710	0	1,626	1,800	0	29,536	19,802	1.5	0.4	25.7	64.3	1.7	0.7
1983	29,500	0	812	1,221	0	29,091	20,363	1.4	0.4	25.3	63.6	1.7	0.7
1984	29,790	0	178	27	0	29,941	20,918	1.4	0.4	20.4	58.5	2.1	0.7
1985	26,320	0	940	0	0	27,260	21,462	1.3	0.4	24.0	64.5	1.6	0.6
1986	23,983	0	120	300	0	23,803	21,996	1.1	0.3	22.8	64.6	1.4	0.5
1987	27,243	0	418	192	0	27,469	22,520	1.2	0.4	21.7	63.9	1.7	0.6
1988	29,245	0	448	1,392	0	28,301	23,036	1.2	0.4	21.5	64.2	1.7	0.6
1989	30,300	0	1,886	60	0	32,126	23,549	1.4	0.4	22.5	65.5	1.8	0.6
1990	31,734	0	1,800	60	0	33,474	24,061	1.4	0.4	22.8	65.6	1.8	0.6
1991	33,303	0	1,560	75	0	34,788	24,574	1.4	0.4	22.9	66.9	1.8	0.6
1992	35,200	0	1,600	90	0	36,710	25,088	1.5	0.4	23.0	67.7	1.9	0.6
1993	40,200	0	1,600	146	0	41,654	25,610	1.6	0.5	23.3	70.1	2.1	0.7
1994	44,245	0	1,640	155	0	45,730	26,148	1.8	0.5	23.7	72.1	2.2	0.7
1995	45,000	0	1,706	100	0	46,606	26,707	1.8	0.5	23.8	72.3	2.2	0.7

207 SURINAME	- FOOD BALANCE SHEET OF FISH AND FISHERY PRODUCTS IN LIVE WEIGHT AND FISH CONTRIBUTION TO PROTEIN SUPPLY -	FAO-FIDI

YEAR	PRODUCTION	NON-FOOD USES	IMPORTS	EXPORTS	STOCKS VARIATIONS	TOTAL FOOD SUPPLY	POPULATION	PER CAPUT SUPPLY	FISH PROTEINS	ANIMAL PROTEINS	TOTAL PROTEINS	FISH/ANIMAL PROTEINS	FISH/TOTAL PROTEINS
	(..		metric tons in live weight		....)		(thousands)	(kilograms)	(grams per caput per day)			(%)	(%)
1961	4,700	0	4,697	300	0	9,097	298	30.5	9.4	23.8	50.4	39.6	18.7
1962	3,400	0	4,794	500	0	7,694	306	25.1	7.9	21.6	48.4	36.3	16.2
1963	3,600	0	4,897	629	0	7,868	315	25.0	7.9	23.3	51.2	33.8	15.4
1964	3,700	0	4,230	775	0	7,155	323	22.1	7.8	23.0	56.3	34.0	13.9
1965	4,400	0	4,565	1,000	0	7,965	332	24.0	7.7	22.0	55.0	35.0	14.0
1966	4,500	0	4,180	1,335	0	7,345	342	21.5	7.7	22.2	53.9	34.9	14.4
1967	4,300	0	3,841	1,150	0	6,991	351	19.9	6.9	21.6	54.2	32.0	12.7
1968	5,000	0	4,064	1,570	0	7,494	360	20.8	7.2	22.9	55.6	31.5	13.0
1969	6,000	0	6,630	1,500	0	11,130	368	30.3	9.9	27.1	59.7	36.5	16.6
1970	7,200	0	3,521	1,500	0	9,221	372	24.8	8.1	23.8	56.0	34.2	14.5
1971	6,500	0	4,521	2,850	0	8,171	374	21.8	7.0	23.3	55.0	30.2	12.8
1972	7,700	0	5,021	1,600	0	11,121	374	29.8	9.1	24.7	56.8	37.0	16.1
1973	6,800	0	3,618	1,800	0	8,618	371	23.2	7.3	23.6	54.8	30.9	13.3
1974	6,444	0	2,964	1,000	0	8,408	368	22.9	7.0	24.9	55.0	28.0	12.7
1975	7,361	0	1,822	1,500	0	7,683	365	21.1	6.2	23.2	54.1	26.6	11.4
1976	4,592	0	1,800	471	0	5,921	361	16.4	5.0	25.1	56.9	19.7	8.7
1977	5,068	0	1,856	673	0	6,251	358	17.5	5.1	25.4	58.6	20.2	8.7
1978	3,547	0	4,259	184	-1,000	6,622	356	18.6	6.0	28.3	60.1	21.3	10.0
1979	3,491	0	2,895	295	0	6,091	355	17.2	5.6	26.6	58.2	21.0	9.6
1980	2,960	0	3,509	372	0	6,097	355	17.2	5.7	28.9	61.4	19.7	9.2
1981	3,354	0	4,353	1,493	-200	6,014	357	16.8	5.6	29.7	63.2	18.8	8.8
1982	2,890	0	3,763	1,537	0	5,116	361	14.2	4.8	29.3	64.0	16.3	7.5
1983	3,592	0	4,714	2,132	-600	5,574	366	15.2	5.0	30.0	67.6	16.8	7.5
1984	4,102	0	392	2,038	600	3,056	372	8.2	2.4	25.0	65.2	9.6	3.7
1985	4,141	0	291	2,372	700	2,760	377	7.3	2.1	26.1	65.1	8.2	3.3
1986	3,651	0	20	1,105	-400	2,166	382	5.7	1.6	23.7	60.8	6.9	2.7
1987	5,187	0	50	2,851	900	3,286	386	8.5	2.4	23.9	62.8	10.1	3.9
1988	3,684	0	93	1,210	0	2,567	391	6.6	2.3	23.3	57.7	9.9	4.0
1989	6,188	0	18	1,420	0	4,786	395	12.1	3.9	26.0	62.5	15.0	6.3
1990	6,500	0	69	1,793	0	4,776	400	11.9	3.8	26.4	63.7	14.3	5.9
1991	7,358	0	1,160	1,066	0	7,452	406	18.4	6.2	30.1	67.2	20.4	9.2
1992	10,933	0	114	1,227	0	9,820	411	23.9	7.4	28.8	68.1	25.7	10.9
1993	9,503	0	120	1,537	0	8,086	416	19.4	6.2	25.6	64.9	24.1	9.5
1994	14,466	0	585	1,545	0	13,506	422	32.0	9.7	26.8	65.2	36.4	14.9
1995	13,000	0	728	3,554	0	10,175	427	23.8	7.3	27.5	65.1	26.4	11.2

209 SWAZILAND	- FOOD BALANCE SHEET OF FISH AND FISHERY PRODUCTS IN LIVE WEIGHT AND FISH CONTRIBUTION TO PROTEIN SUPPLY -	FAO-FIDI

YEAR	PRODUCTION	NON-FOOD USES	IMPORTS	EXPORTS	STOCKS VARIATIONS	TOTAL FOOD SUPPLY	POPULATION	PER CAPUT SUPPLY	FISH PROTEINS	ANIMAL PROTEINS	TOTAL PROTEINS	FISH/ANIMAL PROTEINS	FISH/TOTAL PROTEINS
	(............................ metric tons in live weight ....)						(thousands)	(kilograms)	(grams per caput per day)			(%)	(%)
1961	...	...	...	...	...	...	334	...	...	21.9	68.6	...	...
1962	...	...	...	...	...	...	343	...	...	23.4	67.9	...	...
1963	...	...	...	...	...	...	352	...	...	22.5	64.9	...	...
1964	...	...	...	...	...	...	361	...	...	22.7	64.5	...	...
1965	...	...	...	...	...	...	370	...	...	21.6	64.6	...	...
1966	...	...	...	...	...	...	379	...	...	21.5	63.5	...	...
1967	...	...	...	...	...	...	389	...	...	22.1	60.9	...	...
1968	...	...	...	...	...	...	398	...	...	22.1	66.6	...	...
1969	...	...	...	...	...	...	409	...	...	20.1	64.8	...	...
1970	...	...	...	...	...	...	419	...	...	20.4	65.8	...	...
1971	...	...	...	...	...	...	431	...	...	19.6	64.5	...	...
1972	...	...	...	...	...	...	442	...	...	21.2	66.7	...	...
1973	...	...	...	...	...	...	455	...	...	18.2	64.9	...	...
1974	...	...	...	...	...	...	468	...	...	20.1	65.6	...	...
1975	...	...	...	...	...	...	482	...	...	21.2	65.9	...	...
1976	...	...	...	...	...	...	496	...	...	21.6	68.0	...	...
1977	53	0	...	...	0	53	511	0.1	0.0	20.2	62.2	0.2	0.1
1978	44	0	...	...	0	44	527	0.1	0.0	21.2	63.6	0.1	0.0
1979	60	0	...	...	0	60	543	0.1	0.0	21.3	65.1	0.2	0.1
1980	70	0	...	...	0	70	560	0.1	0.0	22.3	64.0	0.2	0.1
1981	70	0	...	...	0	70	577	0.1	0.0	23.6	65.1	0.2	0.1
1982	80	0	...	...	0	80	595	0.1	0.0	22.0	63.6	0.2	0.1
1983	80	0	...	...	0	80	613	0.1	0.0	22.9	64.1	0.2	0.1
1984	90	0	...	...	0	90	631	0.1	0.0	19.7	63.1	0.2	0.1
1985	90	0	...	...	0	90	649	0.1	0.0	19.3	61.3	0.2	0.1
1986	100	0	...	...	0	100	667	0.2	0.0	18.1	63.6	0.3	0.1
1987	100	0	...	...	0	100	686	0.2	0.0	20.4	66.3	0.2	0.1
1988	110	0	...	...	0	110	705	0.2	0.1	19.7	67.6	0.2	0.1
1989	110	0	...	...	0	110	724	0.2	0.1	19.9	67.7	0.2	0.1
1990	110	0	...	...	0	110	744	0.2	0.0	18.8	64.9	0.2	0.1
1991	105	0	...	...	0	105	765	0.1	0.0	19.6	64.1	0.2	0.1
1992	100	0	...	...	0	100	787	0.1	0.0	21.3	64.7	0.2	0.1
1993	110	0	...	...	0	110	810	0.1	0.0	19.8	59.0	0.2	0.1
1994	125	0	...	...	0	125	833	0.2	0.0	18.4	62.8	0.2	0.1
1995	148	0	...	...	0	148	857	0.2	0.1	17.8	61.3	0.3	0.1

Note: trade statistics included with those of South Africa.

210 SWEDEN	- FOOD BALANCE SHEET OF FISH AND FISHERY PRODUCTS IN LIVE WEIGHT AND FISH CONTRIBUTION TO PROTEIN SUPPLY -	FAO-FIDI

YEAR	PRODUCTION	NON-FOOD USES	IMPORTS	EXPORTS	STOCKS VARIATIONS	TOTAL FOOD SUPPLY	POPULATION	PER CAPUT SUPPLY	FISH PROTEINS	ANIMAL PROTEINS	TOTAL PROTEINS	FISH/ANIMAL PROTEINS	FISH/TOTAL PROTEINS
	(..................... metric tons in live weight					.....................)	(thousands)	(kilograms)	(grams per caput per day)			(%)	(%)
1961	268,900	26,000	86,864	133,909	0	195,855	7,526	26.0	8.6	57.4	86.2	14.9	9.9
1962	294,700	37,500	89,757	145,608	0	201,349	7,574	26.6	8.5	55.3	84.7	15.3	10.0
1963	342,600	41,500	91,463	200,740	0	191,823	7,623	25.2	7.9	53.6	84.4	14.8	9.4
1964	377,898	42,500	99,169	240,490	0	194,077	7,677	25.3	8.3	54.6	86.0	15.2	9.6
1965	370,556	44,500	106,860	234,456	0	198,460	7,734	25.7	8.6	56.3	85.9	15.3	10.0
1966	320,490	46,000	122,840	188,376	0	208,954	7,796	26.8	8.5	56.1	87.5	15.2	9.7
1967	341,668	49,500	119,234	203,513	0	207,889	7,863	26.4	8.5	56.1	86.7	15.2	9.8
1968	318,960	40,500	126,718	189,111	0	216,067	7,929	27.3	8.4	56.5	86.2	14.9	9.8
1969	269,758	33,000	131,334	152,997	0	215,095	7,990	26.9	8.2	56.1	85.7	14.6	9.6
1970	285,869	37,000	143,776	162,250	0	230,395	8,043	28.7	8.6	57.7	87.7	14.9	9.8
1971	229,486	29,000	148,888	120,260	0	229,114	8,085	28.3	8.4	59.1	88.9	14.3	9.5
1972	218,105	45,000	158,139	104,649	0	226,595	8,113	27.9	8.1	56.7	85.9	14.3	9.4
1973	220,248	53,000	153,291	103,693	0	216,846	8,144	26.6	8.0	57.4	87.0	13.9	9.2
1974	205,282	54,000	150,750	80,869	-1,000	220,163	8,168	27.0	7.9	60.0	89.6	13.2	8.8
1975	206,894	53,027	155,945	78,131	0	231,681	8,193	28.3	8.1	62.8	92.3	12.9	8.8
1976	204,852	40,062	152,274	88,367	512	229,208	8,219	27.9	9.2	65.4	96.4	14.0	9.5
1977	182,346	20,050	153,934	84,539	-100	231,590	8,246	28.1	9.7	66.8	96.4	14.5	10.1
1978	184,982	47,457	149,518	67,766	412	219,689	8,273	26.6	8.6	65.0	94.3	13.2	9.1
1979	197,527	25,259	162,022	110,519	-175	223,595	8,295	27.0	8.9	66.7	96.4	13.4	9.3
1980	232,733	38,586	165,123	129,119	263	230,414	8,310	27.7	9.1	67.8	98.1	13.5	9.3
1981	257,275	43,208	140,881	140,345	72	214,675	8,319	25.8	8.1	66.8	96.3	12.1	8.4
1982	255,790	44,224	153,235	139,608	-160	225,033	8,323	27.0	8.3	66.4	96.9	12.4	8.5
1983	261,642	39,833	145,005	144,342	175	222,647	8,325	26.7	8.3	65.5	95.8	12.6	8.6
1984	274,617	54,987	148,138	143,984	-222	223,562	8,333	26.8	8.4	66.1	98.1	12.7	8.6
1985	239,721	52,280	145,009	115,973	-800	215,677	8,350	25.8	8.0	65.1	96.1	12.3	8.4
1986	215,315	46,070	152,645	95,472	-1,511	224,907	8,379	26.8	8.4	66.0	96.4	12.8	8.7
1987	214,512	65,021	156,425	82,027	-5,449	218,440	8,416	26.0	8.2	64.1	94.1	12.8	8.7
1988	250,993	80,019	153,618	101,726	2,786	225,652	8,461	26.7	8.2	64.0	95.1	12.8	8.6
1989	257,720	66,023	162,340	123,076	927	231,888	8,510	27.3	8.6	66.3	97.1	12.9	8.8
1990	260,124	57,235	169,842	139,843	600	233,487	8,559	27.3	8.4	63.7	95.6	13.2	8.8
1991	245,016	50,134	168,804	135,771	2,480	230,395	8,608	26.8	8.5	63.2	94.4	13.4	9.0
1992	314,686	63,241	178,692	202,376	1,200	228,962	8,658	26.4	8.2	64.9	99.0	12.7	8.3
1993	347,820	53,360	177,761	226,464	-167	245,591	8,706	28.2	9.2	66.4	101.3	13.8	9.1
1994	394,242	67,369	207,698	277,955	167	256,783	8,750	29.4	9.3	69.1	102.8	13.4	9.0
1995	412,153	20,067	211,790	359,028	0	244,847	8,788	27.9	9.6	66.9	100.7	14.4	9.6

211 SWITZERLAND	- FOOD BALANCE SHEET OF FISH AND FISHERY PRODUCTS IN LIVE WEIGHT AND FISH CONTRIBUTION TO PROTEIN SUPPLY -	FAO-FIDI

YEAR	PRODUCTION	NON-FOOD USES	IMPORTS	EXPORTS	STOCKS VARIATIONS	TOTAL FOOD SUPPLY	POPULATION	PER CAPUT SUPPLY	FISH PROTEINS	ANIMAL PROTEINS	TOTAL PROTEINS	FISH/ANIMAL PROTEINS	FISH/TOTAL PROTEINS
	(.. metric tons in live weight ....)						(thousands)	(kilograms)	(grams per caput per day)			(%)	(%)
1961	2,800	0	39,366	1,000	0	41,166	5,458	7.5	1.9	49.3	93.7	3.8	2.0
1962	2,800	0	42,928	1,333	0	44,394	5,560	8.0	2.0	50.2	89.0	4.0	2.3
1963	3,300	0	46,042	1,000	0	48,342	5,664	8.5	2.1	49.5	90.0	4.2	2.3
1964	2,800	0	47,327	633	0	49,494	5,765	8.6	2.1	50.5	89.7	4.2	2.3
1965	2,700	0	49,346	367	0	51,680	5,857	8.8	2.2	51.1	90.9	4.2	2.4
1966	3,000	0	46,457	400	0	49,057	5,940	8.3	2.0	50.7	88.9	4.0	2.3
1967	3,000	0	48,521	567	0	50,955	6,013	8.5	2.1	50.0	86.5	4.2	2.4
1968	3,500	0	54,538	467	0	57,571	6,078	9.5	2.3	50.9	88.0	4.5	2.6
1969	3,500	0	56,447	467	0	59,480	6,135	9.7	2.3	51.6	88.6	4.4	2.6
1970	2,400	0	62,350	567	0	64,183	6,187	10.4	2.4	54.0	90.8	4.5	2.7
1971	2,800	0	63,346	667	0	65,479	6,232	10.5	2.5	53.3	89.3	4.6	2.8
1972	3,600	0	67,605	467	0	70,739	6,272	11.3	2.6	54.1	91.5	4.9	2.9
1973	3,600	0	65,275	480	0	68,396	6,304	10.9	2.5	55.5	92.5	4.5	2.7
1974	3,800	0	64,241	609	0	67,432	6,327	10.7	2.5	55.1	89.7	4.5	2.8
1975	3,900	0	60,175	681	0	63,395	6,339	10.0	2.3	54.7	88.0	4.2	2.6
1976	3,915	0	44,608	689	0	47,834	6,339	7.6	2.4	57.2	91.4	4.1	2.6
1977	3,982	0	47,660	759	0	50,883	6,329	8.0	2.5	59.6	95.2	4.2	2.6
1978	3,820	0	48,410	350	0	51,880	6,317	8.2	2.5	60.2	96.9	4.2	2.6
1979	3,890	0	52,167	463	0	55,594	6,312	8.8	2.6	59.8	94.1	4.4	2.8
1980	3,500	0	59,102	429	0	62,173	6,319	9.8	2.9	61.7	97.2	4.7	3.0
1981	3,676	0	57,652	303	0	61,024	6,342	9.6	2.8	61.3	98.2	4.6	2.9
1982	3,369	0	57,891	442	0	60,818	6,380	9.5	2.8	61.8	97.0	4.6	2.9
1983	3,540	0	62,668	534	0	65,674	6,427	10.2	3.0	61.2	94.2	4.8	3.1
1984	4,113	0	65,145	909	0	68,350	6,481	10.6	3.1	62.4	95.4	5.0	3.2
1985	4,564	0	68,555	1,080	0	72,039	6,536	11.0	3.2	62.2	96.0	5.2	3.3
1986	4,657	0	74,277	891	0	78,043	6,592	11.8	3.5	61.9	94.9	5.6	3.6
1987	4,523	0	79,676	794	0	83,405	6,649	12.5	3.6	61.9	95.9	5.9	3.8
1988	4,227	38	83,114	1,388	0	85,916	6,708	12.8	3.4	60.1	94.0	5.7	3.6
1989	4,483	39	85,571	1,483	0	88,531	6,770	13.1	3.5	60.1	94.0	5.8	3.7
1990	4,228	27	87,893	1,397	0	90,697	6,834	13.3	3.5	60.3	94.5	5.9	3.7
1991	4,829	42	92,254	1,334	0	95,707	6,901	13.9	3.7	60.9	95.2	6.0	3.8
1992	3,945	52	89,323	1,183	0	92,034	6,970	13.2	3.5	59.5	95.1	5.8	3.6
1993	2,957	42	91,182	996	0	93,101	7,038	13.2	3.5	58.3	93.0	6.0	3.7
1994	2,716	44	96,676	867	0	98,481	7,104	13.9	3.6	57.2	92.4	6.3	3.9
1995	2,749	50	95,825	720	0	97,804	7,166	13.7	3.6	56.3	91.1	6.3	3.9

212 SYRIA	- FOOD BALANCE SHEET OF FISH AND FISHERY PRODUCTS IN LIVE WEIGHT AND FISH CONTRIBUTION TO PROTEIN SUPPLY -	FAO-FIDI

YEAR	PRODUCTION	NON-FOOD USES	IMPORTS	EXPORTS	STOCKS VARIATIONS	TOTAL FOOD SUPPLY	POPULATION	PER CAPUT SUPPLY	FISH PROTEINS	ANIMAL PROTEINS	TOTAL PROTEINS	FISH/ANIMAL PROTEINS	FISH/TOTAL PROTEINS
	(.. metric tons in live weight					....)	(thousands)	(kilograms)	(grams per caput per day)			(%)	(%)
1961	1,500	0	5,200	1,100	0	5,600	4,700	1.2	0.3	12.5	61.9	2.7	0.6
1962	1,500	0	5,100	800	0	5,800	4,846	1.2	0.3	12.8	62.9	2.7	0.5
1963	1,500	0	4,100	800	0	4,800	5,000	1.0	0.3	13.4	64.4	2.0	0.4
1964	1,500	0	5,100	800	0	5,800	5,159	1.1	0.3	12.9	60.8	2.5	0.5
1965	1,600	0	4,700	900	0	5,400	5,325	1.0	0.3	13.1	60.0	2.2	0.5
1966	1,600	0	3,900	900	0	4,600	5,496	0.8	0.2	12.5	58.8	1.9	0.4
1967	1,600	0	1,700	700	0	2,600	5,673	0.5	0.1	12.1	62.1	1.1	0.2
1968	1,700	0	4,400	500	0	5,600	5,857	1.0	0.3	12.4	63.0	2.2	0.4
1969	1,700	0	5,800	600	0	6,900	6,052	1.1	0.3	13.1	61.7	2.5	0.5
1970	1,700	0	4,700	200	0	6,200	6,258	1.0	0.3	12.0	63.4	2.4	0.5
1971	1,700	0	6,900	100	0	8,500	6,478	1.3	0.4	12.6	64.1	3.0	0.6
1972	1,600	0	4,200	50	0	5,750	6,710	0.9	0.2	11.7	65.8	2.1	0.4
1973	1,400	0	10,600	100	0	11,900	6,951	1.7	0.5	11.4	64.7	4.3	0.8
1974	1,650	0	9,900	150	0	11,400	7,195	1.6	0.5	13.6	69.5	3.3	0.7
1975	2,001	0	11,100	50	0	13,051	7,438	1.8	0.5	14.1	69.4	3.5	0.7
1976	3,307	0	4,576	65	0	7,818	7,678	1.0	0.3	14.6	72.3	1.9	0.4
1977	3,617	0	9,459	86	0	12,989	7,918	1.6	0.5	15.8	73.0	2.9	0.6
1978	3,755	0	12,440	117	0	16,078	8,163	2.0	0.6	18.2	75.5	3.1	0.7
1979	3,826	0	15,051	77	0	18,800	8,423	2.2	0.6	20.0	76.6	3.1	0.8
1980	3,971	0	14,801	279	0	18,493	8,704	2.1	0.6	19.9	78.9	3.0	0.8
1981	3,838	0	15,749	61	0	19,526	9,006	2.2	0.6	23.9	85.0	2.5	0.7
1982	4,063	0	11,444	26	0	15,481	9,328	1.7	0.5	24.0	88.0	2.0	0.5
1983	4,410	0	16,775	105	0	21,080	9,668	2.2	0.6	23.7	89.2	2.7	0.7
1984	5,414	0	8,703	100	0	14,017	10,026	1.4	0.4	22.1	86.8	1.8	0.5
1985	5,887	0	13,440	103	-6	19,218	10,397	1.9	0.5	21.2	87.9	2.5	0.6
1986	5,299	0	182	0	-6	5,475	10,785	0.5	0.2	21.3	86.0	0.7	0.2
1987	5,381	0	2,106	35	0	7,452	11,186	0.7	0.2	19.3	84.7	1.0	0.2
1988	5,533	0	109	57	-9	5,576	11,594	0.5	0.1	20.3	83.3	0.7	0.2
1989	5,095	0	64	65	0	5,094	11,997	0.4	0.1	18.1	82.2	0.7	0.2
1990	5,811	0	62	0	-20	5,853	12,388	0.5	0.1	19.2	85.1	0.7	0.2
1991	7,721	0	33	48	0	7,706	12,765	0.6	0.2	18.5	84.9	1.0	0.2
1992	9,654	0	102	42	20	9,734	13,128	0.7	0.2	18.6	85.5	1.2	0.3
1993	9,189	0	197	67	20	9,339	13,484	0.7	0.2	17.9	84.4	1.1	0.2
1994	10,041	0	44	56	0	10,029	13,840	0.7	0.2	17.7	86.6	1.2	0.3
1995	10,540	0	2,026	101	0	12,465	14,203	0.9	0.3	19.0	88.9	1.4	0.3

208 TAJIKISTAN	- FOOD BALANCE SHEET OF FISH AND FISHERY PRODUCTS IN LIVE WEIGHT AND FISH CONTRIBUTION TO PROTEIN SUPPLY -												FAO-FIDI
YEAR	PRODUCTION	NON-FOOD USES	IMPORTS	EXPORTS	STOCKS VARIATIONS	TOTAL FOOD SUPPLY	POPULATION	PER CAPUT SUPPLY	FISH PROTEINS	ANIMAL PROTEINS	TOTAL PROTEINS	FISH/ANIMAL PROTEINS	FISH/TOTAL PROTEINS
	(..		metric tons in live weight		....)		(thousands)	(kilograms)		(grams per caput per day)		(%)	(%)
1961	0	0	0	0	0	0	2,162	0.0	0.0	0.0	0.0	0.0	0.0
1962	0	0	0	0	0	0	2,252	0.0	0.0	0.0	0.0	0.0	0.0
1963	0	0	0	0	0	0	2,338	0.0	0.0	0.0	0.0	0.0	0.0
1964	0	0	0	0	0	0	2,422	0.0	0.0	0.0	0.0	0.0	0.0
1965	0	0	0	0	0	0	2,507	0.0	0.0	0.0	0.0	0.0	0.0
1966	0	0	0	0	0	0	2,589	0.0	0.0	0.0	0.0	0.0	0.0
1967	0	0	0	0	0	0	2,669	0.0	0.0	0.0	0.0	0.0	0.0
1968	0	0	0	0	0	0	2,755	0.0	0.0	0.0	0.0	0.0	0.0
1969	0	0	0	0	0	0	2,848	0.0	0.0	0.0	0.0	0.0	0.0
1970	0	0	0	0	0	0	2,942	0.0	0.0	0.0	0.0	0.0	0.0
1971	0	0	0	0	0	0	3,040	0.0	0.0	0.0	0.0	0.0	0.0
1972	0	0	0	0	0	0	3,144	0.0	0.0	0.0	0.0	0.0	0.0
1973	0	0	0	0	0	0	3,240	0.0	0.0	0.0	0.0	0.0	0.0
1974	0	0	0	0	0	0	3,340	0.0	0.0	0.0	0.0	0.0	0.0
1975	0	0	0	0	0	0	3,442	0.0	0.0	0.0	0.0	0.0	0.0
1976	0	0	0	0	0	0	3,545	0.0	0.0	0.0	0.0	0.0	0.0
1977	0	0	0	0	0	0	3,649	0.0	0.0	0.0	0.0	0.0	0.0
1978	0	0	0	0	0	0	3,750	0.0	0.0	0.0	0.0	0.0	0.0
1979	0	0	0	0	0	0	3,851	0.0	0.0	0.0	0.0	0.0	0.0
1980	0	0	0	0	0	0	3,954	0.0	0.0	0.0	0.0	0.0	0.0
1981	0	0	0	0	0	0	4,064	0.0	0.0	0.0	0.0	0.0	0.0
1982	0	0	0	0	0	0	4,180	0.0	0.0	0.0	0.0	0.0	0.0
1983	0	0	0	0	0	0	4,303	0.0	0.0	0.0	0.0	0.0	0.0
1984	0	0	0	0	0	0	4,433	0.0	0.0	0.0	0.0	0.0	0.0
1985	0	0	0	0	0	0	4,558	0.0	0.0	0.0	0.0	0.0	0.0
1986	0	0	0	0	0	0	4,712	0.0	0.0	0.0	0.0	0.0	0.0
1987	0	0	0	0	0	0	4,874	0.0	0.0	0.0	0.0	0.0	0.0
1988	0	0	0	0	0	0	5,023	0.0	0.0	0.0	0.0	0.0	0.0
1989	0	0	0	0	0	0	5,168	0.0	0.0	0.0	0.0	0.0	0.0
1990	0	0	0	0	0	0	5,303	0.0	0.0	0.0	0.0	0.0	0.0
1991	3,945	0	...	0	0	3,945	5,423	0.7	0.2	18.6	61.3	1.2	0.4
1992	2,129	0	...	0	0	2,129	5,530	0.4	0.1	17.0	68.0	0.7	0.2
1993	2,704	0	...	0	0	2,704	5,629	0.5	0.1	14.9	65.0	1.0	0.2
1994	1,228	0	288	0	0	1,516	5,726	0.3	0.1	14.1	63.9	0.6	0.1
1995	544	0	223	3	0	764	5,828	0.1	0.0	12.9	62.4	0.3	0.1

Note: for years before 1991 see USSR.

215 TANZANIA	- FOOD BALANCE SHEET OF FISH AND FISHERY PRODUCTS IN LIVE WEIGHT AND FISH CONTRIBUTION TO PROTEIN SUPPLY -												FAO-FIDI
YEAR	PRODUCTION	NON-FOOD USES	IMPORTS	EXPORTS	STOCKS VARIATIONS	TOTAL FOOD SUPPLY	POPULATION	PER CAPUT SUPPLY	FISH PROTEINS	ANIMAL PROTEINS	TOTAL PROTEINS	FISH/ANIMAL PROTEINS	FISH/TOTAL PROTEINS
	(..		metric tons in live weight		....)		(thousands)	(kilograms)		(grams per caput per day)		(%)	(%)
1961	70,700	0	2,950	5,986	0	67,664	10,495	6.5	1.9	9.4	42.2	20.2	4.5
1962	70,800	0	3,400	6,615	0	67,585	10,798	6.3	1.8	9.2	41.8	19.8	4.4
1963	76,600	100	3,900	6,012	0	74,388	11,112	6.7	2.0	9.3	42.5	21.1	4.6
1964	100,200	100	3,350	5,500	0	97,950	11,439	8.6	2.5	9.8	41.0	25.9	6.2
1965	102,900	100	4,100	6,110	0	100,790	11,781	8.6	2.5	9.7	41.7	26.0	6.1
1966	102,000	100	4,250	7,011	0	99,139	12,138	8.2	2.4	9.9	44.0	24.2	5.5
1967	128,400	100	4,800	6,107	0	126,993	12,509	10.2	3.0	10.9	42.1	27.3	7.1
1968	152,100	100	5,950	7,609	0	150,341	12,893	11.7	3.5	11.1	42.5	31.1	8.1
1969	150,200	0	4,900	13,131	0	141,969	13,289	10.7	3.2	11.0	42.1	28.7	7.5
1970	186,700	300	5,100	25,468	0	166,032	13,694	12.1	3.6	11.5	42.4	30.9	8.4
1971	181,800	100	6,150	12,827	0	175,023	14,108	12.4	3.7	11.5	41.5	31.8	8.8
1972	157,300	100	4,650	6,967	0	154,883	14,533	10.7	3.1	10.5	41.4	29.9	7.6
1973	162,200	0	7,000	4,353	0	164,847	14,971	11.0	3.2	10.5	43.7	30.7	7.4
1974	174,058	24	5,000	2,222	0	176,812	15,426	11.5	3.3	10.5	46.7	31.8	7.2
1975	211,018	14	4,600	922	0	214,682	15,900	13.5	3.9	11.4	53.1	34.7	7.4
1976	232,393	0	3,378	561	0	235,210	16,395	14.4	4.1	11.6	54.7	35.5	7.5
1977	271,371	37	758	499	0	271,592	16,910	16.1	4.6	12.5	56.2	37.2	8.3
1978	211,348	49	1,376	298	0	212,377	17,446	12.2	3.5	11.5	56.1	30.5	6.2
1979	180,633	40	525	477	0	180,641	18,003	10.0	2.9	10.6	56.5	27.5	5.1
1980	227,894	71	1,120	347	0	228,597	18,581	12.3	3.5	10.6	56.0	33.5	6.3
1981	230,673	26	1,188	385	0	231,450	19,180	12.1	3.5	10.9	54.6	31.9	6.4
1982	227,779	50	617	149	0	228,197	19,803	11.5	3.4	10.8	54.7	31.3	6.1
1983	239,185	29	800	1,028	-625	238,303	20,446	11.7	3.4	10.5	55.3	32.2	6.1
1984	277,287	68	321	744	-125	276,672	21,106	13.1	3.8	10.9	56.7	34.9	6.7
1985	300,644	0	721	585	-500	300,280	21,783	13.8	4.0	11.4	58.2	35.2	6.9
1986	309,855	0	403	1,380	525	309,403	22,471	13.8	4.0	11.5	55.4	34.8	7.2
1987	342,338	0	85	1,618	375	341,181	23,171	14.7	4.3	11.6	56.9	36.8	7.5
1988	392,958	0	985	2,633	350	391,660	23,896	16.4	4.8	12.3	55.6	38.7	8.6
1989	377,097	4	150	1,452	0	375,791	24,663	15.2	4.4	11.9	58.0	37.1	7.6
1990	414,040	0	231	2,581	0	411,690	25,483	16.2	4.7	12.0	53.4	38.9	8.8
1991	326,753	186	255	2,365	0	324,457	26,365	12.3	3.5	11.0	53.1	32.2	6.7
1992	331,585	235	363	2,147	0	329,565	27,301	12.1	3.5	10.8	49.7	32.2	7.0
1993	331,467	430	500	9,363	0	322,174	28,253	11.4	3.3	10.5	49.1	31.6	6.8
1994	288,549	711	500	14,135	0	274,204	29,172	9.4	2.7	9.6	46.9	28.2	5.8
1995	360,000	1,020	1,050	21,598	0	338,432	30,026	11.3	3.3	10.2	49.2	32.0	6.6

216 THAILAND - FOOD BALANCE SHEET OF FISH AND FISHERY PRODUCTS IN LIVE WEIGHT AND FISH CONTRIBUTION TO PROTEIN SUPPLY - FAO-FIDI

YEAR	PRODUCTION	NON-FOOD USES	IMPORTS	EXPORTS	STOCKS VARIATIONS	TOTAL FOOD SUPPLY	POPULATION	PER CAPUT SUPPLY	FISH PROTEINS	ANIMAL PROTEINS	TOTAL PROTEINS	FISH/ANIMAL PROTEINS	FISH/TOTAL PROTEINS
	(........ metric tons in live weight					........)	(thousands)	(kilograms)	(grams per caput per day)			(%)	(%)
1961	305,600	74,200	17,679	13,033	0	236,046	27,189	8.7	2.7	10.5	41.5	25.4	6.4
1962	339,700	61,000	14,226	10,676	0	282,249	28,006	10.1	3.2	11.1	43.0	28.8	7.4
1963	418,700	90,200	12,364	10,200	0	330,664	28,848	11.5	3.5	11.5	45.4	30.8	7.8
1964	577,000	208,900	14,903	12,233	0	370,770	29,725	12.5	4.0	12.4	46.2	32.0	8.5
1965	615,100	186,200	18,929	16,800	0	431,030	30,641	14.1	4.5	12.8	45.7	35.3	9.9
1966	708,100	209,900	15,827	20,940	0	493,087	31,596	15.6	4.9	13.6	48.3	36.1	10.1
1967	847,100	184,600	15,289	19,333	0	658,456	32,588	20.2	5.9	14.6	48.0	40.4	12.3
1968	1,088,800	311,300	14,038	19,067	0	772,471	33,612	23.0	6.9	15.9	50.1	43.5	13.8
1969	1,269,600	501,300	13,252	22,304	0	759,249	34,666	21.9	6.6	15.9	50.0	41.6	13.2
1970	1,437,301	541,200	27,170	40,063	0	883,208	35,745	24.7	7.1	16.6	50.8	42.8	14.0
1971	1,586,949	637,700	35,098	49,436	0	934,911	36,847	25.4	7.0	15.7	49.8	44.2	14.0
1972	1,678,236	744,300	40,066	69,441	0	904,562	37,967	23.8	7.4	16.1	52.1	45.9	14.2
1973	1,678,936	763,500	42,629	114,905	0	843,159	39,098	21.6	6.3	16.0	52.0	39.7	12.2
1974	1,515,482	691,081	52,234	83,564	0	793,071	40,231	19.7	5.9	14.8	50.2	40.0	11.8
1975	1,548,347	636,477	33,604	90,884	-1,333	853,256	41,359	20.6	6.1	15.1	49.9	40.3	12.2
1976	1,660,386	652,090	30,173	111,855	-667	925,947	42,478	21.8	6.1	15.0	49.5	40.7	12.4
1977	2,188,885	1,093,587	22,712	128,644	2,143	991,509	43,587	22.8	6.4	15.6	49.9	40.9	12.7
1978	2,099,310	1,010,401	31,602	164,186	-2,609	953,717	44,673	21.4	6.2	16.8	50.3	36.8	12.2
1979	1,950,676	950,694	82,996	191,848	267	891,397	45,720	19.5	5.1	14.7	47.9	34.7	10.6
1980	1,797,960	825,080	48,132	176,274	540	845,278	46,718	18.1	5.3	14.6	48.4	36.6	11.0
1981	1,989,025	842,474	50,939	232,184	-5,753	959,552	47,656	20.1	5.7	15.4	49.0	37.4	11.7
1982	2,120,133	875,461	50,641	265,764	-8,997	1,020,553	48,539	21.0	5.8	15.1	47.4	38.5	12.2
1983	2,260,024	865,605	64,029	283,826	-13,335	1,161,287	49,391	23.5	6.5	16.1	50.1	40.2	12.9
1984	2,139,606	849,383	123,848	390,184	16,009	1,039,895	50,246	20.7	6.4	16.1	48.9	40.1	13.2
1985	2,229,142	853,794	161,951	470,267	13,992	1,081,023	51,128	21.1	6.3	16.8	49.0	37.2	12.8
1986	2,540,835	1,041,743	271,665	642,925	-16,312	1,111,520	52,050	21.4	6.0	15.9	48.8	37.7	12.3
1987	2,783,644	1,183,450	229,933	737,732	145	1,092,539	52,996	20.6	5.7	16.1	47.5	35.6	12.0
1988	2,651,266	1,047,935	330,435	827,014	-18,081	1,088,672	53,931	20.2	5.7	16.2	48.1	35.3	11.9
1989	2,704,076	1,071,054	447,871	974,958	5,267	1,111,202	54,804	20.3	5.8	16.8	49.8	34.7	11.7
1990	2,789,953	1,152,455	497,335	1,042,310	23,213	1,115,736	55,580	20.1	5.9	17.8	49.7	33.5	12.0
1991	2,972,113	1,119,931	684,323	1,191,269	3,333	1,348,569	56,246	24.0	7.2	19.4	52.1	37.2	13.9
1992	3,246,408	1,082,914	663,772	1,233,371	3,678	1,597,574	56,816	28.1	8.2	20.8	53.4	39.6	15.4
1993	3,385,118	1,112,573	640,773	1,221,682	-1,345	1,690,290	57,316	29.5	8.8	21.1	52.9	41.5	16.6
1994	3,522,233	1,131,212	780,513	1,363,050	-1,583	1,806,901	57,782	31.3	9.3	22.0	54.0	42.3	17.2
1995	3,756,066	1,307,555	739,395	1,295,434	2,369	1,894,842	58,242	32.5	9.6	22.9	55.0	41.9	17.4

217 TOGO	- FOOD BALANCE SHEET OF FISH AND FISHERY PRODUCTS IN LIVE WEIGHT AND FISH CONTRIBUTION TO PROTEIN SUPPLY -												FAO-FIDI

YEAR	PRODUCTION	NON-FOOD USES	IMPORTS	EXPORTS	STOCKS VARIATIONS	TOTAL FOOD SUPPLY	POPULATION	PER CAPUT SUPPLY	FISH PROTEINS	ANIMAL PROTEINS	TOTAL PROTEINS	FISH/ANIMAL PROTEINS	FISH/TOTAL PROTEINS
	(.......... metric tons in live weight ....)						(thousands)	(kilograms)	(grams per caput per day)			(%)	(%)
1961	4,000	0	1,200	0	0	5,200	1,528	3.4	1.0	5.3	44.6	19.6	2.3
1962	4,300	0	1,300	0	0	5,600	1,540	3.6	1.1	5.5	50.4	20.4	2.2
1963	4,500	0	5,800	0	0	10,300	1,556	6.6	2.0	6.4	49.3	30.4	4.0
1964	5,000	0	7,517	0	0	12,517	1,583	7.9	2.4	7.0	51.7	34.3	4.6
1965	6,000	0	8,692	0	0	14,692	1,627	9.0	2.8	7.3	54.0	37.8	5.1
1966	7,300	0	10,244	0	0	17,544	1,690	10.4	3.1	7.7	51.7	40.8	6.0
1967	7,500	0	13,744	0	0	21,244	1,769	12.0	3.6	8.0	55.4	44.8	6.5
1968	10,000	0	8,000	0	0	18,000	1,858	9.7	2.9	7.0	50.7	41.4	5.7
1969	10,000	0	15,085	0	0	25,085	1,944	12.9	3.8	8.0	47.6	47.7	8.1
1970	8,900	0	12,485	0	0	21,385	2,020	10.6	3.1	7.0	49.0	43.8	6.3
1971	10,600	0	10,496	0	0	21,096	2,084	10.1	3.0	6.6	53.1	45.0	5.6
1972	10,600	0	9,800	0	0	20,400	2,137	9.5	2.9	6.7	52.7	42.3	5.4
1973	10,900	0	13,566	0	0	24,466	2,185	11.2	3.3	7.3	50.3	44.9	6.5
1974	11,150	0	15,460	0	0	26,610	2,233	11.9	3.5	7.5	48.7	47.0	7.2
1975	14,420	0	15,813	0	0	30,233	2,285	13.2	3.9	7.9	46.0	49.3	8.5
1976	12,820	0	18,590	7	-4,444	26,958	2,343	11.5	3.5	7.4	44.6	47.4	7.9
1977	10,847	0	13,288	40	-1,111	22,985	2,406	9.6	2.8	6.9	42.9	40.5	6.5
1978	15,363	0	12,036	49	5,556	32,905	2,473	13.3	3.9	7.7	45.3	50.6	8.6
1979	8,096	0	18,519	12	-5,556	21,047	2,543	8.3	2.3	5.9	46.8	39.5	5.0
1980	9,134	0	20,458	135	0	29,458	2,615	11.3	3.3	6.8	48.3	48.3	6.8
1981	10,325	0	23,720	24	0	34,021	2,691	12.6	3.7	7.6	50.4	48.7	7.3
1982	14,530	0	21,851	43	-2,222	34,116	2,770	12.3	3.7	8.2	48.2	45.6	7.8
1983	14,556	0	13,495	58	3,333	31,327	2,853	11.0	3.3	7.4	46.6	44.9	7.1
1984	14,587	0	11,839	36	4,444	30,835	2,939	10.5	3.2	7.8	48.8	41.3	6.6
1985	15,574	0	22,414	72	0	37,916	3,028	12.5	3.8	8.5	53.3	45.0	7.2
1986	14,838	0	19,818	16	0	34,640	3,121	11.1	3.3	8.6	47.4	38.9	7.0
1987	15,189	0	29,108	265	0	44,032	3,217	13.7	4.1	9.4	49.1	44.1	8.4
1988	15,469	0	35,588	283	0	50,775	3,316	15.3	4.5	10.1	50.9	44.6	8.9
1989	16,458	0	34,755	163	-17	51,033	3,418	14.9	4.4	9.7	54.6	45.7	8.1
1990	15,800	0	37,665	230	0	53,235	3,524	15.1	4.4	9.3	56.7	47.8	7.8
1991	12,524	0	34,953	167	-167	47,143	3,632	13.0	3.9	8.7	48.9	45.1	8.0
1992	10,899	0	33,962	251	117	44,727	3,743	12.0	3.5	8.6	49.1	40.9	7.2
1993	17,114	0	20,697	641	0	37,170	3,856	9.6	3.4	8.1	51.3	41.5	6.5
1994	13,202	0	42,851	1,341	67	54,778	3,970	13.8	4.1	8.9	49.7	46.2	8.2
1995	12,222	0	40,166	1,086	0	51,302	4,085	12.6	3.8	8.7	45.2	44.0	8.5

218 TOKELAU	- FOOD BALANCE SHEET OF FISH AND FISHERY PRODUCTS IN LIVE WEIGHT AND FISH CONTRIBUTION TO PROTEIN SUPPLY -												FAO-FIDI
YEAR	PRODUCTION	NON-FOOD USES	IMPORTS	EXPORTS	STOCKS VARIATIONS	TOTAL FOOD SUPPLY	POPULATION	PER CAPUT SUPPLY	FISH PROTEINS	ANIMAL PROTEINS	TOTAL PROTEINS	FISH/ANIMAL PROTEINS	FISH/TOTAL PROTEINS
	(........ metric tons in live weight ....)						(thousands)	(kilograms)	(grams per caput per day)			(%)	(%)
1961	...	0	...	...	0	...	2	...	...	1.4	8.5	...	...
1962	...	0	...	...	0	...	2	...	...	1.4	10.6	...	...
1963	...	0	...	...	0	...	2	...	...	1.3	8.2	...	...
1964	...	0	...	...	0	...	2	...	...	3.0	13.9	...	...
1965	...	0	...	...	0	...	2	...	...	3.0	18.0	...	...
1966	...	0	...	...	0	...	2	...	...	3.0	17.2	...	...
1967	...	0	...	...	0	...	2	...	...	3.0	18.2	...	...
1968	...	0	...	...	0	...	2	...	...	3.0	18.9	...	...
1969	...	0	...	...	0	...	2	...	...	3.1	19.2	...	...
1970	100	0	...	...	0	100	2	50.0	14.1	17.4	33.8	81.2	41.7
1971	100	0	...	...	0	100	2	50.0	14.1	17.4	35.2	80.9	40.1
1972	100	0	...	...	0	100	2	50.0	14.1	17.5	34.9	80.6	40.5
1973	100	0	...	...	0	100	2	50.0	14.1	17.5	35.5	80.5	39.8
1974	100	0	...	...	0	100	2	50.0	14.1	17.7	23.4	80.0	60.2
1975	100	0	...	...	0	100	2	50.0	14.1	17.8	23.7	79.2	59.6
1976	100	0	...	...	0	100	2	50.0	14.1	17.5	23.1	80.6	61.1
1977	100	0	...	...	0	100	2	50.0	14.1	17.5	23.0	80.4	61.3
1978	100	0	...	...	0	100	2	50.0	14.1	18.0	23.5	78.6	60.1
1979	100	0	...	...	0	100	2	50.0	14.1	18.1	23.8	77.8	59.4
1980	100	0	...	...	0	100	2	50.0	14.1	18.2	24.0	77.4	58.8
1981	100	0	...	...	0	100	2	50.0	14.1	18.4	24.2	76.8	58.3
1982	100	0	...	...	0	100	2	50.0	14.1	19.1	25.1	74.0	56.3
1983	100	0	...	...	0	100	2	50.0	14.1	19.2	25.3	73.6	55.8
1984	150	0	...	...	0	150	2	75.0	21.2	26.0	32.2	81.3	65.8
1985	150	0	...	...	0	150	2	75.0	21.2	26.1	32.4	81.0	65.4
1986	150	0	...	...	0	150	2	75.0	21.2	26.2	32.6	80.7	65.0
1987	200	0	...	...	0	200	2	100.0	28.2	33.4	39.8	84.5	70.9
1988	200	0	...	...	0	200	2	100.0	28.2	33.5	40.0	84.4	70.6
1989	200	0	...	...	0	200	2	100.0	28.2	33.5	40.1	84.2	70.3
1990	200	0	...	...	0	200	2	100.0	28.2	33.6	40.3	84.1	70.0
1991	231	0	...	...	0	231	2	115.5	32.6	37.7	44.6	86.4	73.1
1992	191	0	...	...	0	191	2	95.5	27.0	32.1	39.0	84.0	69.1
1993	200	0	...	...	0	200	2	100.0	28.2	33.3	40.4	84.6	69.9
1994	190	0	...	...	0	190	2	95.0	26.8	31.9	39.0	84.0	68.8
1995	200	0	...	...	0	200	2	100.0	28.2	33.3	40.4	84.6	69.9

219 TONGA - FOOD BALANCE SHEET OF FISH AND FISHERY PRODUCTS IN LIVE WEIGHT AND FISH CONTRIBUTION TO PROTEIN SUPPLY - FAO-FIDI

YEAR	PRODUCTION	NON-FOOD USES	IMPORTS	EXPORTS	STOCKS VARIATIONS	TOTAL FOOD SUPPLY	POPULATION	PER CAPUT SUPPLY	FISH PROTEINS	ANIMAL PROTEINS	TOTAL PROTEINS	FISH/ANIMAL PROTEINS	FISH/TOTAL PROTEINS
	(........ metric tons in live weight					........)	(thousands)	(kilograms)	(grams per caput per day)			(%)	(%)
1961	400	0	0	0	0	400	66	6.1	1.7	7.0	45.3	24.3	3.8
1962	400	0	0	0	0	400	68	5.9	1.7	7.8	42.1	21.3	3.9
1963	400	0	0	0	0	400	70	5.7	1.6	8.2	42.6	19.6	3.8
1964	400	0	0	0	0	400	72	5.6	1.6	9.3	45.6	16.8	3.4
1965	500	0	0	0	0	500	74	6.8	1.9	9.8	46.8	19.5	4.1
1966	500	0	0	0	0	500	76	6.6	1.9	10.4	46.6	17.9	4.0
1967	500	0	200	0	0	700	77	9.0	2.6	11.0	48.2	23.2	5.3
1968	500	0	200	0	0	700	79	8.9	2.5	11.3	49.1	22.2	5.1
1969	500	0	386	0	0	886	81	11.0	3.1	11.3	52.4	27.7	5.9
1970	400	0	200	0	0	600	82	7.3	2.1	10.7	50.5	19.3	4.1
1971	500	0	48	0	0	548	83	6.6	1.9	10.9	49.6	17.1	3.7
1972	500	0	100	0	0	600	85	7.1	2.0	11.7	53.1	17.1	3.8
1973	600	0	212	0	0	812	86	9.5	2.7	12.8	55.7	21.0	4.8
1974	726	0	298	0	0	1,024	87	11.8	3.3	17.9	58.3	18.6	5.7
1975	901	0	100	0	0	1,001	88	11.4	3.2	19.4	58.8	16.6	5.5
1976	1,019	0	0	0	0	1,019	89	11.4	3.2	20.1	58.0	16.1	5.6
1977	1,197	0	0	0	0	1,197	90	13.3	3.8	20.1	59.2	18.7	6.3
1978	1,210	0	351	70	0	1,491	91	16.4	4.9	23.5	63.9	20.9	7.7
1979	2,000	0	352	0	0	2,352	92	25.7	7.5	24.4	64.1	30.9	11.7
1980	1,994	0	582	2	0	2,574	92	28.0	8.7	26.9	66.5	32.2	13.0
1981	2,093	0	543	3	0	2,633	92	28.6	8.8	27.7	66.7	31.9	13.2
1982	2,310	0	926	5	0	3,231	92	35.3	11.1	28.5	67.7	39.1	16.5
1983	2,413	0	423	8	0	2,828	91	31.0	9.5	26.9	69.7	35.4	13.6
1984	2,557	0	568	49	0	3,076	91	33.8	10.4	29.5	70.8	35.4	14.7
1985	2,734	0	453	321	0	2,866	91	31.5	9.7	29.6	71.1	32.8	13.6
1986	2,984	0	300	334	0	2,950	92	32.2	9.8	30.2	71.7	32.4	13.6
1987	2,756	0	527	1,157	0	2,126	93	23.0	7.1	29.8	71.4	23.9	10.0
1988	2,717	0	319	468	0	2,569	94	27.4	8.2	33.9	73.6	24.1	11.1
1989	2,691	0	437	379	0	2,750	95	28.9	8.7	34.2	71.5	25.5	12.2
1990	1,656	0	277	308	0	1,625	96	16.9	5.2	29.2	68.0	17.7	7.6
1991	1,959	0	573	413	0	2,119	97	21.9	6.8	30.6	70.0	22.2	9.7
1992	2,283	0	492	451	0	2,324	97	23.9	7.3	31.6	72.3	23.2	10.1
1993	2,481	0	504	461	0	2,524	97	25.9	8.0	33.2	70.0	24.2	11.5
1994	2,540	0	781	477	0	2,845	98	29.1	9.1	36.4	76.2	25.1	12.0
1995	2,596	0	750	275	-13	3,058	98	31.2	9.5	36.9	69.5	25.8	13.7

220 TRINIDAD and TOBAGO	- FOOD BALANCE SHEET OF FISH AND FISHERY PRODUCTS IN LIVE WEIGHT AND FISH CONTRIBUTION TO PROTEIN SUPPLY -	FAO-FIDI

YEAR	PRODUCTION	NON-FOOD USES	IMPORTS	EXPORTS	STOCKS VARIATIONS	TOTAL FOOD SUPPLY	POPULATION	PER CAPUT SUPPLY	FISH PROTEINS	ANIMAL PROTEINS	TOTAL PROTEINS	FISH/ANIMAL PROTEINS	FISH/TOTAL PROTEINS
	(.. metric tons in live weight					....)	(thousands)	(kilograms)	(grams per caput per day)			(%)	(%)
1961	2,400	0	9,282	100	0	11,582	858	13.5	4.4	22.4	57.8	19.7	7.6
1962	2,400	0	10,991	100	0	13,291	869	15.3	5.0	23.0	60.1	21.6	8.3
1963	2,400	0	10,688	200	0	12,888	877	14.7	4.8	24.0	60.8	19.9	7.9
1964	2,400	0	11,091	100	0	13,391	886	15.1	4.9	25.9	63.1	19.0	7.8
1965	3,000	0	10,385	100	0	13,285	896	14.8	4.8	24.5	60.6	19.5	7.9
1966	3,000	0	10,888	400	0	13,488	910	14.8	4.8	25.3	61.6	19.1	7.8
1967	3,200	0	9,679	400	0	12,479	926	13.5	4.4	25.0	60.7	17.5	7.2
1968	3,400	0	8,673	500	0	11,573	942	12.3	4.0	24.0	57.2	16.6	7.0
1969	3,400	0	9,576	703	0	12,273	958	12.8	4.1	26.0	62.3	15.9	6.6
1970	3,600	0	8,970	800	0	11,770	971	12.1	3.9	26.7	63.4	14.7	6.2
1971	4,100	0	7,961	700	0	11,361	981	11.6	3.7	26.8	64.9	14.0	5.8
1972	3,700	0	8,764	500	0	11,964	989	12.1	3.9	27.1	65.0	14.3	5.9
1973	4,000	0	6,439	1,100	0	9,339	996	9.4	3.0	25.6	64.4	11.5	4.6
1974	3,725	0	7,713	948	-350	10,140	1,003	10.1	3.2	27.4	66.1	11.6	4.8
1975	4,417	0	7,417	1,232	175	10,776	1,012	10.7	3.4	30.0	69.3	11.2	4.8
1976	4,417	0	8,759	958	30	12,248	1,023	12.0	3.8	31.7	72.9	11.9	5.2
1977	4,303	0	6,689	833	145	10,304	1,035	10.0	3.2	33.9	74.2	9.3	4.3
1978	4,823	0	10,881	571	-29	15,105	1,049	14.4	4.5	37.4	80.3	11.9	5.6
1979	3,840	0	8,834	549	-21	12,104	1,065	11.4	3.6	31.6	73.3	11.3	4.9
1980	4,461	0	10,282	1,147	50	13,646	1,082	12.6	4.0	37.8	81.3	10.5	4.9
1981	3,804	0	14,322	1,188	0	16,938	1,101	15.4	4.8	36.8	79.6	13.1	6.0
1982	4,574	0	12,301	904	0	15,971	1,121	14.3	4.5	40.3	82.2	11.1	5.4
1983	4,350	0	14,473	311	-30	18,482	1,141	16.2	5.1	41.5	81.4	12.3	6.3
1984	3,723	8	15,250	174	-30	18,760	1,161	16.2	5.1	41.3	80.1	12.3	6.3
1985	5,538	15	13,707	374	-38	18,819	1,178	16.0	5.0	40.6	77.4	12.4	6.5
1986	4,347	3	15,111	1,250	-25	18,179	1,193	15.2	4.7	38.8	80.1	12.2	5.9
1987	6,971	3	11,086	952	-3	17,099	1,206	14.2	4.3	33.5	74.5	12.9	5.8
1988	7,587	2	7,242	1,298	0	13,529	1,216	11.1	3.5	26.1	64.7	13.2	5.3
1989	8,217	5	7,619	1,363	41	14,509	1,226	11.8	3.6	25.8	63.6	14.1	5.7
1990	8,519	4	3,938	1,392	-10	11,051	1,236	8.9	2.7	24.7	62.9	10.8	4.3
1991	12,407	5	6,124	1,392	4	17,138	1,247	13.8	4.1	27.8	64.4	14.8	6.4
1992	15,557	5	4,514	3,456	65	16,675	1,257	13.3	4.0	26.1	61.0	15.3	6.5
1993	11,043	5	4,866	4,809	10	11,105	1,267	8.8	2.7	25.4	58.9	10.5	4.5
1994	14,307	8	5,364	4,890	15	14,788	1,277	11.6	3.5	25.3	58.0	13.9	6.0
1995	13,261	5	6,853	5,232	-30	14,847	1,287	11.5	3.5	25.7	61.7	13.6	5.7

222 TUNISIA	- FOOD BALANCE SHEET OF FISH AND FISHERY PRODUCTS IN LIVE WEIGHT AND FISH CONTRIBUTION TO PROTEIN SUPPLY -	FAO-FIDI

YEAR	PRODUCTION	NON-FOOD USES	IMPORTS	EXPORTS	STOCKS VARIATIONS	TOTAL FOOD SUPPLY	POPULATION	PER CAPUT SUPPLY	FISH PROTEINS	ANIMAL PROTEINS	TOTAL PROTEINS	FISH/ANIMAL PROTEINS	FISH/TOTAL PROTEINS
	(............ metric tons in live weight ....)						(thousands)	(kilograms)	(grams per caput per day)			(%)	(%)
1961	21,500	0	811	5,900	0	16,411	4,297	3.8	1.1	10.4	56.5	10.5	1.9
1962	20,600	0	400	5,400	0	15,600	4,376	3.6	1.0	10.1	55.0	10.2	1.9
1963	25,800	0	583	5,900	0	20,483	4,457	4.6	1.3	9.9	55.1	13.2	2.4
1964	21,400	0	361	5,645	0	16,116	4,541	3.6	1.0	10.1	55.1	9.8	1.8
1965	23,000	0	667	2,569	0	21,098	4,630	4.6	1.3	11.1	56.8	12.1	2.4
1966	25,900	0	444	3,461	0	22,884	4,723	4.8	1.4	11.7	58.4	12.2	2.4
1967	28,000	0	333	2,979	0	25,355	4,822	5.3	1.5	12.0	58.4	12.7	2.6
1968	28,000	0	444	1,965	0	26,480	4,923	5.4	1.6	11.7	57.8	13.6	2.8
1969	29,700	0	333	3,267	0	26,767	5,025	5.3	1.6	11.8	57.6	13.4	2.8
1970	24,034	0	0	3,300	0	20,734	5,127	4.0	1.1	10.6	56.5	10.7	2.0
1971	27,475	0	1,000	2,200	0	26,275	5,227	5.0	1.4	11.4	63.6	12.5	2.2
1972	28,095	0	222	2,468	-200	25,649	5,327	4.8	1.5	11.9	66.6	12.7	2.3
1973	32,102	0	889	3,177	0	29,813	5,431	5.5	1.7	13.1	68.4	13.0	2.5
1974	42,532	0	1,444	3,064	-250	40,663	5,543	7.3	2.2	14.4	70.3	15.5	3.2
1975	45,128	0	1,254	3,603	-200	42,580	5,668	7.5	2.1	15.2	70.5	13.9	3.0
1976	49,833	0	515	4,151	-1,350	44,848	5,805	7.7	2.2	16.7	71.9	13.4	3.1
1977	55,531	0	562	4,936	100	51,257	5,954	8.6	2.5	17.8	75.7	13.9	3.3
1978	56,193	0	810	6,911	200	50,292	6,112	8.2	2.4	16.1	77.6	14.8	3.1
1979	59,497	0	300	7,499	1,300	53,597	6,278	8.5	2.5	15.7	76.6	15.7	3.2
1980	62,362	5	474	6,581	-150	56,100	6,448	8.7	2.5	16.3	79.8	15.5	3.2
1981	58,982	505	405	6,843	100	52,139	6,622	7.9	2.3	16.7	75.6	13.5	3.0
1982	64,753	2,000	500	3,724	100	59,630	6,801	8.8	2.5	17.2	74.5	14.6	3.4
1983	68,710	1,500	1,706	7,027	-863	61,026	6,980	8.7	2.5	17.6	78.5	14.3	3.2
1984	76,594	1,302	904	8,678	503	68,021	7,159	9.5	2.7	18.1	79.3	15.0	3.4
1985	91,147	2,000	1,112	9,466	150	80,943	7,334	11.0	3.2	19.1	82.3	16.9	3.9
1986	92,843	1,252	11	13,217	160	78,545	7,505	10.5	3.1	20.1	83.2	15.2	3.7
1987	99,357	1,000	56	15,066	264	83,611	7,672	10.9	3.3	19.7	85.1	16.6	3.8
1988	103,056	1,002	1,694	17,558	141	86,331	7,836	11.0	3.2	19.9	84.2	16.3	3.9
1989	95,491	3,500	958	15,255	194	77,888	7,998	9.7	2.8	19.5	84.3	14.5	3.4
1990	88,923	1,181	1,373	14,716	-204	74,195	8,162	9.1	2.6	18.2	83.5	14.5	3.2
1991	87,994	...	376	14,027	300	74,643	8,325	9.0	2.7	18.2	85.8	14.7	3.1
1992	88,824	...	486	12,147	-994	76,170	8,489	9.0	2.7	19.5	84.8	13.7	3.1
1993	83,962	...	1,011	15,324	505	70,153	8,654	8.1	2.4	18.6	85.1	12.7	2.8
1994	87,010	...	5,877	13,758	529	79,658	8,820	9.0	2.7	19.2	82.8	13.8	3.2
1995	84,405	...	13,145	10,073	0	87,477	8,987	9.7	2.8	19.7	83.5	14.4	3.4

223 TURKEY	- FOOD BALANCE SHEET OF FISH AND FISHERY PRODUCTS IN LIVE WEIGHT AND FISH CONTRIBUTION TO PROTEIN SUPPLY -	FAO-FIDI

YEAR	PRODUCTION	NON-FOOD USES	IMPORTS	EXPORTS	STOCKS VARIATIONS	TOTAL FOOD SUPPLY	POPULATION	PER CAPUT SUPPLY	FISH PROTEINS	ANIMAL PROTEINS	TOTAL PROTEINS	FISH/ANIMAL PROTEINS	FISH/TOTAL PROTEINS
	(...... metric tons in live weight ......)						(thousands)	(kilograms)	(grams per caput per day)			(%)	(%)
1961	81,100	1,100	0	11,736	0	68,264	28,233	2.4	0.8	25.9	89.0	3.1	0.9
1962	60,100	500	0	7,976	0	51,624	28,948	1.8	0.6	25.6	87.3	2.2	0.7
1963	130,200	800	0	6,835	0	122,565	29,664	4.1	1.4	25.7	87.9	5.4	1.6
1964	121,000	3,200	0	7,553	0	110,247	30,394	3.6	1.2	25.2	87.4	4.9	1.4
1965	134,400	6,200	0	11,525	0	116,675	31,151	3.8	1.3	24.9	87.6	5.1	1.5
1966	116,800	1,800	0	11,829	0	103,171	31,934	3.2	1.1	25.3	88.2	4.3	1.2
1967	187,900	11,000	0	7,800	0	169,100	32,741	5.2	1.7	25.7	88.1	6.6	1.9
1968	133,200	5,900	0	6,353	0	120,947	33,573	3.6	1.2	24.9	87.3	4.6	1.3
1969	169,700	1,700	0	7,447	0	160,553	34,433	4.7	1.5	25.3	89.2	6.1	1.7
1970	180,483	1,000	0	7,424	0	172,059	35,321	4.9	1.6	24.8	89.8	6.4	1.8
1971	161,327	1,700	0	4,523	0	155,104	36,243	4.3	1.4	24.6	89.8	5.8	1.6
1972	174,733	3,800	0	4,425	0	166,508	37,196	4.5	1.5	24.5	91.0	6.1	1.6
1973	145,426	4,100	0	7,475	0	133,851	38,161	3.5	1.2	24.4	88.5	4.9	1.3
1974	128,583	13,971	0	6,843	0	107,769	39,109	2.8	0.9	24.5	90.3	3.8	1.0
1975	121,688	16,530	0	2,566	0	102,592	40,025	2.6	0.8	25.1	91.6	3.4	0.9
1976	152,650	14,096	4	2,496	0	136,062	40,895	3.3	1.1	25.2	93.4	4.4	1.2
1977	164,464	16,520	0	4,849	0	143,095	41,731	3.4	1.1	25.9	94.5	4.4	1.2
1978	244,118	36,038	0	6,881	0	201,199	42,569	4.7	1.5	26.3	93.5	5.8	1.6
1979	349,681	89,406	0	9,321	0	250,954	43,460	5.8	1.9	26.4	96.5	7.3	2.0
1980	426,845	94,718	0	9,671	0	322,456	44,438	7.3	2.5	26.8	97.1	9.5	2.6
1981	470,156	106,500	0	15,381	0	348,275	45,515	7.7	2.5	27.4	96.9	9.2	2.6
1982	503,457	99,040	0	15,076	0	389,341	46,674	8.3	2.8	26.1	95.8	10.6	2.9
1983	557,286	178,225	0	17,148	0	361,913	47,890	7.6	2.5	24.5	97.1	10.1	2.6
1984	566,930	240,450	31	24,203	0	302,308	49,123	6.2	2.0	26.9	97.9	7.4	2.0
1985	578,068	119,339	91	21,658	-996	436,166	50,345	8.7	2.9	27.8	101.9	10.3	2.8
1986	582,864	134,625	571	25,710	0	423,100	51,550	8.2	2.7	28.7	102.0	9.5	2.7
1987	627,907	212,963	2,707	29,041	-910	387,700	52,741	7.4	2.4	26.9	100.7	8.9	2.4
1988	674,177	190,928	2,261	28,698	-400	456,412	53,905	8.5	2.7	27.0	101.3	10.0	2.7
1989	455,940	92,361	5,740	44,732	-200	324,387	55,027	5.9	1.9	25.8	100.6	7.3	1.9
1990	384,948	33,212	16,862	34,084	400	334,913	56,098	6.0	1.9	25.8	101.9	7.2	1.8
1991	364,628	66,155	23,925	25,331	747	297,814	57,109	5.2	1.6	25.9	102.6	6.2	1.6
1992	454,345	37,647	30,517	23,308	375	424,282	58,065	7.3	2.2	26.2	101.4	8.6	2.2
1993	559,234	108,810	18,724	16,783	200	452,565	58,986	7.7	2.3	26.5	102.4	8.6	2.2
1994	603,333	111,898	28,886	38,248	0	482,072	59,903	8.1	2.4	26.4	102.1	9.2	2.4
1995	652,193	64,267	37,558	44,213	0	581,271	60,838	9.6	3.1	27.0	104.7	11.4	2.9

213 TURKMENISTAN	- FOOD BALANCE SHEET OF FISH AND FISHERY PRODUCTS IN LIVE WEIGHT AND FISH CONTRIBUTION TO PROTEIN SUPPLY -	FAO-FIDI

YEAR	PRODUCTION	NON-FOOD USES	IMPORTS	EXPORTS	STOCKS VARIATIONS	TOTAL FOOD SUPPLY	POPULATION	PER CAPUT SUPPLY	FISH PROTEINS	ANIMAL PROTEINS	TOTAL PROTEINS	FISH/ANIMAL PROTEINS	FISH/TOTAL PROTEINS
	(............ metric tons in live weight					............)	(thousands)	(kilograms)	(grams per caput per day)			(%)	(%)
1961	0	0	0	0	0	0	1,653	0.0	0.0	0.0	0.0	0.0	0.0
1962	0	0	0	0	0	0	1,713	0.0	0.0	0.0	0.0	0.0	0.0
1963	0	0	0	0	0	0	1,774	0.0	0.0	0.0	0.0	0.0	0.0
1964	0	0	0	0	0	0	1,834	0.0	0.0	0.0	0.0	0.0	0.0
1965	0	0	0	0	0	0	1,891	0.0	0.0	0.0	0.0	0.0	0.0
1966	0	0	0	0	0	0	1,944	0.0	0.0	0.0	0.0	0.0	0.0
1967	0	0	0	0	0	0	2,002	0.0	0.0	0.0	0.0	0.0	0.0
1968	0	0	0	0	0	0	2,062	0.0	0.0	0.0	0.0	0.0	0.0
1969	0	0	0	0	0	0	2,125	0.0	0.0	0.0	0.0	0.0	0.0
1970	0	0	0	0	0	0	2,189	0.0	0.0	0.0	0.0	0.0	0.0
1971	0	0	0	0	0	0	2,251	0.0	0.0	0.0	0.0	0.0	0.0
1972	0	0	0	0	0	0	2,316	0.0	0.0	0.0	0.0	0.0	0.0
1973	0	0	0	0	0	0	2,380	0.0	0.0	0.0	0.0	0.0	0.0
1974	0	0	0	0	0	0	2,449	0.0	0.0	0.0	0.0	0.0	0.0
1975	0	0	0	0	0	0	2,520	0.0	0.0	0.0	0.0	0.0	0.0
1976	0	0	0	0	0	0	2,588	0.0	0.0	0.0	0.0	0.0	0.0
1977	0	0	0	0	0	0	2,655	0.0	0.0	0.0	0.0	0.0	0.0
1978	0	0	0	0	0	0	2,724	0.0	0.0	0.0	0.0	0.0	0.0
1979	0	0	0	0	0	0	2,793	0.0	0.0	0.0	0.0	0.0	0.0
1980	0	0	0	0	0	0	2,864	0.0	0.0	0.0	0.0	0.0	0.0
1981	0	0	0	0	0	0	2,936	0.0	0.0	0.0	0.0	0.0	0.0
1982	0	0	0	0	0	0	3,009	0.0	0.0	0.0	0.0	0.0	0.0
1983	0	0	0	0	0	0	3,083	0.0	0.0	0.0	0.0	0.0	0.0
1984	0	0	0	0	0	0	3,156	0.0	0.0	0.0	0.0	0.0	0.0
1985	0	0	0	0	0	0	3,225	0.0	0.0	0.0	0.0	0.0	0.0
1986	0	0	0	0	0	0	3,310	0.0	0.0	0.0	0.0	0.0	0.0
1987	0	0	0	0	0	0	3,405	0.0	0.0	0.0	0.0	0.0	0.0
1988	0	0	0	0	0	0	3,482	0.0	0.0	0.0	0.0	0.0	0.0
1989	0	0	0	0	0	0	3,581	0.0	0.0	0.0	0.0	0.0	0.0
1990	0	0	0	0	0	0	3,668	0.0	0.0	0.0	0.0	0.0	0.0
1991	43,041	0	...	...	0	43,041	3,753	11.5	3.4	30.4	83.7	11.3	4.1
1992	33,701	0	...	...	0	33,701	3,835	8.8	2.6	30.5	83.1	8.6	3.2
1993	16,901	0	...	...	0	16,901	3,915	4.3	1.3	28.7	72.9	4.5	1.8
1994	16,331	0	830	1,205	0	15,956	3,995	4.0	1.2	26.2	71.7	4.5	1.7
1995	10,207	0	179	726	0	9,660	4,075	2.4	0.7	25.9	73.1	2.7	1.0

Note. for years before 1991 see USSR.

224 TURKS and CAICOS ISLANDS	- FOOD BALANCE SHEET OF FISH AND FISHERY PRODUCTS IN LIVE WEIGHT AND FISH CONTRIBUTION TO PROTEIN SUPPLY -	FAO-FIDI

YEAR	PRODUCTION	NON-FOOD USES	IMPORTS	EXPORTS	STOCKS VARIATIONS	TOTAL FOOD SUPPLY	POPULATION	PER CAPUT SUPPLY	FISH PROTEINS	ANIMAL PROTEINS	TOTAL PROTEINS	FISH/ANIMAL PROTEINS	FISH/TOTAL PROTEINS
	(...... metric tons in live weight					......)	(thousands)	(kilograms)	(grams per caput per day)			(%)	(%)
1961	400	110	0	0	0	290	6	47.7	12.6	...	...	...	...
1962	500	210	0	0	0	290	6	47.6	12.5	...	...	...	...
1963	600	300	0	0	0	300	6	49.4	13.1	...	...	...	...
1964	600	310	0	0	0	290	6	48.1	12.7	...	...	...	...
1965	600	310	0	0	0	290	6	48.3	12.7	...	...	...	...
1966	600	300	0	0	0	300	6	50.1	13.2	...	...	...	...
1967	700	380	0	0	0	320	6	53.4	13.7	...	...	...	...
1968	700	390	0	0	0	310	6	51.6	13.2	...	...	...	...
1969	700	300	0	100	0	300	6	49.9	13.2	...	...	...	...
1970	700	300	0	100	0	300	6	50.0	12.7	...	...	...	...
1971	800	400	0	100	0	300	6	50.2	12.8	...	...	...	...
1972	900	300	0	300	0	300	6	50.4	12.9	...	...	...	...
1973	900	300	0	300	0	300	6	50.6	12.9	...	...	...	...
1974	890	202	0	380	0	308	6	51.8	14.6	...	...	...	...
1975	1,132	504	0	300	0	328	6	54.7	15.3	...	...	...	...
1976	1,269	775	0	180	0	314	6	51.4	13.9	...	...	...	...
1977	946	483	0	140	0	323	6	51.6	13.9	...	...	...	...
1978	1,128	657	0	129	0	342	6	52.9	14.4	...	...	...	...
1979	1,385	847	0	159	0	379	7	56.4	15.3	...	...	...	...
1980	1,190	630	0	165	0	395	7	56.4	15.3	...	...	...	...
1981	989	0	0	589	0	400	7	54.7	15.1	...	...	...	...
1982	1,061	145	0	503	0	413	8	53.9	14.3	...	...	...	...
1983	1,236	0	0	795	0	441	8	54.8	15.0	...	...	...	...
1984	1,212	90	0	685	0	437	9	51.4	13.8	...	...	...	...
1985	1,349	100	0	820	0	429	9	47.7	12.8	...	...	...	...
1986	1,379	0	0	1,010	0	369	10	38.6	10.7	...	...	...	...
1987	1,248	0	0	868	0	380	10	37.3	10.3	...	...	...	...
1988	1,185	0	24	820	0	389	11	35.9	9.9	...	...	...	...
1989	1,152	428	33	306	0	451	11	39.4	10.9	...	...	...	...
1990	941	281	40	239	0	461	12	38.4	10.7	...	...	...	...
1991	1,120	331	29	313	0	505	12	40.6	11.1	...	...	...	...
1992	1,193	207	3	434	0	555	13	43.3	11.7	...	...	...	...
1993	1,459	409	12	577	0	485	13	36.8	10.1	...	...	...	...
1994	1,420	360	17	600	0	477	14	35.2	9.7	...	...	...	...
1995	1,400	208	47	660	0	579	14	41.4	11.3	...	...	...	...

227 TUVALU	- FOOD BALANCE SHEET OF FISH AND FISHERY PRODUCTS IN LIVE WEIGHT AND FISH CONTRIBUTION TO PROTEIN SUPPLY -	FAO-FIDI

YEAR	PRODUCTION	NON-FOOD USES	IMPORTS	EXPORTS	STOCKS VARIATIONS	TOTAL FOOD SUPPLY	POPULATION	PER CAPUT SUPPLY	FISH PROTEINS	ANIMAL PROTEINS	TOTAL PROTEINS	FISH/ANIMAL PROTEINS	FISH/TOTAL PROTEINS
	(..		metric tons in live weight		....)		(thousands)	(kilograms)	(grams per caput per day)			(%)	(%)
1961	...	0	...	...	0	...	5	...	...	2.3	8.6	...	...
1962	...	0	...	...	0	...	5	...	...	2.2	8.3	...	...
1963	...	0	...	...	0	...	6	...	...	2.1	7.9	...	...
1964	...	0	...	...	0	...	6	...	...	2.0	9.8	...	...
1965	...	0	...	...	0	...	6	...	...	2.0	9.6	...	...
1966	...	0	...	...	0	...	6	...	...	2.0	9.9	...	...
1967	...	0	...	...	0	...	6	...	...	2.0	9.0	...	...
1968	...	0	...	...	0	...	6	...	...	2.0	9.0	...	...
1969	...	0	...	...	0	...	6	...	...	2.0	9.1	...	...
1970	...	0	...	...	0	...	6	...	...	2.0	9.9	...	...
1971	...	0	...	...	0	...	6	...	...	2.0	8.7	...	...
1972	...	0	...	...	0	...	6	...	...	2.0	9.2	...	...
1973	...	0	...	...	0	...	6	...	...	2.1	10.0	...	...
1974	...	0	...	...	0	...	6	...	...	2.0	9.2	...	...
1975	...	0	...	...	0	...	6	...	...	2.0	9.0	...	...
1976	...	0	...	...	0	...	6	...	...	1.9	10.2	...	...
1977	...	0	...	...	0	...	7	...	...	13.5	36.3	...	...
1978	80	0	...	...	0	80	7	11.0	3.1	19.3	43.6	16.1	7.1
1979	100	0	...	...	0	100	8	13.0	3.7	19.6	43.5	18.7	8.4
1980	150	0	...	...	0	150	8	18.8	5.3	23.4	49.1	22.6	10.8
1981	180	0	...	...	0	180	8	22.1	6.2	17.5	38.5	35.7	16.2
1982	429	100	...	...	0	329	8	40.5	14.8	23.9	48.0	61.9	30.9
1983	784	400	...	...	0	384	8	47.7	17.4	28.3	49.4	61.2	35.1
1984	840	440	...	...	0	400	8	50.1	16.0	23.8	45.4	67.3	35.3
1985	313	0	...	...	0	313	8	39.1	11.1	20.7	51.8	53.6	21.5
1986	660	230	...	...	0	430	8	53.1	16.7	24.2	50.2	68.9	33.2
1987	933	480	...	...	0	453	8	54.7	17.3	25.9	43.0	66.8	40.2
1988	1,409	940	...	...	0	469	9	55.1	17.8	28.3	57.2	62.6	31.0
1989	519	20	...	...	0	499	9	57.0	17.1	26.3	57.2	64.9	29.8
1990	518	0	...	...	0	518	9	57.6	17.0	26.7	52.2	63.6	32.5
1991	526	0	...	...	0	526	9	56.9	16.4	29.2	54.6	55.9	30.0
1992	499	0	...	...	0	499	9	52.6	15.0	27.4	53.0	54.6	28.2
1993	1,460	900	...	...	0	560	10	57.7	17.6	29.5	52.2	59.9	33.8
1994	561	0	...	269	0	292	10	29.6	9.5	25.2	49.9	37.7	19.0
1995	399	0	...	269	0	130	10	13.0	3.8	21.5	47.2	17.9	8.1

226 UGANDA	- FOOD BALANCE SHEET OF FISH AND FISHERY PRODUCTS IN LIVE WEIGHT AND FISH CONTRIBUTION TO PROTEIN SUPPLY -												FAO-FIDI
YEAR	PRODUCTION	NON-FOOD USES	IMPORTS	EXPORTS	STOCKS VARIATIONS	TOTAL FOOD SUPPLY	POPULATION	PER CAPUT SUPPLY	FISH PROTEINS	ANIMAL PROTEINS	TOTAL PROTEINS	FISH/ANIMAL PROTEINS	FISH/TOTAL PROTEINS
	(.......... metric tons in live weight ....)						(thousands)	(kilograms)	(grams per caput per day)			(%)	(%)
1961	61,200	0	299	6,412	0	55,087	6,821	8.1	2.4	11.5	54.3	20.6	4.4
1962	64,500	100	0	2,960	0	61,440	7,100	8.7	2.5	11.0	53.3	23.0	4.8
1963	69,600	100	0	585	0	68,915	7,398	9.3	2.7	11.0	52.7	24.7	5.2
1964	70,500	200	200	592	0	69,908	7,715	9.1	2.6	10.6	52.8	24.9	5.0
1965	76,300	100	0	301	0	75,899	8,047	9.4	2.8	10.6	52.2	26.1	5.3
1966	83,300	100	200	2,096	0	81,304	8,397	9.7	2.8	10.5	48.7	27.1	5.8
1967	99,600	0	902	1,503	0	98,999	8,762	11.3	3.3	10.9	53.6	30.5	6.2
1968	108,400	0	600	1,500	0	107,500	9,130	11.8	3.5	10.7	55.1	32.4	6.3
1969	125,300	0	200	1,500	0	124,000	9,482	13.1	3.9	11.3	54.4	34.1	7.1
1970	129,000	0	2,250	2,400	0	128,850	9,806	13.1	3.9	11.5	55.7	33.9	7.0
1971	137,000	0	3,750	2,004	0	138,746	10,096	13.7	4.1	11.7	58.6	34.8	7.0
1972	166,000	0	1,250	1,715	0	165,535	10,356	16.0	4.7	12.3	60.0	38.4	7.8
1973	169,500	0	1,750	1,514	0	169,736	10,607	16.0	4.7	12.5	53.8	37.6	8.7
1974	167,500	0	1,250	1,008	0	167,742	10,876	15.4	4.5	12.4	56.8	36.8	8.0
1975	188,000	0	...	702	0	187,298	11,183	16.8	5.0	12.7	55.1	39.2	9.0
1976	152,400	0	...	...	0	152,400	11,535	13.2	4.0	11.4	56.8	34.6	7.0
1977	219,220	0	...	...	0	219,220	11,925	18.4	5.5	13.2	56.5	41.6	9.7
1978	223,830	0	...	...	0	223,830	12,336	18.1	5.4	13.6	57.7	39.8	9.4
1979	179,930	0	...	...	0	179,930	12,740	14.1	4.2	12.2	51.3	34.5	8.2
1980	165,840	0	...	...	0	165,840	13,120	12.6	3.8	11.2	46.9	33.8	8.1
1981	166,590	0	...	...	0	166,590	13,472	12.4	3.7	11.1	46.9	33.4	7.9
1982	170,035	0	...	...	0	170,035	13,804	12.3	3.7	10.9	48.1	33.7	7.6
1983	172,000	0	...	...	0	172,000	14,123	12.2	3.6	10.9	51.8	33.4	7.0
1984	212,300	0	...	...	0	212,300	14,439	14.7	4.4	11.6	50.9	37.7	8.6
1985	160,800	0	...	...	0	160,800	14,762	10.9	3.3	10.5	45.3	31.1	7.2
1986	197,600	0	...	...	0	197,600	15,089	13.1	3.9	10.2	46.6	38.3	8.4
1987	200,000	0	...	...	0	200,000	15,421	13.0	3.9	9.9	46.1	39.3	8.4
1988	214,291	0	...	...	0	214,291	15,778	13.6	4.1	10.2	49.5	39.9	8.2
1989	212,205	0	...	...	0	212,205	16,182	13.1	3.9	10.7	53.2	36.6	7.4
1990	245,223	0	...	4,165	0	241,058	16,649	14.5	4.3	11.6	55.4	37.3	7.8
1991	254,900	0	...	11,723	0	243,178	17,190	14.2	4.2	11.5	54.1	36.8	7.8
1992	250,000	0	...	12,128	0	237,873	17,796	13.4	4.0	11.2	52.7	35.6	7.6
1993	219,814	0	...	16,823	0	202,992	18,440	11.0	3.3	10.5	52.1	31.3	6.3
1994	213,145	0	...	18,412	0	194,733	19,080	10.2	3.1	9.8	52.1	31.1	5.9
1995	208,807	0	...	21,580	0	187,227	19,689	9.5	2.8	9.5	51.3	29.8	5.5

230 UKRAINE	- FOOD BALANCE SHEET OF FISH AND FISHERY PRODUCTS IN LIVE WEIGHT AND FISH CONTRIBUTION TO PROTEIN SUPPLY -	FAO-FIDI

YEAR	PRODUCTION	NON-FOOD USES	IMPORTS	EXPORTS	STOCKS VARIATIONS	TOTAL FOOD SUPPLY	POPULATION	PER CAPUT SUPPLY	FISH PROTEINS	ANIMAL PROTEINS	TOTAL PROTEINS	FISH/ANIMAL PROTEINS	FISH/TOTAL PROTEINS
	(..		metric tons in live weight		....)		(thousands)	(kilograms)	(grams per caput per day)			(%)	(%)
1961	0	0	0	0	0	0	43,406	0.0	0.0	0.0	0.0	0.0	0.0
1962	0	0	0	0	0	0	43,902	0.0	0.0	0.0	0.0	0.0	0.0
1963	0	0	0	0	0	0	44,456	0.0	0.0	0.0	0.0	0.0	0.0
1964	0	0	0	0	0	0	44,979	0.0	0.0	0.0	0.0	0.0	0.0
1965	0	0	0	0	0	0	45,375	0.0	0.0	0.0	0.0	0.0	0.0
1966	0	0	0	0	0	0	45,806	0.0	0.0	0.0	0.0	0.0	0.0
1967	0	0	0	0	0	0	46,237	0.0	0.0	0.0	0.0	0.0	0.0
1968	0	0	0	0	0	0	46,628	0.0	0.0	0.0	0.0	0.0	0.0
1969	0	0	0	0	0	0	46,987	0.0	0.0	0.0	0.0	0.0	0.0
1970	0	0	0	0	0	0	47,317	0.0	0.0	0.0	0.0	0.0	0.0
1971	0	0	0	0	0	0	47,705	0.0	0.0	0.0	0.0	0.0	0.0
1972	0	0	0	0	0	0	48,089	0.0	0.0	0.0	0.0	0.0	0.0
1973	0	0	0	0	0	0	48,423	0.0	0.0	0.0	0.0	0.0	0.0
1974	0	0	0	0	0	0	48,726	0.0	0.0	0.0	0.0	0.0	0.0
1975	0	0	0	0	0	0	49,016	0.0	0.0	0.0	0.0	0.0	0.0
1976	0	0	0	0	0	0	49,270	0.0	0.0	0.0	0.0	0.0	0.0
1977	0	0	0	0	0	0	49,483	0.0	0.0	0.0	0.0	0.0	0.0
1978	0	0	0	0	0	0	49,667	0.0	0.0	0.0	0.0	0.0	0.0
1979	0	0	0	0	0	0	49,854	0.0	0.0	0.0	0.0	0.0	0.0
1980	0	0	0	0	0	0	49,961	0.0	0.0	0.0	0.0	0.0	0.0
1981	0	0	0	0	0	0	50,137	0.0	0.0	0.0	0.0	0.0	0.0
1982	0	0	0	0	0	0	50,336	0.0	0.0	0.0	0.0	0.0	0.0
1983	0	0	0	0	0	0	50,531	0.0	0.0	0.0	0.0	0.0	0.0
1984	0	0	0	0	0	0	50,726	0.0	0.0	0.0	0.0	0.0	0.0
1985	0	0	0	0	0	0	50,914	0.0	0.0	0.0	0.0	0.0	0.0
1986	0	0	0	0	0	0	51,095	0.0	0.0	0.0	0.0	0.0	0.0
1987	0	0	0	0	0	0	51,280	0.0	0.0	0.0	0.0	0.0	0.0
1988	0	0	0	0	0	0	51,441	0.0	0.0	0.0	0.0	0.0	0.0
1989	0	0	0	0	0	0	51,765	0.0	0.0	0.0	0.0	0.0	0.0
1990	0	0	0	0	0	0	51,891	0.0	0.0	0.0	0.0	0.0	0.0
1991	865,227	0	50,000	300,000	0	615,227	51,961	11.8	3.5	47.3	98.5	7.5	3.6
1992	525,801	0	150,444	200,000	0	476,245	51,977	9.2	2.6	40.3	92.5	6.5	2.9
1993	371,343	0	107,405	129,772	-5,263	343,713	51,944	6.6	2.0	36.0	89.4	5.6	2.3
1994	310,722	0	27,612	133,957	0	204,377	51,869	3.9	1.2	36.0	85.7	3.4	1.4
1995	413,882	175	38,965	163,916	5,263	294,019	51,757	5.7	1.7	35.8	82.5	4.9	2.1

Note: for years before 1991 see USSR.

228 USSR	- FOOD BALANCE SHEET OF FISH AND FISHERY PRODUCTS IN LIVE WEIGHT AND FISH CONTRIBUTION TO PROTEIN SUPPLY -	FAO-FIDI

YEAR	PRODUCTION	NON-FOOD USES	IMPORTS	EXPORTS	STOCKS VARIATIONS	TOTAL FOOD SUPPLY	POPULATION	PER CAPUT SUPPLY	FISH PROTEINS	ANIMAL PROTEINS	TOTAL PROTEINS	FISH/ANIMAL PROTEINS	FISH/TOTAL PROTEINS
	(..		metric tons in live weight			....)	(thousands)	(kilograms)		(grams per caput per day)		(%)	(%)
1961	3,236,000	97,800	32,037	113,697	-556	3,055,985	217,904	14.0	5.5	36.4	95.9	15.1	5.7
1962	3,601,800	97,400	64,660	134,191	-4,444	3,430,426	221,411	15.5	5.9	38.2	98.0	15.4	6.0
1963	3,963,300	128,300	79,725	188,656	6,111	3,732,180	224,794	16.6	6.4	40.2	99.7	15.9	6.4
1964	4,475,700	233,600	83,048	191,387	-3,333	4,130,429	227,984	18.1	6.6	37.4	97.3	17.5	6.7
1965	5,099,900	364,100	57,628	257,939	3,333	4,538,822	230,936	19.7	6.6	38.2	97.0	17.4	6.8
1966	5,348,800	441,100	47,799	299,025	-4,444	4,652,028	233,621	19.9	6.5	39.8	97.3	16.4	6.7
1967	5,777,200	520,300	53,235	282,909	1,667	5,028,891	236,061	21.3	6.9	42.5	99.2	16.2	6.9
1968	6,082,100	782,500	43,924	281,924	0	5,061,600	238,334	21.2	6.9	44.0	100.1	15.8	6.9
1969	6,498,400	823,300	40,423	258,325	-4,444	5,452,754	240,555	22.7	7.3	44.9	101.0	16.2	7.2
1970	7,207,929	1,058,400	45,018	314,646	1,667	5,881,567	242,806	24.2	7.7	46.2	102.4	16.6	7.5
1971	7,298,840	1,369,800	26,908	342,481	3,889	5,617,355	245,117	22.9	7.4	46.7	102.2	15.8	7.2
1972	7,860,605	1,133,200	25,199	304,593	6,667	6,454,678	247,472	26.1	8.1	47.8	102.2	16.9	7.9
1973	8,636,567	1,713,400	29,141	318,052	-556	6,633,700	249,844	26.6	8.6	47.6	102.3	18.1	8.4
1974	9,200,158	1,718,600	42,952	428,712	-1,663	7,094,134	252,187	28.1	9.1	51.0	105.2	17.8	8.6
1975	10,025,040	2,254,552	56,461	573,796	3,885	7,257,034	254,469	28.5	9.3	52.7	106.2	17.6	8.7
1976	10,167,870	2,733,650	29,867	542,686	0	6,921,398	256,687	27.0	8.0	50.3	104.1	16.0	7.7
1977	9,229,514	2,227,250	40,721	501,779	0	6,541,206	258,864	25.3	7.7	49.5	102.5	15.6	7.5
1978	9,046,691	1,877,400	85,027	541,507	0	6,712,811	261,025	25.7	8.1	50.7	103.9	16.0	7.8
1979	9,049,666	1,690,700	116,670	562,556	0	6,913,080	263,207	26.3	8.4	51.2	104.1	16.3	8.0
1980	9,515,640	2,106,100	189,426	563,584	0	7,035,382	265,436	26.5	8.5	50.4	102.7	16.8	8.3
1981	9,566,299	1,716,400	71,012	455,410	0	7,465,501	267,698	27.9	8.6	50.2	102.0	17.2	8.5
1982	9,990,642	2,199,537	55,952	369,112	0	7,477,945	269,981	27.7	8.8	50.0	102.4	17.6	8.6
1983	9,816,651	2,334,932	333,222	573,378	0	7,241,564	272,315	26.6	8.5	50.7	102.4	16.9	8.4
1984	10,592,920	2,487,536	387,869	591,598	0	7,901,658	274,737	28.8	9.2	52.4	103.9	17.5	8.8
1985	10,522,830	2,140,514	443,464	677,026	0	8,148,755	277,261	29.4	9.5	53.6	105.3	17.7	9.0
1986	11,259,960	2,330,228	467,020	861,395	0	8,535,352	279,913	30.5	9.5	54.6	106.0	17.4	9.0
1987	11,159,620	2,740,330	514,231	816,591	0	8,116,926	282,644	28.7	9.2	54.9	106.6	16.7	8.6
1988	11,332,060	3,065,035	637,764	1,034,533	0	7,870,260	285,296	27.6	8.9	55.4	106.3	16.0	8.4
1989	11,310,090	3,188,670	600,872	877,863	0	7,844,436	287,659	27.3	8.9	56.4	106.4	15.9	8.4
1990	10,389,030	2,653,265	572,880	1,250,862	0	7,057,782	289,586	24.4	8.4	55.9	106.5	15.0	7.9

Note: data refer to the area that was formerly the Union of Soviet Socialist Republics. Starting with 1991 information for each independent Republic is shown separately.

225 UNITED ARAB EMIRATES	- FOOD BALANCE SHEET OF FISH AND FISHERY PRODUCTS IN LIVE WEIGHT AND FISH CONTRIBUTION TO PROTEIN SUPPLY -													FAO-FIDI
YEAR	PRODUCTION	NON-FOOD USES	IMPORTS	EXPORTS	STOCKS VARIATIONS	TOTAL FOOD SUPPLY	POPULATION	PER CAPUT SUPPLY	FISH PROTEINS	ANIMAL PROTEINS	TOTAL PROTEINS	FISH/ANIMAL PROTEINS	FISH/TOTAL PROTEINS
	(.. metric tons in live weight ....)						(thousands)	(kilograms)	(grams per caput per day)			(%)	(%)
1961	30,000	28,129	...	...	0	1,871	98	19.1	5.9	54.8	89.4	10.7	6.6
1962	35,000	32,971	...	...	0	2,029	109	18.7	5.8	50.0	84.0	11.5	6.8
1963	35,000	32,686	...	...	0	2,314	121	19.2	5.9	46.6	81.4	12.7	7.2
1964	35,000	32,543	...	...	0	2,457	133	18.5	5.7	43.6	77.9	13.0	7.3
1965	38,000	35,143	...	...	0	2,857	144	19.8	6.1	41.9	76.2	14.6	8.0
1966	38,000	35,143	...	...	0	2,857	154	18.5	5.7	42.3	81.8	13.5	7.0
1967	40,000	37,029	...	...	0	2,971	164	18.1	5.6	42.9	83.9	13.0	6.6
1968	40,000	31,070	...	5,730	0	3,200	177	18.1	5.6	43.8	87.4	12.7	6.4
1969	40,000	30,870	...	5,730	0	3,400	195	17.4	5.4	43.4	87.4	12.4	6.1
1970	40,000	28,857	...	7,143	0	4,000	223	18.0	5.5	42.9	89.0	12.7	6.1
1971	43,000	24,014	...	14,286	0	4,700	260	18.1	5.0	40.2	82.9	12.5	6.1
1972	43,000	22,157	587	15,714	0	5,715	306	18.7	5.1	39.9	81.6	12.8	6.3
1973	43,000	20,557	339	16,000	0	6,782	362	18.8	5.1	44.2	90.6	11.5	5.6
1974	67,800	41,857	3,995	21,714	0	8,223	428	19.2	5.4	48.9	99.2	11.1	5.5
1975	67,800	57,880	...	...	0	9,920	505	19.7	5.7	48.0	96.6	11.9	5.9
1976	64,600	52,600	...	...	0	12,000	592	20.3	6.2	50.6	102.7	12.2	6.0
1977	64,600	50,476	...	...	0	14,124	690	20.5	6.3	49.0	102.2	12.8	6.1
1978	64,600	56,046	13,184	6,206	0	15,533	795	19.6	6.6	50.1	103.9	13.2	6.4
1979	64,600	47,928	12,353	9,912	0	19,113	904	21.1	6.9	54.7	104.0	12.5	6.6
1980	64,600	57,578	24,714	8,119	0	23,617	1,015	23.3	7.9	59.3	108.3	13.3	7.3
1981	67,760	56,417	21,001	7,696	0	24,648	1,128	21.9	7.5	58.3	105.9	12.9	7.1
1982	69,755	49,887	11,559	5,550	0	25,878	1,240	20.9	6.0	59.3	107.0	10.1	5.6
1983	72,716	50,494	10,931	5,929	0	27,224	1,351	20.2	5.6	56.0	102.2	10.0	5.5
1984	72,716	49,934	9,892	3,145	0	29,530	1,455	20.3	5.4	55.1	101.7	9.9	5.3
1985	72,260	51,278	12,700	2,877	0	30,805	1,552	19.9	5.4	54.0	100.5	10.1	5.4
1986	79,321	52,639	10,605	4,185	0	33,102	1,639	20.2	5.6	52.2	98.3	10.6	5.6
1987	85,248	53,488	10,860	7,206	0	35,414	1,718	20.6	6.0	51.0	97.3	11.8	6.2
1988	89,500	49,544	9,399	13,947	0	35,408	1,789	19.8	5.9	51.1	97.5	11.6	6.1
1989	91,160	44,253	9,929	20,777	0	36,059	1,856	19.4	5.5	51.1	98.4	10.7	5.5
1990	95,129	46,835	9,586	19,088	0	38,792	1,921	20.2	5.4	51.8	99.0	10.3	5.4
1991	92,300	52,324	16,381	13,400	0	42,957	1,983	21.7	6.1	52.2	99.3	11.6	6.1
1992	95,046	50,879	10,830	10,735	0	44,262	2,044	21.7	5.6	52.2	100.7	10.8	5.6
1993	95,000	47,099	9,543	9,269	0	48,176	2,102	22.9	6.0	54.1	111.9	11.1	5.4
1994	108,000	59,804	13,438	7,548	0	54,085	2,157	25.1	6.6	54.3	104.2	12.2	6.4
1995	105,554	55,800	13,511	6,767	0	56,498	2,210	25.6	6.7	54.8	104.3	12.3	6.5

229 UNITED KINGDOM — - FOOD BALANCE SHEET OF FISH AND FISHERY PRODUCTS IN LIVE WEIGHT AND FISH CONTRIBUTION TO PROTEIN SUPPLY - — FAO-FIDI

YEAR	PRODUCTION	NON-FOOD USES	IMPORTS	EXPORTS	STOCKS VARIATIONS	TOTAL FOOD SUPPLY	POPULATION	PER CAPUT SUPPLY	FISH PROTEINS	ANIMAL PROTEINS	TOTAL PROTEINS	FISH/ANIMAL PROTEINS	FISH/TOTAL PROTEINS
	(....................................		metric tons in live weight			....................................)	(thousands)	(kilograms)	(grams per caput per day)			(%)	(%)
1961	881,700	45,100	280,044	64,676	0	1,051,968	52,735	20.0	4.9	54.1	91.5	9.0	5.3
1962	931,100	63,000	317,118	75,046	0	1,110,172	53,140	20.9	5.2	55.2	93.1	9.4	5.6
1963	950,500	74,400	285,457	78,154	0	1,083,403	53,564	20.2	5.0	54.6	92.6	9.2	5.4
1964	960,600	61,500	337,267	69,412	0	1,166,956	53,975	21.6	5.6	55.9	93.2	10.1	6.1
1965	1,031,000	90,100	342,945	87,032	0	1,196,814	54,350	22.0	5.6	55.1	92.3	10.2	6.1
1966	1,053,400	143,600	334,994	122,713	0	1,122,082	54,680	20.5	5.1	54.4	91.4	9.3	5.6
1967	1,010,500	94,200	351,248	94,112	0	1,173,436	54,968	21.4	5.2	54.8	91.3	9.6	5.7
1968	1,025,600	102,900	384,090	96,437	0	1,210,352	55,218	21.9	5.5	55.3	91.7	10.0	6.0
1969	1,069,500	100,100	342,057	132,673	0	1,178,784	55,437	21.3	5.3	54.8	91.8	9.6	5.7
1970	1,092,200	64,800	331,449	176,812	-14,000	1,168,037	55,632	21.0	5.2	55.5	92.9	9.4	5.6
1971	1,094,600	92,200	303,519	200,652	-10,000	1,095,268	55,803	19.6	5.0	55.2	91.7	9.1	5.5
1972	1,073,900	122,600	315,813	195,400	-24,000	1,047,713	55,948	18.7	4.9	55.3	91.2	8.8	5.4
1973	1,128,800	145,600	367,133	294,721	-3,000	1,052,612	56,066	18.8	4.8	53.7	90.2	9.0	5.3
1974	1,075,700	161,300	328,844	263,996	-1,000	978,248	56,159	17.4	4.4	53.8	89.7	8.3	5.0
1975	969,426	126,461	335,924	226,343	-3,000	949,633	56,226	16.9	4.3	53.3	89.0	8.0	4.8
1976	1,027,015	168,944	354,760	203,881	-4,000	1,004,967	56,269	17.9	4.4	52.8	89.0	8.2	4.9
1977	999,086	176,716	360,154	215,862	-2,054	964,608	56,293	17.1	4.2	51.8	89.2	8.2	4.7
1978	1,031,899	144,524	434,747	402,854	-6,000	913,268	56,305	16.2	3.9	52.5	87.8	7.4	4.4
1979	909,127	63,159	517,428	477,363	30,000	916,044	56,315	16.3	3.7	53.3	90.0	7.0	4.2
1980	847,835	91,000	553,492	405,116	4	905,215	56,330	16.1	3.5	52.0	87.5	6.8	4.0
1981	882,989	114,589	563,322	324,715	-8,000	999,007	56,351	17.7	4.1	52.5	87.6	7.8	4.7
1982	912,246	123,769	560,375	349,313	-8,950	990,589	56,380	17.6	4.2	52.3	87.4	8.0	4.8
1983	852,042	93,891	587,501	429,861	50,980	966,771	56,427	17.1	4.0	52.0	87.7	7.7	4.6
1984	845,948	72,418	598,825	376,821	-19,367	976,166	56,504	17.3	4.3	53.1	90.1	8.1	4.7
1985	901,474	88,340	646,208	406,080	-33,312	1,019,950	56,618	18.0	4.4	53.9	92.6	8.2	4.8
1986	859,866	55,962	698,834	457,752	-24,193	1,020,792	56,774	18.0	4.3	53.4	92.8	8.1	4.7
1987	953,899	93,630	691,329	480,791	-40,902	1,029,905	56,967	18.1	4.4	53.8	92.6	8.2	4.8
1988	945,479	104,956	737,892	493,395	-37,567	1,047,454	57,178	18.3	4.6	53.4	94.1	8.6	4.9
1989	911,974	71,608	813,337	539,159	-39,165	1,075,378	57,382	18.7	4.8	53.1	92.0	9.1	5.3
1990	817,319	56,678	857,680	524,504	13,633	1,107,451	57,561	19.2	4.8	53.2	93.6	9.1	5.2
1991	858,920	6,543	820,365	622,333	32,082	1,082,493	57,708	18.8	4.8	52.4	92.1	9.2	5.2
1992	877,273	6,272	834,478	653,350	59,597	1,111,726	57,827	19.2	4.9	53.0	92.3	9.2	5.3
1993	936,754	6,494	790,187	684,236	63,626	1,099,841	57,923	19.0	4.8	50.1	91.0	9.6	5.3
1994	971,527	52,576	860,505	686,410	28,533	1,121,579	58,005	19.3	5.1	51.2	92.3	9.9	5.5
1995	1,011,474	71,289	834,227	661,373	0	1,113,042	58,079	19.2	5.0	52.1	92.0	9.7	5.5

231 U.S.A.	- FOOD BALANCE SHEET OF FISH AND FISHERY PRODUCTS IN LIVE WEIGHT AND FISH CONTRIBUTION TO PROTEIN SUPPLY -	FAO-FIDI

YEAR	PRODUCTION	NON-FOOD USES	IMPORTS	EXPORTS	STOCKS VARIATIONS	TOTAL FOOD SUPPLY	POPULATION	PER CAPUT SUPPLY	FISH PROTEINS	ANIMAL PROTEINS	TOTAL PROTEINS	FISH/ANIMAL PROTEINS	FISH/TOTAL PROTEINS
	(.. metric tons in live weight					....)	(thousands)	(kilograms)	(grams per caput per day)			(%)	(%)
1961	2,807,285	1,135,877	821,509	37,242	-1,008	2,454,668	189,091	13.0	3.1	65.1	98.4	4.7	3.1
1962	2,865,323	1,175,406	911,886	45,881	338	2,556,260	191,954	13.3	3.2	64.8	97.7	4.9	3.3
1963	2,648,475	944,520	878,286	54,893	673	2,528,021	194,714	13.0	3.0	65.8	98.6	4.6	3.1
1964	2,512,682	835,996	899,267	76,072	0	2,499,881	197,336	12.7	3.0	67.2	100.2	4.5	3.0
1965	2,624,276	906,150	909,364	89,592	0	2,537,898	199,796	12.7	3.0	65.1	96.8	4.7	3.1
1966	2,407,551	725,705	1,068,854	90,209	0	2,660,491	202,077	13.2	3.3	66.5	98.3	4.9	3.3
1967	2,292,044	666,775	1,025,738	89,863	0	2,561,144	204,194	12.5	2.9	66.8	98.8	4.3	2.9
1968	2,391,744	730,524	1,304,490	74,692	-8,011	2,883,007	206,195	14.0	3.2	67.8	99.9	4.7	3.2
1969	2,410,909	828,993	1,258,570	106,680	8,188	2,741,995	208,150	13.2	3.1	66.9	99.4	4.6	3.1
1970	2,809,700	1,004,200	1,323,312	121,766	-1,400	3,005,646	210,111	14.3	3.4	65.5	97.7	5.2	3.5
1971	2,879,900	1,111,000	1,324,682	145,659	-3,172	2,944,751	212,099	13.9	3.2	68.9	101.3	4.6	3.1
1972	2,767,000	1,024,900	1,782,588	145,063	0	3,379,626	214,108	15.8	3.6	69.2	100.8	5.2	3.6
1973	2,803,700	1,056,375	1,725,380	193,145	5,350	3,284,912	216,131	15.2	3.7	66.6	99.0	5.5	3.7
1974	2,852,634	1,147,070	1,611,041	143,776	-1,120	3,171,710	218,153	14.5	3.5	67.1	98.8	5.2	3.5
1975	2,848,356	1,097,739	1,451,927	164,075	-774	3,037,696	220,165	13.8	3.1	65.2	99.2	4.8	3.1
1976	3,057,560	1,184,505	1,659,522	177,467	401	3,355,511	222,168	15.1	3.5	68.4	102.4	5.1	3.4
1977	2,994,487	1,053,392	1,648,328	245,770	0	3,343,652	224,177	14.9	3.3	67.8	100.8	4.9	3.3
1978	3,429,642	1,309,033	1,752,863	307,261	1,735	3,567,946	226,208	15.8	3.5	67.3	100.4	5.3	3.5
1979	3,527,655	1,358,308	1,758,129	390,730	0	3,536,747	228,279	15.5	3.5	66.7	101.0	5.2	3.4
1980	3,651,095	1,289,753	1,538,705	451,377	-2,500	3,446,169	230,406	15.0	3.5	66.9	99.1	5.2	3.5
1981	3,790,312	1,107,596	1,790,930	572,289	-3,000	3,898,356	232,597	16.8	3.5	66.9	101.2	5.2	3.5
1982	4,029,213	1,409,334	1,801,878	665,698	-1,000	3,755,058	234,847	16.0	3.4	65.9	99.9	5.1	3.4
1983	4,315,896	1,459,385	1,990,857	802,175	2,500	4,047,693	237,149	17.1	3.6	67.7	101.3	5.3	3.5
1984	4,987,567	1,421,642	2,037,339	1,009,439	1,000	4,594,828	239,488	19.2	3.9	68.2	102.3	5.8	3.8
1985	4,947,566	1,353,882	2,353,751	1,295,466	-4,000	4,647,970	241,855	19.2	4.0	69.6	106.2	5.8	3.8
1986	5,183,138	1,210,612	2,433,768	1,749,348	-3,500	4,653,447	244,238	19.1	4.2	70.1	105.7	5.9	3.9
1987	5,988,691	1,360,574	2,635,350	1,969,166	-14,589	5,279,712	246,640	21.4	4.6	70.9	108.6	6.4	4.2
1988	5,952,881	1,208,324	2,405,373	2,105,752	-11,978	5,032,200	249,074	20.2	4.3	70.5	108.8	6.1	4.0
1989	5,774,846	1,056,949	2,566,432	1,658,747	-50,122	5,575,460	251,559	22.2	4.9	70.6	108.7	7.0	4.5
1990	5,867,749	1,105,944	2,329,115	1,718,790	-23,279	5,348,851	254,106	21.1	4.7	70.7	109.5	6.6	4.3
1991	5,485,824	1,151,493	2,412,320	1,497,606	-2,666	5,246,378	256,722	20.4	4.6	71.1	111.4	6.5	4.2
1992	5,598,547	936,275	2,310,791	1,438,707	-2,330	5,532,025	259,389	21.3	4.8	72.4	112.0	6.6	4.3
1993	5,931,575	1,053,926	2,352,112	1,414,326	45,000	5,860,436	262,054	22.4	5.0	72.0	112.5	7.0	4.5
1994	5,940,761	1,173,356	2,413,301	1,421,376	10,000	5,769,330	264,646	21.8	4.9	73.2	114.5	6.6	4.3
1995	5,614,534	987,425	2,331,829	1,239,149	60,000	5,779,789	267,115	21.6	4.9	72.8	112.8	6.8	4.4

240 U.S. VIRGIN ISLANDS	- FOOD BALANCE SHEET OF FISH AND FISHERY PRODUCTS IN LIVE WEIGHT AND FISH CONTRIBUTION TO PROTEIN SUPPLY -	FAO-FIDI

YEAR	PRODUCTION	NON-FOOD USES	IMPORTS	EXPORTS	STOCKS VARIATIONS	TOTAL FOOD SUPPLY	POPULATION	PER CAPUT SUPPLY	FISH PROTEINS	ANIMAL PROTEINS	TOTAL PROTEINS	FISH/ANIMAL PROTEINS	FISH/TOTAL PROTEINS
	(........................ metric tons in live weight ....)						(thousands)	(kilograms)	(grams per caput per day)			(%)	(%)
1961	500	0	...	...	0	500	35	14.4	4.1	41.6	41.6	9.8	9.8
1962	500	0	...	...	0	500	37	13.7	3.9	38.7	38.7	10.0	10.0
1963	500	0	...	...	0	500	39	12.9	3.7	24.2	24.2	15.1	15.1
1964	500	0	...	...	0	500	41	12.2	3.4	38.2	38.2	9.0	9.0
1965	500	0	...	...	0	500	44	11.4	3.2	38.1	38.1	8.4	8.4
1966	600	0	...	...	0	600	47	12.7	3.6	34.3	34.8	10.5	10.3
1967	600	0	...	...	0	600	50	11.9	3.4	45.4	45.8	7.4	7.3
1968	600	0	...	...	0	600	54	11.1	3.1	52.9	53.3	5.9	5.9
1969	600	0	...	...	0	600	59	10.2	2.9	53.7	54.0	5.4	5.3
1970	600	0	...	...	0	600	64	9.4	2.7	56.6	56.9	4.7	4.7
1971	600	0	...	...	0	600	70	8.5	2.4	50.3	50.6	4.8	4.8
1972	500	0	...	...	0	500	77	6.5	1.8	35.0	35.2	5.2	5.2
1973	500	0	...	...	0	500	84	5.9	1.7	33.1	33.4	5.0	5.0
1974	500	0	...	...	0	500	91	5.5	1.6	39.8	40.0	3.9	3.9
1975	503	0	...	...	0	503	95	5.3	1.5	41.4	41.6	3.6	3.6
1976	500	0	...	...	0	500	98	5.1	1.4	35.2	35.4	4.1	4.1
1977	602	0	...	...	0	602	98	6.1	1.7	13.7	13.7	12.5	12.5
1978	557	0	...	...	0	557	98	5.7	1.6	24.2	24.5	6.5	6.5
1979	547	0	...	...	0	547	97	5.6	1.6	23.8	24.0	6.6	6.5
1980	669	0	...	...	0	669	97	6.9	1.9	24.3	24.4	7.6	7.6
1981	603	0	...	...	0	603	97	6.2	1.7	6.0	6.0	28.2	28.2
1982	891	0	...	...	0	891	97	9.2	2.6	7.1	7.1	36.4	36.4
1983	612	0	...	...	0	612	98	6.3	1.8	6.2	6.2	28.2	28.2
1984	687	0	...	...	0	687	98	7.0	1.9	6.4	6.4	29.7	29.7
1985	607	0	...	...	0	607	99	6.1	1.7	6.3	6.3	26.8	26.8
1986	931	0	...	...	0	931	100	9.4	2.6	7.2	7.2	36.1	36.1
1987	895	0	...	...	0	895	100	8.9	2.5	4.9	4.9	50.7	50.7
1988	686	0	...	...	0	686	101	6.8	1.9	5.1	5.1	37.3	37.3
1989	793	0	...	...	0	793	101	7.8	2.2	5.3	5.3	40.8	40.8
1990	684	0	...	...	0	684	102	6.7	1.8	4.7	4.7	38.4	38.4
1991	880	0	...	...	0	880	103	8.6	2.3	5.4	5.4	42.9	42.9
1992	880	0	...	...	0	880	103	8.5	2.3	5.4	5.4	42.6	42.6
1993	890	0	...	...	0	890	104	8.6	2.3	5.4	5.4	42.9	42.9
1994	900	0	...	...	0	900	104	8.6	2.3	5.4	5.4	43.1	43.1
1995	920	0	...	...	0	920	105	8.8	2.4	5.4	5.4	43.6	43.6

234 URUGUAY	- FOOD BALANCE SHEET OF FISH AND FISHERY PRODUCTS IN LIVE WEIGHT AND FISH CONTRIBUTION TO PROTEIN SUPPLY -	FAO-FIDI

YEAR	PRODUCTION	NON-FOOD USES	IMPORTS	EXPORTS	STOCKS VARIATIONS	TOTAL FOOD SUPPLY	POPULATION	PER CAPUT SUPPLY	FISH PROTEINS	ANIMAL PROTEINS	TOTAL PROTEINS	FISH/ANIMAL PROTEINS	FISH/TOTAL PROTEINS
	(..		metric tons in live weight	....		)	(thousands)	(kilograms)		(grams per caput per day)		(%)	(%)
1961	10,500	0	200	0	0	10,700	2,571	4.2	0.9	59.1	87.3	1.6	1.1
1962	7,700	0	1,309	200	0	8,809	2,603	3.4	0.8	56.8	84.4	1.4	0.9
1963	8,100	0	2,206	400	0	9,906	2,634	3.8	0.9	57.0	85.8	1.6	1.0
1964	12,200	0	703	800	0	12,103	2,664	4.5	1.0	60.1	92.2	1.7	1.1
1965	15,800	0	0	600	0	15,200	2,693	5.6	1.2	55.7	85.0	2.2	1.4
1966	12,500	0	0	1,000	0	11,500	2,722	4.2	0.9	54.2	85.4	1.6	1.0
1967	10,900	0	0	1,000	0	9,900	2,749	3.6	0.8	50.1	80.4	1.5	0.9
1968	11,000	0	200	1,100	0	10,100	2,773	3.6	0.8	57.6	87.1	1.4	0.9
1969	12,700	0	600	1,700	0	11,600	2,794	4.2	1.0	58.7	90.6	1.7	1.1
1970	13,200	0	600	3,900	0	9,900	2,808	3.5	0.9	62.7	94.4	1.4	0.9
1971	14,400	0	200	3,500	0	11,100	2,817	3.9	0.9	55.6	87.7	1.6	1.0
1972	20,600	0	0	1,400	0	19,200	2,820	6.8	1.6	54.4	84.8	2.9	1.9
1973	17,500	0	78	2,300	0	15,278	2,821	5.4	1.3	52.3	83.3	2.4	1.5
1974	16,000	0	814	3,699	0	13,115	2,823	4.7	1.1	56.8	87.6	1.9	1.3
1975	26,248	0	179	13,550	-984	11,893	2,829	4.2	1.0	57.8	89.8	1.7	1.1
1976	33,604	0	471	18,181	-1,071	14,822	2,840	5.2	1.1	55.7	87.3	2.1	1.3
1977	48,274	2,100	744	28,317	-447	18,154	2,855	6.4	1.5	54.9	84.9	2.7	1.8
1978	74,223	2,241	1,392	51,863	-1,767	19,744	2,874	6.9	1.7	53.6	83.7	3.1	2.0
1979	108,133	2,150	5,050	83,278	-8,728	19,027	2,394	6.6	1.7	53.6	85.8	3.1	2.0
1980	120,399	1,499	9,578	103,958	-5,235	19,285	2,914	6.6	1.6	54.5	87.1	2.9	1.8
1981	146,951	6,170	2,827	134,341	4,968	14,235	2,933	4.9	0.9	51.8	84.2	1.8	1.1
1982	118,983	1,088	3,049	111,386	1,114	10,672	2,952	3.6	0.8	53.3	83.6	1.4	0.9
1983	143,403	18,382	1,295	112,977	-3,944	9,396	2,971	3.2	0.7	52.2	82.0	1.4	0.9
1984	132,958	16,832	1,005	111,355	4,890	10,665	2,990	3.6	0.8	48.4	79.5	1.6	1.0
1985	138,369	828	1,540	122,522	-7,476	9,083	3,008	3.0	0.7	47.2	78.7	1.5	0.9
1986	140,737	1,360	2,340	137,184	7,547	12,080	3,026	4.0	0.9	47.4	78.9	2.0	1.2
1987	137,799	1,710	7,054	116,850	-7,119	19,174	3,043	6.3	1.3	51.6	82.5	2.6	1.6
1988	107,348	1,870	2,592	108,825	13,113	12,359	3,060	4.0	1.0	50.6	80.4	2.0	1.3
1989	121,716	618	3,324	110,808	4,071	17,684	3,077	5.8	1.6	48.1	77.3	3.3	2.1
1990	90,832	77	3,520	78,691	600	16,185	3,094	5.2	1.3	44.9	75.6	2.9	1.7
1991	143,714	797	3,732	123,303	-4,935	18,411	3,112	5.9	1.5	49.8	80.2	3.0	1.9
1992	125,758	461	7,078	119,532	3,045	15,889	3,130	5.1	1.3	52.1	84.5	2.6	1.6
1993	118,820	1,000	4,504	97,652	-1,542	23,129	3,149	7.4	1.9	56.7	85.8	3.3	2.2
1994	120,742	1,680	5,485	102,333	2,588	24,803	3,167	7.8	2.1	57.5	87.4	3.6	2.4
1995	126,466	2,015	6,073	109,441	111	21,194	3,186	6.7	1.7	60.1	90.2	2.8	1.9

235 UZBEKISTAN	- FOOD BALANCE SHEET OF FISH AND FISHERY PRODUCTS IN LIVE WEIGHT AND FISH CONTRIBUTION TO PROTEIN SUPPLY -											FAO-FIDI
YEAR	PRODUCTION	NON-FOOD USES	IMPORTS	EXPORTS	STOCKS VARIATIONS	TOTAL FOOD SUPPLY	POPULATION	PER CAPUT SUPPLY	FISH PROTEINS	ANIMAL PROTEINS	TOTAL PROTEINS	FISH/ANIMAL PROTEINS	FISH/TOTAL PROTEINS
	(............ metric tons in live weight ....)						(thousands)	(kilograms)	(grams per caput per day)			(%)	(%)
1961	0	0	0	0	0	0	8,937	0.0	0.0	0.0	0.0	0.0	0.0
1962	0	0	0	0	0	0	9,280	0.0	0.0	0.0	0.0	0.0	0.0
1963	0	0	0	0	0	0	9,619	0.0	0.0	0.0	0.0	0.0	0.0
1964	0	0	0	0	0	0	9,952	0.0	0.0	0.0	0.0	0.0	0.0
1965	0	0	0	0	0	0	10,303	0.0	0.0	0.0	0.0	0.0	0.0
1966	0	0	0	0	0	0	10,629	0.0	0.0	0.0	0.0	0.0	0.0
1967	0	0	0	0	0	0	10,965	0.0	0.0	0.0	0.0	0.0	0.0
1968	0	0	0	0	0	0	11,336	0.0	0.0	0.0	0.0	0.0	0.0
1969	0	0	0	0	0	0	11,704	0.0	0.0	0.0	0.0	0.0	0.0
1970	0	0	0	0	0	0	11,973	0.0	0.0	0.0	0.0	0.0	0.0
1971	0	0	0	0	0	0	12,355	0.0	0.0	0.0	0.0	0.0	0.0
1972	0	0	0	0	0	0	12,756	0.0	0.0	0.0	0.0	0.0	0.0
1973	0	0	0	0	0	0	13,156	0.0	0.0	0.0	0.0	0.0	0.0
1974	0	0	0	0	0	0	13,570	0.0	0.0	0.0	0.0	0.0	0.0
1975	0	0	0	0	0	0	13,981	0.0	0.0	0.0	0.0	0.0	0.0
1976	0	0	0	0	0	0	14,390	0.0	0.0	0.0	0.0	0.0	0.0
1977	0	0	0	0	0	0	14,786	0.0	0.0	0.0	0.0	0.0	0.0
1978	0	0	0	0	0	0	15,184	0.0	0.0	0.0	0.0	0.0	0.0
1979	0	0	0	0	0	0	15,578	0.0	0.0	0.0	0.0	0.0	0.0
1980	0	0	0	0	0	0	15,936	0.0	0.0	0.0	0.0	0.0	0.0
1981	0	0	0	0	0	0	16,349	0.0	0.0	0.0	0.0	0.0	0.0
1982	0	0	0	0	0	0	16,788	0.0	0.0	0.0	0.0	0.0	0.0
1983	0	0	0	0	0	0	17,241	0.0	0.0	0.0	0.0	0.0	0.0
1984	0	0	0	0	0	0	17,708	0.0	0.0	0.0	0.0	0.0	0.0
1985	0	0	0	0	0	0	18,112	0.0	0.0	0.0	0.0	0.0	0.0
1986	0	0	0	0	0	0	18,634	0.0	0.0	0.0	0.0	0.0	0.0
1987	0	0	0	0	0	0	19,174	0.0	0.0	0.0	0.0	0.0	0.0
1988	0	0	0	0	0	0	19,611	0.0	0.0	0.0	0.0	0.0	0.0
1989	0	0	0	0	0	0	20,051	0.0	0.0	0.0	0.0	0.0	0.0
1990	0	0	0	0	0	0	20,515	0.0	0.0	0.0	0.0	0.0	0.0
1991	27,439	0	...	...	0	27,439	20,973	1.3	0.4	28.2	80.4	1.4	0.5
1992	28,097	0	...	...	0	28,097	21,424	1.3	0.4	26.2	77.4	1.6	0.5
1993	23,401	2,600	...	...	0	20,801	21,871	1.0	0.3	26.8	76.9	1.1	0.4
1994	17,654	1,500	...	811	-615	14,728	22,317	0.7	0.2	28.5	76.8	0.7	0.3
1995	13,808	0	2,556	2,248	616	14,731	22,762	0.7	0.2	28.6	79.5	0.7	0.3

Note: for years before 1991 see USSR.

155 VANUATU	- FOOD BALANCE SHEET OF FISH AND FISHERY PRODUCTS IN LIVE WEIGHT AND FISH CONTRIBUTION TO PROTEIN SUPPLY -												FAO-FIDI
YEAR	PRODUCTION	NON-FOOD USES	IMPORTS	EXPORTS	STOCKS VARIATIONS	TOTAL FOOD SUPPLY	POPULATION	PER CAPUT SUPPLY	FISH PROTEINS	ANIMAL PROTEINS	TOTAL PROTEINS	FISH/ANIMAL PROTEINS	FISH/TOTAL PROTEINS
	(.. metric tons in live weight					....)	(thousands)	(kilograms)	(grams per caput per day)			(%)	(%)
1961	1,400	0	2,965	2,500	0	1,865	66	28.3	8.6	27.2	58.0	31.7	14.9
1962	1,400	0	3,075	2,438	0	2,038	68	30.0	10.6	29.2	60.3	36.2	17.5
1963	1,500	0	1,819	1,219	0	2,100	70	30.0	10.5	30.9	62.7	34.0	16.8
1964	1,500	0	2,938	1,950	0	2,488	72	34.5	10.2	31.1	61.1	32.6	16.6
1965	1,800	0	2,300	1,800	0	2,300	75	30.9	9.4	30.0	61.1	31.4	15.4
1966	1,800	0	7,200	6,564	0	2,436	77	31.7	9.7	30.1	60.6	32.3	16.1
1967	1,800	0	6,800	5,977	0	2,623	79	33.0	10.2	30.1	61.1	34.0	16.8
1968	1,800	0	7,700	6,627	0	2,873	82	35.1	11.0	31.2	62.2	35.2	17.6
1969	1,800	0	9,180	7,988	0	2,992	85	35.4	10.8	31.0	62.9	35.0	17.2
1970	2,100	0	11,909	10,242	0	3,767	87	43.1	12.7	33.9	65.7	37.5	19.3
1971	2,200	0	17,032	14,829	0	4,403	90	48.8	14.3	36.9	68.5	38.8	20.9
1972	2,300	0	19,407	17,331	0	4,376	93	47.0	13.7	35.6	67.8	38.6	20.3
1973	2,400	0	19,059	16,812	0	4,647	96	48.4	14.1	35.1	66.5	40.3	21.2
1974	2,510	0	12,646	10,916	0	4,240	99	42.8	12.5	37.3	69.2	33.6	18.1
1975	2,610	0	6,593	5,798	0	3,405	102	33.4	9.9	31.2	64.1	31.5	15.4
1976	2,710	0	9,589	6,773	0	5,526	105	52.6	15.3	36.3	68.3	42.2	22.4
1977	2,816	0	13,922	11,126	0	5,612	108	51.9	15.1	33.7	66.0	44.9	22.9
1978	2,825	0	13,033	10,223	0	5,635	111	50.7	14.7	32.7	64.5	45.0	22.8
1979	2,819	0	10,964	8,610	0	5,172	114	45.3	13.1	35.2	69.4	37.3	18.9
1980	2,937	0	10,693	9,329	0	4,301	117	36.7	10.4	30.1	64.2	34.7	16.3
1981	2,715	0	7,221	5,392	0	4,544	120	37.8	10.7	26.9	61.4	39.7	17.4
1982	2,715	0	5,687	4,305	0	4,097	123	33.3	9.3	25.7	61.5	36.2	15.2
1983	2,494	0	6,543	5,079	0	3,958	126	31.4	8.8	26.0	62.9	33.9	14.0
1984	2,932	0	6,193	4,417	0	4,708	129	36.4	8.9	27.6	65.6	32.4	13.6
1985	3,555	0	6,240	4,557	0	5,238	132	39.6	10.4	28.4	65.4	36.5	15.9
1986	3,221	0	4,548	2,790	0	4,979	136	36.7	9.4	26.4	64.0	35.6	14.7
1987	3,253	0	1,136	4	0	4,385	139	31.6	8.0	25.3	63.2	31.5	12.6
1988	3,360	0	1,671	60	0	4,971	142	34.9	8.9	23.8	61.0	37.4	14.6
1989	3,340	0	900	40	0	4,200	146	28.8	7.1	22.0	58.9	32.5	12.1
1990	5,634	0	1,776	8	0	7,402	149	49.5	13.7	29.6	66.5	46.3	20.6
1991	3,270	0	996	0	-5	4,261	153	27.8	6.9	23.7	59.7	29.1	11.6
1992	2,521	0	1,744	7	-5	4,253	157	27.1	6.9	23.6	59.7	29.3	11.6
1993	2,922	0	1,807	30	-5	4,694	161	29.2	7.8	23.7	59.1	33.1	13.3
1994	2,733	0	1,844	33	-3	4,541	165	27.5	7.4	23.6	58.6	31.6	12.7
1995	2,727	0	2,253	37	-204	4,739	169	28.0	7.6	22.7	58.5	33.2	12.9

236 VENEZUELA	- FOOD BALANCE SHEET OF FISH AND FISHERY PRODUCTS IN LIVE WEIGHT AND FISH CONTRIBUTION TO PROTEIN SUPPLY -	FAO-FIDI

YEAR	PRODUCTION	NON-FOOD USES	IMPORTS	EXPORTS	STOCKS VARIATIONS	TOTAL FOOD SUPPLY	POPULATION	PER CAPUT SUPPLY	FISH PROTEINS	ANIMAL PROTEINS	TOTAL PROTEINS	FISH/ANIMAL PROTEINS	FISH/TOTAL PROTEINS
	(.. metric tons in live weight ....)						(thousands)	(kilograms)	(grams per caput per day)			(%)	(%)
1961	84,900	4,500	7,124	4,740	0	82,784	7,871	10.5	2.8	23.4	54.5	12.1	5.2
1962	94,900	4,500	6,365	11,109	0	85,655	8,171	10.5	3.3	23.5	53.6	14.1	6.2
1963	97,300	4,700	5,283	7,316	0	90,567	8,476	10.7	3.5	23.6	54.2	14.8	6.4
1964	110,600	3,100	5,370	6,576	0	106,294	8,784	12.1	3.9	24.9	58.9	15.5	6.6
1965	119,400	7,400	6,753	7,820	0	110,933	9,094	12.2	3.8	24.0	56.5	15.9	6.8
1966	116,900	7,200	4,188	2,462	0	111,526	9,404	11.9	3.6	23.6	56.2	15.1	6.4
1967	107,500	9,800	3,660	4,386	0	96,974	9,717	10.0	3.1	23.0	55.9	13.6	5.6
1968	126,100	16,400	2,781	4,230	0	108,251	10,038	10.8	3.3	23.4	56.1	14.2	5.9
1969	134,100	11,800	3,120	5,772	0	119,648	10,371	11.5	3.5	25.1	58.7	14.1	6.0
1970	126,720	15,900	2,648	8,490	0	104,978	10,721	9.8	3.1	24.7	58.8	12.6	5.3
1971	139,046	15,900	3,130	11,918	0	114,358	11,089	10.3	3.2	25.1	58.4	12.7	5.5
1972	152,220	16,600	3,487	14,914	0	124,193	11,473	10.8	3.2	25.4	58.1	12.4	5.4
1973	162,600	19,800	3,604	10,600	0	135,803	11,874	11.4	3.6	27.9	58.0	12.8	6.2
1974	145,059	12,710	5,256	7,940	0	129,665	12,295	10.6	3.0	26.9	57.1	11.0	5.2
1975	152,833	23,000	4,791	9,980	0	124,644	12,734	9.8	2.7	28.7	61.6	9.3	4.4
1976	145,810	20,050	5,134	11,447	0	119,447	13,195	9.1	2.7	29.3	62.8	9.1	4.2
1977	145,919	15,048	13,009	6,862	0	137,019	13,675	10.0	2.9	33.0	68.0	8.8	4.3
1978	166,118	7,579	16,307	11,190	0	163,656	14,161	11.6	2.9	34.2	66.9	8.3	4.3
1979	169,016	25,080	22,991	7,452	0	159,476	14,637	10.9	3.1	33.7	66.8	9.2	4.6
1980	185,293	29,000	27,810	4,243	0	179,860	15,091	11.9	3.6	33.2	71.0	10.7	5.0
1981	179,725	20,000	35,511	9,197	0	186,039	15,518	12.0	3.4	35.2	71.1	9.6	4.8
1982	221,259	26,078	36,465	19,811	-1,000	210,835	15,921	13.2	4.0	36.3	69.0	11.0	5.8
1983	230,445	22,000	6,867	24,725	400	190,988	16,314	11.7	3.6	32.8	67.0	11.1	5.4
1984	260,229	24,027	5,644	32,816	200	209,230	16,715	12.5	3.7	33.4	69.7	11.1	5.3
1985	264,512	28,027	3,463	50,154	200	189,994	17,138	11.1	3.4	32.0	66.6	10.5	5.1
1986	284,640	33,009	112	50,691	200	201,253	17,585	11.4	3.4	29.2	59.3	11.7	5.8
1987	297,958	23,505	231	22,484	0	252,200	18,054	14.0	4.1	30.7	64.3	13.3	6.4
1988	286,187	23,201	598	11,600	0	251,984	18,536	13.6	4.0	33.6	70.0	12.0	5.8
1989	329,485	21,302	118	48,906	0	259,396	19,021	13.6	4.0	27.7	56.9	14.2	6.9
1990	339,287	47,429	5,164	38,107	0	258,914	19,502	13.3	3.7	24.8	56.0	15.0	6.6
1991	344,447	36,509	7,394	39,267	0	276,066	19,976	13.8	4.2	27.3	61.3	15.3	6.8
1992	333,589	25,161	8,919	32,079	0	285,268	20,446	14.0	4.1	28.1	63.1	14.7	6.6
1993	397,257	35,009	14,427	41,915	-15	334,744	20,913	16.0	4.5	28.9	60.5	15.4	7.4
1994	441,279	14,805	12,214	57,169	15	381,534	21,378	17.9	5.1	27.6	59.4	18.6	8.6
1995	505,471	15,003	20,819	56,422	0	454,865	21,844	20.8	6.1	28.2	61.2	21.6	10.0

237 VIET NAM | **- FOOD BALANCE SHEET OF FISH AND FISHERY PRODUCTS IN LIVE WEIGHT AND FISH CONTRIBUTION TO PROTEIN SUPPLY -** | **FAO-FIDI**

YEAR	PRODUCTION	NON-FOOD USES	IMPORTS	EXPORTS	STOCKS VARIATIONS	TOTAL FOOD SUPPLY	POPULATION	PER CAPUT SUPPLY	FISH PROTEINS	ANIMAL PROTEINS	TOTAL PROTEINS	FISH/ANIMAL PROTEINS	FISH/TOTAL PROTEINS
	(.......... metric tons in live weight					..........)	(thousands)	(kilograms)	(grams per caput per day)			(%)	(%)
1961	472,900	0	150	600	0	472,450	35,393	13.4	4.0	8.7	45.2	45.8	8.8
1962	543,700	0	...	700	0	543,000	36,081	15.1	4.5	9.2	47.5	48.5	9.4
1963	628,600	0	...	1,000	0	627,600	36,803	17.1	4.8	9.6	49.4	50.3	9.7
1964	647,000	0	...	1,153	0	645,847	37,557	17.2	5.0	9.8	50.0	50.9	10.0
1965	675,000	0	...	1,195	0	673,805	38,341	17.6	5.1	10.0	49.1	50.8	10.3
1966	680,500	0	1,350	900	0	680,950	39,152	17.4	5.0	10.3	48.7	48.3	10.2
1967	710,700	0	1,200	700	0	711,200	39,991	17.8	5.1	10.2	49.3	50.0	10.3
1968	710,000	0	16,550	100	0	726,450	40,864	17.8	5.9	10.9	50.6	53.9	11.6
1969	763,800	0	26,316	100	0	790,016	41,775	18.9	6.1	11.0	50.6	55.5	12.0
1970	617,400	0	2,456	0	0	619,856	42,729	14.5	4.9	9.7	50.0	50.1	9.7
1971	687,500	0	3,480	400	0	690,580	43,728	15.8	5.2	9.9	50.5	52.6	10.3
1972	777,700	0	19,445	1,900	0	795,245	44,767	17.8	5.8	10.4	51.8	55.5	11.1
1973	813,500	0	5,614	5,015	0	814,100	45,838	17.8	5.7	10.1	49.9	56.1	11.3
1974	572,000	0	5,614	3,391	0	574,223	46,928	12.2	4.2	8.4	48.5	49.7	8.6
1975	546,000	0	2,632	3,380	0	545,252	48,030	11.4	4.0	7.9	47.6	50.2	8.3
1976	610,000	0	...	2,477	0	607,523	49,139	12.4	4.2	8.1	48.2	52.1	8.8
1977	588,000	0	...	2,902	0	585,099	50,259	11.6	4.0	7.9	46.3	50.4	8.6
1978	583,000	0	...	2,535	0	580,465	51,390	11.3	3.9	8.0	45.8	48.6	8.5
1979	600,000	0	...	4,514	0	595,486	52,540	11.3	3.9	8.2	45.5	47.9	8.6
1980	558,660	0	1,554	3,241	0	556,973	53,711	10.4	2.9	7.2	46.9	40.0	6.1
1981	596,866	30,000	...	6,044	0	560,822	54,903	10.2	2.6	7.3	48.1	35.5	5.4
1982	661,208	21,400	...	12,987	0	626,821	56,116	11.2	2.8	7.9	48.7	35.7	5.8
1983	757,138	24,100	...	13,108	0	719,930	57,351	12.6	3.2	8.7	49.8	36.6	6.4
1984	776,308	26,700	...	24,118	0	725,490	58,611	12.4	3.1	8.8	50.7	35.7	6.2
1985	808,010	24,400	...	31,433	0	752,177	59,898	12.6	3.2	8.9	50.1	35.5	6.3
1986	824,743	26,800	...	34,900	0	763,043	61,212	12.5	3.2	9.1	50.6	34.9	6.3
1987	871,404	32,900	...	37,886	0	800,618	62,550	12.8	3.3	9.6	49.4	34.1	6.6
1988	900,000	50,700	...	51,024	0	798,276	63,910	12.5	3.2	9.4	49.4	34.1	6.5
1989	930,000	12,200	...	39,669	0	878,132	65,291	13.5	3.5	9.9	49.6	35.7	7.1
1990	960,000	9,300	...	55,826	0	894,874	66,689	13.4	3.4	9.8	50.4	35.1	6.8
1991	1,020,000	32,900	...	79,336	0	907,764	68,103	13.3	3.8	10.0	50.4	37.6	7.5
1992	1,080,279	27,400	...	97,669	0	955,210	69,530	13.7	4.1	11.0	52.7	37.5	7.8
1993	1,100,000	27,500	...	128,345	0	944,155	70,960	13.3	4.0	11.1	54.4	36.1	7.4
1994	1,150,000	30,000	...	140,809	0	979,191	72,384	13.5	4.1	11.9	55.5	34.3	7.4
1995	1,100,000	30,000	959	118,904	0	952,055	73,793	12.9	3.9	12.2	57.5	32.2	6.9

243 WALLIS and FUTUNA ISLANDS	- FOOD BALANCE SHEET OF FISH AND FISHERY PRODUCTS IN LIVE WEIGHT AND FISH CONTRIBUTION TO PROTEIN SUPPLY -	FAO-FIDI

YEAR	PRODUCTION	NON-FOOD USES	IMPORTS	EXPORTS	STOCKS VARIATIONS	TOTAL FOOD SUPPLY	POPULATION	PER CAPUT SUPPLY	FISH PROTEINS	ANIMAL PROTEINS	TOTAL PROTEINS	FISH/ANIMAL PROTEINS	FISH/TOTAL PROTEINS
	(.. metric tons in live weight ....)						(thousands)	(kilograms)	(grams per caput per day)			(%)	(%)
1961	30	0	0	0	0	30	8	3.7	1.1	15.1	36.9	7.0	2.9
1962	30	0	0	0	0	30	8	3.7	1.1	15.1	37.1	7.0	2.8
1963	30	0	0	0	0	30	8	3.8	1.1	15.3	37.4	7.0	2.8
1964	30	0	0	0	0	30	8	3.8	1.1	15.3	37.6	7.0	2.8
1965	30	0	0	0	0	30	8	3.8	1.1	15.4	37.5	6.9	2.8
1966	30	0	0	0	0	30	8	3.7	1.0	15.3	37.0	6.8	2.8
1967	30	0	0	0	0	30	8	3.6	1.0	16.2	41.2	6.3	2.5
1968	30	0	0	0	0	30	9	3.5	1.0	15.7	40.0	6.2	2.5
1969	30	0	0	0	0	30	9	3.4	1.0	15.3	38.9	6.2	2.5
1970	30	0	0	0	0	30	9	3.3	0.9	15.3	38.5	6.2	2.4
1971	30	0	0	0	0	30	9	3.3	0.9	15.2	40.4	6.2	2.3
1972	30	0	0	0	0	30	9	3.4	0.9	15.3	40.7	6.2	2.3
1973	30	0	0	0	0	30	9	3.4	1.0	15.4	41.0	6.2	2.3
1974	30	0	0	0	0	30	9	3.4	1.0	16.1	39.2	5.9	2.4
1975	30	0	0	0	0	30	9	3.3	0.9	17.9	45.8	5.3	2.1
1976	40	0	0	0	0	40	9	4.3	1.2	14.9	38.7	8.2	3.1
1977	40	0	0	0	0	40	10	4.1	1.2	16.1	42.4	7.3	2.8
1978	50	0	0	0	0	50	10	4.9	1.4	20.8	50.1	6.7	2.8
1979	50	0	0	0	0	50	11	4.7	1.3	17.5	39.3	7.6	3.4
1980	60	0	0	0	0	60	11	5.5	1.5	17.9	43.5	8.6	3.5
1981	60	0	0	0	0	60	11	5.3	1.5	20.5	48.5	7.3	3.1
1982	70	0	0	0	0	70	11	6.1	1.7	24.8	40.0	7.0	4.3
1983	70	0	0	0	0	70	12	6.1	1.7	26.8	41.6	6.4	4.1
1984	70	0	0	0	0	70	12	6.0	1.7	37.5	61.7	4.5	2.7
1985	80	0	0	0	0	80	12	6.7	1.9	26.4	51.6	7.1	3.7
1986	80	0	0	0	0	80	12	6.5	1.8	21.5	43.2	8.5	4.2
1987	80	0	0	0	0	80	13	6.2	1.8	19.2	40.2	9.2	4.4
1988	100	0	0	0	0	100	13	7.5	2.1	21.2	41.5	10.0	5.1
1989	100	0	0	0	0	100	14	7.3	2.1	19.8	39.6	10.4	5.2
1990	80	0	0	0	0	80	14	5.7	1.6	18.6	35.9	8.7	4.5
1991	70	0	0	0	0	70	14	5.0	1.4	24.2	45.5	5.8	3.1
1992	143	0	0	0	0	143	14	10.1	2.9	24.6	46.0	11.6	6.2
1993	150	0	0	0	0	150	14	10.7	3.0	25.1	45.5	12.0	6.6
1994	193	0	0	0	0	193	14	13.8	3.9	33.1	56.8	11.7	6.8
1995	170	0	0	0	0	170	14	12.1	3.4	39.8	65.6	8.6	5.2

249 YEMEN	- FOOD BALANCE SHEET OF FISH AND FISHERY PRODUCTS IN LIVE WEIGHT AND FISH CONTRIBUTION TO PROTEIN SUPPLY -	FAO-FIDI

YEAR	PRODUCTION	NON-FOOD USES	IMPORTS	EXPORTS	STOCKS VARIATIONS	TOTAL FOOD SUPPLY	POPULATION	PER CAPUT SUPPLY	FISH PROTEINS	ANIMAL PROTEINS	TOTAL PROTEINS	FISH/ANIMAL PROTEINS	FISH/TOTAL PROTEINS
	(.. metric tons in live weight					....)	(thousands)	(kilograms)	(grams per caput per day)			(%)	(%)
1961	22,000	1,850	2,000	8,750	0	13,400	5,365	2.5	0.8	7.5	50.7	10.0	1.5
1962	22,000	7,500	2,000	4,750	0	11,750	5,488	2.1	0.7	7.5	50.4	8.7	1.3
1963	22,000	3,950	3,700	8,700	0	13,050	5,612	2.3	0.7	7.4	51.1	9.4	1.4
1964	25,000	2,300	5,350	10,700	0	17,350	5,731	3.0	0.9	7.8	50.7	11.7	1.8
1965	25,000	2,200	7,500	15,700	0	14,600	5,843	2.5	0.7	7.8	50.3	9.6	1.5
1966	25,000	2,870	7,000	14,800	0	14,330	5,946	2.4	0.7	7.4	50.3	9.9	1.5
1967	27,000	1,640	3,250	12,150	0	16,460	6,043	2.7	0.8	7.5	50.2	10.9	1.6
1968	27,000	1,500	4,300	11,500	0	18,300	6,137	3.0	0.9	6.6	49.0	13.7	1.8
1969	27,000	510	6,200	11,000	-9,000	12,690	6,232	2.0	0.6	6.1	48.6	10.1	1.3
1970	27,400	...	8,450	20,150	800	16,500	6,332	2.6	0.8	5.8	49.7	13.5	1.6
1971	26,400	...	4,619	21,452	6,200	15,767	6,437	2.5	0.8	5.9	51.0	12.7	1.5
1972	31,600	...	8,448	19,949	1,000	21,099	6,548	3.2	0.9	6.0	51.3	15.2	1.8
1973	43,200	...	6,362	13,585	-3,000	32,977	6,672	4.9	1.4	6.5	51.2	21.7	2.7
1974	51,244	...	2,940	21,992	-400	31,792	6,818	4.7	1.3	6.6	51.5	20.0	2.6
1975	50,829	...	3,600	10,111	-2,800	41,518	6,991	5.9	1.7	7.4	52.4	22.6	3.2
1976	65,022	5,880	2,526	7,533	3,000	57,135	7,194	7.9	2.5	8.2	53.8	30.8	4.7
1977	66,096	9,080	2,526	8,906	3,600	54,236	7,426	7.3	2.0	9.9	57.0	19.7	3.4
1978	67,047	7,570	5,595	5,415	600	60,257	7,680	7.9	2.3	12.2	59.1	18.8	3.9
1979	69,600	3,680	6,597	4,103	-4,500	63,914	7,946	8.0	2.4	13.3	59.5	17.7	4.0
1980	75,349	7,740	9,474	6,273	2,000	72,810	8,219	8.9	2.5	12.6	57.7	19.9	4.4
1981	68,226	8,100	4,411	2,272	500	62,765	8,498	7.4	2.3	14.2	58.9	16.3	3.9
1982	58,131	5,540	6,317	3,963	500	55,445	8,787	6.3	1.9	13.9	58.6	13.4	3.2
1983	63,663	4,830	6,437	2,520	-800	61,950	9,084	6.8	2.0	13.4	57.4	15.1	3.5
1984	65,710	8,040	5,186	6,423	1,500	57,933	9,388	6.2	1.7	14.5	57.0	11.9	3.0
1985	71,310	7,480	4,875	5,684	200	63,221	9,699	6.5	1.9	14.6	57.5	13.1	3.3
1986	72,747	13,310	4,512	6,186	0	57,762	10,012	5.8	1.7	13.9	59.2	11.9	2.8
1987	71,869	12,040	2,570	3,654	0	58,745	10,331	5.7	1.6	12.0	57.2	13.1	2.8
1988	73,156	12,250	2,665	3,300	0	60,271	10,680	5.6	1.6	13.7	61.9	11.7	2.6
1989	71,881	12,200	3,164	3,900	600	59,545	11,094	5.4	1.5	12.2	61.8	12.3	2.4
1990	77,860	12,770	1,799	3,942	0	62,947	11,592	5.4	1.6	11.0	56.3	14.4	2.8
1991	83,431	...	2,769	2,700	0	83,500	12,188	6.9	2.2	11.4	55.0	19.2	4.0
1992	80,387	...	2,699	2,311	0	80,775	12,868	6.3	1.9	9.5	55.8	20.4	3.5
1993	86,828	...	3,712	2,300	0	88,240	13,598	6.5	2.0	10.9	56.8	18.1	3.5
1994	83,458	...	5,021	1,400	0	87,079	14,329	6.1	1.9	9.5	54.8	19.5	3.4
1995	103,964	...	9,085	2,613	0	110,436	15,027	7.4	2.3	9.1	55.1	25.2	4.2

186 YUGOSLAVIA	- FOOD BALANCE SHEET OF FISH AND FISHERY PRODUCTS IN LIVE WEIGHT AND FISH CONTRIBUTION TO PROTEIN SUPPLY -	FAO-FIDI

YEAR	PRODUCTION	NON-FOOD USES	IMPORTS	EXPORTS	STOCKS VARIATIONS	TOTAL FOOD SUPPLY	POPULATION	PER CAPUT SUPPLY	FISH PROTEINS	ANIMAL PROTEINS	TOTAL PROTEINS	FISH/ANIMAL PROTEINS	FISH/TOTAL PROTEINS
	(...... metric tons in live weight ......)						(thousands)	(kilograms)	(grams per caput per day)			(%)	(%)
1961	0	0	0	0	0	0	8,123	0.0	0.0	0.0	0.0	0.0	0.0
1962	0	0	0	0	0	0	8,196	0.0	0.0	0.0	0.0	0.0	0.0
1963	0	0	0	0	0	0	8,266	0.0	0.0	0.0	0.0	0.0	0.0
1964	0	0	0	0	0	0	8,333	0.0	0.0	0.0	0.0	0.0	0.0
1965	0	0	0	0	0	0	8,396	0.0	0.0	0.0	0.0	0.0	0.0
1966	0	0	0	0	0	0	8,455	0.0	0.0	0.0	0.0	0.0	0.0
1967	0	0	0	0	0	0	8,512	0.0	0.0	0.0	0.0	0.0	0.0
1968	0	0	0	0	0	0	8,568	0.0	0.0	0.0	0.0	0.0	0.0
1969	0	0	0	0	0	0	8,627	0.0	0.0	0.0	0.0	0.0	0.0
1970	0	0	0	0	0	0	8,691	0.0	0.0	0.0	0.0	0.0	0.0
1971	0	0	0	0	0	0	8,761	0.0	0.0	0.0	0.0	0.0	0.0
1972	0	0	0	0	0	0	8,837	0.0	0.0	0.0	0.0	0.0	0.0
1973	0	0	0	0	0	0	8,916	0.0	0.0	0.0	0.0	0.0	0.0
1974	0	0	0	0	0	0	9,000	0.0	0.0	0.0	0.0	0.0	0.0
1975	0	0	0	0	0	0	9,085	0.0	0.0	0.0	0.0	0.0	0.0
1976	0	0	0	0	0	0	9,174	0.0	0.0	0.0	0.0	0.0	0.0
1977	0	0	0	0	0	0	9,264	0.0	0.0	0.0	0.0	0.0	0.0
1978	0	0	0	0	0	0	9,355	0.0	0.0	0.0	0.0	0.0	0.0
1979	0	0	0	0	0	0	9,441	0.0	0.0	0.0	0.0	0.0	0.0
1980	0	0	0	0	0	0	9,522	0.0	0.0	0.0	0.0	0.0	0.0
1981	0	0	0	0	0	0	9,597	0.0	0.0	0.0	0.0	0.0	0.0
1982	0	0	0	0	0	0	9,667	0.0	0.0	0.0	0.0	0.0	0.0
1983	0	0	0	0	0	0	9,732	0.0	0.0	0.0	0.0	0.0	0.0
1984	0	0	0	0	0	0	9,792	0.0	0.0	0.0	0.0	0.0	0.0
1985	0	0	0	0	0	0	9,848	0.0	0.0	0.0	0.0	0.0	0.0
1986	0	0	0	0	0	0	9,898	0.0	0.0	0.0	0.0	0.0	0.0
1987	0	0	0	0	0	0	9,941	0.0	0.0	0.0	0.0	0.0	0.0
1988	0	0	0	0	0	0	9,989	0.0	0.0	0.0	0.0	0.0	0.0
1989	0	0	0	0	0	0	10,111	0.0	0.0	0.0	0.0	0.0	0.0
1990	0	0	0	0	0	0	10,156	0.0	0.0	0.0	0.0	0.0	0.0
1991	0	0	0	0	0	0	10,184	0.0	0.0	0.0	0.0	0.0	0.0
1992	7,635	10	18,837	8,719	0	17,743	10,200	1.7	0.5	47.8	89.0	1.0	0.5
1993	6,455	105	3,034	279	0	9,105	10,209	0.9	0.3	45.6	88.9	0.6	0.3
1994	6,755	45	3,368	53	0	10,025	10,224	1.0	0.3	46.8	84.8	0.6	0.3
1995	6,540	0	4,782	413	-222	10,686	10,251	1.0	0.3	50.7	93.5	0.6	0.3

Note: for years before 1992 see Yugoslavia, S.F.R.

248 YUGOSLAVIA, S.F.R.	- FOOD BALANCE SHEET OF FISH AND FISHERY PRODUCTS IN LIVE WEIGHT AND FISH CONTRIBUTION TO PROTEIN SUPPLY -	FAO-FIDI

YEAR	PRODUCTION	NON-FOOD USES	IMPORTS	EXPORTS	STOCKS VARIATIONS	TOTAL FOOD SUPPLY	POPULATION	PER CAPUT SUPPLY	FISH PROTEINS	ANIMAL PROTEINS	TOTAL PROTEINS	FISH/ANIMAL PROTEINS	FISH/TOTAL PROTEINS
	(..		metric tons in live weight	....)			(thousands)	(kilograms)		(grams per caput per day)		(%)	(%)
1961	37,300	0	14,072	24,900	0	26,472	18,406	1.4	0.4	22.5	90.4	1.9	0.5
1962	30,300	0	12,222	25,261	0	17,261	18,567	0.9	0.3	22.4	93.0	1.2	0.3
1963	34,400	0	15,056	28,155	0	21,301	18,728	1.1	0.3	21.3	92.4	1.6	0.4
1964	38,300	0	20,361	28,700	0	29,961	18,882	1.6	0.5	22.5	94.4	2.1	0.5
1965	41,900	0	16,694	26,600	0	31,994	19,026	1.7	0.5	24.2	94.3	2.0	0.5
1966	45,500	0	13,380	11,905	0	46,975	19,157	2.5	0.9	24.8	96.6	3.6	0.9
1967	47,900	0	10,130	13,258	0	44,772	19,280	2.3	0.9	26.7	98.5	3.2	0.9
1968	44,900	0	10,098	11,499	0	43,499	19,399	2.2	0.8	27.8	97.9	2.9	0.8
1969	43,800	0	16,378	12,087	0	48,091	19,525	2.5	0.9	27.4	97.0	3.1	0.9
1970	46,129	0	26,029	12,086	0	60,073	19,663	3.1	1.0	28.5	96.3	3.7	1.1
1971	49,272	0	22,990	11,996	0	60,266	19,814	3.0	1.0	29.6	97.7	3.4	1.0
1972	49,241	0	24,301	14,002	0	59,540	19,978	3.0	1.0	28.9	96.2	3.3	1.0
1973	50,590	0	31,297	16,221	0	65,666	20,151	3.3	1.1	30.8	97.6	3.7	1.2
1974	54,161	0	40,073	17,529	0	76,705	20,331	3.8	1.3	33.4	102.0	3.8	1.2
1975	56,559	0	27,210	17,026	0	66,744	20,513	3.3	1.1	34.7	101.5	3.1	1.1
1976	58,674	0	27,996	26,749	0	59,921	20,699	2.9	1.0	36.6	104.3	2.6	0.9
1977	60,750	0	28,512	21,555	-200	67,507	20,887	3.2	1.1	38.1	105.4	2.8	1.0
1978	62,895	0	34,230	23,233	-200	73,692	21,075	3.5	1.2	39.4	105.3	3.0	1.1
1979	56,403	0	29,024	27,714	0	57,712	21,259	2.7	0.9	39.9	106.1	2.3	0.9
1980	58,301	0	36,189	21,940	-400	72,150	21,439	3.4	1.2	39.8	106.5	2.9	1.1
1981	71,629	0	27,244	23,821	-200	74,852	21,609	3.5	1.2	39.9	105.6	3.0	1.1
1982	66,560	0	32,833	28,815	-1,500	69,078	21,768	3.2	1.1	39.1	103.1	2.8	1.1
1983	79,685	0	32,670	26,277	2,500	88,578	21,922	4.0	1.4	39.6	105.0	3.4	1.3
1984	73,546	0	27,630	22,077	0	79,099	22,080	3.6	1.2	40.3	103.9	3.1	1.2
1985	75,187	0	28,871	21,845	0	82,213	22,245	3.7	1.3	39.6	101.9	3.2	1.2
1986	77,732	0	36,160	23,889	0	90,004	22,427	4.0	1.3	40.5	104.0	3.3	1.3
1987	81,586	0	33,233	25,466	0	89,354	22,620	4.0	1.3	41.1	102.1	3.2	1.3
1988	72,064	0	47,678	27,350	0	92,392	22,795	4.1	1.4	41.2	101.1	3.5	1.4
1989	71,992	0	44,414	27,280	0	89,126	22,911	3.9	1.3	40.5	101.5	3.3	1.3
1990	65,767	0	63,781	21,492	0	108,056	22,945	4.7	1.6	44.1	105.6	3.7	1.5
1991	35,811	0	14,780	1,118	0	49,473	22,877	2.2	0.8	37.8	99.1	2.1	0.8

Note: data refer to the area that was formerly the Socialist Federal Republic of Yugoslavia. Since 1992 information for each independent Republic is shown separately.

251 ZAMBIA	- FOOD BALANCE SHEET OF FISH AND FISHERY PRODUCTS IN LIVE WEIGHT AND FISH CONTRIBUTION TO PROTEIN SUPPLY -												FAO-FIDI
YEAR	PRODUCTION	NON-FOOD USES	IMPORTS	EXPORTS	STOCKS VARIATIONS	TOTAL FOOD SUPPLY	POPULATION	PER CAPUT SUPPLY	FISH PROTEINS	ANIMAL PROTEINS	TOTAL PROTEINS	FISH/ANIMAL PROTEINS	FISH/TOTAL PROTEINS
	(.................................... metric tons in live weight ....)						(thousands)	(kilograms)	(grams per caput per day)			(%)	(%)
1961	20,400	0	9,301	3,000	0	26,701	3,228	8.3	2.4	10.9	61.1	22.2	3.9
1962	22,200	0	12,001	3,000	0	31,201	3,318	9.4	2.7	10.9	59.1	25.0	4.6
1963	40,100	0	12,001	6,001	0	46,101	3,412	13.5	3.9	11.9	61.0	33.0	6.5
1964	42,300	0	12,701	8,501	0	46,500	3,511	13.3	3.9	11.7	60.2	32.8	6.4
1965	40,100	0	14,901	7,201	0	47,801	3,614	13.2	3.8	12.1	60.7	31.8	6.3
1966	40,100	0	13,901	2,400	0	51,601	3,721	13.9	4.0	13.0	63.3	31.1	6.4
1967	38,500	0	18,001	2,100	0	54,401	3,834	14.2	4.1	13.2	62.2	31.1	6.6
1968	41,300	0	16,201	0	0	57,501	3,951	14.6	4.2	13.9	63.1	30.3	6.7
1969	44,000	0	21,602	0	0	65,602	4,069	16.1	4.7	14.4	64.3	32.3	7.3
1970	52,100	0	20,650	600	0	72,150	4,189	17.2	5.2	15.1	63.8	34.7	8.2
1971	46,800	0	17,000	0	0	63,800	4,309	14.8	4.5	14.4	64.5	31.4	7.0
1972	49,900	0	19,600	0	0	69,500	4,429	15.7	4.8	15.2	66.1	31.4	7.2
1973	50,400	0	10,200	0	0	60,600	4,554	13.3	4.0	13.5	61.9	29.7	6.5
1974	46,929	0	6,150	100	0	52,979	4,690	11.3	3.4	13.6	64.7	25.0	5.2
1975	57,426	0	9,250	200	0	66,476	4,841	13.7	4.1	14.4	64.9	28.3	6.3
1976	54,267	0	3,529	28	0	57,768	5,011	11.5	3.4	13.9	65.1	24.6	5.3
1977	53,732	0	4,006	50	0	57,688	5,195	11.1	3.3	13.0	65.8	25.4	5.0
1978	47,567	0	2,982	0	0	50,549	5,386	9.4	2.8	12.0	60.7	23.3	4.6
1979	49,511	0	2,805	0	0	52,316	5,570	9.4	2.8	11.5	59.7	24.6	4.7
1980	50,987	0	5,641	0	0	56,628	5,738	9.9	3.0	11.8	58.3	25.4	5.1
1981	38,818	0	4,762	4	0	43,577	5,888	7.4	2.2	10.5	54.8	21.4	4.1
1982	55,809	0	5,990	0	0	61,799	6,021	10.3	3.1	11.7	54.8	26.5	5.7
1983	67,234	0	3,468	0	0	70,702	6,146	11.5	3.4	11.1	54.9	31.2	6.3
1984	64,621	0	2,318	21	0	66,918	6,274	10.7	3.2	10.1	54.8	31.4	5.8
1985	67,731	0	625	0	0	68,356	6,411	10.7	3.2	10.6	53.4	29.9	5.9
1986	68,199	0	342	942	0	67,599	6,560	10.3	3.0	10.4	52.7	29.1	5.7
1987	63,565	0	355	1,750	0	62,170	6,720	9.3	2.7	10.0	52.7	26.6	5.0
1988	60,584	0	597	1,528	0	59,653	6,887	8.7	2.5	9.9	52.3	25.3	4.8
1989	66,818	0	2,250	1,150	0	67,918	7,056	9.6	2.8	10.3	53.2	27.4	5.3
1990	64,771	0	441	807	0	64,405	7,224	8.9	2.6	10.2	52.0	25.2	4.9
1991	65,353	0	2,250	701	0	66,902	7,389	9.1	2.6	10.4	52.5	25.3	5.0
1992	67,349	0	3,037	613	0	69,772	7,555	9.2	2.7	10.8	50.2	24.8	5.3
1993	65,307	0	5,000	580	0	69,727	7,723	9.0	2.7	11.0	52.1	24.2	5.1
1994	70,056	0	2,171	160	0	72,067	7,897	9.1	2.6	10.4	51.3	25.5	5.2
1995	70,546	0	1,776	162	0	72,160	8,081	8.9	2.6	9.5	50.5	27.2	5.1

181 ZIMBABWE	- FOOD BALANCE SHEET OF FISH AND FISHERY PRODUCTS IN LIVE WEIGHT AND FISH CONTRIBUTION TO PROTEIN SUPPLY -	FAO-FIDI

YEAR	PRODUCTION	NON-FOOD USES	IMPORTS	EXPORTS	STOCKS VARIATIONS	TOTAL FOOD SUPPLY	POPULATION	PER CAPUT SUPPLY	FISH PROTEINS	ANIMAL PROTEINS	TOTAL PROTEINS	FISH/ANIMAL PROTEINS	FISH/TOTAL PROTEINS
	(.................................... metric tons in live weight					....................................)	(thousands)	(kilograms)	(grams per caput per day)			(%)	(%)
1961	1,300	0	4,578	0	0	5,878	3,933	1.5	0.4	12.9	61.6	3.1	0.7
1962	1,300	0	4,911	0	0	6,211	4,059	1.5	0.4	12.7	61.9	3.2	0.7
1963	1,300	0	4,689	0	0	5,989	4,189	1.4	0.4	12.5	60.8	3.1	0.6
1964	1,900	0	6,272	0	0	8,172	4,324	1.9	0.5	11.3	59.3	4.7	0.9
1965	2,000	0	6,619	0	0	8,619	4,466	1.9	0.6	10.6	58.5	5.2	0.9
1966	2,000	0	7,621	0	0	9,621	4,614	2.1	0.6	11.1	58.3	5.3	1.0
1967	2,000	0	7,562	0	0	9,562	4,769	2.0	0.6	9.6	55.3	6.0	1.0
1968	2,000	0	8,054	0	0	10,054	4,929	2.0	0.6	10.1	58.1	5.8	1.0
1969	2,000	0	8,167	0	0	10,167	5,093	2.0	0.6	10.4	55.1	5.5	1.1
1970	2,000	0	6,953	0	0	8,953	5,260	1.7	0.5	10.8	61.2	4.6	0.8
1971	2,500	0	6,747	0	0	9,247	5,431	1.7	0.5	12.4	63.1	4.0	0.8
1972	3,000	0	7,291	0	0	10,291	5,605	1.8	0.5	14.1	62.6	3.8	0.9
1973	3,000	0	6,722	0	0	9,722	5,782	1.7	0.5	15.0	61.0	3.3	0.8
1974	3,087	0	5,433	0	0	8,520	5,961	1.4	0.4	11.9	60.3	3.5	0.7
1975	3,654	0	4,490	0	0	8,144	6,143	1.3	0.4	11.5	54.2	3.4	0.7
1976	4,550	0	3,759	0	0	8,309	6,327	1.3	0.5	12.2	59.8	3.8	0.8
1977	4,771	0	3,981	0	0	8,752	6,514	1.3	0.5	12.3	62.1	3.8	0.8
1978	7,772	0	2,537	0	0	10,309	6,707	1.5	0.5	11.8	62.3	4.2	0.8
1979	9,874	0	1,938	90	0	11,721	6,910	1.7	0.5	11.0	58.8	4.7	0.9
1980	13,288	0	6,422	97	0	19,614	7,126	2.8	0.9	12.1	61.2	7.0	1.4
1981	16,431	0	4,085	185	0	20,331	7,355	2.8	0.8	12.2	58.2	6.9	1.4
1982	17,513	0	2,522	156	0	19,879	7,597	2.6	0.8	12.5	58.3	6.3	1.4
1983	13,609	0	2,019	0	-250	15,378	7,852	2.0	0.6	11.4	54.3	5.2	1.1
1984	16,415	0	753	152	0	17,015	8,117	2.1	0.6	10.7	52.6	5.8	1.2
1985	17,400	0	103	167	250	17,586	8,392	2.1	0.6	10.5	53.7	6.0	1.2
1986	18,952	0	874	53	0	19,772	8,678	2.3	0.7	11.2	53.0	6.1	1.3
1987	19,184	0	951	61	0	20,074	8,972	2.2	0.7	10.9	53.1	6.2	1.3
1988	22,222	0	440	111	0	22,551	9,270	2.4	0.7	10.8	55.2	6.7	1.3
1989	24,058	0	392	0	0	24,450	9,566	2.6	0.8	11.0	55.6	6.9	1.4
1990	25,772	0	328	89	-18	25,994	9,855	2.6	0.8	11.3	54.3	7.0	1.4
1991	22,167	0	307	86	18	22,405	10,136	2.2	0.7	10.1	52.2	6.6	1.3
1992	21,741	0	2,326	190	0	23,877	10,410	2.3	0.7	10.2	50.2	6.7	1.4
1993	21,370	0	5,742	219	0	26,892	10,676	2.5	0.8	7.3	50.4	10.2	1.5
1994	20,333	0	18,284	133	0	38,484	10,936	3.5	1.1	7.6	49.1	14.0	2.2
1995	16,570	0	22,705	588	0	38,687	11,190	3.5	1.0	7.9	46.2	13.2	2.3

REFERENCES

FAO, 1981. Programme for statistics of food consumption and nutrition. Rome, FAO, 72 p.

FAO, 1986. Economic and Social Policy Department, Statistics Division (Comp.), The ICS users' manual. Interlinked Computer Storage and processing system of the food and agricultural commodity data. Rome, FAO, 201 p.

FAO, 1988. Review of food consumption surveys – 1988. Household food consumption by economic groups. FAO Food Nutr.Pap., (44): 197 p.

FAO, 1989. Economic and Social Policy Department, Statistical Analysis Service, Annual series of total population estimates, 1960-88, based on the UN 1988 revision of World population prospects. Rome, FAO, ESS/Misc/89-1: 44 p.

FAO, 1989a. Comprehensive demographic estimates and populations 1950-2025. Rome, FAO, ESS/Misc/89-2: 194 p.

FAO, 1993. Compendium of food consumption statistics from household surveys in developing countries. FAO Econ. Soc. Dev. Pap., (116/1): 383 p. (116/2): 294 p.

FAO, 1996. Statistics Division/Division des Statistiques/Dirección de Estadística, Food Balance Sheets 1993-1995 average. Bilans alimentaires, Moyenne 1993-1995. Hojas de balance de alimentos, Promedio 1993-1995. Rome, FAO. 497 p.

Laureti, E. (comp.), 1990. Fish and fishery products. World apparent consumption statistics based on food balance sheets (1975-86), FAO Fish. Circ., (821): 523 p.

Laureti, E. (comp.), 1991. Fish and fishery products. World apparent consumption statistics based on food balance sheets (1961-89), FAO Fish. Circ., (821), Rev.1: 423 p.

Laureti, E. (comp.), 1992. Fish and fishery products. World apparent consumption statistics based on food balance sheets (1961-90), FAO Fish. Circ., (821), Rev.2: 477 p.

Laureti, E. (comp.), 1996. Fish and fishery products. World apparent consumption statistics based on food balance sheets (1961-93), FAO Fish. Circ., (821), Rev.3: 235 p.

Torry Research Station, Aberdeen (UK), 1989. Yield and nutritional value of the commercial more important fish species. FAO Fish. Tech. Pap., (309): 187 p.

United Nations, 1996. Population Division, Revision of world population prospects. New York, UN.